Evolutionary Biology: Mechanisms and Applications

Evolutionary Biology: Mechanisms and Applications

Editor: Noah Rodriguez

www.callistoreference.com

Callisto Reference,
118-35 Queens Blvd., Suite 400,
Forest Hills, NY 11375, USA

Visit us on the World Wide Web at:
www.callistoreference.com

ISBN: 978-1-64116-302-6 (Hardback)

Cataloging-in-publication Data

Evolutionary biology : mechanisms and applications / edited by Noah Rodriguez.
    p. cm.
Includes bibliographical references and index.
ISBN 978-1-64116-302-6
1. Evolution (Biology). 2. Biology. 3. Evolution. I. Rodriguez, Noah.
QH366.2 .E86 2020
575--dc23

Table of Contents

Preface

The world is advancing at a fast pace like never before. Therefore, the need is to keep up with the latest developments. This book was an idea that came to fruition when the specialists in the area realized the need to coordinate together and document essential themes in the subject. That's when I was requested to be the editor. Editing this book has been an honour as it brings together diverse authors researching on different streams of the field. The book collates essential materials contributed by veterans in the area which can be utilized by students and researchers alike.

The sub-field of biology concerned with the in-depth study of evolutionary processes is known as evolutionary biology. These processes are responsible for the diversity of life on this planet. Speciation, natural selection and common descent are some common evolutionary processes. Diverse topics and ideas are incorporated in the current research in evolutionary biology such as computer science and molecular genetics. The research has further widened to cover the genetic architecture of adaptation, molecular evolution, and the different forces that contribute to evolution such as biogeography, sexual selection and genetic drift. This book brings forth some of the most innovative concepts and elucidates the unexplored aspects of evolutionary biology. It is a valuable compilation of topics, ranging from the basic to the most complex advancements in the field of evolutionary biology. This book is appropriate for students seeking detailed information in this area as well as for experts.

Each chapter is a sole-standing publication that reflects each author's interpretation. Thus, the book displays a multi-facetted picture of our current understanding of application, resources and aspects of the field. I would like to thank the contributors of this book and my family for their endless support.

Editor

Hybridization and the spread of the apple maggot fly, *Rhagoletis pomonella* (Diptera: Tephritidae), in the northwestern United States

Tracy Arcella,[1,§] Glen R. Hood,[1,§] Thomas H. Q. Powell,[1,*,§] Sheina B. Sim,[1,†] Wee L. Yee,[2] Dietmar Schwarz,[3] Scott P. Egan,[1,4,‡] Robert B. Goughnour,[5] James J. Smith[6] and Jeffrey L. Feder[1,4,7]

1 Department of Biological Sciences, University of Notre Dame, Notre Dame, IN, USA
2 USDA-ARS, Yakima Agricultural Research Laboratory, Wapato, WA, USA
3 Department of Biology, Western Washington University, Bellingham, WA, USA
4 Advanced Diagnostics and Therapeutics, University of Notre Dame, Notre Dame, IN, USA
5 Washington State University Extension, Vancouver, WA, USA
6 Departments of Entomology & Lyman Briggs College, Michigan State University, E. Lansing, MI, USA
7 Environmental Change Initiative, University of Notre Dame, Notre Dame, IN, USA
*Present address: Department of Entomology and Nematology, University of Florida, Gainesville, FL 32611, USA
†Present address: USDA-ARS US PBARC, 64 Nowelo Street, Hilo, HI 96720, USA
‡Present address: Department of Biosciences, Rice University, Houston, TX 77005, USA

Keywords
introgression, insect pest, microsatellites, *Rhagoletis zephyria*, snowberries, Washington state.

Correspondence
Glen R. Hood, Department of Biological Sciences, Galvin Life Sciences Building, University of Notre Dame, Notre Dame, IN 46556, USA.

e-mail: ghood@nd.edu

§Denotes equal contribution

Abstract

Hybridization may be an important process interjecting variation into insect populations enabling host plant shifts and the origin of new economic pests. Here, we examine whether hybridization between the native snowberry-infesting fruit fly *Rhagoletis zephyria* (Snow) and the introduced quarantine pest *R. pomonella* (Walsh) is occurring and may aid the spread of the latter into more arid commercial apple-growing regions of central Washington state, USA. Results for 19 microsatellites implied hybridization occurring at a rate of 1.44% per generation between the species. However, there was no evidence for increased hybridization in central Washington. Allele frequencies for seven microsatellites in *R. pomonella* were more 'R. zephyria-like' in central Washington, suggesting that genes conferring resistance to desiccation may be adaptively introgressing from *R. zephyria*. However, in only one case was the putatively introgressing allele from *R. zephyria* not found in *R. pomonella* in the eastern USA. Thus, many of the alleles changing in frequency may have been prestanding in the introduced *R. pomonella* population. The dynamics of hybridization are therefore complex and nuanced for *R. pomonella*, with various causes and factors, including introgression for a portion, but not all of the genome, potentially contributing to the pest insect's spread.

Introduction

Hybridization between closely related species can provide insights into the nature of species boundaries and the speciation process (Barton and Hewitt 1985; Arnold 1992; Harrison 1993; Mallet 2005). Hybrid individuals can have lower fitness due to sterility, the breakup of locally adapted gene complexes (Dobzhansky and Pavlovsky 1958; Haddon 1984; Hubbard et al. 1992; Maeher and Caddick 1995; Levin et al. 1996), or a mismatch between hybrid phenotype and parental environments (Rundle 2002; Egan and Funk 2009). In other circumstances, hybrids may have higher fitness, which can lead to the creation of a hybrid swarm or a new, genetically distinct evolutionary lineage or species (Gallez and Gottlieb 1982; Rieseberg 1997; Seehausen 2004; Grant et al. 2005; Schwarz et al. 2005; Mavárez et al. 2006; Abbott et al. 2013). In other cases, hybrids themselves may not have higher fitness, but certain genes may be favored, resulting in the adaptive introgression of a subset of favorable alleles into the genetic background of the alternate parental population (Seehausen 2004; Mallet 2005).

Hybridization and introgression can also affect biological communities and ecosystems, particularly in light of global climate change where rapid adaptation to novel environmental conditions is prevalent (Scriber 2011, 2014; Pauls et al. 2013; Chown et al. 2014; Chunco 2014; Moran and Alexander 2014). The process may be particularly detrimental to ecosystems when one species involved is invasive. In this case, hybridization can contribute to the genetic extirpation of native species (Echelle and Connor 1989; Rhymer and Simberloff 1996; Huxel 1999) or help facilitate the spread of a modified form of the invader into previously unpopulated habitats (Ellstrand and Schierenbeck 2000; Perry et al. 2001; Arcella et al. 2013). In addition, if hybrids alter the local ecology or experience an escape from biotic factors normally constraining population densities, they can cause the loss of endemic biodiversity and ecosystem functions (Lodge et al. 2012).

One area where the detrimental consequences of hybridization and introgression may be underappreciated and understudied concerns the evolution of new pest insects (Diehl and Bush 1984; Kirk et al. 2013). In this instance, hybridization may increase levels of genetic and phenotypic variation to help enable the creation of new biotypes or races capable of shifting and differentially adapting to novel host plants of agricultural importance. Also, hybridization need not involve the evolution of novel host plant-related traits, *per se*, to create a new pest. Instead, it may facilitate adaptation to nonhost related biotic or abiotic conditions associated with the agricultural setting or climate change, allowing an insect to expand its ecology to become an economic threat.

Here, we investigate the spread of the apple maggot fly, *Rhagoletis pomonella* Walsh, into the commercial apple-growing region of central Washington (WA) state, a $2.25 billion annual industry accounting for 75% of apple production in the United States (Mertz et al. 2013). Specifically, we examine whether hybridization of *R. pomonella* with its sibling species *R. zephyria* Snow is occurring and may be aiding the apple maggot in spreading into more arid and hotter central WA from mesic habitats west of the Cascade Mountains. The apple maggot fly was likely introduced to the Pacific Northwest (PNW) from its native range in the eastern United States (AliNiazee and Penrose 1981; AliNiazee and Westcott 1986; Brunner 1987; Tracewski et al. 1987; Hood et al. 2013; Sim 2013) where it is native to ancestral host hawthorn (*Crataegus* spp.). In the eastern United States, the fly shifted to introduced, domesticated apple *Malus domestica* ~160 ya, forming a new host race, the initial step in ecological speciation with

gene flow (Bush 1966, 1969; Feder et al. 1988, 1993; Egan et al. 2015). In the process, the apple-infesting race of *R. pomonella* became a major frugivorous pest of commercially grown apple. Female flies oviposit into ripening fruit growing in trees and, following egg hatch, larvae feed within fruit, causing damage and making the fruit unmarketable. More recently, *R. pomonella* has been detected in the PNW. It is believed the fly was originally introduced via larval-infested apples into the Portland, Oregon (OR) area (arrow 1 in Fig. 1), where the first report of apple infestation was made in 1979 (AliNiazee and Penrose 1981). Subsequently, *R. pomonella* spread north and south from Portland into WA and OR on the western side of the Cascade Mountains (arrow 2 in Fig. 1). The fly also moved eastward into the Columbia River Gorge and other passages in the Cascades (arrow 3 in Fig. 1) and has been encroaching on the commercial apple-growing region of central WA since the mid-1990s (Yee et al. 2012). Here, there is a zero infestation policy for apple export to foreign markets and for domestic consumption (WSDA 2001, Yee et al. 2012).

The spread of *R. pomonella* in the PNW is complicated by two factors of evolutionary and economic significance. First, *R. pomonella* in the PNW also attacks native black hawthorn, *C. douglasii* Lindley, and to a lesser degree native *C. suksdorffii* Sarg. and *C. douglasii* × *C. suksdorffii* hybrids, as well as the introduced ornamental hawthorn, *C. monogyna* Jacquin (Tracewski et al. 1987; Yee 2008; Yee

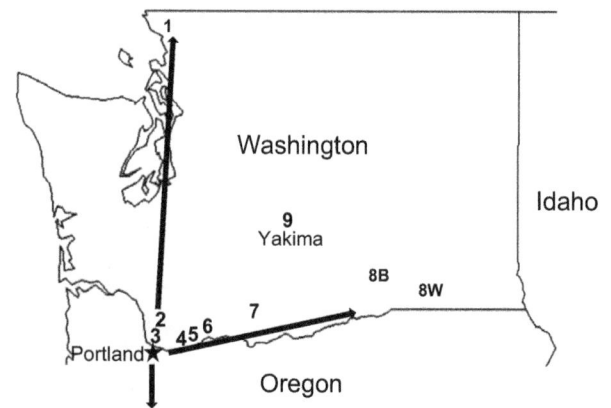

Figure 1 Map of the nine paired collection sites in Washington (WA) state genetically analyzed in the study. 1 = Bellingham; 2 = Vancouver, Washington State University campus; 3 = Vancouver; Burnt Bridge Creek Greenway; 4 = St. Cloud Park; 5 = Beacon Rock State Park; 6 = Home Valley; 7 = Klickitat; 8B = Burbank black hawthorn; 8W = Walla Walla snowberry; 9 = Tampico near Yakima. See Table S1 for site descriptions. Arrows denote spread of *R. pomonella* north and south along the western side of the Cascade Mountains and eastward into the Columbia River gorge following its putative introduction into Portland, OR. Black hawthorn-infesting populations of the fly have now encroached on the commercial apple-growing region of central WA centered in Yakima.

and Goughnour 2008; Yee et al. 2012; Hood et al. 2013). The existence of black hawthorn-infesting populations of *R. pomonella* raises the possibility that the fly is native to the PNW. If this is true, then flies shifted from black hawthorn to apples and ornamental hawthorns when these latter two plants were introduced to the region. However, current knowledge of the geographic distribution of *R. pomonella* in the western United States is consistent with the introduction hypothesis. An extensive field survey of black hawthorns failed to detect *R. pomonella* in the PNW aside from areas where the fly was already known to occur (Hood et al. 2013). A more recent survey found *R. pomonella* infesting black hawthorn at an isolated site in Troy, Montana (Yee et al. 2015). However, flies were found infesting only one of 24 trees surveyed across a five-year period, a pattern consistent with a local introduction. Moreover, *R. pomonella* was first reported to attack *C. douglasii* in central WA in 2003 (Yee 2008; Yee et al. 2012). If the fly was native on black hawthorn and did not recently disperse into the area, then in all likelihood it should have been detected earlier.

Additionally, the pattern of genetic variation within *R. pomonella* conforms to the introduction hypothesis. Genetic diversity is reduced in the PNW compared to the eastern United States, but highest in and around the hypothesized area of introduction in Portland, OR (Sim 2013). Also, a genetic distance network based on microsatellites clustered all PNW populations of *R. pomonella* together and derived from a source in the midwestern United States where *C. monogyna* is absent. Finally, coalescence simulations of microsatellites estimate the age of the split between northwestern and eastern fly populations as only 13.6 years (Sim 2013), consistent with a recent historical timeframe for *R. pomonella* being introduced to the PNW.

Although *R. pomonella* is unlikely to be native on black hawthorn, the host currently appears to be the major conduit for spreading the fly into the apple-growing region of central WA (Yee et al. 2012). Moving eastward in the Columbia River Gorge and other mountain passes into central WA, environmental conditions become increasingly more arid and hotter. Feral apples are rare, and black hawthorn is the principal host for *R. pomonella* along creeks and streams, including those flowing into the Yakima River Valley, which is the center of the commercial apple industry (Yee et al. 2012). As yet, no exported apple from central WA has been found infested with a fly larva. However, computer simulations suggest that if unchecked, all apple-producing areas may be infested by *R. pomonella* in <30 years (Zhao et al. 2007).

The second factor complicating and potentially contributing to the spread of *R. pomonella* concerns *R. zephyria*, which infests snowberries (*Symphoricarpos* spp.). The two flies are parapatric in their distribution across the

northern United States, overlapping extensively through the Midwest in Minnesota and Wisconsin (Bush 1966). In addition, *R. zephyria* is found in the northeastern United States into Canada, but exhibits a more patchy distribution across the region and may be non-native (Gavrilovic et al. 2007). The snowberry fly is also distributed through the northern plains states, where *R. pomonella* is not present, and westward into the PNW, where it is native and co-occurs with *R. pomonella*. In central WA, *R. zephyria* is problematic to commercial apple growers because it is abundant and difficult to definitively distinguish morphologically from the rarer *R. pomonella* when trapped in monitoring surveys (Westcott 1982; Yee et al. 2009, 2011, 2013). Due to the zero tolerance policy, misidentification of *R. zephyria* as *R. pomonella* is of concern because false positives can result in unnecessary quarantine measures being imposed at great cost to stakeholders, including growers and local and federal agencies (St. Jean et al. 2013).

Hybridization of *R. pomonella* with *R. zephyria* potentially poses a threat to the apple industry. Evidence suggests hybridization at a rate of 0.1% per generation between *R. pomonella* and *R. zephyria* in the eastern United States (Feder et al. 1999). In addition, *R. pomonella* and *R. zephyria* can be crossed to produce viable and fertile offspring in the laboratory, although at a reduced rate compared to pure parental matings (Yee and Goughnour 2011). Results from Green et al. (2013) suggest that a higher rate of hybridization is occurring in the PNW than eastern United States, with *R. zephyria* alleles extensively introgressing into *R. pomonella* in central WA. Here, low densities of *R. pomonella* infesting black hawthorn may encourage hybridization with the more abundant *R. zephyria*. Alleles common to *R. zephyria* elsewhere in WA were elevated in frequency in black hawthorn flies in the central apple-growing region of the state. However, the findings of Green et al. (2013) were based on a limited number of populations and individuals scored (60 flies combined from 2 locations in WA) and only 11 genetic markers. Nevertheless, the unidirectional pattern of introgression is consistent with transplant studies indicating that larval survivorship of *R. pomonella* is low in snowberry (Ragland et al. 2015). Hybridization could also help explain the difficulty in morphologically distinguishing *R. pomonella* from *R. zephyria* in central WA while contributing to the spread of a transgressive form (Rieseberg et al. 1999) of the apple maggot fly possessing features outside the normal phenotypic range of *R. pomonella* into the more arid and hotter apple-growing region of the state centered in Yakima.

Here, using a set of 19 microsatellite loci, we genotyped flies infesting black hawthorn and snowberry across nine pairs of sites where the flies co-occur or are in geographic proximity from west of the Cascade Mountains into central

WA to test the hypotheses that: (i) *R. pomonella* and *R. zephyria* are hybridizing; and (ii) alleles from *R. zephyria* are introgressing into *R. pomonella* and potentially aiding its spread into the hotter and more arid central apple-growing region of the state. Comparisons of allele frequencies among these nine sites, as well as with potential source populations in the eastern United States, imply a complex pattern of low level hybridization and asymmetric introgression at some, but not all loci, implying a heterogeneous pattern of introgression throughout the genome.

Materials and methods

Sample collection
Infested fruit from snowberry bushes and black hawthorn trees were collected from nine pairs of 'sympatric' field sites in Washington state from July to September 2009 to 2012 where the flies co-occur less than 200 meters apart (Fig. 1; Table S1). The only exception was the 'Burbank' site where black hawthorn flies were collected in Burbank, WA, while the corresponding sample of snowberry flies was collected 50 km east near Walla Walla, WA. Fruits were transported back to the greenhouse of the Washington State University Extension Services, Clark County 78th street Heritage Farm, Vancouver, Washington, where they were then placed separately by host plant and site onto wire mesh racks held over plastic collecting tubs. Larvae were allowed to emerge from the fruit and pupate in the tubs. Pupae were collected on a daily basis and frozen immediately for later genetic analysis.

Microsatellites
Flies were genotyped for 19 microsatellites originally developed for *R. pomonella* by Velez et al. (2006; see Table S2). The 19 microsatellites constitute a standard set of core loci analyzed for population differentiation in the *R. pomonella* sibling species group because they successfully PCR amplify and can be readily scored by multiplex genotyping for all taxa in the group (Michel et al. 2010; Cha et al. 2012; Powell et al. 2013, 2014). In addition, the 19 microsatellites are distributed across five of the six chromosomes (Michel et al. 2010) of the *Rhagoletis* genome (the small sixth dot chromosome is highly heterochromatic and currently does not contain a marker). Consequently, the genetic survey was not limited to one region of the genome but screened a representative portion for evidence of hybridization and introgression.

A total of 605 individuals collected from the nine sympatric sites ($n = 278$ from black hawthorn and $n = 327$ from snowberry) were genotyped in the study. Genomic DNA was isolated and purified from pupae or adults using PUREGENE extraction kits (Gentra Systems, Minneapolis, MN). PCR amplification and genotyping of microsatellites

were carried out as previously described for *R. pomonella* in Michel et al. (2010) and Powell et al. (2013, 2014).

Population genetic analyses
An unrooted neighbor-joining network was constructed based on overall Nei's (1972) genetic distances for the 19 microsatellites between populations using PowerMarker v3.25 (Liu and Muse 2005). For the Klickitat, WA population, two loci (p3 and p16) could not be scored due to a shortage of material and, thus, only 17 microsatellites were included in pairwise genetic distance measures calculated for this population. Bootstrap values were calculated based on 10 000 replicates across all loci.

To quantify genetic divergence between *R. pomonella* and *R. zephyria* and assess genotypes of individual flies for evidence of interhost migration and hybridization, we conducted a four-stage STRUCTURE analysis (v2.3.4; Pritchard et al. 2000). First, we tested for overall population structure for the entire microsatellite data set by conducting a blind (i.e., without *a priori* population information) STRUCTURE analysis for all 18 populations. For this analysis, we used the admixture and correlated alleles model, investigating $K = 1$–18 as possible numbers of genetically distinct subpopulations existing across sites. Three replicate runs of 250 000 MCMC generations, following a burn in period of 250 000 generations, were assessed for each K value.

Next, we conducted a more thorough second analysis of population subdivision for the 5 K values displaying the highest likelihood estimates in the initial STRUCTURE runs ($K = 1$–5). The follow-up analysis involved five replicate runs of 1 000 000 MCMC repetitions, following a burn in period of 500 000, for each $K = 1$–5 value. We used the ΔK method of Evanno et al. (2005) to identify the best value of K in the secondary STRUCTURE analysis.

Third, we assessed each of the nine sympatric *R. pomonella* and *R. zephyria* sites for evidence of population subdivision by conducting five blind runs for $K = 1$–2 (1 000 000 MCMC repetitions, following a burn in period of 500 000) separately for each paired site. Because the Evanno et al. (2005) method cannot be used for a comparison of $K = 1$ and $K = 2$, we used the mean Ln Likelihood estimates to evaluate the difference between these two values of K.

Fourth, potential migrants and hybrids were identified from STRUCTURE analyses performed at paired sympatric sites using the built-in function for population priors in cases when strong genetic structure exists between populations, as determined by step three above. The analyses for migrants and hybrids were conducted using a correlated allele frequency model and three replicate runs for $K = 2$ (1 000 000 MCMC repetitions, following a burn in period of 500 000) for each of three prior migration rates of 0.05 (the default), 0.01, and 0.1. The analysis produced posterior probabilities for each individual being derived from (i) cross of pure parental genotypes of natal host population origin (i.e., resident *R. pomonella* or *R. zephyria* flies infesting black hawthorn versus snowberry fruit, respectively); (ii) a pure parental cross of non-natal host origin (i.e., a fly whose mother was a migrant from the alternate host); (iii) a hybrid cross between *R. zephyria* and *R. pomonella* (F1 hybrid); and (iv) a cross of a hybrid individual with one or the other parental types (backcross). Migration rates (gene flow levels) per generation between hosts could then be estimated as the number of migrant genotypes plus half the number of F1 hybrids divided by the total number of flies scored. Hybridization rates per generation similarly could be estimated as the number of F1 hybrids divided by the total number of flies scored.

Mantel and partial Mantel tests for associations between Nei's genetic distances D between populations for microsatellites and their physical geographic distance and species identity (as a binary distance) were conducted using the package *vegan* (Oksanen et al. 2015) in R (R Development Core). Geographic distance matrices were produced using the R package *geosphere* (Hijmans 2014). Mantel tests were conducted based on 10 000 permutations and a Pearson correlation coefficient.

The overall mean and individual microsatellite locus allele frequency differences between *R. pomonella* and *R. zephyria* populations at the nine paired sites were also analyzed by linear regression against their geographic distance to the Tampico site near Yakima in central WA in R (R Development Core). To test the hypothesis that the genetic difference between species should decrease with proximity to Yakima given adaptive introgression from *R. zephyria* into *R. pomonella*, nonparametric, Monte Carlo simulations were used to determine significance levels for the regressions. In contrast to the Mantel tests which considered geographic distance between populations *per se* as the determinant of genetic differentiation regardless of ecology (i.e., isolation by distance), the linear regression with distance to Yakima tested for microsatellite convergence of *R. pomonella* with *R. zephyria* along the primary axes of climatic change (drier and hotter environmental conditions) varying from western to central WA hypothesized to underlie the adaptive introgression of alleles into *R. pomonella*. To test for statistical significance, microsatellite genotypes were randomly assigned within each species among the nine sites to determine how often a positive regression coefficient as large as that observed between allele frequency differences between the species at sites and the distance of sites to Yakima could be generated by chance. In addition, linear regressions were performed to test for an association between mean allele frequency

differences and three environmental factors: (i) mean precipitation at sites from July to October, (ii) average high temperature at sites in July, and (iii) average low temperature at sites in January, as measured at recording stations near each site from 1981 to 2010, as listed on the Web sites for US Climate Data (www.usclimatedata.com) and the Western Regional Climate Center (www.wrcc.dri.edu).

Results

Species differences
None of the 19 microsatellites scored displayed a diagnostic difference between *R. pomonella* and *R. zephyria* (i.e., no one allele or set of alleles was fixed at a frequency of 1.0 in one of the species and absent in the other; Table S3). Nevertheless, several microsatellites displayed large frequency differences between the species. For example, allele 226 at locus p7 was present at a frequency of at least 0.50 in all *R. zephyria* populations (range 0.50–0.96), while its highest frequency in any of the nine *R. pomonella* populations surveyed was 0.19. Conversely, allele 150 at locus p27 was present at a frequency of at least 0.18 in all *R. pomonella* populations, while its highest frequency in any *R. zephyria* population was 0.03. There were also a total of 196 private alleles for the 19 microsatellites found in one or more population of one of the species and not the other (Table S3). These private alleles were generally not observed at high frequency within populations (<0.10) and were usually not found across all nine populations of a single species. Indeed, almost half of the private alleles were present in only one of the 18 host-associated populations surveyed in WA. *Rhagoletis pomonella* possessed a higher percentage of private alleles (129 of 196 total = 65.8%) than *R. zephyria* ($X^2 = 19.6$, $P = 0.00001$, df = 1), consistent with the hypothesis of gene flow being primarily in the direction of snowberry into apple maggot fly populations. However, the difference could also be explained by larger effective population sizes and/or higher microsatellite mutation rates in *R. pomonella*, but given the history of recent introduction and generally lower densities of the apple maggot in the PNW, these two hypotheses seem less likely. Moreover, we genotyped fewer black hawthorn ($n = 278$) than snowberry-infesting ($n = 327$) flies, biasing the detection of private alleles in the direction of *R. zephyria*.

Despite the lack of a fixed difference, microsatellite allele frequencies differed sufficiently enough such that all nine *R. zephyria* and nine *R. pomonella* populations clustered distinctly and separately from one another with 100% bootstrap support in the neighbor-joining (NJ) network (Fig. 2). In addition, a blind STRUCTURE analysis conducted by combining flies from all nine paired sympatric sites gave a best fit of $K = 2$ across the PNW, corresponding to the species *R. zephyria* and *R. pomonella*

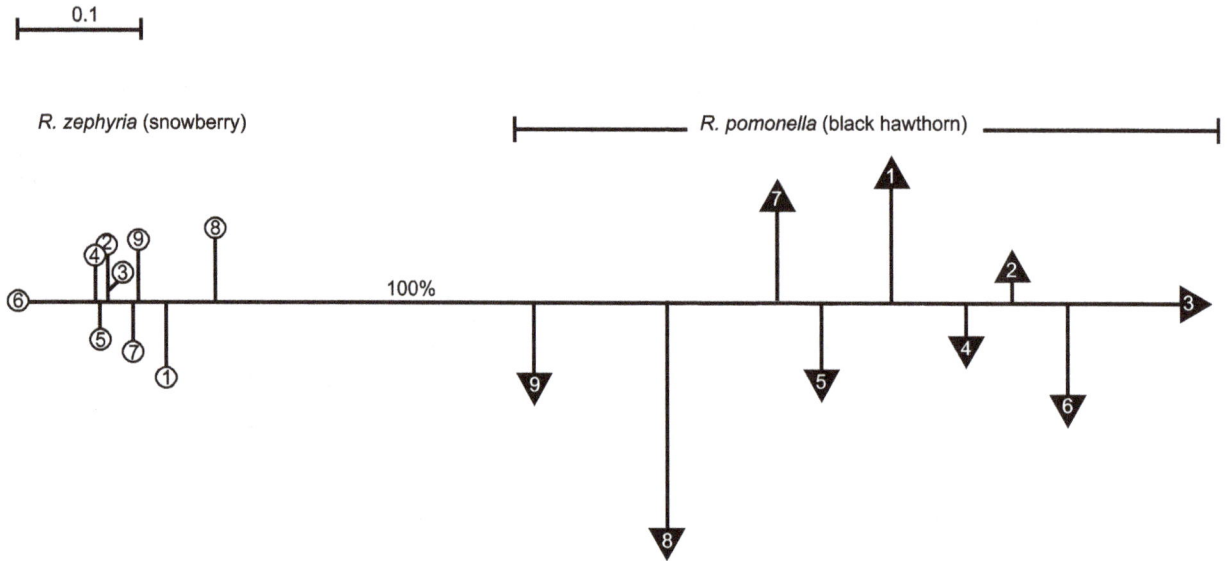

Figure 2 Neighbor-joining network for the nine paired black hawthorn-infesting *R. pomonella* (unfilled circles) and snowberry-infesting *R. zephyria* populations (dark triangles) in WA based on Nei's overall genetic distances for 19 microsatellite loci. See Fig. 1 legend and Table S1 for descriptions of the nine paired sites designated. Bootstrap support values based on 10 000 replicates are given for the node separating *R. pomonella* and *R. zephyria* populations.

(Table S4; Figs S1 and S2). Similarly, a model of $K = 2$ was supported for each of the nine paired sites considered separately, representing the two taxa (see Fig. 3A–D for STRUCTURE plots for the St. Cloud Park, Beacon Rock State Park, Home Valley, and Burbank/Walla Walla, and Fig. S2 for STRUCTURE plots for all sites, and Table S5 for mean Ln Likelihoods).

Inbreeding coefficients within populations

Microsatellite genotypes within populations of both species tended to be slightly heterozygote deficient, with low positive mean inbreeding coefficients (f) across loci (Table S6). There appeared to be considerable variation in f values among loci. However, only one population, *R. pomonella* from Devine (site #3), had a consistently positive inbreeding coefficient such that the standard deviation of f values across loci did not overlap zero. *Rhagoletis zephyria* and *R. pomonella* did not differ in f (paired *t*-test; $t = -0.331$; $P = 0.7491$; df = 8), but *R. zephyria* populations did have lower observed heterozygosity than *R. pomonella* (Table S3; paired *t*-test; $t = 6.68$; $P = 0.00016$; df = 8), consistent with the generally lower level of polymorphism found in the former species (see above). The results imply that deviations from random mating were not significant within local populations of either *R. zephyria* or *R. pomonella* and that the 19 microsatellites scored in the study did not possess high frequencies of null alleles.

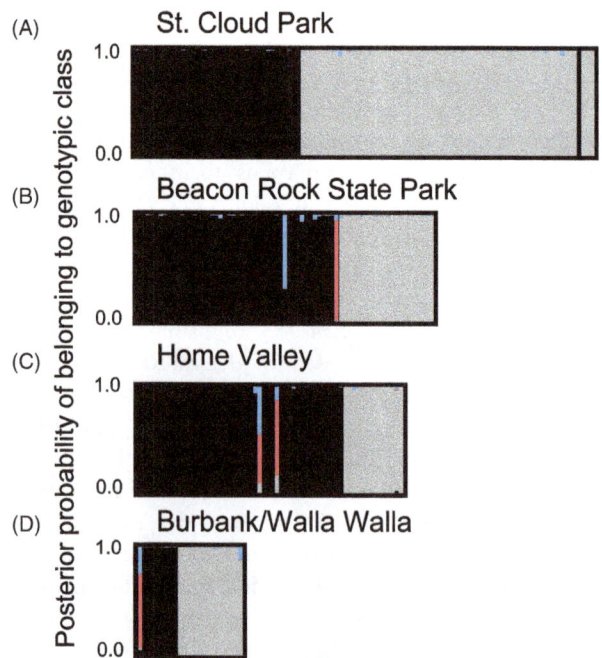

Figure 3 STRUCTURE bar plots for four paired sites at (A) St. Cloud Park; (B) Beacon Rock State Park; (C) Home Valley; and (D) Burbank/Walla Walla, WA, depicting posterior probabilities of individual *R. pomonella* black hawthorn fly genotypes (on left) and *R. zephyria* snowberry fly genotypes (on right) belonging to one of four genotypic classes: pure *R. pomonella* origin (black), pure *R. zephyria* (light gray), F1 hybrid (red), or backcross (blue), based on genotypes at 19 microsatellite loci. Bars along the *x*-axis represent individual flies.

Hybridization and gene flow

The finding of substantial population structure between black hawthorn versus snowberry-infesting fly populations at each of the nine paired sympatric sites allowed for posterior probabilities to be estimated for individuals to assess their genetic ancestry. Based on these probabilities, we found evidence for ongoing migration, gene flow, hybridization, and introgression between *R. pomonella* and *R. zephyria* in WA (Figs 3 and S2). The results were qualitatively similar for the three different priors used for the migration rate ($m = 0.05$, 0.01, and 0.1), with the exception that more backcross individuals were identified for the $m = 0.1$ model. We therefore present the finding generated from the most conservative $m = 0.01$ model. One of the 46 flies reared from black hawthorn at the Beacon Rock (Fig. 3B), two of the 48 black hawthorn-origin flies from the Home Valley (Fig. 3C), and one of the nine black hawthorn flies from the Burbank/Walla Walla (Fig. 3D) possessed multilocus microsatellite genotypes making them most likely to be F1 hybrids. One black hawthorn-origin fly from Beacon Rock had a genotype with a high posterior probability of being a later generation F2 or backcross hybrid (Fig. 3B). In addition, one of 71 individual reared from snowberry at the St. Cloud had a posterior probability for being a parental *R. pomonella* migrant (Fig. 3A). None of the identified migrants or hybrids had greater than two microsatellite loci with missing data. Based on these data, we estimated the hybridization rate of *R. pomonella* and *R. zephyria* on black hawthorn as 1.44% per generation (= 4 F1 hybrids/278 black hawthorn-origin flies genotyped in the study). In contrast, no evidence for hybridization was detected between the two species on snowberries. The estimated rate of migration and gene flow from the snowberry into black hawthorn-infesting fly populations was 0.0072 per generation, while it was 0.0031 in the reverse direction.

Isolation by distance

In a partial Mantel across all populations after species identity was accounted for, geographic distance was not significantly related to population genetic distance (Mantel $r = 0.032$; $P = 0.2785$). Geographic distance was correlated with population genetic distance for *R. zephyria* populations analyzed separately (Mantel $r = 0.7457$; $P = 0.0067$), but not for *R. pomonella* populations (Mantel $r = 0.4924$; $P = 0.11689$).

Genetic divergence in relation to Yakima

Microsatellite allele frequency differences between *R. pomonella* and *R. zephyria* populations across the nine

Table 1. Regression coefficients (*r*) for microsatellite loci of allele frequency differences between co-occurring *R. zephyria* and *R. pomonella* populations at nine paired sites against geographic distance of each location to the Yakima County site at Rosyln, WA. Also given in the column designated 'Chr #' are the chromosome (linkage group) assignments that each microsatellite locus has been mapped to in the *Rhagoletis* genome (Michel et al. 2010) and significance levels (*P* values) for regressions (significant loci are bolded). Significance was determined by nonparametric Mote Carlo simulations as described in the Materials and methods.

Locus	Chr #	r	P
p71	1	0.549	0.085
p37	1	0.169	0.355
p4	**1**	**0.796**	**0.020**
p3	1	0.607	0.107
p70	2	−0.115	0.574
p46	**2**	**0.707**	**0.034**
p73	2	0.097	0.409
p7	**3**	**0.655**	**0.049**
p80	**3**	**0.671**	**0.048**
p16	**3**	**0.766**	**0.045**
p66	3	−0.507	0.864
p11	4	0.149	0.376
p29	4	0.483	0.140
p25	**4**	**0.772**	**0.017**
p50	4	0.200	0.339
p60	4	0.508	0.141
p18	**5**	**0.847**	**0.007**
p9	5	0.001	0.502
p27	5	0.303	0.276
All loci		**0.769**	**0.023**

paired sites were significantly related to geographic distance to Tampico near Yakima in central WA (Table 1, Fig. 4). Seven loci designated p4, p7, p16, p18, p25, p46, and p80, as well as the overall pattern for all 19 loci scored in the study, displayed significant trends for black hawthorn flies to become more 'snowberry-like' in their frequencies with geographic proximity to Yakima (Table 1, Fig. 4). As mean July to October precipitation also decreases with proximity to Yakima, overall microsatellite genetic distance for black hawthorn flies was also significantly correlated with rainfall ($r = 0.729$, $P = 0.026$, df = 8), consistent with the desiccation hypothesis. Mean allele frequency differences were also strongly correlated with summer high temperatures ($r = -0.752$, $P = 0.019$, df = 8) and winter low temperatures ($r = 0.958$, $P < 0.0001$, df = 8). The seven significant loci displaying a relationship with distance to Yakima were distributed across all five of the chromosomes for which genetic markers were scored (Table 1), implying that the pattern was not restricted to just one particular region of the genome. The increased genetic similarity of *R. pomonella* to *R. zephyria* was also reflected in the Burbank and Yakima black hawthorn populations (sites 8 and 9) being closest to snowberry flies in the genetic

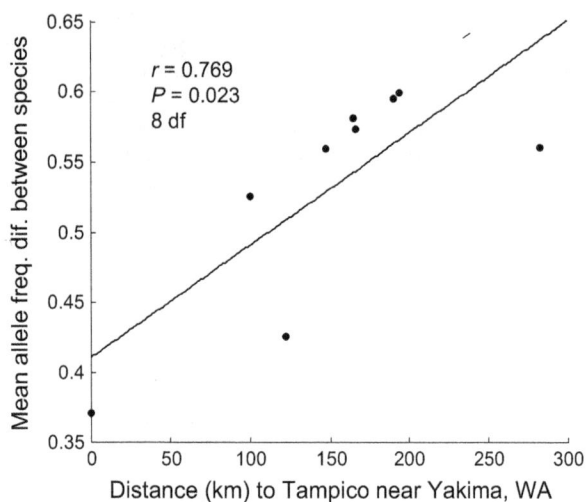

Figure 4 Association between mean allele frequency difference for 19 microsatellite loci between *R. zephyria* and *R. pomonella* populations at each of the nine paired sites plotted against each pairs geographic distance to the Tampico unincorporated community near Yakima, WA. Best fit line added to illustrate association. Note: the mean allele frequency difference at the Klickitat site was based on only 17 loci.

distance network (Fig. 2). However, allele frequencies for the remaining 12 microsatellites did not differ significantly with geographic distance, with two loci (p66 and p70) having negative, but not significant, regression coefficients (Table 1). Also, the seven significant loci described above also differed significantly between *R. pomonella* and *R. zephyria* Burbank and Yakima in central WA (Table S3). Moreover, with the exception of locus p16, every microsatellite possessed at least one private allele present in *R. zephyria*, but not in any of the nine co-occurring *R. pomonella* populations scored, even for those loci displaying significant frequency convergence between the two species with geographic proximity to Yakima. In addition, there were a total of 41 instances in which an allele was present at a microsatellite locus in only two of the total of 18 host-associated populations surveyed in the study. If introgression was appreciable between local *R. pomonella* and *R. zephyria* populations, then we would expect to observe several instances in which the rare allele was shared between the taxa at sympatric sites and not present anywhere else. However, in only one of the 41 cases was this so, which was lower than the null expectation of 1.2 (= [9/306] × 41). Consequently, while the microsatellites implied hybridization, gene flow, and a degree of genetic converge between *R. zephyria* and *R. pomonella* approaching central WA, there was also evidence for introgression being restricted between the species despite hybridization.

Discussion

Our results imply a more nuanced scenario than envisioned by Green et al. (2013) for the potential role that hybridization may be playing in the spread of *R. pomonella* into the more arid apple-growing region of central Washington. We found no evidence for a hybrid swarm between *R. pomonella* and *R. zephyria* in central WA. However, the STRUCTURE analysis implied mainly unidirectional hybridization between *R. zephyria* into *R. pomonella* on black hawthorn occurring at a rate of 1.44% per generation (4 F1 hybrids of a total of 278 black hawthorn-infesting flies scored), which is higher than that of 0.1% inferred from previous work (Feder et al. 1999). Nevertheless, there was no indication for a dramatically increased rate of hybridization in central WA compared to elsewhere. One of the likely F1 hybrids was detected at the Burbank site in central Washington, but none were evident at the Yakima site in the heart of apple production region of the state (Fig. S1). Moreover, one putative F1 hybrid was found at the Beacon Rock site and two putative hybrids were present at the Home Valley site in the Columbia River Gorge. Thus, while hybridization appears to be occurring and may disproportionately result in an influx of *R. zephyria* alleles into *R. pomonella*, the microsatellites imply that the rate is not greatly elevated in central WA.

There was a general tendency for *R. pomonella* to become genetically more similar to, but still distinct from, *R. zephyria*, approaching Yakima in central WA. There was no overall pattern of isolation by distance for *R. pomonella*, however. Thus, geographic distance *per se* between black hawthorn sites was not significantly related to the degree to which they were genetically diverged. Rather, it was the distance of sites from Yakima along the axis of varying environmental conditions that differentiated fly populations. The results therefore suggest that alleles associated with the snowberry fly may be favored in black hawthorn populations in the central WA region, possibly due to selection for increase desiccation resistance, but not in black hawthorn flies in western WA. However, this trend was not apparent for every locus. Indeed, many microsatellites suggest rather restricted hybridization and little gene flow locally between the two species. It may therefore be that much of the genome is not freely introgressing between *R. pomonella* and *R. zephyria*. Thus, although hybridization between the two species in the PNW may generally be higher than previously appreciated (1.44%), the effective introgression rate may be low, while potentially contributing to the adaptive process for *R. pomonella*. Such a scenario seems more in line with studies showing that larval survivorship is not high in reciprocal host transplant experiments (Ragland et al. 2015) and that a degree of postzygotic isolation also exists between *R. pomonella* and *R. zephyria* (Yee and Goughnour 2011). Future studies investigating the feeding performance the genetic patterns of surviving individuals from reciprocally transplanted F1 *R. pomonella* × *R. zephyria* hybrids and backcrossed individuals would be of particular interest.

Comparisons of microsatellite variation between *R. pomonella* and *R. zephyria* in the PNW to that for *R. pomonella* in its native range in the eastern United States also implies more subtle dynamics of introgression between the two species. Combining microsatellite data from Michel et al. (2010) and Powell et al. (2013, 2014) with those from the current study, we identified a total of 50 alleles in western *R. zephyria* not found in *R. pomonella* populations in the midwestern United States, the area from where the apple maggot fly was likely introduced (Sim 2013). These *R. zephyria* alleles 'private' to the northwest are insightful because they provide a means to gauge the degree and effectiveness of gene flow from *R. zephyria* into *R. pomonella* in the region. If these *R. zephyria* alleles are also present in western *R. pomonella* populations but not in the Midwest, then it is likely they introgressed from *R. zephyria*. It would seem improbable that if a high proportion of *R. zephyria* variants are present in western but not eastern *R. pomonella*, that they could all have been independently derived in western apple maggot flies following their recent introduction, but rather represent homoplasy, especially for alleles found at high frequencies in black hawthorn-infesting fly populations. Nor could the pattern be due to founder effects associated with the introduction of *R. pomonella* to the western United States because the alleles are not present in the midwestern United States.

In total, we detected 19 of 50 'unique *R. zephyria*' alleles (38%) in black hawthorn populations in WA, implying introgression. However, most of these shared alleles between the two species in WA were present at low frequencies (<0.05%) and in only a subset of the nine *R. pomonella* populations surveyed. Indeed, in only eight of the 19 instances was the putatively introgressed *R. zephyria* allele found at modest to high frequencies in a majority of the *R. pomonella* populations sampled (allele 143 at locus p3, alleles 301 and 315 at locus p18, allele 156 at locus p50, alleles 212 and 232 at locus p66, and alleles 207 and 211 at locus p80). Moreover, of these eight variants, only allele 301 at microsatellite p18 showed a clear pattern of increased frequency in central WA, although this was the locus displaying the most pronounced convergence with *R. zephyria* in the study. For the other significant microsatellites positively varying with proximity to Yakima, patterns were diverse. First, no shared and presumably introgressed unique *R. zephyria* allele was present at four loci (p4, p7, p25, and p46); second, for one locus (p16), a shared allele (292) was present at low frequency (0.0132)

and only in one *R. pomonella* population (WSU); third, for the remaining locus (p80), there were several shared alleles distributed widely across black hawthorn fly populations, but not obviously increasing in frequency in central WA. Thus, for the majority of loci showing convergence with *R. zephyria* in central WA, the alleles can also be found in *R. pomonella* populations in the midwestern United States. As such, these genes may represent standing variation in *R. pomonella* that was present at the time of the fly's introduction prior to its spread into central WA. The origin of these alleles in *R. pomonella* could conceivably trace to past historical gene flow between *R. zephyria* and *R. pomonella* predating the western introduction. However, there is currently no evidence from the microsatellites that these genes recently introgressed into *R. pomonella* to facilitate the fly's movement into central WA. Proof or refutation of recent introgression will require much more detailed DNA sequence analysis. Consequently, certain variants at certain microsatellites appear to have introgressed and may be adaptive in *R. pomonella* (in particular, allele 301 at microsatellite p18), while other instances of observed geographic variation in WA could involve standing genetic variation.

The story of hybridization and its role in the adaptive spread of the apple maggot fly in WA may therefore be more complicated than one involving widespread introgression of *R. zephyria* alleles into introduced *R. pomonella* populations. In central WA, larvae finish feeding, exit fruit, burrow into the soil, and form puparia during the hottest and driest period of the year. Studies suggest that *R. zephyria* pupae have a greater tolerance to desiccation than *R. pomonella* (Neilson 1964; Tracewski and Brunner 1987). Consistent with the desiccation hypothesis, we found an overall trend for *R. pomonella* populations to become more '*R. zephyria*-like' in their allele frequencies with decreased rainfall in proximity to Yakima. It is therefore possible that selection for desiccation resistance alleles introgressing from *R. zephyria* are actively adapting *R. pomonella* to the harsher environmental conditions for pupae in central WA, facilitating its spread into the commercial apple regions of the state. In addition, genetic similarity between *R. pomonella* and *R. zephyria* also increases with increasing high temperature at sites in July and with lower mean temperature in January, suggesting that diapause life history timing or another phenotype related to temperature rather than desiccation *per se* could alternatively be the source of selection. It remains to be determined, however, whether the observed relationships are causative with respect to desiccation resistance or merely represent correlations with aridity. Resolving these issues will require coupling selection experiments in *R. pomonella* on desiccation resistance and diapause traits to connect allelic variants with survivorship differences to these evolutionary histories. Such studies are also necessary to discount neutral isolation by distance due to variable gene flow and drift among populations generating the observed pattern of introgression. In this regard, there is a general decrease of genetic diversity in *R. pomonella* populations from the putative site of introduction in Portland, OR eastward that could reflect decreasing population sizes associated with the fly's spread toward central WA (Sim 2013). However, allele frequencies of the seven microsatellites showing a significant relationship with proximity to Yakima in central WA (Table 1; Fig. 4) were all in the direction of being more '*R. zephyria*-like' compared to populations in the western portion of the state, implying a deterministic rather than demographic explanation for the observed patterns of introgression.

Strong population genetic differentiation is maintained between *R. pomonella* and *R. zephyria*, despite a gross hybridization rate (~1.44%) that should lead to the eventual erosion of differentiation at equilibrium. The data presented here therefore add to the evidence that many 'good' species persist in the face of considerable hybridization (Mallet 2005) and that population genetic differentiation is not always a straightforward function of the migration rate. Factors including positive and negative selection acting differentially on genes, and structural features of the genome, such as inversion polymorphism in *R. pomonella* (Feder et al. 2003a) and the general lack of recombination in Dipteran males, can all disassociate rates of hybridization and effective migration (introgression) variably across the genome between populations.

In conclusion, there are a growing number of examples documenting differential adaptation of insect populations to commercially grown host plants (Kirk et al. 2013). Standing genetic variation is often cited as one important factor facilitating insect shifts to novel host species (Barrett and Schluter 2008; Kirk et al. 2013). Hybridization could therefore be an important and understudied process interjecting genetic variation into insect populations enabling host shifts and the genesis of new economic pests. Adaptive introgression could be of particular concern in an era of rapid global climate change and increasing species introductions, as populations encounter novel environmental conditions on an accelerated basis (Balint et al. 2011; Scriber 2011, 2014; Pauls et al. 2013; Chown et al. 2014; Chunco 2014; Moran and Alexander 2014). Previous studies of hawthorn-infesting populations of *R. pomonella* have suggested that past cycles of geographic isolation, genetic differentiation, secondary contact, and subsequent hybridization have contributed to the adaptive radiation of members of the *R. pomonella* sibling species complex onto a variety of novel host plants (Feder et al. 2003b, 2005; Xie et al. 2008). In addition, hybridization between *R. zephyria* and *R. mendax* may have given rise to a new honeysuckle-infesting population of *Rhagoletis* attacking *Lonicera* (Schwarz et al. 2005). Here, we investigated whether ongoing hybridization between *R. zephyria* and the recently introduced *R. pomonella* is contributing to the spread and adaptation of the apple maggot fly to harsher environmental conditions found in the commercial apple-growing region of central WA. Our results imply that standing variation may be important for *R. pomonella*'s establishment in central WA. We found evidence that some of this variation could have its roots in *R. zephyria* and have recently introgressed into western *R. pomonella* populations, while a significant portion may have a deeper history in *R. pomonella* predating its introduction. Thus, a diversity of causes and factors, including hybridization, likely underlie the story of *R. pomonella*'s invasion and spread in the western United States.

Acknowledgements

The authors would like to thank the Clark County Washington 78th street Heritage Farm, the Washington State University Research and Extension Unit, Vancouver, Blair Wolfley, Doug Stienbarger, Terry Porter, and Kathleen Rogers for their support and assistance on the project. This work was supported in part by grants to JLF from the NSF and the USDA and to WLY by the Washington Tree Fruit Research Commission and Washington State Commission on Pesticide Registration.

Literature Cited

Abbott, R., D. Albach, S. Ansell, J. W. Arntzen, S. J. E. Baird, N. Bierne, J. Boughman et al. 2013. Hybridization and speciation. Journal of Evolutionary Biology **26**:229–246.

AliNiazee, M. T., and R. L. Penrose 1981. Apple maggot in Oregon: a possible new threat to the northwest apple industry. Bulletin of the Entomological Society of America **27**:245–246.

AliNiazee, M. T., and R. L. Westcott 1986. Distribution of the apple maggot *Rhagoletis pomonella* (Diptera: Tephritidae) in Oregon. Journal of the Entomological Society of British Columbia **83**:54–56.

Arcella, T. E., W. L. Perry, D. M. Lodge, and J. L. Feder 2013. The role of hybridization in a species invasion and extirpation of resident fauna: hybrid vigor and breakdown in the rusty crayfish (*Orconectes rusticus*). Journal of Crustacean Biology **34**:157–164.

Arnold, M. L. 1992. Natural hybridization as an evolutionary process. Annual Review of Ecology and Systematics **23**:237–261.

Balint, M., S. Domisch, C. H. M. Engelhardt, P. Haase, S. Lehrian, J. Sauer, K. Theissinger et al. 2011. Cryptic biodiversity loss linked to global climate change. Nature Climate Change **1**:313–318.

Barrett, R. D., and D. Schluter 2008. Adaptation from standing genetic variation. Trends in Ecology and Evolution **23**:38–44.

Barton, N. H., and G. M. Hewitt 1985. Analysis of hybrid zones. Annual Review of Ecology and Systematics **16**:113–148.

Brunner, J. F. 1987. Apple maggot in Washington state: a review with special reference to its status in western states. Melanderia **45**:33–51.

Bush, G. L. 1966. The Taxonomy, Cytology, and Evolution of the Genus

Rhagoletis in North America (Diptera: Tephritidae). Museum of Comparative Zoology, Cambridge, MA.

Bush, G. L. 1969. Sympatric host race formation and speciation in frugivorous flies of the genus *Rhagoletis* (Diptera, Tephritidae). Evolution 23:237–251.

Cha, D. H., T. H. Q. Powell, J. L. Feder, and C. E. Linn 2012. Geographic variation in fruit volatiles emitted by the hawthorn *Crataegus mollis* and its consequences for host race formation in the apple maggot fly, *Rhagoletis pomonella*. Entomologia Experimentalis Et Applicata 143:254–268.

Chown, S. L., K. A. Hodgins, P. C. Griffin, J. G. Oakeshott, M. Bryne, and A. A. Hoffmann 2014. Biological invasions, climate change and genomics. Evolutionary Applications 8:23–46.

Chunco, A. J. 2014. Hybridization in a warmer world. Ecology and Evolution 4:2019–2031.

Diehl, S. R., and G. L. Bush 1984. An evolutionary and applied perspective of insect biotypes. Annual Review of Entomology 29: 471–504.

Dobzhansky, T., and O. Pavlovsky 1958. Interracial hybridization and breakdown of coadapted gene complexes in *Drosophila paulistorum* and *Drosophila willistoni*. Proceedings of the National Academy of Sciences USA 44:622–629.

Echelle, A. A., and P. J. Connor 1989. Rapid, geographically extensive genetic introgression after secondary contact between two pupfish species (Cyprinodon, Cyprinodontidae). Evolution 43:717–727.

Egan, S. P., and D. J. Funk 2009. Ecologically dependent postmating isolation between sympatric 'host forms' of *Neochlamisus bebbianae* leaf beetles. Proceedings of the National Academy of Sciences 106:19426–19431.

Egan, S. P., G. R. Ragland, L. Assour, T. H. Q. Powell, G. R. Hood, S. Emrich, P. Nosil et al. 2015. Experimental evidence of genome-wide impact of ecological speciation during early stages of speciation-with-gene-flow. Ecology Letters 18:817–825.

Ellstrand, N. C., and K. A. Schierenbeck 2000. Hybridization as a stimulus for the evolution of invasiveness in plants? Proceedings of the National Academy of Sciences USA 97:7043–7050.

Evanno, G., S. Regnaut, and J. Goudet 2005. Detecting the number of clusters of individuals using the software STRUCTURE: a simulation study. Molecular Ecology 14:2611–2620.

Feder, J. L., C. A. Chilcote, and G. L. Bush 1988. Genetic differentiation between sympatric host races of the apple maggot fly *Rhagoletis pomonella*. Nature 336:61–64.

Feder, J. L., T. A. Hunt, and G. L. Bush 1993. The effects of climate, host plant phenology and host fidelity on the genetics of apple and hawthorn infesting races of *Rhagoletis pomonella*. Entomologia Experimentalis et Applicata 69:117–135.

Feder, J. L., S. M. Williams, S. H. Berlocher, B. A. McPheron, and G. L. Bush 1999. The population genetics of the apple maggot fly, *Rhagoletis pomonella* and the snowberry maggot, *R. zephyria*: implications for models of sympatric speciation. Entomologia Experimentalis et Applicata 90:9–24.

Feder, J. L., J. B. Roethele, K. E. Filchak, J. Niedbalski, and J. Romero-Severson 2003a. Evidence for inversions related to sympatric host race formation in the apple maggot fly, *Rhagoletis pomonella* (Diptera: Tephritidae). Genetics 163:939–953.

Feder, J. L., S. H. Berlocher, J. B. Roethele, J. J. Smith, W. L. Perry, V. Gavrilovic, K. E. Filchak et al. 2003b. Allopatric genetic origins for sympatric host race formation in *Rhagoletis*. Proceedings of the National Academy of Sciences USA 100:10314–10319.

Feder, J. L., X. Xie, J. Rull, S. Velez, A. Forbes, H. Dambroski, K. Filchak

et al. 2005. Mayr, Dobzhansky, Bush and the complexities of sympatric speciation in *Rhagoletis*. Proceedings of the National Academy of Sciences USA 102:6573–6580.

Gallez, G. P., and L. D. Gottlieb 1982. Genetic evidence for the hybrid origin of the diploid plant *Stephanomeria diegensis*. Evolution 36:1158–1167.

Gavrilovic, V., G. L. Bush, D. Schwarz, J. E. Crossno, and J. J. Smith 2007. *Rhagoletis zephyria* Snow (Diptera: Tephritidae) in the Great Lakes basin: a native insect on native hosts? Annals of the Entomological Society of America 100:474–482.

Grant, P. R., B. R. Grant, and K. Petren 2005. Hybridization in the recent past. American Naturalist 166:56–67.

Green, E., K. Almskaar, S. B. Sim, T. Arcella, W. L. Yee, J. L. Feder, and D. Schwarz 2013. Molecular species identification of cryptic apple and snowberry maggots (Diptera: Tephritidae) in Western and central Washington. 2013. Environmental Entomology 42:1100–1109.

Haddon, M. 1984. A re-analysis of hybridization between mallards and grey ducks in New Zealand. The Auk 101:190–191.

Harrison, R. G. 1993. Hybrids and hybrid zones: historical perspectives. In: R. G. Harrison, ed. Hybrid Zones and the Evolutionary Process, pp. 3–12. Oxford University Press, Oxford.

Hijmans, R. J. 2014. Geosphere: Spherical Trigonmetry. R package version 1.3-11. http:/CRAN.R-project.org/package=geosphere (accessed on 23 May 2015).

Hood, G. R., W. Yee, R. B. Goughnour, S. B. Sim, S. P. Egan, T. Arcella, G. St. Jean et al. 2013. The geographic distribution of *Rhagoletis pomonella* (Diptera: Tephritidae) in the western United States: introduced species or native population? Annals of the Entomological Society of America 106:59–65.

Hubbard, A. L., S. McOrist, T. W. Jones, R. Boid, R. Scott, and N. Easterbee 1992. Is the survival of European wildcats *Felis silvestris* in Britain threatened by interbreeding with domestic cats? Biological Conservation 61:203–208.

Huxel, G. R. 1999. Rapid displacement of native species by invasive species: effects of hybridization. Biological Conservation 89:143–152.

Kirk, H., S. Dorn, and D. Mazzi 2013. Molecular genetics and genomics generate new insights into invertebrate pest invasions. Evolutionary Applications 6:842–856.

Levin, D. A., J. Francisco-Ortega, and R. K. Jansen 1996. Hybridization and the extinction of rare plant species. Conservation Biology 10:10–16.

Liu, K. J., and S. V. Muse 2005. PowerMarker: an integrated analysis environment for genetic marker analysis. Bioinformatics 21:2128–2129.

Lodge, D. M., T. Arcella, A. K. Baldridge, M. A. Barnes, L. Chadderton, A. Deines, J. L. Feder et al. 2012. Global introductions of crayfishes: evaluating the impact of species invasions on ecosystem services. Annual Review of Ecology Evolution and Systematics 43:449–472.

Maeher, D., and G. Caddick 1995. Demographics and genetic introgression in the Florida panther. Conservation Biology 9:1295–1298.

Mallet, J. 2005. Hybridization as an invasion of the genome. Trends in Ecology and Evolution 20:229–237.

Mavárez, J., C. A. Salazar, E. Bermingham, C. Salcedo, C. D. Jiggins, and M. Linares 2006. Speciation by hybridization in *Heliconius* butterflies. Nature 441:868–871.

Mertz, C., D. Koong, and S. Anderson 2013. Washington Annual Agriculture Bulletin. United States Department of Agriculture National Agricultural Statistics Service Northwest Regional Field Office, Olympia, WA.

Michel, A. P., S. B. Sim, T. H. Q. Powell, M. S. Taylor, P. Nosil, and J. L.

Feder 2010. Widespread genomic divergence during sympatric speciation. Proceedings of the National Academy of Sciences USA
107:9724–9729.

Moran, E. V., and J. M. Alexander 2014. Evolutionary responses to global change: lessons from invasive species. Ecology Letters 17:637–649.

Nei, M. 1972. Genetic distance between populations. American Naturalist 106:283–292.

Neilson, W. T. A. 1964. Some effects of relative humidity on development of the apple maggot, Rhagoletis pomonella (Walsh). Canadian
Entomologist 96:810–811.

Oksanen, J., F. G. Blancet, R. Kindt, P. Legendre, P. R. Minchin, R. B.
O'hara, G. L. Simpson et al. 2015. vegan: Community Ecology Package. R package version 2.2-1. http:/CRAN.R-project.org/package=ve-
gan (accessed on 23 May 2015).

Pauls, S. U., C. Nowak, M. Balint, and M. Pfenniger 2013. The impact of
global climate change on genetic diversity within populations and species. Molecular Ecology 22:925–946.

Perry, W. L., J. L. Feder, G. Dwyer, and D. M. Lodge 2001. Hybrid zone
dynamics and species replacement between Orconectes crayfishes in a
northern Wisconsin lake. Evolution 55:1153–1166.

Powell, T. H. Q., G. R. Hood, M. O. Murphy, J. S. Heilveil, S. H. Berlocher, P. Nosil, and J. L. Feder 2013. Genetic divergence along the
speciation continuum: the transition from host race to species in
Rhagoletis (Diptera: Tephritidae). Evolution 67:2561–2576.

Powell, T. H. Q., A. A. Forbes, G. R. Hood, and J. L. Feder 2014. Ecological adaptation and reproductive isolation in sympatry: genetic and
phenotypic evidence for native host races of Rhagoletis pomonella.
Molecular Ecology 23:688–704.

Pritchard, J. K., M. Stephens, and P. Donnelly 2000. Inference of population structure using multilocus genotype data. Genetics 155:945–959.

R Core Team 2014. R: A Language and Environment for Statistical Computing Computer Program, Version 3.1.2. Team, R. C., Vienna, Austria.

Ragland, G. J., K. Almskaar, K. L. Vertacnik, H. M. Gough, J. L. Feder,
D. A. Hahn, and D. Schwartz 2015. Differences in performance and
transcriptome-wide gene expression associated with Rhagoletis (Dip-
tera:Tephritidae) larvae feeding in alternate host fruit environments.
Molecular Ecology. 24:2759–2776.

Rhymer, J. M., and D. Simberloff 1996. Extinction by hybridization and
introgression. Annual Review of Ecology and Systematics 27:83–109.

Rieseberg, L. H. 1997. Hybrid origins of plant species. Annual Review of
Ecology and Systematics 28:359–389.

Rieseberg, L. H., M. A. Archer, and R. K. Wayne 1999. Transgressive segregation, adaptation and speciation. Heredity 83:363–372.

Rundle, H. D. 2002. A test of ecologically dependent postmating isolation between sympatric sticklebacks. Evolution 56:322–329.

Schwarz, D., B. M. Matta, N. L. Shakir-Botteri, and B. A. McPheron
2005. Host shift to an invasive plant triggers rapid animal speciation.
Nature 436:546–549.

Scriber, J. M. 2011. Impacts of climate warming on hybrid zone movement: geographically diffuse and biologically porous "species boundaries". Insect Science 18:121–159.

Scriber, J. M. 2014. Climate-driven reshuffling of species and genes:
potential conservation roles for species translocations and recombi-
nant hybrid genotypes. Insects 5:1–61.

Seehausen, O. 2004. Hybridization and adaptive radiation. Trends in
Ecology and Evolution 19:198–207.

Sim, S. B. 2013. The frontier of ecological speciation: investigating western populations of Rhagoletis pomonella. Ph.D. Dissertation. University of Notre Dame, Notre Dame, IN.

St. Jean, G., S. P. Egan, W. L. Yee, and J. L. Feder 2013. Genetic identification of an unknown Rhagoletis fruit fly (Diptera: Tephritidae)
infesting Chinese Crabapple: implications for apple pest management.
Journal of Economic Entomology 106:1511–1515.

Tracewski, K. T., and J. F. Brunner 1987. Seasonal and diurnal activity of
Rhagoletis zephyria Snow. Melanderia 45:27–32.

Tracewski, K. T., J. F. Brunner, S. C. Hoyt, and S. R. Dewey 1987. Occurrence of Rhagoletis pomonella (Walsh) in hawthorns, Crataegus, of the
PNW. Melanderia 45:19–25.

Velez, S., M. S. Taylor, M. A. F. Noor, N. F. Lobo, and J. L. Feder 2006.
Isolation and characterization of microsatellite loci from the apple
maggot fly Rhagoletis pomonella (Diptera: Tephritidae). Molecular
Ecology Notes 6:90–92.

Westcott, R. L. 1982. Differentiating adults of apple maggot, Rhagoletis
pomonella (Walsh) from snowberry maggot, Rhagoletis zephyria Snow
(Diptera: Tephritidae) in Oregon. Pan-Pacific Entomologist 58:25–30.

WSDA 2001. Washington Administrative Code 16-470-108, Distribution
of Infested or Damaged Fruit is Prohibited. Washington State Department of Agriculture, Olympia, WA.

Xie, X., A. P. Michel, D. Schwarz, J. Rull, S. Velez, A. A. Forbes, M. Aluja
et al. 2008. Radiation and divergence in the Rhagoletis pomonella species complex: inferences from DNA sequence data. Journal of Evolutionary Biology 21:900–913.

Yee, W. L. 2008. Host plant use by apple maggot, western cherry fruit
fly, and other Rhagoletis species (Diptera: Tephritidae) in central
Washington state. Pan-Pacific Entomologist 84:163–178.

Yee, W. L., and R. B. Goughnour 2008. Host plant use by and new host
records of apple maggot, western cherry fruit fly, and other Rhagoletis
species (Diptera: Tephritidae) in western Washington state. Pan-Pacific Entomologist 84:179–193.

Yee, W. L., and R. B. Goughnour 2011. Mating frequencies and production of hybrids by Rhagoletis pomonella and Rhagoletis zephyria
(Diptera: Tephritidae) in the laboratory. Canadian Entomologist
143:82–90.

Yee, W. L., P. S. Chapman, H. D. Sheets, and T. R. Unruh 2009. Analysis
of body measurements and wing shape to discriminate Rhagoletis pomonella and Rhagoletis zephyria (Diptera: Tephritidae) in Washington
State. Annals of the Entomological Society of America 102:1013–1028.

Yee, W. L., H. D. Sheets, and P. S. Chapman 2011. Analysis of surstylus
and aculeus shape and size using geometric morphometrics to discriminate Rhagoletis pomonella and Rhagoletis zephyria (Diptera:
Tephritidae). Annals of the Entomological Society of America
104:105–114.

Yee, W. L., M. W. Klaus, D. H. Cha, C. E. Linn, R. B. Goughnour, and J.
L. Feder 2012. Abundance of apple maggot, Rhagoletis pomonella,
across different areas in central Washington, with special reference to
black-fruited hawthorns. Journal of Insect Science 12:124.

Yee, W. L., P. S. Chapman, and H. D. Sheets 2013. Comparative body
size and shape analyses of F1 hybrid Rhagoletis pomonella and Rhagoletis zephyria (Diptera: Tephritidae). Annals of the Entomological
Society of America 106:410–423.

Yee, W. L., T. W. Lawrence, G. R. Hood, and J. L. Feder 2015. New
records of Rhagoletis Leow, 1862 (Diptera: Tephritidae) and their host
plants in western Montana, U.S.A. Pan-Pacific Entomologist
91:39–57.

Zhao, Z., T. Wahl, and T. Marsh 2007. Economic effects of mitigating
apple maggot spread. Canadian Journal of Agricultural Economics-
Revue Canadienne D Agroeconomie 55:499–514.

The contribution of phenotypic plasticity to the evolution of insecticide tolerance in amphibian populations

Jessica Hua,[1],* Devin K. Jones,[2],* Brian M. Mattes,[2] Rickey D. Cothran,[3] Rick A. Relyea[2] and Jason T. Hoverman[1]

1 Department of Forestry and Natural Resources, Purdue University, West Lafayette, IN, USA
2 Department of Biological Sciences, Rensselaer Polytechnic Institute, Troy, NY, USA
3 Department of Biological Sciences, Southwestern Oklahoma State University, Weatherford, OK, USA

Keywords

acetylcholine esterase inhibitor, amphibian declines, genetic accommodation, *Lithobates sylvaticus*, phenotypic plasticity, toxicology.

Correspondence

Jessica Hua, Department of Forestry and Natural Resources, Purdue University, West Lafayette, IN 47907, USA.

e-mail: jhua13@gmail.com

*Both authors contributed equally to this manuscript.

Abstract

Understanding population responses to rapid environmental changes caused by anthropogenic activities, such as pesticides, is a research frontier. Genetic assimilation (GA), a process initiated by phenotypic plasticity, is one mechanism potentially influencing evolutionary responses to novel environments. While theoretical and laboratory research suggests that GA has the potential to influence evolutionary trajectories, few studies have assessed its role in the evolution of wild populations experiencing novel environments. Using the insecticide, carbaryl, and 15 wood frog populations distributed across an agricultural gradient, we tested whether GA contributed to the evolution of pesticide tolerance. First, we investigated the evidence for evolved tolerance to carbaryl and discovered that population-level patterns of tolerance were consistent with evolutionary responses to pesticides; wood frog populations living closer to agriculture were more tolerant than populations living far from agriculture. Next, we tested the potential role of GA in the evolution of pesticide tolerance by assessing whether patterns of tolerance were consistent with theoretical predictions. We found that populations close to agriculture displayed constitutive tolerance to carbaryl whereas populations far from agriculture had low naïve tolerance but high magnitudes of induced tolerance. These results suggest GA could play a role in evolutionary responses to novel environments in nature.

Introduction

Human activities have dramatically altered the environment through climate change, habitat fragmentation, introduced species, and pollution (Goudie 2005). A significant concern is how populations will respond to such rapid and novel environmental changes (Sutherland et al. 2013). The traditional paradigm predicts that novel environments can enact selection upon existing constitutive traits (i.e. a trait that is constantly expressed regardless of environment; Pigliucci et al. 2006) driving populations toward an optimum (Hoffmann and Sgrò 2011; Lawrence et al. 2012). However, this process depends on existing levels of genetic variation and mutation rates, which often can limit evolution (Le Rouzic and Carlborg 2008). Because of these limitations, there has been a surge of interest in understanding the role of phenotypic plasticity in evolutionary innovation (Pigliucci et al.

2006; Crispo 2007; Moczek et al. 2011; Wund 2012). Phenotypic plasticity, defined as the capacity of a single genotype to produce different phenotypes in different environments, represents a rapid alternative solution in response to novel environments (West-Eberhard 2003; Schlichting 2008). Novel environments can induce organisms to exhibit cryptic genetic variation that may code for adaptive traits within a single generation and allow a population to persist (Lande 2009; Bondduriansky et al. 2012). Thus, phenotypic plasticity has the potential to influence evolutionary outcomes and shape adaptations in populations that experience novel environments. Despite the unprecedented rate of environmental change, our understanding of plasticity's role in evolutionary responses to novel environmental changes remains limited in wild populations (Moczek et al. 2011).

There has been considerable controversy over whether phenotypic plasticity can facilitate adaptation to novel environments, especially in wild populations (De Jong 2005; Pigliucci et al. 2006; Crispo 2007). At the core of this controversy are two fundamental questions: 1) Does exposure to a novel environment reveal cryptic genetic variation in a population through phenotypic plasticity? and 2) Does phenotypic plasticity impede or promote evolutionary change (Braendle and Flatt 2006; Wund 2012)? Controlled laboratory experiments have routinely shown that exposure to novel environments can reveal cryptic genetic variation in populations that sets the stage for evolutionary processes to operate (Schlichting 2008). Moreover, theoretical research and laboratory selection experiments have demonstrated that phenotypic plasticity can promote evolutionary change via genetic accommodation (evolution of environmentally induced phenotypes) including the evolution of constitutive traits from initially plastic traits (i.e. genetic assimilation; Schmalhauzen 1949; Waddington 1956; West-Eberhard 2003; Crispo 2007; Moczek et al. 2011). Through the process of genetic assimilation, selection acts upon the reaction norm leading to a loss of plasticity. Thus, over time, a trait that was previously induced via an environmental cue no longer requires the cue to be expressed (i.e. constitutive expression; Pigliucci et al. 2006). While this research has demonstrated the potential role of phenotypic plasticity in evolutionary responses to novel environments, there have been relatively few attempts to determine whether it actually occurs in nature (Braendle and Flatt 2006; Moczek et al. 2011).

There are challenges to testing whether and how phenotypic plasticity contributes to evolutionary responses to novel environments in wild populations (Scoville and Pfrender 2010; Moczek et al. 2011). In particular, research can be hampered by an incomplete knowledge of the ancestral environment and the initial phenotypic responses of ancestral populations. For instance, both selection on existing constitutive trait and the loss of plasticity through genetic assimilation can result in the same evolutionary end point of constitutive trait expression. Although it is usually impossible to assess ancestral conditions (except in resurrection studies; Franks 2011), a viable solution is to use populations existing along spatial environmental gradients to infer evolutionary processes and ancestral conditions (Scoville and Pfrender 2010). By substituting space for time, one can investigate evidence for genetic assimilation by examining ancestral populations that have not consistently experienced a novel environment to determine whether they express phenotypic plasticity when exposed to a novel environment and whether derived populations that have been consistently exposed to a novel environment express constitutive traits that represent adaptations to the novel environment. While the space for time approach provides a useful proxy for assessing evolutionary responses within a single generation, multigenerational studies that track populations across time are essential next steps to corroborating discoveries that use the space for time approach (Moczek et al. 2011; Wund 2012).

Pesticides are useful tools for addressing the mechanisms underlying evolutionary processes because the agent of selection is known and populations can be easily manipulated (Mallet 1989). Moreover, pesticide use varies across the landscape from high usage in areas close to agriculture to low usage in areas far from agriculture (Declerck et al. 2006). Because spatiotemporal variation in pesticide exposure can lead to rapid environmental changes (Odenkirchen and Wente 2007), the ability to rapidly induce tolerance could play a significant role not only in population persistence within a single generation but also in the evolution of constitutive tolerance across multiple generations through genetic assimilation.

By choosing populations distributed along an agricultural gradient, it is possible to substitute space for time with populations far from agriculture representing more ancestral populations (i.e. with respect to pesticide use) and populations closer to agriculture representing derived populations (Cothran et al. 2013). We can examine individuals that have never been exposed during their lifetime (hereafter termed 'naïve' individuals) and individuals that have been exposed earlier in their life (hereafter termed 'exposed' individuals) to test the potential role of plasticity in evolutionary responses to pesticides. For instance, if pesticides select for constitutive tolerance in populations, we would predict that naïve individuals from derived populations (i.e. populations close to agriculture) should express higher naïve tolerance while individuals from ancestral populations (i.e. populations far from agriculture) should express lower naïve tolerance (Cothran et al. 2013). Thus, we would predict a negative relationship between distance to agriculture and the naïve tolerance of populations (Fig. 1A). Additionally, if genetic assimilation is the mechanism underlying the evolution of constitutive tolerance, ancestral populations should express a greater magnitude of induced tolerance in response to a novel pesticide environment whereas derived populations should express a lower magnitude of induced tolerance, perhaps to the point of expressing only high levels of constitutive pesticide tolerance (Waddington 1956; Pigliucci et al. 2006; Crispo 2007; Wund 2012; Hua et al. 2013b, 2014). Thus, we would predict a positive relationship between distance to agriculture and the magnitude of induced pesticide tolerance in a population (Fig. 1B). Finally, the role of genetic assimilation in evolutionary innovation (i.e. the acquisition of novel morphologies and/or behaviors that open new niches, providing new ways to successfully exploit the environment; Allen and Holling 2010) also can be evaluated by testing whether populations expressing the highest naïve tolerance express the lowest magnitudes of inducible tolerance, perhaps due

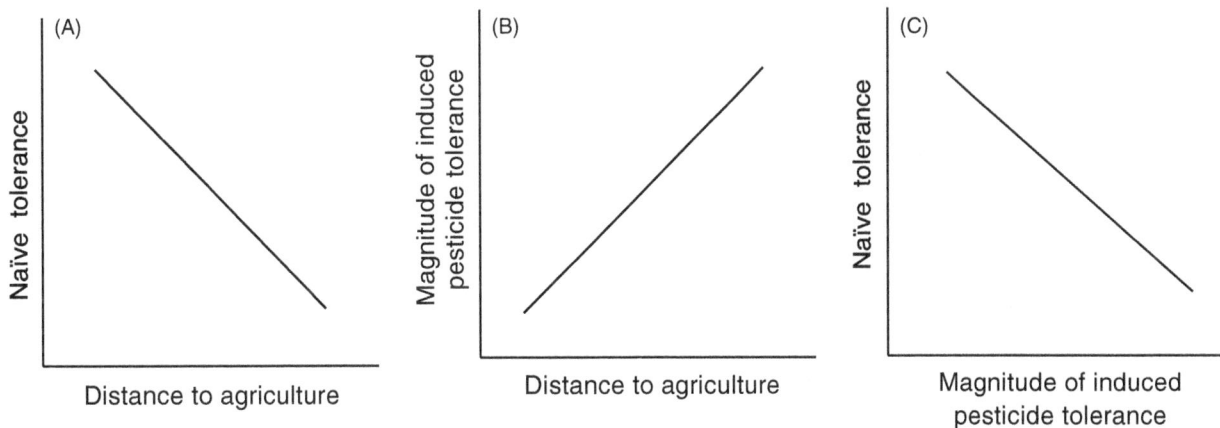

Figure 1 A conceptual framework for the predictions of evolved pesticide tolerance through the process of genetic assimilation. (A) If pesticides select for increased tolerance in populations over time, a negative relationship is predicted between distance to agriculture and naïve tolerance to pesticides. (B) If genetic assimilation is a mechanism for achieving the evolution of constitutive tolerance, a positive relationship is predicted between distance to agriculture and the magnitude of induced pesticide tolerance among populations. (C) If populations consistently exposed to pesticides incur costs associated with the expression of plasticity, then a negative relationship is predicted between naïve tolerance to insecticides and the magnitude of induced pesticide tolerance.

to the costs associated with the expression of plasticity in constant environments (Crispo 2007). Thus, we would predict a negative relationship between the naïve tolerance of populations and the magnitude of induced pesticide tolerance (Fig. 1C). Here, we test these predictions using a commonly used insecticide, carbaryl, and larval wood frogs from 15 populations distributed along an agricultural gradient.

Study system

Ponds provide an excellent study system to explore the role of phenotypic plasticity in evolutionary innovation. Ponds are abundant, have well-defined populations, and are broadly distributed across agricultural gradients that pose variable amounts of insecticide exposure risk due to direct application and runoff (De Meester et al. 2005; Declerck et al. 2006; Gilliom 2007). Using ponds distributed along an agricultural gradient, we can assess whether patterns of pesticide tolerance are consistent with predictions of genetic assimilation. Furthermore, ponds are important habitats for many species, including amphibians. We have recently shown that nine amphibian populations vary in their constitutive tolerance to pesticides with populations living close to agriculture having higher constitutive tolerance compared to populations far from agriculture (Cothran et al. 2013). Additionally, amphibians have the ability to respond plastically to insecticides by inducing increased tolerance to lethal concentrations of insecticides following early exposure to sublethal concentrations of insecticides (Hua et al. 2013b, 2014). Using four wood frog (*Lithobates sylvaticus*) populations, Hua et al. (2013b) demonstrated that patterns of insecticide tolerance were consistent with

genetic assimilation theory with the two ancestral (far from agriculture) wood frog populations exhibiting plasticity to a common insecticide by inducing increased tolerance but the two derived populations (close to agriculture) exhibiting constitutive tolerance. While our past work suggests that phenotypic plasticity and genetic assimilation played a role in evolutionary responses to pesticides, the essential next step is to expand beyond four populations to specifically test the theoretical predictions needed to provide support for the possible role of genetic assimilation in wild populations (Fig. 1; Waddington 1956; Pigliucci et al. 2006; Crispo 2007; Wund 2012).

Methods

Insecticide background

Our focal pesticide was the insecticide carbaryl (commercial formulation Sevin©, 22.5% active ingredient CAS 63-25-2), which is a common insecticide used for pest control and disease prevention in agricultural and residential settings (Grube et al. 2011). The half-life of carbaryl at a pH of 7 is 10 days, and environmental concentrations in aquatic systems range from 0.73 to 1.5 mg/L (USEPA 2008). Carbaryl operates by reversibly binding to acetylcholine esterase (AChE) ultimately leading to the accumulation of acetylcholine and mortality (Lajmanovich et al. 2010). Reported carbaryl LC50 values for amphibians range from 1.2 to 22 mg/L (Boone and Bridges 1999; Relyea 2005).

Animal collection and husbandry

We tested for induced tolerance to carbaryl in 15 wood frog populations collected across western Pennsylvania, USA

(Table S1). We chose wood frog populations that were separated by at least 4 km. The genetic neighborhood of wood frogs is generally within ~1 km of the breeding pond which means that we were most likely using 15 distinct populations (Berven and Grudzien 1990; Semlitsch 1998). Animals from all 15 wood frog populations were collected as early-stage embryos (Gosner 1960) within a 7-days period (Table S1). To control for the effects of developmental stage and size on sensitivity to insecticides, we manipulated temperature to standardize hatching time (Cothran et al. 2013; Hua et al. 2013a). Initially, all clutches were raised outdoors in 100-L pools filled with 90 L of aged well water (air temperature ranged from 1 to 20.6°C). On 13 April 2014, clutches collected before 7 April 2014 were chilled (1.6°C) by placing them in a walk-in cooler to slow development while clutches collected after 7 April 2014 remained outdoors in 100-L pools where they experienced warmer conditions (air temperature ranged from 10.5 to 26.1°C). After 34 h, embryonic development of clutches collected after 7 April 2014 converged with those collected before 7 April 2014 and we moved all egg masses collected prior to 7 April 2014 back into outdoor 100-L pools filled with 90 L of aged well water. On 21 April 2014, wood frogs from all 15 populations hatched within a 20-h period (Gosner stage 20).

Distance to agriculture

To determine each population's proximity to agriculture, we measured the linear distance from each pond (at the location, egg masses were collected) to the nearest agricultural field using Google Earth (2013, v. 7.1.2). We defined an agricultural field as any plot of land from 1993 to 2013 that was used for growing crops, raising livestock and small farm animals for domestic and commercial uses, or feedlots intended for game animals. We did not differentiate among the various types of agriculture as farmers in the area rotate crops planted from year to year. For each population, we confirmed agricultural status of all fields by visiting the field and talking to landowners or local USDA agents. However, we were not able to assess the amount or frequency of pesticide application or historical land use prior to 1993. Additionally, we note that other factors, such as topography, canopy cover, water depth, and surface area, may influence the amount of pesticides that runoff into each pond (Schriever and Liess 2007). However, these additional factors would be expected to add variation to our data and potentially obscure spatial patterns of tolerance across the 15 populations.

Experimental design overview

We tested for induced tolerance using a two-phase experiment similar to that of Hua et al. (2013b). In Phase 1, we exposed wood frog hatchlings (Gosner stage 20) from all 15 populations to either a pesticide-free control or a sublethal carbaryl treatment to induce tolerance. In Phase 2 of the experiment, we tested whether exposure to a sublethal concentration of carbaryl during the hatchling stage induced an increase in tolerance to carbaryl during the tadpole stage (Gosner stage 25). We assessed tolerance using a time to death (TTD) assay, a common measure of relative tolerance among different experimental groups (Bridges and Semlitsch 2000; Cothran et al. 2013).

Phase 1 – Inducing higher tolerance

For all 15 populations, we haphazardly chose 300 hatchlings from each population once animals reached Gosner stage 20 on 21 April 2014. Using 14-L plastic containers as our experimental unit, we assigned 150 hatchlings from each population to 7 L of a pesticide-free control (UV-irradiated, carbon-filtered well water) or 7 L of a sublethal carbaryl solution (nominal concentration: 0.5 mg/L of carbaryl). These two groups represented our naïve and exposed tadpoles, respectively. We chose 0.5 mg/L as the sublethal concentration because past studies have demonstrated that this concentration induces tolerance without causing mortality (Hua et al. 2013b). Hatchlings were held in the laboratory at a constant temperature of 21°C on a 16:8 light dark cycle, and the insecticide solutions were not renewed. After 72 h of exposure, we transferred all individuals (Gosner 24) to 14-L containers filled with 7 L of pesticide-free well water. The hatchlings were held in clean water and were not fed because they were still living on their yolk reserves for 24 h until all individuals reached Gosner stage 25. We euthanized (MS-222 overdose) and preserved 10 randomly selected tadpoles from each treatment to assess the effect of sublethal exposure to carbaryl on tadpole mass at the end of Phase I (Appendix S1).

Phase 2 – TTD assay: Lethal exposure to assess induced tolerance

Once tadpoles from all populations reached Gosner stage 25 on 25 April 2014, we began Phase 2 of the experiment by crossing the two Phase 1 treatments with a pesticide-free control and a lethal carbaryl treatment in the TTD assay. Our objective for the TTD assays was to cause moderate mortality over time (Newman 2010). Based on a pilot toxicity study, we chose to use 20 mg/L of carbaryl. TTD assays commonly use relatively high concentrations as a tool to assess the relative sensitivities of different groups with the expectation that these differences in mortality also provide information regarding relative differences in sublethal effects between groups (Newman 2010). Using a factorial, completely randomized design, this produced 60 treatments (15 populations × two Phase 1 treatments × two Phase 2 treatments) that were each replicated five times for a total of 300 experimental units.

The experimental units were 100-mL glass Petri dishes filled with either 70 mL of water (control) or 70 mL of the lethal carbaryl solution (20 mg/L). Keeping individuals from each population together, we haphazardly assigned 10 tadpoles to each experimental unit. We conducted water changes every 24 h with a renewal of the pesticide concentration. To assess tadpole tolerance using TTD, we monitored tadpole mortality every 2 h for the first 12 h, every 4 h after 12 h, and terminated the experiment at 96 h. In accordance with standard toxicity tests, tadpoles were not fed during the test (ASTM 2014). The tadpoles had food reserves in the form of yolk as evidenced by the low mortality observed in the pesticide-free controls from the TTD assay (0.2% mortality).

Insecticide applications

To create working solutions, we mixed commercial grade carbaryl with UV-irradiated, carbon-filtered well water (pH = 7.5). For Phase 1, we added 15 µL of commercial grade carbaryl to 7 L of filtered water in plastic 14-L containers to achieve 0.5 mg/L of carbaryl. Hatchlings from all 15 populations were added within 10 min of dosing. For Phase 2, we added 1.185 mL of commercial grade carbaryl to 14 L of filtered water in a 45.5 L glass aquarium to achieve 20 mg/L. We then added 70 mL of the 20 mg/L carbaryl solution to each Petri dish. After adding the insecticide solutions, we added ten tadpoles to each Petri dish within 20 min of dosing. Finally, we used UV-irradiated, carbon-filtered water to create the control solutions and added 70 mL and ten tadpoles to each Petri dish within 30 min of their carbaryl counterparts.

Insecticide testing

To determine the actual concentrations of insecticides used in this study, we collected a 500-mL sample of the 0.5 mg/L treatment after hatchlings were added during Phase 1 and a 500-mL sample of the 20 mg/L treatment after tadpoles were added into Petri dishes during Phase 2. Because we used filtered water from the same source for the control treatments in both Phase 1 and 2, we collected a single 500 mL sample from this source to be tested. All samples were sent to the University of Connecticut's Center for Environmental Sciences and Engineering (Storrs, CT). For nominal concentration 0, 0.5, and 20 mg/L, actual concentrations were 0, 0.7, and 21 mg/L, respectively (reporting limit = 0.5 µg/L).

Statistical analysis

To explore the potential contribution of plasticity to the evolution of constitutive tolerance, we tested the following predictions: (i) the naïve tolerance of populations to carbaryl will be negatively related to distance to agriculture, (ii) the magnitude of induced tolerance will be positively related to distance to agriculture, and (iii) the magnitude of induced tolerance in the populations will be negatively related to the naïve tolerance of the populations.

For our measure of naïve pesticide tolerance for each of the populations, we focused on tadpoles that were not exposed to pesticides during Phase 1 but were exposed to carbaryl during the TTD assay of Phase 2. Naïve tolerance was calculated as the proportion of tadpoles that survived the lethal carbaryl exposure at 96 h for each of the 15 populations. We used a univariate analysis of variance (UNIANOVA; SPSS 21) to assess population-level differences in survival at 96 h. Because individuals within a Petri dish are not independent of each other, we included dish as a random effect. Also, as the normality assumption was not met, we ranked transformed the data and used SNK post hoc analysis which is an appropriate pairwise analysis for ranked cases (Quinn and Keough 2002).

To test for the presence of induced tolerance to carbaryl in the populations, we used Cox regression analyses (SPSS 21) to calculate a hazard regression coefficient (b) for each population that compared the survival of naïve individuals to the survival of exposed individuals. We conducted a separate Cox regression analysis for each population using individual tadpole time to death values with Petri dish included as a covariate (Hua et al. 2013b, 2014). The value of the coefficient indicates the relative probability that an exposed tadpole will experience mortality when exposed to a lethal concentration of carbaryl compared to naïve tadpoles (Walters 2009). When b < 0, exposed tadpoles are less likely to experience mortality from a lethal dose compared to naïve tadpoles (i.e. there is induced tolerance; Walters 2009). In contrast, when b > 0, exposed tadpoles are more likely to experience mortality from a lethal dose compared to naïve tadpoles (i.e. there is induced susceptibility).

Using our measures of naïve and induced tolerance to carbaryl, we conducted three linear regression analyses (SPSS 21) to assess the predictions of genetic assimilation theory. Our first analysis assessed the relationship between distance to agriculture and naïve tolerance (average survival at 96 h). Our second analyses assessed the relationship between distance to agriculture and the magnitude of induced tolerance to carbaryl (b). Our third analysis assessed the relationship between naïve tolerance (average survival at 96 h) and the magnitude of induced tolerance to carbaryl (b). In two populations (RR and STB), a pre-exposure induced lower tolerance (i.e. b > 0). To assess whether these populations influenced the interpretation of the regression analyses, we conducted two additional regression analyses that assessed the relationship between induced tolerance to carbaryl versus distance to agriculture and induced tolerance to carbaryl versus constitutive tolerance (average survival at 96 h) with the two populations excluded. For all analyses, as we had *a priori* predictions about the direction of ach relationship, we present just the one-tailed results (all other results reported in Table S2).

Results

Patterns of naïve carbaryl tolerance with distance to agriculture

Wood frog populations displayed significant variation in their naïve tolerance to lethal carbaryl concentrations ($F_{14,56} = 4.7$; $P < 0.001$). Average mortality at 96 h ranged from 4% to 56% (Fig. S2). Population-level variation in naïve tolerance was negatively related to distance to agriculture ($r = -0.47$, $P = 0.04$; Fig. 2A); the naïve tolerance of populations living close to agriculture was higher than populations living far from agriculture.

Patterns of the magnitude of induced tolerance with distance to agriculture

We found significant population-level variation in the magnitude of induced tolerance to carbaryl (Fig. 3; Table 1). Cox regression analyses indicated four wood frog populations (HOP, LOG, REE, and XTI ponds), which are all relatively far from agriculture, displayed significant, induced tolerance to carbaryl. In these populations, exposed tadpoles had significantly higher tolerance to a lethal dose of carbaryl later in life than naïve tadpoles from

the same population (all $P < 0.05$). In contrast, early exposure to 0.5 mg/L of carbaryl resulted in a decrease in tolerance for tadpoles from RR and STB pond, which are relatively close to agriculture, although only significant for STB (Fig. 3; Table 1).

The magnitude of induced pesticide tolerance in a population was related to distance to agriculture. Using all 15 populations, we found a positive relationship between distance to agriculture and inducible tolerance ($r = 0.56$, $P = 0.01$; Fig. 2B). The relationship between distance to agriculture and induced tolerance was stronger when the two populations with positive hazard regression coefficient (b) were excluded ($r = 0.73$, $P = 0.002$). Collectively, these results demonstrate that populations farther from agriculture were more inducible for carbaryl tolerance than populations closer to agriculture.

Relationship between the magnitude of induced tolerance and naïve tolerance

Our final analysis examined the relationship between naïve pesticide tolerance and the magnitude of induced pesticide tolerance. Regardless of whether populations with positive b-values were included or not (Fig. 2C), we found a negative relationship between the naïve tolerance of populations and the magnitude of induced tolerance in these populations ($r = -0.73$, $P = 0.001$ and $r = -0.63$, $P = 0.01$, respectively).

Reaction norms of tadpoles exposed to sublethal vs. no carbaryl across 15 populations

Finally, using mean mortality of each population at the end of the TTD assay, we present our data in reaction norm format. We first indicate predicted reaction norms for

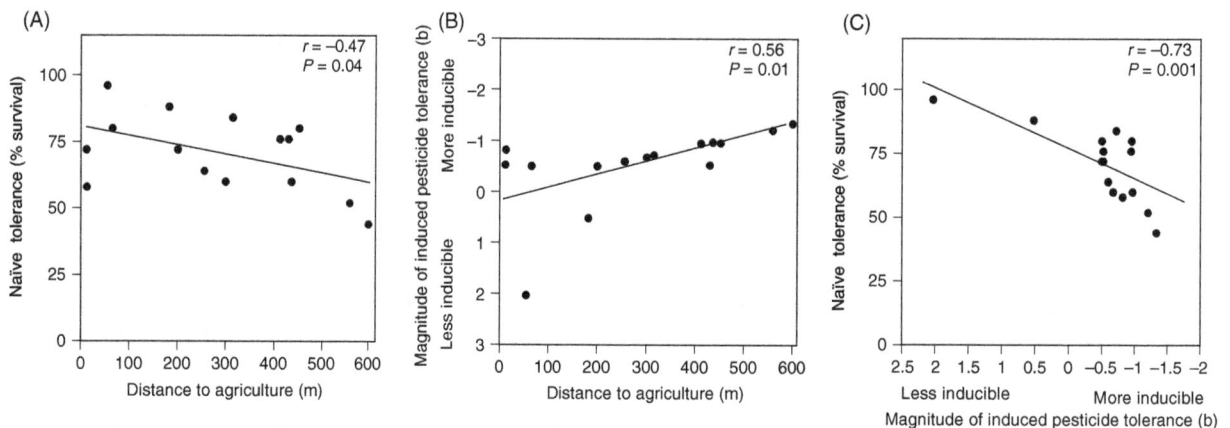

Figure 2 The relationships between (A) distance to agriculture and mean naïve tolerance, (B) distance to agriculture and the magnitude of induced tolerance to carbaryl, and (C) the magnitude of induced tolerance and naïve tolerance to carbaryl. All three relationships were consistent with predictions of genetic assimilation.

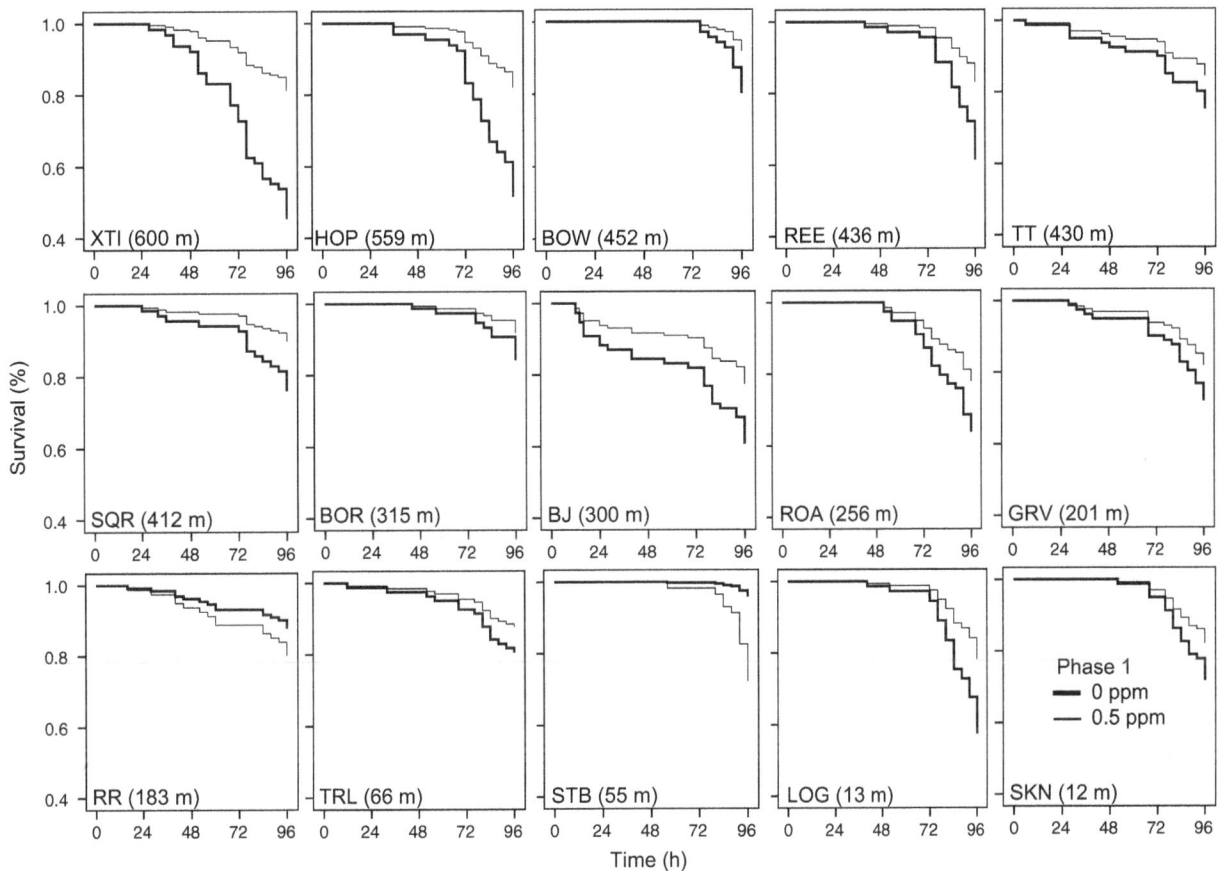

Figure 3 The proportion survival of wood frog tadpoles from 15 populations over time after being exposed to 0 vs. 0.5 mg/L of carbaryl at the hatchling stage and a lethal concentration of carbaryl (20 mg/L) as tadpoles. Values in parenthesis indicate the population's distance from agriculture.

ancestral and derived population if genetic assimilation (GA) is occurring (Fig. 4A). Next, we present the reaction norm of each of the 15 wood frog populations (Fig. 4B). Given the large number of populations, we report the reaction norms of all 15 populations by splitting the populations into three groups: Populations that are >415 m from agriculture (TT, BOW, REE, HOP, XTI), populations that are 201–415 m from agriculture (GRV, ROA, BJ, BOR, and SQR), and populations that are <200 m from agriculture (SKN, LOG, STB, TRL, and RR). In addition to reporting the reaction norm line for each population (dotted lines), we also report the average reaction norm of all populations within each group (solid lines). We find that the patterns of reaction norms of populations that vary in distance to agriculture are consistent with predictions of GA.

Discussion

As human populations grow, there is a critical need to understand how anthropogenic stressors will influence the ecology and evolution of wild populations (Goudie 2005;

Lawrence et al. 2012). In particular, there has been an increasing emphasis on identifying the underlying mechanisms that drive evolutionary responses of populations to anthropogenic influences, including the role of plasticity (Crispo et al. 2010). We found that population-level patterns of pesticide tolerance in wood frogs were consistent with evolutionary responses to agricultural pesticide use and that the patterns of induced versus constitutive tolerance were consistent with predictions of genetic assimilation. Our results suggest that the ability to induce tolerance to pesticides is a common phenomenon across multiple wood frog populations allowing amphibians to rapidly induce increased pesticide tolerance within a single generation. Additionally, patterns of inducible tolerance suggest that genetic assimilation may play a role in the evolution of constitutive tolerance in wild populations.

As our reliance on pesticides intensifies, there is a great need to understand the mechanisms that drive evolutionary responses of wild populations that are often inadvertently exposed to contaminant stressors. Toward this goal, one challenge in tracking evolutionary responses in wild popu-

Table 1. Hazard regression coefficient (b) of tadpoles exposed to 0 mg/L vs. 0.5 mg/L in Phase 1 determined by Cox regression analysis for tadpoles from each population. Censored values indicate % tadpoles that did not experience mortality by 96 h in the TTD assay. Bold values indicate a significant difference in the TTD of tadpoles exposed to 0 mg/L vs 0.5 mg/L.

Population	% Censored	Hazard regression coefficient (P-value)
BJ	69	−0.7 (0.07)
BOR	88	−0.72 (0.23)
BOW	86	−0.95 (0.09)
GRV	77	−0.49 (0.24)
HOP	67	−1.20 **(0.001)**
LOG	68	−0.81 **(0.03)**
REE	72	−0.97 **(0.02)**
ROA	71	−0.59 (0.12)
RR	84	0.53 (0.30)
SKN	77	−0.52 (0.22)
SQR	83	−0.94 (0.06)
STB	84	2.04 **(0.001)**
TRL	84	−0.49 (0.33)
TT	80	−0.52 (0.25)
XTI	63	−1.3 **(<0.001)**

lations is that the process can occur across broad temporal scales. However, using a space for time approach, it is possible to infer evolutionary processes. For instance, evolutionary theory predicts that if the ancestral state of the populations is to have inducible tolerance, then populations consistently exposed to pesticides (i.e. populations close to agriculture) should evolve high, constitutive tolerance whereas those not consistently exposed to pesticides (i.e. populations far from agriculture) should retain a low level of naïve tolerance that can be increased via induction by a sublethal exposure to the pesticide (Crispo 2007; Brausch and Smith 2009; Cothran et al. 2013). In support of this theory, we provide evidence demonstrating that wood frog populations living closer to agriculture had higher naïve tolerance to the insecticide carbaryl compared to those living far from agriculture. A growing number of studies have also demonstrated similar patterns in target and nontarget species (Brausch and Smith 2009; Cothran et al. 2013; Bendis and Relyea 2014; Nkya et al. 2014). For instance, a negative relationship between distance to agriculture and naïve tolerance was detected in fairy shrimp in response to cyfluthrin, methyl parathion, and DDT (Brausch and Smith 2009). A similar relationship was documented in mosquito populations in response to the pyrethroid insecticide, and deltamethrin (Nkya et al. 2014). Thus, despite the initially detrimental effects of pesticides on nontarget populations, evidence suggests that populations can evolve in response to these stressors. Identifying the mechanisms that drive these processes is an essential next step in understanding the broader implications of evolved tolerance in nontarget populations.

Evolved pesticide tolerance can be achieved through selection for existing constitutive traits or by the evolution of constitutive traits from initially plastic traits (i.e. genetic assimilation). If genetic assimilation plays a role in evolved constitutive tolerance, theory predicts that 1) ancestral populations (i.e. populations far from agriculture) should induce increase tolerance when exposed to pesticides, thereby making their formerly cryptic variation for pesticide tolerance become apparent to selection (i.e. noncryptic), 2) derived populations (i.e. populations close to agriculture) should express high naïve tolerance to pesticides, and 3) the degree of naïve tolerance will be negatively related with the magnitude of induced tolerance. Our analyses indicated that early exposure to sublethal carbaryl indeed induced increased tolerance to lethal concentrations of carbaryl in tadpoles from populations relatively far from agriculture (HOP, LOG, REE, and XTI ponds). Tadpoles with induced tolerance to carbaryl have previously been shown to have increased AChE concentration (Hua et al. 2013b). However, the molecular mechanisms allowing organism to achieve tolerance can be highly variable (i.e. modification of ACh binding sites, metabolic detoxification; Feyereisen 1995), underscoring the need for future studies exploring the mechanisms underlying induced tolerance. We also found that the magnitude of induced tolerance was positively correlated with proximity to agriculture. Consistent with our predictions of genetic assimilation, we found a negative relationship between induced pesticide tolerance and naïve tolerance; populations with high naïve tolerance to carbaryl expressed low amounts on induced tolerance to carbaryl. The results are consistent with our past study, which examined only four populations and found that the two populations closer to agriculture were unable to induce tolerance while those farther from agriculture were able to induce increase tolerance (Hua et al. 2013b). Thus, our results suggest that the ability to induce tolerance to pesticides may be a common phenomenon across multiple wood frog populations and patterns of inducible tolerance were consistent with predictions of genetic assimilation.

Within the past few decades, the field of evolutionary biology has seen a surge of interest in understanding the role of plasticity and genetic assimilation in evolutionary innovation (Wund 2012; Schlichting and Wund 2014). Understanding the ability for populations to induce pesticide tolerance and develop constitutive tolerance through the process of genetic assimilation may have broad conservation implications. Amphibian populations are declining worldwide, and pesticides are often implicated as a contributor to these declines (Blaustein et al. 2011). We suggest that genetic assimilation offers an alternative perspective to understanding the role of pesticide contaminants in amphibian populations declines. With growing human

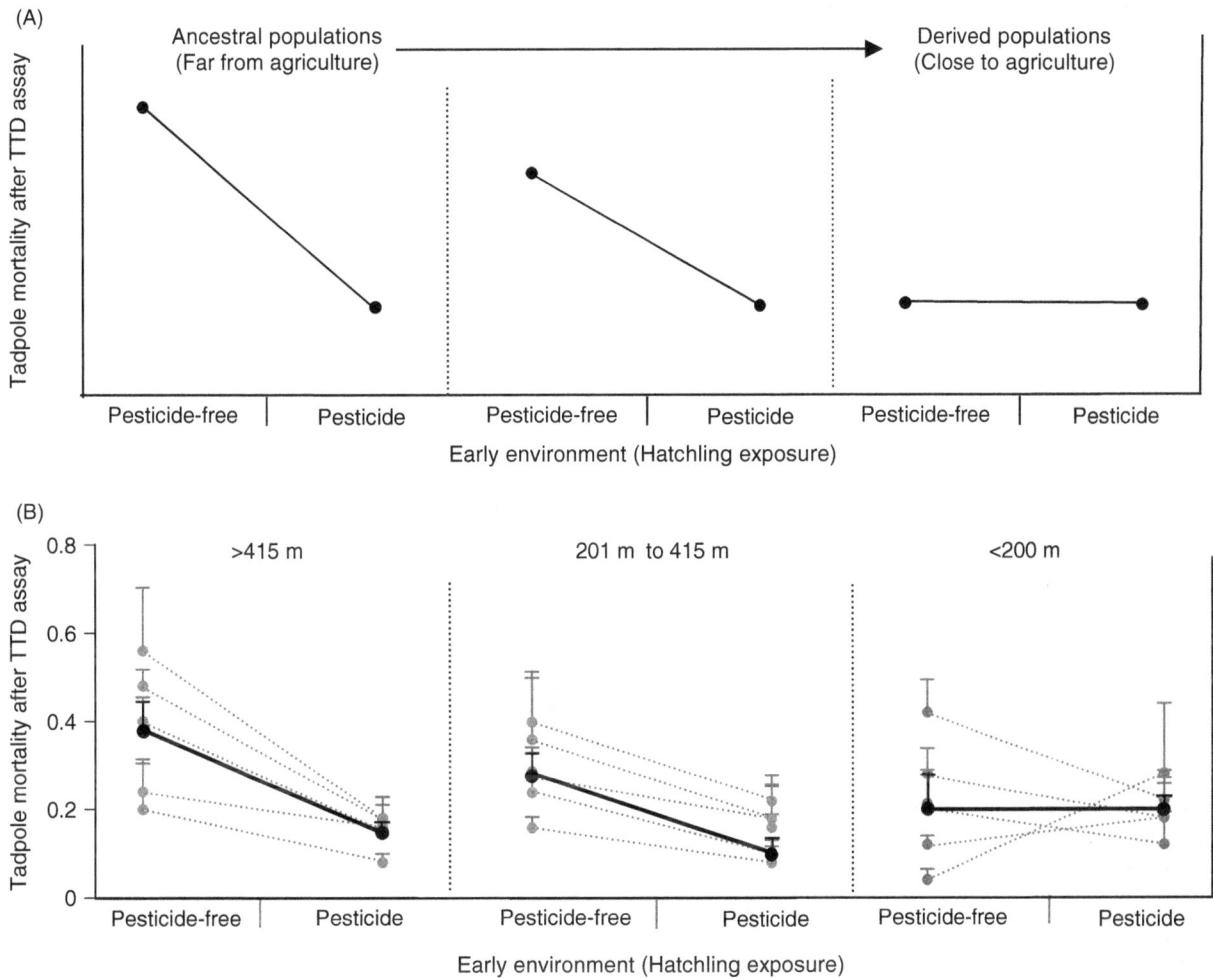

Figure 4 (A) Predicated reaction norms for ancestral and derived population if genetic assimilation (GA) is occurring. (B) Reaction norm of 15 wood frog populations that vary in distance to agriculture is consistent with predictions of GA. Dotted lines indicate the reaction norm for each individual population within a distance to agriculture range (i.e. >415 m, 201–415 m, or <200 m). Solid lines indicate the average reaction norm of all populations within a distance to agricultural range. Pesticide free = 0 ppm carbaryl and Pesticide = 0.5 ppm carbaryl.

populations, genetic assimilation may allow wild populations to rapidly evolve in response to anthropogenic stressors in their environment. However, although genetic assimilation offers a potentially optimistic outlook for populations faced with anthropogenic stressors, consideration of potential costs associated with genetic assimilation will be crucial.

Although the evolution of constitutive trait expression can be achieved through selection for existing constitutive traits and selection on plastic traits (i.e. genetic assimilation), the role of genetic assimilation has largely been ignored. To date, our study provides one of the most comprehensive approaches examining the role of plasticity and genetic assimilation in the evolution of pesticide tolerance in wild populations. However, we emphasize that additional studies are imperative. The present study focused on

a single species and pesticide; future work needs to expand to other taxa and pesticides to determine whether the identified patterns are generalizable. Further, our work does not consider the potential for nongenetic inheritance due to epigenetic inheritance (transmission of DNA methylation variants), somatic inheritance (parentally derived somatic resources that affect development), or behavioral inheritance (parental influence on developmental environment) and future work considering these contributions are necessary. Additionally, despite the many benefits of space for time substitutions for species with relatively longer generation times, future studies utilizing alternative model organisms with shorter generation times in controlled laboratory or mesocosms settings will be helpful in providing direct mechanistic evidence for genetic assimilation (Fox and Wolf 2006). Thus, while our study provides evidence consistent with the predictions of genetic assimilation in nature, a critical next step is to experimentally track and document this process across multiple generations. In addition to multigeneration research, future studies should also consider the standing intra-population-level variation, which can play a large role in determining patterns of local adaptation. Further, predictions of genetic assimilation infer that costs of plasticity drive selection for constitutive tolerance across time (Crispo 2007). Thus, an important next step is to identify these potential costs. Finally, future studies that identify the underlying molecular mechanisms associated with tolerance can potentially lead to the development of effective detection tools for predicting the contribution of genetic assimilation to the evolution of wild populations. In conclusion, plasticity and genetic assimilation offer an exciting and novel perspective for exploring how anthropogenic stressors influence the evolution of populations in nature (Schlichting and Smith 2002; Schlichting and Wund 2014). Studies exploring these mechanisms are imperative and will likely have broad evolutionary and conservation implications.

Acknowledgements

We thank Erika Yates for her help during the experiments. This work was funded by a National Science Foundation grant to RAR (DEB 11-19430), grants from the University of Pittsburgh's G. Murray McKinley Research Fund to DKJ, a Summer Faculty Grant from the Purdue Research Foundation to JTH, and the Purdue Postdoctoral Scholars in Natural Resources fellowship to JH. All methods were approved by the University of Pittsburgh's IACUC (protocol 12050451).

Literature cited

Allen, C., and C. S. Holling 2010. Novelty, adaptive capacity, and resilience. Ecology and Society 15:24–39.

ASTM E729-96. 2014. Standard Guide for Conducting Acute Toxicity Tests on Test Materials with Fishes, Macroinvertebrates, and Amphibians. ASTM International, West Conshohocken, PA, 2014, www.astm.org. doi: 10.1520/E0729-96R14.

Bendis, R. J., and R. A. Relyea 2014. Living on the edge: populations of two zooplankton species living closer to agricultural fields are more resistant to a common insecticide. Environmental Toxicology and Chemistry/SETAC 33:2835–2841.

Berven, K. A., and T. A. Grudzien 1990. Dispersal in the wood frog (*Rana sylvatica*): implications for genetic population structure. Evolution 44:2047–2056.

Blaustein, A. R., B. A. Han, R. A. Relyea, P. T. J. Johnson, J. C. Buck, S. S. Gervasi, and L. B. Kats 2011. The complexity of amphibian population declines: understanding the role of cofactors in driving amphibian losses. Annals of the New York Academy of Sciences 1223:108–119.

Bonduriansky, R., A. J. Crean, and T. Day 2012. The implications of nongenetic inheritance for evolution in changing environments. Evolutionary Applications 5:192–201.

Boone, M. D., and C. M. Bridges 1999. The effect of temperature on the potency of carbaryl for survival of tadpoles of the green frog (*Rana clamitans*). Environmental Toxicology and Chemistry 18:1482–1484.

Braendle, C., and T. Flatt 2006. A role for genetic accommodation in evolution? BioEssays: news and reviews in molecular. Cellular and Developmental Biology 28:868–873.

Brausch, J. M., and P. N. Smith 2009. Mechanisms of resistance and cross-resistance to agrochemicals in the fairy shrimp *Thamnocephalus platyurus* (Crustacea: Anostraca). Aquatic Toxicology 92:140–145.

Bridges, C. M., and R. D. Semlitsch 2000. Variation in pesticide tolerance of tadpoles among and within species of Ranidae and patterns of amphibian decline. Conservation Biology 14:1490–1499.

Cothran, R. D., J. M. Brown, and R. A. Relyea 2013. Proximity to agriculture is correlated with pesticide tolerance: evidence for the evolution of amphibian resistance to modern pesticides. Evolutionary Applications 6:832–841.

Crispo, E. 2007. The Baldwin effect and genetic assimilation: revisiting two mechanisms of evolutionary change mediated by phenotypic plasticity. Evolution 61:2469–2479.

Crispo, E., J. D. DiBattista, C. Correa, X. Thirbert-Plante, A. E. McKellar, A. K. Schwartz, D. Berner et al. 2010. The evolution of phenotypic plasticity in response to anthropogenic disturbance. Evolutionary Ecology Research 12:47–66.

Declerck, S., T. De Bie, D. Ercken, H. Hampel, S. Schrijvers, J. Van Wichelen, V. Gillard et al. 2006. Ecological characteristics of small farmland ponds: associations with land use practices at multiple spatial scales. Biological Conservation 131:523–532.

Feyereisen, R. 1995. Molecular biology of insecticide resistance. Toxicology Letters 82–3:83–90.

Fox, C. W., and J. B. Wolf 2006. Evolutionary Genetics: Concepts and Case Studies, 1st edn. Oxford University Press, New York, NY.

Franks, S. J. 2011. Plasticity and evolution in drought avoidance and escape in the annual plant *Brassica rapa*. New Phytologist 190:249–257.

Gilliom, R. J. 2007. Pesticides in U.S. streams and groundwater. Environmental Science and Technology 41:3408–3414.

Gosner, K. L. 1960. A simplified table for staging anuran embryos and larvae with notes on identification. Herpetologica 16:183–190.

Goudie, A. S. 2005. The Human Impact on the Natural Environment: Past, Present, and Future, 6th edn. Wiley-Blackwell, Malden, MA.

Grube, A., D. Donaldson, T. Kiely, and L. Wu 2011. Pesticides Industry Sales and Usage: 2006 and 2007 Market Estimates. U.S. EPA, Washington, DC.

Hoffmann, A. A., and C. M. Sgrò 2011. Climate change and evolutionary adaptation. Nature 470:479–485.

Hua, J., R. Cothran, A. Stoler, and R. Relyea 2013a. Cross tolerance in amphibians: wood frog (*Lithobates sylvatica*) mortality when exposed to three insecticides with a common mode of action. Environmental Toxicology and Chemistry 32:932–936.

Hua, J., D. K. Jones, and R. A. Relyea 2014. Induced tolerance from a sublethal insecticide leads to cross-tolerance to other insecticides. Environmental Science and Technology 48:4078–4085.

Hua, J., N. I. Morehouse, and R. Relyea 2013b. Pesticide tolerance in amphibians: induced tolerance in susceptible populations, constitutive tolerance in tolerant populations. Evolutionary Applications 6:1028–1040.

De Jong, G. 2005. Evolution of phenotypic plasticity: patterns of plasticity and the emergence of ecotypes. The New Phytologist 166:101–117.

Lajmanovich, R. C., A. M. Attademo, P. M. Peltzer, C. M. Junges, and M. C. Cabagna 2010. Toxicity of four herbicide formulations with glyphosate on *Rhinella arenarum* (Anura: Bufonidae) tadpoles: B-esterases and glutathione S-transferase inhibitors. Archives of Environmental Contamination and Toxicology 60:681–689.

Lande, R. 2009. Adaptation to an extraordinary environment by evolution of phenotypic plasticity and genetic assimilation. Journal of Evolutionary Biology 22:1435–1446.

Lawrence, D., F. Fiegna, V. Behrends, J. G. Bundy, A. B. Phillimore, T. Bell, and T. G. Barraclough 2012. Species interactions alter evolutionary responses to a novel environment. PLoS Biology 10: e1001330.

Mallet, J. 1989. The evolution of insecticide resistance: have the insects won? Trends in Ecology and Evolution 4:336–340.

De Meester, L., S. Declerck, R. Stoks, G. Louette, F. Van De Meutter, T. De Bie, E. Michels et al. 2005. Ponds and pools as model systems in conservation biology, ecology and evolutionary biology. Aquatic Conservation: Marine and Freshwater Ecosystems 15:715–725.

Moczek, A. P., S. Sultan, S. Foster, C. Ledón-Rettig, I. Dworkin, H. F. Nijhout, E. Abouheif et al. 2011. The role of developmental plasticity in evolutionary innovation. Proceedings of the Royal Society of London B: Biological Sciences 278:2705–2713.

Newman, M. C. 2010. Fundamentals of Ecotoxicology. CRC Press, Boca Raton, FL.

Nkya, T. E., I. Akhouayri, R. Poupardin, B. Batengana, F. Mosha, S. Magesa, W. Kisinza et al. 2014. Insecticide resistance mechanisms associated with different environments in the malaria vector *Anopheles gambiae*: a case study in Tanzania. Malaria Journal 13:1–15.

Odenkirchen, E., and S. Wente 2007. Risks of Malathion Use to Federally Listed California Red-Legged Frog (*Rana Aurora draytonii*), pp. 1–272. Environmental Fate and Effects Division Office of Pesticide Programs, Washington, DC.

Pigliucci, M., C. J. Murren, and C. D. Schlichting 2006. Phenotypic plasticity and evolution by genetic assimilation. The Journal of Experimental Biology 209:2362–2367.

Quinn, G. P., and M. J. Keough 2002. Experimental Design and Data Analysis for Biologists. Cambridge University Press, New York, NY.

Relyea, R. A. 2005. The impact of insecticides and herbicides on the biodiversity and productivity of aquatic communities. Ecological Applications 15:618–627.

Le Rouzic, A., and O. Carlborg 2008. Evolutionary potential of hidden genetic variation. Trends in Ecology & Evolution 23:33–37.

Schlichting, C. D. 2008. Hidden reaction norms, cryptic genetic variation, and evolvability. Annals of the New York Academy of Sciences 1133:187–203.

Schlichting, C. D., and H. Smith 2002. Phenotypic plasticity: linking molecular mechanisms with evolutionary outcomes. Evolutionary Ecology 16:189–211.

Schlichting, C. D., and M. A. Wund 2014. Phenotypic plasticity and epigenetic marking: an assessment of evidence for genetic accommodation. Evolution 68:656–672.

Schmalhauzen, I. I. 1949. Factors of Evolution: The Theory of Stabilizing Selection. Blakiston Company, Philadelphia, PA.

Schriever, C. A., and M. Liess 2007. Mapping ecological risk of agricultural pesticide runoff. Science of The Total Environment 384:264–279.

Scoville, A. G., and M. E. Pfrender 2010. Phenotypic plasticity facilitates recurrent rapid adaptation to introduced predators. Proceedings of the National Academy of Sciences of the United States of America 107:4260–4263.

Semlitsch, R. D. 1998. Biological delineation of terrestrial buffer zones for pond-breeding salamanders. Conservation Biology 12:1113–1119.

Sutherland, W. J., R. P. Freckleton, H. C. J. Godfray, S. R. Beissinger, T. Benton, D. D. Cameron, Y. Carmel et al. 2013. Identification of 100 fundamental ecological questions. Journal of Ecology 101:58–67.

USEPA 2008. Amended Reregistration Eligibility Decision (RED) for Carbaryl. USEPA, Washington, DC.

Waddington, C. H. 1956. Genetic assimilation of the bithorax phenotype. Evolution 10:1–13.

Walters, S. 2009. What is a Cox model? http://www.whatisseries. co.uk. (accessed on 30 June 2012).

West-Eberhard, M. J. 2003. Developmental Plasticity and Evolution. Oxford University Press, New York, NY.

Wund, M. A. 2012. Assessing the impacts of phenotypic plasticity on evolution. Integrative and Comparative Biology 52:5–15.

Experimental evolution reveals high insecticide tolerance in *Daphnia* inhabiting farmland ponds

Mieke Jansen,[1] Anja Coors,[2,3] Joost Vanoverbeke,[1] Melissa Schepens,[1] Pim De Voogt,[4] Karel A. C. De Schamphelaere[5] and Luc De Meester[1]

1 Laboratory of Aquatic Ecology, Evolution and Conservation, KU Leuven, Leuven, Belgium
2 ECT Oekotoxikologie GmbH, Flörsheim a.M., Germany
3 Biodiversity and Climate Research Centre (BiK-F), Frankfurt a.M., Germany
4 Institute for Biodiversity and Ecosystem Dynamics (IBED), Universiteit Amsterdam, Amsterdam, The Netherlands
5 Laboratory for Environmental Toxicology and Aquatic Ecology, Environmental Toxicology Unit (GhEnToxLab), Ghent University, Ghent, Belgium

Keywords

adaptation, carbaryl, *Daphnia magna*, evolutionary potential, insecticide tolerance.

Correspondence

Mieke Jansen, Laboratory of Aquatic Ecology, Evolution and Conservation, KU Leuven, Charles Deberiotstraat 32, 3000 Leuven, Belgium.

e-mail: mieke.jansen@bio.kuleuven.be

Abstract

Exposure of nontarget populations to agricultural chemicals is an important aspect of global change. We quantified the capacity of natural *Daphnia magna* populations to locally adapt to insecticide exposure through a selection experiment involving carbaryl exposure and a control. Carbaryl tolerance after selection under carbaryl exposure did not increase significantly compared to the tolerance of the original field populations. However, there was evolution of a decreased tolerance in the control experimental populations compared to the original field populations. The magnitude of this decrease was positively correlated with land use intensity in the neighbourhood of the ponds from which the original populations were sampled. The genetic change in carbaryl tolerance in the control rather than in the carbaryl treatment suggests widespread selection for insecticide tolerance in the field associated with land use intensity and suggests that this evolution comes at a cost. Our data suggest a strong impact of current agricultural land use on nontarget natural *Daphnia* populations.

Introduction

Pesticides are widely used in agriculture to increase crop yield (Waterfield and Ziberman 2012). Spray drift or run-off of pesticides can affect natural, nontarget populations inhabiting surface waters situated in the vicinity of the treated land area, impacting community structure and ecosystem functioning (Parker et al. 1999). Several studies have shown that exposure to toxic substances may induce micro-evolutionary responses in natural populations, leading to genetic differences in the concentration–response curve between exposed and nonexposed populations (e.g. Raymond et al. 2001; Medina et al. 2007; Jansen et al. 2011). Genetic adaptation to pesticides may also come at a cost, however, such as an increased susceptibility to parasites (Jansen et al. 2011) and loss of genetic diversity (Coors et al. 2009).

The water flea *Daphnia* (Crustacea) belongs to the freshwater zooplankton and is a key model organism in ecology and evolution (Lampert 2006; Decaestecker et al. 2007;

Van Doorslaer et al. 2009a; Colbourne et al. 2011; Jansen et al. 2011; Miner et al. 2012) and in ecotoxicology (OECD 2004; Walker et al. 2006). Their cyclic parthenogenetic reproduction allows the exposure of single genotypes to a range of environmental conditions using clonal replicates, to quantify genotype-dependent responses to environmental gradients. The short generation time allows monitoring responses in experimental evolution trials (Van Doorslaer et al. 2009b; Jansen et al. 2011), while the production of dormant stages (ephippia, Hebert 1978) that accumulate in dormant egg banks allows obtaining representative samples of the gene pool of natural populations. Large-bodied *Daphnia* species such as *D. magna* play a central role in the food web of eutrophic ponds and shallow lakes. Such habitats are often quite abundant in agriculture landscapes, where exposure of *Daphnia* populations to pesticides through drainage, spray drift or run-off may be common.

The aim of this study was to investigate whether natural *Daphnia magna* populations inhabiting ponds in agricultural landscapes possess evolutionary potential for adaptation to changes in the level of pesticide exposure, using carbaryl as a model acetylcholinesterase-inhibiting insecticide. To that end, we carried out a selection experiment starting from standing genetic variation sampled by hatching dormant eggs from the egg bank of seven natural populations. After the selection experiment, we performed acute toxicity tests to compare the carbaryl tolerance of the experimental populations that had been exposed to high concentration pulses of carbaryl, the experimental populations that had been exposed to control conditions, and the original field populations. We tested two hypotheses. First, we tested the hypothesis that natural populations of this nontarget species harbour evolutionary potential to respond to selection exerted by pulse exposure to carbaryl. Second, we tested the hypothesis that the selection background of the natural populations (i.e. the selection conditions in their original habitat) influences their response to insecticide selection.

Materials and methods

Our overall research strategy involved a selection experiment on carbaryl tolerance in outdoor containers followed by an assessment of the genetic changes in tolerance to carbaryl in laboratory short-term exposure experiments.

Daphnia populations

We exposed dormant eggs contained in the sediments of seven ponds and small shallow lakes in Flanders known to be inhabited by *D. magna* (Blankaart, Uitkerke, Moorsel, OM 1, OM 2, Oud-Heverlee and Tersaart; see Coors et al. 2009) to favourable hatching stimuli (i.e. exposure to 'spring conditions': fresh noncontaminated dechlorated tap water, long day photoperiod 16L/8D and 20°C, after storage of the eggs for > 1 month at 4°C in the dark; see De Meester and De Jager 1993), and randomly isolated 125 hatchlings from each population. Dormant eggs in *D. magna* are produced sexually, so all hatchlings are genetically unique. These hatchlings were cultured in isolation to establish clonal lineages. The seven study populations here are a subset of the populations studied by Coors et al. (2009) in their analysis of the association between carbaryl tolerance of 10 *D. magna* populations and land use intensity in the immediate neighbourhood of their habitats.

Selection factor carbaryl

The pesticide carbaryl (1-naphthyl methylcarbamate, CAS 63-25-2, purity 99.8%, Sigma-Aldrich, Germany) belongs to the group of carbamates, which together with the organophosphorous insecticides act as acetylcholinesterase inhibitors.

Carbaryl prevents the breakdown of the chemical messenger acetylcholine between the synapses in the nervous system. This results in an overload of acetylcholine in the synaptic cleft leading to overstimulation of the postsynaptic receptors. The receptors are then no longer able to contract or relax in response to a synaptic stimulus (Walker et al. 2006). Since 2007, carbaryl has no marketing authorisation as agricultural insecticide in Belgium anymore, but already sold products may still be used and the VMM (Vlaamse Milieu Maatschappij) still reports carbaryl concentrations in the field (Vlaamse Milieu Maatschappij 2010).

Experimental evolution trials

Using the hatchlings obtained from the sampled dormant egg bank of each of the seven natural populations, we carried out a selection experiment in outdoor containers. We applied two treatments: a carbaryl exposure treatment and an ethanol control treatment, with two replicate containers for every population x treatment combination, giving 28 experimental units. We used 28, 225 L containers filled with 180 L of dechlorinated tap water and covered by mosquito netting to keep away predatory midge larvae. Two weeks before the inoculation of the *D. magna* juveniles, we added an inoculum of 6×10^8 cells of the unicellular green alga *Scenedesmus obliquus* to each container (i.e. 3.3×10^3 cells mL^{-1}). The growth of these algae during the 2 weeks before inoculation of the *Daphnia* provided sufficient food for rapid population growth of the *Daphnia* upon their inoculation. The alga community was not replenished during the experiment and was self-maintained.

In each container, we inoculated clonal descendants of the 125 genetically unique hatchlings of a given *D. magna* population. The inoculate used can be considered a representative sample of the standing genetic variation present in the natural population at the start of a growing season, when populations in nature hatch from their dormant egg banks. Each container received one individual (< 5 days old) from each of 125 clones derived from a single population, resulting in a start density of 0.70 individuals per litre; the inoculated individuals were first-generation and second-clutch offspring of the animals that hatched from the dormant egg bank. For each population, all containers (two replicates x two treatments) received the same, standardized set of unique clones. Carbaryl exposure was achieved by exposing the populations to three pulses of carbaryl (nominal concentration of 32 μg L^{-1}) that were given at an interval of 14 days to mimic spray season, starting 2 weeks after stocking the containers with *D. magna* juveniles. Exposure to 32 μg L^{-1} can be considered a strong selection pressure, as the EC_{50} value for carbaryl for those populations has been estimated to be around 8 μg L^{-1} for juveniles (Coors et al. 2009). At the same time as the carbaryl pulses, ethanol pulses were given in the control treatments, exposing the animals to the same concentration of ethanol used as carrier solvent as in the carbaryl exposure treatment (1 mL ethanol (pure, 99%) per pulse and per container i.e. 0.004 mL L^{-1}).

The container selection experiment ran in total for 80 days, with 55 days after the first pulse of carbaryl, representing approximately 5–8 generations; 27 days after the last pulse was given, the selection experiment ended.

Clones used for the acute toxicity tests

At the end of the selection experiment, we randomly isolated 10 individuals from each population × treatment combination (i.e. pooled over both replicate containers) to start clonal lineages in the laboratory. Random extinction of clones after isolation reduced the number of available lineages to 3–6 per population × treatment combination (see Table S1). These losses of clones were caused by the prolonged period (approx. 82 generations) of culture under suboptimal stock conditions following isolation of the lineages. During this period, the cultures were only fed twice a week. Extinction of clones was random, that is independent of population of origin nor selection history (verified by one-way ANOVA on number of clones that survived: population: $F = 0.547$, df = 6, 14 and $P = 0.77$; selection history: $F = 1.98$, df = 2, 18 and $P = 0.17$).

After the selection experiment, we obtained three sets of clonal cultures for each of the seven sampled natural populations: (i) a set of genotypes directly hatched from the dormant egg banks (called original populations; they are a subset of the lineages that were used to inoculate the containers) and never exposed to selection in the containers, (ii) a set of genotypes obtained from the control treatment in the container experiment (control populations), and (iii) a set of genotypes obtained from the outdoor containers exposed to carbaryl (carbaryl-selected population). Of the 102 lineages tested for their sensitivity to carbaryl, the 33 clonal lineages directly hatched from the original populations are genetically unique as they were hatched from dormant eggs, which are the result of sexual recombination. The remaining 69 clonal lineages were isolated from populations that were reproducing parthenogenetically, so that in principle several isolates may belong to the same clone; of these, 34 were screened for their variation at seven polymorphic microsatellite markers (Jansen et al. 2011) and proved to be unique clones. Unfortunately, the remaining 35 clones could not be genotyped because they were accidently lost before we could genotype them using microsatellites. The fact that all 34 genotyped clones proved genetically unique, however, suggests that the likelihood that our results would be influenced by different isolates belonging to the same genotype is small.

Acute toxicity tests

We carried out standard acute toxicity tests for a total of 102 clonal lineages ($n = 4$–6 lineages per population, for two populations $n = 3$; Table S1) based on the OECD 202 guideline acute *Daphnia* test (Organisation for Economic Co-Operation and Development, OECD 2004).

For two generations prior to the start of the experiment, all isolates were grown under standardized conditions in terms of density and medium to minimize maternal effects. For each clonal lineage, four to six cultures were maintained to generate enough juveniles for the toxicity experiments. In each culture, ten individuals are raised as a cohort in 500-mL jars filled with Aachener Daphnien medium (ADaM, Klüttgen et al. 1994) in a climate-controlled room with a constant temperature (20 ± 1°C) and a fixed day/night regime (16 h light/8 h dark), and daily fed 1×10^5 cells mL^{-1} of the unicellular green alga *Scenedesmus obliquus* until maturation, after which food levels were doubled.

For the acute toxicity tests, we exposed second or third clutch neonates < 24 h old to eight different carbaryl concentrations ranging from 4 to 25.1 μg L^{-1} (spacing factor of 1.3 between concentrations) and an ethanol control.

The analysis of the actual concentrations of carbaryl in test solutions was carried out according to the method described by Cerbin et al. (2010). Briefly, water samples for chemical analyses of carbaryl were preserved with formic acid until analysis. A LC–MS/MS [4000QTrap (Applied Biosystems, Gent, Belgium) operated with electrospray ionization in the positive ion mode (ESI)] was used for the determination of the actual concentrations of the pesticide. The separation was done on a C18 column [Pathfinder (Shimadzu, Duisburg, Germany): 4.6 × 50 mm, silica 300, particle size 3.5 μm], using a mobile phase consisting of methanol and water, both with 5 mM of NH_4Ac. The transitions $202 \rightarrow 145$ and $202 \rightarrow 127$ were monitored.

For each clonal lineage and test concentration, four cohorts of five neonates randomly collected from the juveniles produced in the 500-mL cultures were inoculated in 5 mL ADaM in a 30-mL glass jar. All juveniles were first pooled after which we randomly picked out five individuals for each exposure unit (see guideline OECD 202: OECD 2004). Stock solutions of carbaryl were prepared at eight concentrations in ethanol and stored at −20°C.

Given the size of the experiment, with parallel full scale standardized acute toxicity tests on 102 clonal lineages, tests were split over 86 different days. The order in which clonal lineages were tested was randomized across populations. We minimized the time between inoculating the experimental animals and the carbaryl pulse and ensured that all animals tested on 1 day were exposed to their test concentration at the same moment. Each test included a solvent control with 0.005% ethanol, the same ethanol concentration as in all carbaryl test solutions. At the start of the experiment, we also included a blank control containing only ADaM, but as there was no difference (between 0 and 10%) in mortality in both controls (blank and ethanol control), blank controls were not included in further exposure assays. In case there was more than 10% mortality in the control treatment, we decided to remove that specific test from the analysis according the OECD guideline 202. Experimental animals were not fed during the acute toxicity test and tested in a dark/light cycle (16 h light/8 h dark). After 48 h, all individuals were scored for immobility. On each test day, four different variables were measured in the exposure medium: pH, conductivity (μS cm^{-1}), oxygen level and temperature (°C) (Table S2).

Statistical analysis

Concentration–response curves and response to selection
Generalized linear modelling (GLiM) was applied to analyse acute toxicity response data (proportion immobilized daphnids after 48 h exposure versus log$_{10}$-transformed carbaryl concentrations) and to estimate the median effect concentration (EC$_{50}$) for each population × treatment combination. We used GLiM with a probit link and assumed binomial error distribution as recommended by McCullagh and Nelder (1989) in the statistical software R (R Development Core Team, 2007). To account for extrabinomial variation (i.e. overdispersion expressed as the residual deviance being more than 1.5 greater than the degrees of freedom of the residual deviance), we used as model family 'quasibinomial' and replaced χ^2 tests by F-tests as recommended by Crawley (2007). The results of the quasibinomial model differ from the binomial model only with regard to the standard errors of the parameter estimates, with the quasibinomial model yielding larger and therefore more conservative confidence intervals for point estimates such as the EC$_{50}$. The 95% confidence intervals of the EC$_{50}$ were calculated from the variances and covariances of the model parameters using Fieller's theorem as outlined by Kerr and Meador (1996) and Wheeler et al. (2006).

As the accuracy of the estimate of the EC$_{50}$ is greater when replicate observations are combined in one analysis (OECD 2004), we fitted one concentration–response curve for each population × treatment combination (original, control and carbaryl-selected for each of the seven populations) using the results of all replicate trials with the different clones of this population. A backward elimination process (Crawley 2007) was applied to analyse the influence of origin of the populations (factor 'population') and carbaryl exposure during the selection experiment (factor 'selection history') with regard to the toxicity of carbaryl

on the clonal lineages. Starting with the most complex model explaining the proportion of dead daphnids (y~ carbaryl × population × selection history, i.e. taking all possible interactions into account), the term that caused the least reduction in deviance was deleted in each step. This stepwise reduction process was applied until the least complex model was identified that significantly reduced deviance in comparison with the next more complex model. Following the analysis on all populations, we contrasted selection histories to analyse specifically the influence of differential selection due to carbaryl exposure within the experiment (by comparing 'control' and 'carbaryl-selected' populations) of the overall selection experienced by the carbaryl-selected populations (by comparing only 'original' and 'carbaryl-selected' populations) and of the 'background' selection caused by the containers (by including only 'original' and 'control' populations). In addition to this GLiM analysis, we also carried out a general linear model (ANOVA) on the EC$_{50}$ values calculated from concentration–response curves estimated for each clonal isolate of the different treatments and populations separately, with population and selection history as independent variables (see Table S3; run using Statistica v12, Statsoft 2014: http://www.statsoft.com). This ANOVA approach uses the EC$_{50}$ values estimated for each clone separately, and while it incorporates variation among clones for this variable, it has the disadvantage compared to the GLiM analysis that the single clonal EC$_{50}$ values are less informative than considering the whole concentration–response curve per population because they only represent one point estimate of the response and are calculated from a smaller data base (i.e. on the four replicate cohorts of five individuals tested per clone and concentration).

Relationship with land use
To test for the hypothesis that the response to selection is dependent on the selection background of the study populations, that is the degree to which they are impacted by intensive agriculture, we quantified the relationship between the relative change in carbaryl tolerance during the selection experiment and the land use intensity in the immediate vicinity of the habitats the populations were derived from. The relative change in carbaryl tolerance during the selection experiment was described by the log$_2$ of the ratio of the EC$_{50}$ of each container population ('control' or 'carbaryl-selected') over the EC$_{50}$ in the respective original population [log$_2$ (EC$_{50}$ treatment/EC$_{50}$ original)]. This was calculated for both control and carbaryl-selected populations separately. A value above 0 indicates an increase in tolerance in comparison with the original population and a value below 0 a loss in tolerance, respectively. Using this ratio of EC$_{50}$ values, the change in carbaryl tolerance was analysed by Spearman rank correlation for its

association with land use intensity in the vicinity of the pond of origin of the populations. In earlier work, we documented that carbaryl tolerance of *Daphnia* populations tends to increase with increasing intensity of land use in the immediate vicinity of their habitat (Coors et al. 2009). The seven ponds from which the experimental populations were selected were ranked for land use intensity (1 = lowest and 7 = highest land use intensity) based on the visual screening of the percentage of crop (corn and cereals), pasture and garden area in a radius of 50 m around the pond, and using the percentage of cropland as the main criterion (as described in Coors et al. 2009; see Table S4). This assessment was based on earlier work that established that water quality in farmland ponds is largely determined by land use nearby (< 200 m) the ponds (Declerck et al. 2006).

Results

Concentration–response curves and response to selection

The concentration–response curves of acute exposure to carbaryl for the original field populations and for the experimental populations that were ('carbaryl-selected') or were not ('control') exposed to carbaryl in our selection experiment are shown in Fig. 1 (i.e. carbaryl-selected and control experimental populations of each of the seven original field populations; see also Table S1 for summary statistics). There is a significant main effect of acute carbaryl concentration, population and selection history as well as a carbaryl × population and a population × selection history interaction effect on mortality in the overall analysis (Table 1A). Carbaryl concentration is also significant in each of the three comparisons tested (control versus original; carbaryl-selected versus original; carbaryl-selected versus control; Table 1B). The population from which the dormant egg bank was derived also significantly shaped the response to carbaryl in all three comparisons (Table 1B), illustrating that the populations indeed differ in their concentration–response curves. Selection history (original – control – carbaryl-selected) had a significant effect on the response to carbaryl in the comparison of the control treatment with the original population as well as in the comparison of the control with the carbaryl-selected populations, but not in the comparison of the carbaryl-selected with the original populations (Table 1B). Control populations showed less carbaryl tolerance compared to carbaryl-selected populations and compared to the original populations (Fig. 1). There is a significant interaction effect between population and selection history in the comparison between carbaryl-selected and control populations (Table 1B). This implies that the populations reacted differently to the selection pressures imposed during the selection experiment. This is clearly illustrated in Fig. 2D,

which shows that while most populations have a higher tolerance to carbaryl in the carbaryl-selected compared to the control selection treatment, one population (OM 1) had lower tolerance in the carbaryl-selected than in the control population. The ANOVA on the EC_{50} values calculated from concentration–response curves estimated for each clonal isolate of the different treatments and populations separately, yielded results in line with the GLiM analysis, with significant main effects of population (reflecting genetic differences among the populations) and selection history (reflecting evolutionary potential). This analysis did not reveal, however, a population × selection history interaction effect (see Table S3).

Relationship with land use

Similar to the results presented by Coors et al. (2009), the Spearman rank correlation between land use intensity and carbaryl tolerance (represented by the EC_{50}) in the original populations shows a marginally nonsignificant positive association ($r = 0.739$; $P = 0.058$; Fig. 2A). There is a significant negative correlation between the degree to which populations became less tolerant to carbaryl in the control treatment compared to the original populations and land use intensity in the vicinity of the ponds from which the original populations were obtained ($r = -0.93$; $P < 0.01$; Fig. 2B). There is no association between EC_{50} ratio of carbaryl-selected over original populations and land use intensity in the vicinity of the ponds ($r = 0.21$; $P = 0.66$; Fig. 2C). There is a tendency for a positive correlation ($r = 0.71$, marginally nonsignificant: $P = 0.08$) between the difference in carbaryl tolerance between the carbaryl-selected and the control treatment populations and land use intensity (Fig. 2D).

Discussion

Overall, our results show that natural populations of the water flea *D. magna* isolated from a gradient in land use intensity differ in their sensitivity to the pesticide carbaryl (cf. significant effect of population in Table 1), show evolutionary potential to respond to changes in carbaryl exposure (cf. significant effect of selection history in Table 1, our first hypothesis) and differ in the magnitude of this evolutionary potential (cf. significant effect of population × selection history in Table 1, our second hypothesis 2, Fig. 2).

Evolutionary potential for carbaryl tolerance

The overall GLiM analysis indicates significant main effects of carbaryl, population and selection history as well as significant   population × selection   history   and

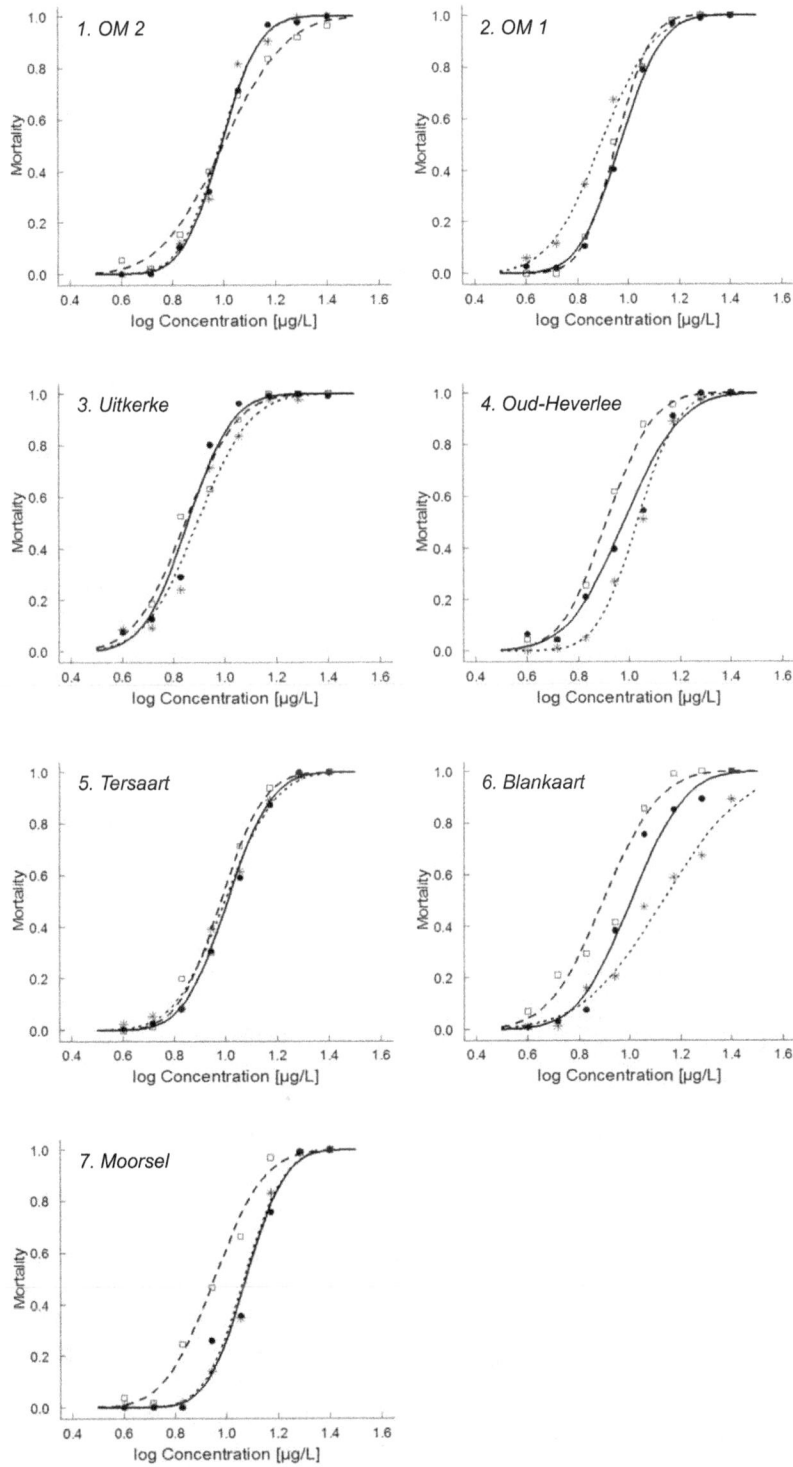

Figure 1 Concentration-response curves for carbaryl for the study populations as sampled from the dormant egg banks (original field populations, full line and circles), the control populations (control condition in experimental evolution trial; dashed line and squares) and the carbaryl-selected populations (from the experimental evolution trial; dotted line and stars) for each of seven natural *Daphnia magna* populations isolated from Flemish farmland ponds situated in areas that differ in land use intensity. The figures of the different populations are ranked based on land use intensity in the neighbourhood of the pond (OM2: low land use – Moorsel: high land use; see Table S4).

Table 1. Overview of the remaining GLiM models following the stepwise backward elimination process of nonsignificant terms ($\alpha = 0.05$) testing for the effect of carbaryl, population, selection history and their interactions on mortality of *D. magna* in standardized acute toxicity experiments. (A) Results of general model analysing the data of all population $\times$ selection treatments; (B) Results of targeted comparisons between selection histories: control versus original, carbaryl-selected versus original, and carbaryl-selected versus control. The two populations obtained in the selection experiment (control and carbaryl-selected population) were compared with each other, and each of them was also compared with the original population in a paired analysis. 'Carbaryl' refers to different carbaryl concentration levels in the acute toxicity tests, 'population' refers to the pond of origin of the populations, and 'selection history' refers to the selection history of the populations (original = prior to selection experiment; control = experimental selection in the absence of carbaryl; carbaryl-selected = experimental selection in the presence of carbaryl).

	df	Residual df	Residual deviance	F	P
(A)					
All clonal lineages					
NULL		775	15634.1		
Carbaryl	1	774	4976.3	1612.0	<0.0001
Population	6	768	4467.8	12.8	<0.0001
Selection history	2	766	4345.5	9.2	0.0001
Carbaryl × Population	6	760	4242.4	2.6	0.017
Population × selection history	12	748	3985.4	3.2	<0.0001
(B)					
Control population versus original population					
NULL		494	10017.6		
Carbaryl	1	493	2916.8	970.3	<0.001
Population	6	487	2616.3	6.8	<0.001
Selection history	1	486	2541.4	10.2	<0.01
Carbaryl-selected versus original population					
NULL		530	10646.9		
Carbaryl	1	529	3509	990.5	<0.001
Population	6	523	2931.5	13.5	<0.001
Carbaryl-selected versus control population					
NULL		525	10556.8		
Carbaryl	1	524	3470.2	1173.4	<0.001
Population	6	518	3181.4	8	<0.001
Selection history	1	517	3069.6	18.54	<0.001
Carbaryl × Population	6	511	2958.5	3.1	<0.01
Population × selection history	6	505	2742.6	6	<0.001

carbaryl × selection history interaction effects (Table 1A). The significant effect of carbaryl indicates that carbaryl exposure impacted mortality, while the carbaryl × population interaction effect reflects that the slope of the concentration–response curves differs among populations. The significant effect of selection history reveals that the selection experiment impacted carbaryl tolerance of the resulting populations, while the selection history × population interaction effect shows that this effect is dependent on the population. In all three targeted contrasts (Table 1B) between different selection histories, carbaryl and population significantly affected mortality. The comparison between the carbaryl-selected and the control populations directly tests for differential evolution in our containers due to carbaryl exposure. In this comparison, the population × selection history interaction is significant, implying that the difference in response to the imposed selection regimes depends on the population of origin. While most populations had a higher EC_{50} value in the carbaryl-selected than in the control populations, one population (OM 1) showed a lower EC_{50} value in the carbaryl-selected than in the control population. Overall, our findings that natural populations can genetically differ in tolerance to pollutants and can show evolutionary potential to adapt to pollutants is in line with the results of several earlier studies that used pesticides (e.g. Tanaka and Tatsuta 2013) or other pollutants (e.g. Muyssen et al. 2005; Agra et al. 2010), and involved *Daphnia* or other organisms (Carriere et al. 1994; Goussen et al. 2013) as model systems.

Relaxation versus enhancement of tolerance

The comparison of carbaryl tolerance between the control and the original populations showed a reduction in carbaryl tolerance in the control populations. Carbaryl tolerance thus rapidly (within 5–8 generations) decreases in the absence of carbaryl. It is indeed striking that the genetically determined reduction in carbaryl tolerance following a release in exposure to carbaryl was stronger than the genetically determined increase in carbaryl tolerance when exposed to pulses of a high concentration of the pesticide in the selection experiment. Selection history (i.e. treatment during the selection experiment) had a highly significant effect on carbaryl tolerance in the control treatment of our selection experiment, whereas exposure to carbaryl pulses did not result in an overall increase in carbaryl tolerance, even though the concentrations we applied in our selection experiment were quite high. We exposed our experimental populations to three pulses of 32 μg carbaryl L^{-1}, while the EC_{50} value in the original populations for neonate survival when exposed to carbaryl has been estimated to be between 6 and 13 μg L^{-1} (present study, Coors et al. 2009). While such high exposure levels in the carbaryl treatment must have strongly impacted mortality in juveniles, the populations persisted. The absence of an evolutionary response upon exposure to the standardized and high concentration carbaryl pulses during the selection experiment may either reflect that there is no evolutionary

(A) *Spearman Rank R = 0.739; P = 0.058*

(C) *Spearman Rank R = 0.21; P = 0.66*

(B) *Spearman Rank R = 0.93; P < 0.01*

(D) *Spearman Rank R = 0.71; P = 0.08*

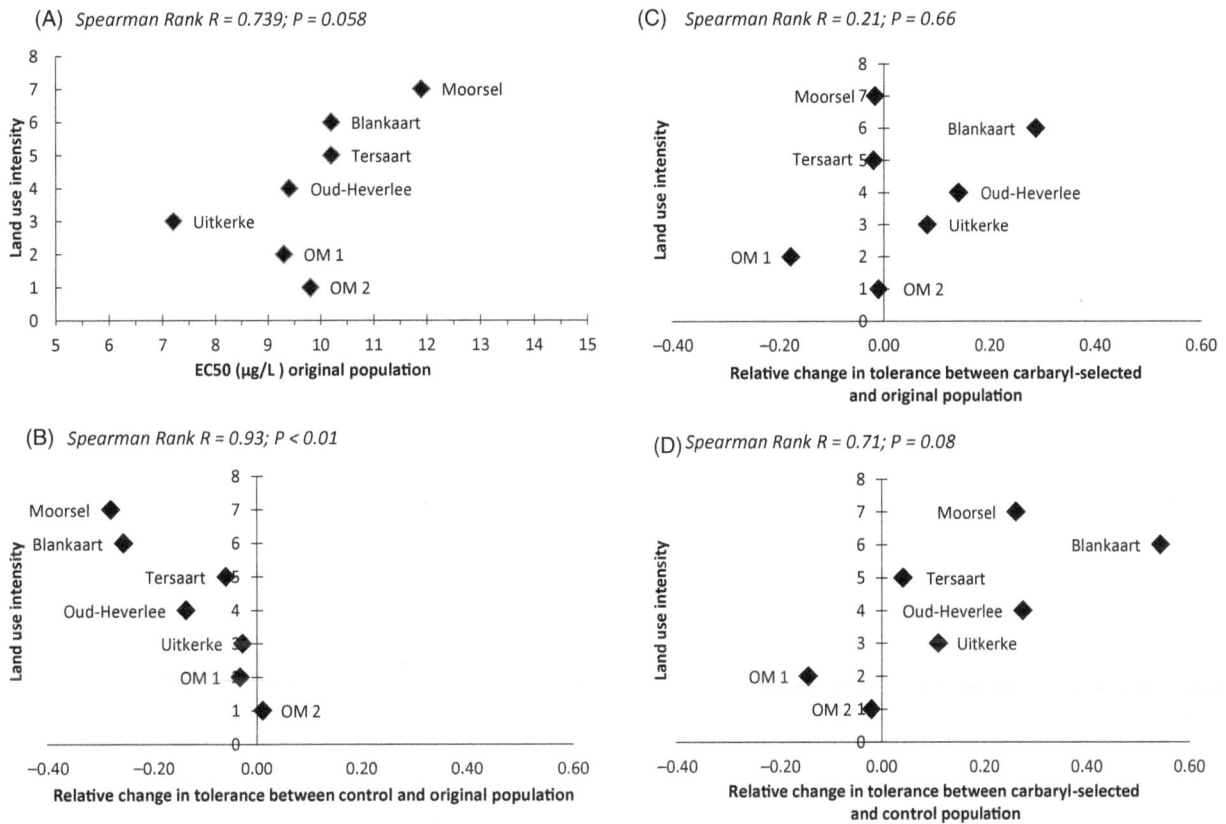

Figure 2 Spearman rank correlations between land use intensity (determined following Coors et al. 2009) and (A) absolute value of the EC_{50} ($\mu g \; L^{-1}$) for carbaryl of the original populations, (B, C) the change in EC_{50} during the selection experiment [$\log_2$ (EC_{50} treatment/EC_{50} original)], and (D) the change in carbaryl tolerance between the control and the carbaryl-selected populations [$\log_2$ (EC_{50} carbaryl-selected/EC_{50} control)]. Panel B shows that the loss in carbaryl tolerance in control populations compared to the original populations increases with land use intensity around the pond [$\log_2$ (EC_{50} control/EC_{50} original)]; this correlation is significant. Panel C shows the change in carbaryl tolerance in carbaryl-selected populations compared to the original populations in relation to land use intensity around the pond [$\log_2$ (EC_{50} carbaryl-selected/EC_{50} original)]; this correlation is not significant. Panel D shows the change in carbaryl tolerance between the control and the carbaryl-selected populations in relation to land use intensity [$\log_2$ (EC_{50} carbaryl-selected/EC_{50} control)]; this relationship is marginally nonsignificant.

potential to further increase tolerance compared to that in the original populations or that the selection pressure we imposed was not significantly different from that experienced in the ponds of origin.

Rapid loss of tolerance to pollutants has been observed before. For instance, Levinton et al. (2003) showed a reduction in Cd resistance after 9–18 generations in the oligochaete *Limnodrilus hoffmeisteri* upon restoration of a heavily metal-polluted site. The rapid decrease in pollutant tolerance in the absence of pollutant exposure is generally viewed as an indication that the tolerance has a cost (Sibly and Calow 1989; Van Straalen and Timmermans 2002; Medina et al. 2007; Saro et al. 2012). A cost of tolerance may be related to an increase in energy demand associated with the mechanism to cope with the pollutant (Sibly and Calow 1989; Agra et al. 2011). In earlier work using controlled laboratory experiments, we have shown that

carbaryl tolerance has a cost in terms of disease development upon exposure of the *Daphnia* to the endoparasitic bacterium *Pasteuria ramosa*. While our results are very suggestive of an overall cost of carbaryl tolerance acting across populations, our data do not allow us to identify the mechanism of this cost.

Relationship with land use

Our observation that the tolerance to carbaryl of the original populations tends to increase with land use intensity in the immediate neighbourhood of the habitats they were isolated from (marginally nonsignificant) is in line with previous results obtained with a different set of clones from the very same populations (Coors et al. 2009). As carbaryl concentrations in the field are difficult to quantify, we used land use intensity as a proxy of exposure. Even though the amount of arable land is a crude measure for the use of insecticides, our results combined with those of Coors et al. (2009) are suggestive of a relationship. It could be that the pattern reflects adaptation to the use of acetylcholinesterase-inhibiting insecticides as a class rather than just carbaryl.

The reduction in carbaryl tolerance we observe in our control treatment is significantly associated with land use intensity around the habitats the populations were isolated from. Overall, the patterns shown in Fig. 2B,D suggest that populations exposed to intensive land use in the vicinity of their habitat show the highest potential to exhibit an evolutionary response to the experimental treatments. This is in line with expectations, as populations that are more strongly selected for carbaryl tolerance can be expected to show the strongest response upon release from this stressor.

Methodological considerations

While our study involved an analysis of concentration–response curves of not less than 102 D. magna clones, the concentration–response curves of each of the populations are based on an analysis of only 3–6 randomly picked out clones per original and experimental population. The maximum of six clones was set for the sake of feasibility, while the lower number in a number of populations resulted from the fact that several clonal lineages were lost during the relatively long period (> 82 generations) of culture under stock conditions prior to the experiment. While a selection of 3–6 clones is at the lower end to obtain a precise estimate of the concentration–response curve of a population, there are several arguments that support the notion that our concentration–response curves capture real patterns. First, our results relating EC_{50} values of the original populations to land use intensity are nearly identical to those obtained by Coors et al. (2009) using a different set of clones and a slightly larger set of populations (the seven populations studied here are a subset of the ten populations studied by Coors et al. 2009). Secondly, in our analysis of the changes in carbaryl tolerance between the control and original populations in relation to land use, the significant and positive association we observed is very unlikely to originate by chance and reflects that the power of our analysis must have been sufficient to capture the essence of interpopulational differences. Indeed, the key conclusions of our study are based on the general patterns across populations (stronger response in the absence than in the presence of selection by carbaryl, relationship with land use), where less precise estimates of responses of specific populations result in some noise but are not detrimental as long as the overall patterns are strong. In these analyses across populations,

our database indeed involves seven times 3–6 clones. In brief, while we want to be careful in our interpretation of the details of the concentration–response curves obtained for each population separately, we feel confident in interpreting the patterns across populations, which was the key aim and approach of this study.

Our observation of an absence of a genetic increase in carbaryl tolerance upon exposure to high doses of carbaryl in our selection experiment is in contrast with previous results, where we provided evidence for the evolution of increased carbaryl tolerance in a standardized 21-day life history test with acute exposure to carbaryl during the first 4 days (Jansen et al. 2011). There are two possible explanations for this discrepancy. First, in our previous study (Jansen et al. 2011), we quantified tolerance using a pulsed exposure test during the first 4 days, while in the present study we carried out static acute toxicity tests. These two approaches can indeed give different outcomes (Naddy and Klaine 2001; Angel et al. 2010). Second, the study of Jansen et al. (2011) involved four rather than seven populations and included one population that showed a strong increased tolerance upon selection (Knokke In). This latter population was not included in the current study due to accidental loss of a number of clones. While this discrepancy underscores that there might be a capacity to evolve an increased tolerance to carbaryl in some populations, the main results of our current study remain unchanged, being that there is evolutionary potential to respond to carbaryl selection, that the reduction in carbaryl tolerance upon release from carbaryl exposure is stronger than the increase in carbaryl tolerance when the animals are exposed to carbaryl in the selection experiment, and that the amplitude of this reduction is related to land use intensity.

Consequences and perspectives

Our observation that the effect of a release from carbaryl exposure under experimental conditions results in a rapid, genetically determined loss of carbaryl tolerance and that this release is associated with land use intensity, has a number of implications and potential applications. First, our results suggest that the intensive agriculture in the Flemish landscape has exerted an important selection pressure on nontarget taxa, resulting in changes in carbaryl tolerance compared to control conditions. While one may conclude that the capacity to genetically adapt allowed the studied Daphnia populations to cope with the stress caused by intensive agriculture, there are two aspects that need to be considered to put our results in context. Obviously, we were only able to test Daphnia populations that in one way or another have been able to cope with the stress caused by agriculture in their natural environment. The populations that may have failed to do so have gone extinct and are therefore absent from our data set, as we only sampled habitats that harboured active *D. magna* populations. Also, we only tested a single nontarget species, and a relatively common one that is typically associated with eutrophied waters. This species may therefore be less sensitive to disturbance than more specialized taxa. Our results suggest that even this species is impacted in its evolutionary trajectory by the intensive agriculture that prevails in the study region.

Second, our results also indicate that genetic adaptation to pollution is ecologically costly, given the fast loss of carbaryl tolerance upon release from the selection and earlier measures of costs upon exposure to additional stressors such as parasites (Jansen et al. 2011). The overall picture is that the *Daphnia* populations, even though they are nontarget species for pesticide use, are exposed to continuous selection pressure by pesticides in ponds in the vicinity of crops.

Third, our observation that genetic adaptation to insecticide presence is widespread and (partly) reversible, suggests that one should be aware of the selection history of the *Daphnia* clones used to carry out toxicity tests. Several authors have already suggested to use a genetically diverse battery of *Daphnia* clones to better capture the variation of toxicological response (Barata and Soares 2002; Messiaen et al. 2013). We suggest that one could also use the degree of genetic adaptation to pollutants as a bio-assay to assess past exposure to pesticides in natural populations. Monitoring pesticide concentrations in the field is often highly laborious and costly because of the short temporal scales of exposure and the rapid degradation of the residues (e.g. a half-life of only a few days in the case of carbaryl). Often, nontarget populations will only be briefly exposed to the pesticide during or just following spraying, but concentrations may be very high. If genetic adaptation to such pesticide exposure is common, as suggested by this study as well as by some other reports on genetic adaptation to pollution stress that were published in the last decade (Hoffman and Fisher 1994; Medina et al. 2007; Agra et al. 2011), the level of genetic adaptation to pollution in natural populations can be used to capture the degree of pollution at an intermediate temporal scale and in an ecologically relevant way. While such experiments would involve quite an investment, they might be the most direct way to assess effective pollution exposure at the population level. Moreover, methodological adjustments, such as directly determining concentration–response curves on the population of hatchlings rather than on the resulting clonal lineages, can reduce workload considerably. In developing such an approach, however, it will be necessary to conduct further studies to quantify the degree to which maternal effects and bio-accumulation of pollutants influence the results directly obtained from hatchlings of eggs derived from nature.

Acknowledgements

MJ is a postdoctoral researcher with the Fund for Scientific Research Flanders (FWO). AC was supported by a postdoctoral fellowship grant from the KU Leuven Research Fund during the experimental phase and by BiK-F through the research funding programme 'LOEWE – Landes-Offensive zur Entwicklung Wissenschaftlich-ökonomischer Exzellenz' of Hesse's Ministry of Higher Education, Research, and the Arts during data analysis and manuscript preparation. We thank Frans van der Wielen (IBED) for his support with the analysis of carbaryl. This research was financially supported by project G.229.09 of the National Science Fund Flanders (FWO), projects GOA/08/06 and PF/2010/07 of the KU Leuven Research Fund and Belspo IAP projects SPEEDY and AQUASTRESS. We thank the editor and two anonymous reviewers for very helpful comments on an earlier version of this manuscript.

Author contributions

LDM, AC and MJ designed the research. MJ, AC and MS performed the experiments and collected the data. Data analysis was performed by MJ and AC with inputs of LDM, JV and KDS. PDV was responsible for the analysis of carbaryl concentrations. MJ wrote the first draft of the manuscript, all authors contributed to editing and revisions.

Literature cited

Agra, A. R., L. Guilhermino, and A. M. V. M. Soares 2010. Genetic costs of tolerance to metals in *Daphnia longispina* populations historically exposed to a copper mine drainage. Environmental Toxicology and Chemistry **29**:939–946.

Agra, A. R., A. M. V. M. Soares, and C. Barata 2011. Life-history consequences of adaptation to pollution '*Daphnia longispina* clones historically exposed to copper'. Ecotoxicology **20**:552–562.

Angel, B. M., S. L. Simpson, and D. F. Jolley 2010. Toxicity to *Melita plumulosa* from intermittent and continuous exposures to dissolved copper. Environmental Toxicology and Chemistry **29**:2823–2830.

Barata, B. C., A. M. V. M. Soares 2002. Determining genetic variability in the distribution of sensitivities to toxic stress among and within field populations of *Daphnia magna*. Environmental Science and Technology **36**:3045–3049.

Carriere, Y., J. P. Deland, D. A. Roff, and C. Vincent 1994. Life-history costs associated with the evolution of insecticide resistance. Proceedings of the Royal Society B-Biological Sciences **258**:35–40.

Cerbin, S., M. H. S. Kraak, P. de Voogt, P. M. Visser, and E. Van Donk 2010. Combined and single effects of pesticide carbaryl and toxic Microcystis aeruginosa on the life history of *Daphnia pulicaria*. Hydrobiologia **643**:129–138.

Colbourne, J. K., M. E. Pfrender, D. Gilbert, W. K. Thomas, A. Tucker, T. H. Oakley, S. Tokishita et al. 2011. The ecoresponsive genome of

Daphnia pulex. Science **331**:555–561.

Coors, A., J. Vanoverbeke, T. De Bie, and L. De Meester 2009. Land use, genetic diversity and toxicant tolerance in natural populations of *Daphnia magna*. Aquatic Toxicology **95**:71–79.

Crawley, M. J. 2007. The R Book. John Wiley & Sons, Chichester, UK.

De Meester, L., and H. De Jager 1993. Hatching of *Daphnia* sexual eggs: 1: intraspecific differences in the hatching responses of *Daphnia magna* eggs. Freshwater Biology **30**:219–226.

Decaestecker, E., S. Gaba, J. A. M. Raeymaekers, R. Stoks, L. Van Kerckhoven, D. Ebert, and L. De Meester 2007. Host-parasite 'Red Queen' dynamics archived in pond sediment. Nature **450**:870–873.

Declerck, S., T. De Bie, D. Ercken, H. Hampel, S. Schrijvers, J. Van Wichelen, V. Gillard et al. 2006. Ecological characteristic's of small farmland ponds: associations with land use practices at multiple spatial scales. Biological Conservation **131**:523–532.

Goussen, B., F. Parisot, R. Beaudouin, M. Dutilleul, A. Buisset-Goussen, A. R. R. Pery, and J. M. Bonzom 2013. Consequences of a multi-generation exposure to uranium *on Caenorhabditis elegans* life parameters and sensitivity. Ecotoxicology **22**:869–878.

Hebert, P. D. N. 1978. Population biology of *Daphnia* (*Crustacea, Daphnidae*). Biological Reviews of the Cambridge Philosophical Society **53**:387–426.

Hoffman, E. R., and S. W. Fisher 1994. Comparison of a field and laboratory-derived population of *Chironomus riparius* (Diptera: Chironomidae): biochemical and fitness evidence for population divergence. Journal of Economic Entomology **87**:318–325.

Jansen, M., R. Stoks, A. Coors, W. van Doorslaer, and L. de Meester 2011. Collateral damage: rapid exposure-induced evolution of pesticide resistance leads to increased susceptibility to parasites. Evolution **65**:2681–2691.

Kerr, D. R., and J. P. Meador 1996. Modeling dose response using generalized linear models. Environmental Toxicology and Chemistry **15**:395–401.

Klüttgen, B., U. Dülmer, M. Engels, and H. T. Ratte 1994. ADaM, an artificial freshwater for the culture of zooplankton. Water Research **28**:406–414.

Lampert, W. 2006. *Daphnia*: model herbivore, predator, prey. Polish Journal of Ecology **54**:607–620.

Levinton, J. S., E. Suatoni, W. Wallace, R. Junkins, B. Kelaher, and B. J. Allen 2003. Rapid loss of genetically based resistance to metals after the cleanup of a superfund site. Proceedings of the National Academy of Sciences of the United States of America **100**:9889–9891.

McCullagh, P., and J. Nelder 1989. Generalized Linear Models, 2nd edn. CRC: Chapman and Hall, Boca Raton, Florida. ISBN 0-412-31760-5.

Medina, M. H., J. A. Correa, and C. Barata 2007. Micro-evolution due to pollution: possible consequences for ecosystem responses to toxic stress. Chemosphere **67**:2105–2114.

Messiaen, M., C. R. Janssen, L. De Meester, and K. A. C. De schamphelaere 2013. The initial tolerance to sub-lethal Cd exposure is the same among ten naïve pond populations of *Daphnia magna*, but their micro-evolutionary potential to develop resistance is very different. Aquatic Toxicology **144-145**:322–331.

Miner, B. E., L. De Meester, M. E. Pfrender, W. Lampert, N. G. Hairston Jr 2012. Linking genes to communities and ecosystems: *Daphnia* as an ecogenomic model. Proceedings of the Royal Society B-Biological Sciences **279**:1873–1882.

Muyssen, B. T. A., B. T. A. Bossuyt, and C. R. Janssen 2005. Inter- and intra-species variation in acute zinc tolerance of field-collected cladoceran populations. Chemosphere **61**:1159–1167.

Naddy, R. B., and S. J. Klaine 2001. Effects of pulse frequency and interval on the toxicity of chlorpyrifos to *Daphnia magna*. Chemosphere **45**:497–506.

OECD 2004. OECD guideline for testing of chemicals - *Daphnia* sp., acute immobilisation test. http://www.oecd.org.

Parker, E. D., V. E. Forbes, S. L. Nielsen, C. Ritter, C. Barata, D. J. Baird, W. Admiraal et al. 1999. Stress in ecological systems. Oikos **86**:179–184.

R Development Core Team 2007. http://www.r-project.org/.

Raymond, M., C. Berticat, M. Weill, N. Pasteur, and C. Chevillon 2001. Insecticide resistance in the mosquito *Culex pipiens*: what have we learned about adaptation? Genetica **112**:287–296.

Saro, L., I. Lopes, and M. Nelson 2012. Testing hypotheses on the resistance to metals by *Daphnia longispina*: differential acclimation endpoints association, and fitness costs. Environmental Toxicology and Chemistry **31**:909–915.

Sibly, R. M., and P. Calow 1989. A life-cycle theory of responses to stress. Biological Journal of the Linnean Society **37**:101–116.

Tanaka, Y., and H. Tatsuta 2013. Retrospective estimation of population-level effect of pollutants based on local adaptation and fitness cost of tolerance. Ecotoxicology **22**:795–802.

Van Doorslaer, W., R. Stoks, C. Duvivier, A. Bednarska, and L. De Meester 2009a. Population dynamics determine genetic adaptation to temperature in *Daphnia*. Evolution **63**:1867–1878.

Van Doorslaer, W., J. Vanoverbeke, C. Duvivier, S. Rousseaux, M. Jansen, B. Jansen, H. Feuchtmayr et al. 2009b. Local adaptation to higher temperatures reduces immigration success of genotypes from a warmer region in the water flea *Daphnia*. Global Change Biology **15**:3046–3055.

Van Straalen, N. M., and M. J. T. N. Timmermans 2002. Genetic variation in toxicant-stressed populations: an evaluation of the 'genetic erosion' hypothesis. Human and Ecological Risk Assessment **8**:983–1002.

VMM, Vlaamse Milieu Maatschappij, Vlaamse Milieu 2010. MIRA-T. Verspreiding van bestrijdingsmiddelen.

Walker, C. H., S. P. Hopkin, R. M. Sibly, and D. B. Peakall 2006. Principles of Ecotoxicology, 3rd edn. Taylor Francis Group, CRC Press, New York.

Waterfield, G. D., and D. Ziberman 2012. Pest management in food systems: an economic perspective. In A. Gadgil, and D. M. Liverman, eds. Annual Review of Environment and Resources, Vol. **37**, pp. 223–245.

Wheeler, M. W., R. M. Park, and A. J. Bailer 2006. Comparing median lethal concentration values using confidence interval overlap or ratio tests. Environmental Toxicology and Chemistry **25**:1441–1444.

4

Additive genetic variation for tolerance to estrogen pollution in natural populations of Alpine whitefish (*Coregonus* sp., Salmonidae)

Gregory Brazzola,[1] Nathalie Chèvre[2] and Claus Wedekind[1]

1 Department of Ecology and Evolution, Biophore, University of Lausanne, Lausanne, Switzerland
2 Institute of Earth Surface Dynamics, University of Lausanne, Lausanne, Switzerland

Keywords
17α-ethinylestradiol, embryo development, fluconazole, micropollutants, Salmonidae, timing of hatching.

Correspondence
Claus Wedekind, Department of Ecology and Evolution, Biophore, University of Lausanne, 1015 Lausanne, Switzerland.

e-mail: claus.wedekind@unil.ch

Abstract

The evolutionary potential of natural populations to adapt to anthropogenic threats critically depends on whether there exists additive genetic variation for tolerance to the threat. A major problem for water-dwelling organisms is chemical pollution, and among the most common pollutants is 17α-ethinylestradiol (EE2), the synthetic estrogen that is used in oral contraceptives and that can affect fish at various developmental stages, including embryogenesis. We tested whether there is variation in the tolerance to EE2 within Alpine whitefish. We sampled spawners from two species of different lakes, bred them *in vitro* in a full-factorial design each, and studied growth and mortality of embryos. Exposure to EE2 turned out to be toxic in all concentrations we tested (≥ 1 ng/L). It reduced embryo viability and slowed down embryogenesis. We found significant additive genetic variation in EE2-induced mortality in both species, that is, genotypes differed in their tolerance to estrogen pollution. We also found maternal effects on embryo development to be influenced by EE2, that is, some maternal sib groups were more susceptible to EE2 than others. In conclusion, the toxic effects of EE2 were strong, but both species demonstrated the kind of additive genetic variation that is necessary for an evolutionary response to this type of pollution.

Introduction

One major question in conservation biology is whether natural populations can adapt early enough to the various anthropogenic challenges they are exposed to before they go extinct (Ferrière et al. 2004; Hendry et al. 2011). Among the major challenges that water-dwelling organisms have been newly exposed to during the last decades are various sorts of chemical pollution through residues in effluents of sewage treatment plants. Among the most common pharmaceuticals that enter the environment after passing municipal sewage treatment and that have well been identified as aquatic environmental risk are the natural steroid estrogen hormone estrone (E1), 17β-estradiol (E2), and 17α-ethinylestradiol (EE2) (Caldwell et al. 2012). The latter (EE2) is used in most formulations of oral contraceptive pills because it mimics the endogenous hormone E2 and is more stable than its natural counterpart (Kime 1998). In the

aquatic environment, EE2 is also more persistent than natural estrogens (its half-life is about 14 days, Shore et al. 1993). EE2 is now commonly found in surface waters at concentrations around 1 ng/L (e.g., Larsson et al. 1999; Vulliet and Cren-Olive 2011; Zhang et al. 2011), but concentrations of 17.2 ng/L (Beck et al. 2005), 42 ng/L (Ternes et al. 1999), and up to 831 ng/L (Kolpin et al. 2002) have been reported, and concentrations of >1 ng/L are sometimes even found in groundwater (Vulliet and Cren-Olive 2011).

EE2 is a potent endocrine disruptor in fish (Kime 1998; Gutendorf and Westendorf 2001; Lange et al. 2001) and has been shown to influence viability and development of zebra fish embryos (*Danio rerio*), either directly as immediate response to an exposure or indirectly via the effects of parents that had been exposure to EE2 (Soares et al. 2009). Overall, the studies so far suggest that embryos are more susceptible to the immediate toxic effects of EE2, while later life history stages may suffer more from the effects EE2 has on sex determination and reproduction (e.g., Segner et al. 2003a; Soares et al. 2009; Harris et al. 2011). Concentrations around 1 ng/L can induce vitellogenin production in male rainbow trout (*Oncorhynchus mykiss*) and zebra fish (Rose et al. 2002) and significantly reduce fertilization success (Segner et al. 2003b). Higher concentrations are known to affect reproductive behavior or sexual characteristics or lead to intersex in, for example, zebra fish (Larsen et al. 2008), fathead minnow (*Pimephales promelas*) (Lange et al. 2001), three-spined sticklebacks (*Gasterosteus aculeatus*) (Dzieweczynski 2011), or the whitefish *Coregonus lavaretus* (Kipfer et al. 2009). Moreover, exposure to substances with as high an estrogenic potency as EE2 is expected to influence sexual differentiation in fish where sex is genetically determined but can be reversed by environmental factors which is the case in many fishes of various families (Devlin and Nagahama 2002; Stelkens and Wedekind 2010). EE2 could be demonstrated to arrest male differentiation in zebra fish when applied during the period of sexual differentiation (Van den Belt et al. 2003; Fenske et al. 2005). Sex ratio management via exposure to hormones is therefore widely used in aquaculture (e.g., if one sex is preferred for economic reasons) (Baroiller et al. 2009) and has been discussed in the context of conservation management (Wedekind 2002b, 2012; Gutierrez and Teem 2006). Estrogens as pollutants in effluents of sewage treatment plants are therefore likely to induce sex reversal and sex ratio distortion in wild fish populations (Jobling et al. 2006; Scholz and Kluver 2009). Indeed, a field experiment on roach (*Rutilus rutilus*) resulted in 98% phenotypic females after 3.5 years of chronic exposure to treated estrogenic wastewater effluents and still 79% phenotypic females in a 50% dilution of these effluents (Lange et al. 2011). On the long term, a biased sex ratio is a serious threat to natural populations because it can considerably reduce genetically effective population sizes, drive sex chromosomes to extinction, and may affect sex ratios in some counterintuitive ways (Cotton and Wedekind 2009). However, Hamilton et al. (2014) recently found populations of roach (*R. rutilus*) to be self-sustaining in heavily estrogen-polluted waters and despite widespread feminization. Such observations raise the question whether natural populations can adapt in useful time to this rather new type of pollution, that is, whether there can be rapid evolution in response to the pollution (Wedekind 2014).

Despite the possible relevance of estrogen pollution worldwide, it is still unclear whether rapid evolutionary changes are possible within natural populations in response to the potential negative effects that estrogens such as EE2 may have on average viability and growth in natural fish populations. First, it needs to be established whether there is, under controlled conditions, phenotypic variation in

response to this selection pressure. It would then be necessary to understand the nature of such phenotypic variation, that is, whether it is due to genetic differences, individual phenotypic plasticity, maternal environmental effects, epigenetic factors, or any form of nongenetic inheritance (Bondurianksy and Day 2009; Hendry et al. 2011; Vandegehuchte and Janssen 2011).

Here, we sampled two natural whitefish populations (*Coregonus* sp.) to (i) study the toxicity of EE2 to embryos and (ii) test whether there is the kind of phenotypic and genetic variation within populations that would be necessary for a rapid evolutionary response to this type of pollution. Alpine whitefish are plankton feeders and typically keystone species in the larger lakes of the pre-Alpine region. The two whitefish species we chose differ in many respect and may hence cover much of the diversity within the Alpine whitefish species complex: a fast-growing, large-type whitefish from the Lake Geneva (*Coregonus palaea* Fatio 1890) and a slow-growing, small-type whitefish from the Lake Brienz (*Coregonus albellus* Fatio 1890). The two lakes are about 100 km apart and belong to different drainage systems. While Lake Brienz has been described as 'ultra-oligotrophic' (Müller et al. 2007) and can be assumed to be comparatively weakly exposed to municipal effluents (few small communities in the catchment area), the state of eutrophication of Lake Geneva has been ranked as moderate to strong (Vonlanthen et al. 2012), and the spawning place of the *C. palaea* study population is close to city of Lausanne (with >300 000 inhabitants living in the city and its agglomeration), that is, exposure to municipal effluents can be assumed in the upper range within Switzerland. We sampled adult breeders from their spawning sites, used their gametes to produce all possible half-sib groups, and exposed the resulting embryos singly to one of several concentrations of EE2 to study growth and survival until hatching. Full-factorial *in vitro* breeding allowed us to separate additive genetic from maternal environmental effects (variation in egg quality) on the susceptibility or tolerance of embryos to estrogen pollution (Lynch and Walsh 1998; Wedekind et al. 2007b).

Methods

Sampling and experimental treatment of *Coregonus palaea*
Adult large-type whitefish ('Palée'; *C. palaea*) from Lake Geneva, Switzerland, were caught with gill nets during their breeding season in December. Four females and six males were stripped to collect their gametes for *in vitro* fertilizations in a full-factorial breeding design. For this, the eggs of each female were distributed to six new petri dishes in about equal amounts, and milt was added and activated with few millilitre of water to produce all 24 possible half-sib groups. The freshly fertilized eggs were left undisturbed

for at least 1 h to allow for egg hardening. They were then transported to the laboratory were they were immediately washed as in von Siebenthal et al. (2009). In total, 2304 eggs (96 eggs per sib group) were then distributed singly to 24-well cell culture plates (Falcon; Becton Dickinson, Allschwil, Switzerland; 2 mL wells). The wells contained 0, 1, 10, and 100 ng/L of analytical 17α-ethinylestradiol (Sigma-Aldrich, Buchs, Switzerland). All water used here had been chemically standardized, that is, reconstituted according to the OECD guideline No. 203, Annex 2 (OECD 1992), tempered, and aerated before use. The embryos were incubated at constant 6.5°C. Embryo mortality and the timing of hatching were recorded daily from day 13 onward (the last dead embryo was recorded at day 123 after fertilization). *Coregonus* sp. eggs are much more translucent than, for example, typical *Salmo* sp. or *Oncorhynchus* sp. eggs, and embryos are easily recognizable after few days of incubation, but it remains difficult to distinguish dead embryos from unfertilized eggs during the very first days of incubation (because they both turn white after rupture of the yolk membrane; Leitritz and Lewis 1976). Therefore, the first recording of mortality at days 13 could include unfertilized eggs and was therefore separately analyzed and interpreted. Permissions for sampling adults, *in vitro* breeding, and the raising of embryos in the laboratory were granted by the fishery inspectorate of the Vaud canton.

Sampling and experimental treatment of *Coregonus albellus*

Adult small-type whitefish ('Brienzlig'; *C. albellus*) from Lake Brienz, Switzerland, were caught with gill nets during their breeding season in September. To minimize temperature variation (the fish spawn in about 60–80 m depth at about 5°C), the fish were immediately transported in cold water to a refrigerated van (IVECO 3T5) where gamete collection and *in vitro* fertilization were done at 5°C as described above. Four females and five males were stripped and used to produce all possible sib groups. After egg hardening, the freshly fertilized eggs were transported to the laboratory and washed, and in total, 1600 eggs (160 per sib group) were distributed to 24-well plates as described above. They were exposed to 0, 1, 5, 10, 50 or 100 ng/L analytical 17α-ethinylestradiol (Sigma-Aldrich).

Whitefish from Lake Brienz show very low growth rates and body condition as compared to other Alpine whitefish (probably because Lake Brienz is an ultra-oligotrophic lake; Müller et al. 2007) and female *C. albellus* produce comparatively few and small eggs (Kirchhofer and Lindt-Kirchhofer 1998) that may be less resistant to handling as other Alpine whitefish. We therefore ran two further controls treated with antimicrobials to potentially reduce stress-induced embryo mortality in the laboratory (Wedekind et al. 2010), additional to the 0 ng/L EE2 control. These further controls were treated with 10 or 100 ng/L analytical fluconazole (Sigma-Aldrich), a broad-spectrum antifungal drug. We did not combine the antimicrobial and the EE2 treatments. The antimicrobial treatment in the additional controls was solely to learn more about the potential relevance of microbes for embryo mortality of a species that is expected to be difficult to raise under experimental conditions.

Embryo mortality and hatching were recorded daily from day 16 postfertilization onward. As in the upper experiment with *C. palaea*, the first recordings of mortality at day 16 could include unfertilized eggs and were therefore separately analyzed and interpreted. Incubation temperature was planned to be constant at 6°C, but because of technical problems went up to 15°C for few hours at day 10 and again at day 14 postfertilization. Hatchlings (alevins) were photographed (Olympus C-5060; Olympus, Shinjuku, Japan) in a drop of water under a microscope on the first and the tenth day after hatching. The notochord length and the volume of the yolk sac of individual hatchlings were determined from these photographs using the open-access software IMAGEJ 1.42q (http://imagej.nih.gov/ij/). Developmental time was determined as degree days (dd). All measurements were taken blindly with respect to the experimental treatment. The expected notochord length at the time the yolk sac would be used up was linearly extrapolated from loss of yolk sac volume and increase of alevin length during the first 10 days. Permissions for sampling adults, *in vitro* breeding, and the raising of embryos in the laboratory were granted by the fishery inspectorate of the Bern canton.

Statistics

Within each experiment, the exposure to estrogen concentrations was full factorial and balanced with respect to parental origin. Parental effects, main effects of EE2 treatment, and treatment × parent interactions were tested either in generalized linear models (on embryo survival) or three-way ANOVAs on continuous dependent variables such as alevin size and growth. All analyses were based on embryo as independent replicates, with treatment and parental origin as fixed factors (we refrained from including second-order interaction terms and from estimating average sire or dam effects because of limited sample sizes per populations). Two male *C. albellus* were excluded from all analyses because total mortality of their offspring turned out to be 100% and 99.2%, respectively. Main treatment effects were tested in directed heterogeneity tests (Rice and Gaines 1994) based on the expectancy that if estrogens have an effect on embryo survival and life history, the effects would increase with increasing estrogen concentrations. Data were analyzed in JMP 9.0 (SAS Institute Inc., Cary, NC, USA) and R 2.14.1 (R Development Core Team 2011).

Results

Embryo mortality

Increased exposure to estrogens increased embryo mortality until hatching in *C. palaea* ($\chi^2 = 7.5$, df = 3, $r_sP_c = 0.75$, $P < 0.05$; Fig. 1A) and in *C. albellus* ($\chi^2 = 52.1$, df = 5, $r_sP_c = 0.75$, $P < 0.01$; Fig. 2A). The fact that very early dead embryos are difficult to distinguish from nonfertilized eggs did not seem to play a role here, because the respective tests on the earliest recording of mortality, that is, the only mortality recording that could

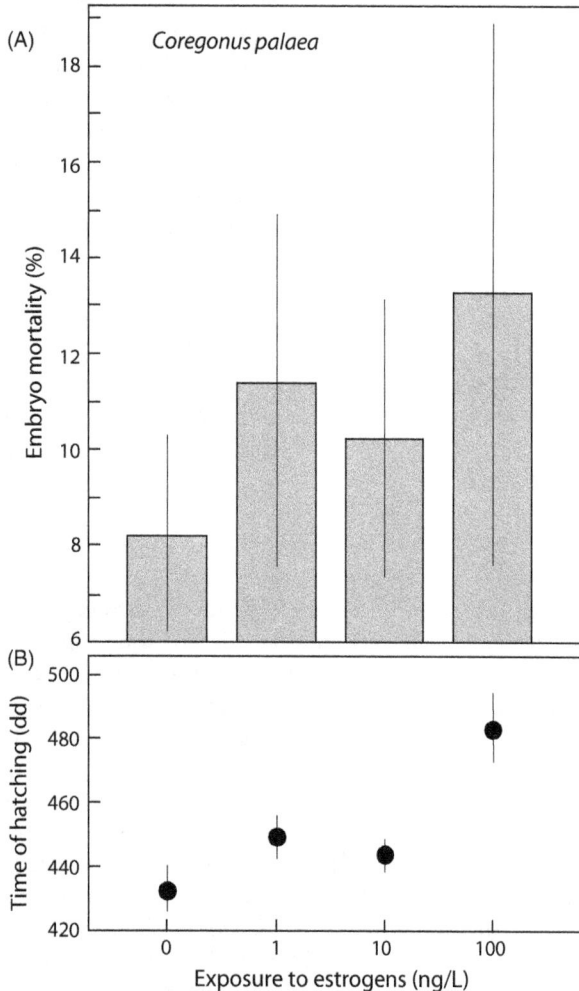

Figure 2 Effects of different experimental stress treatments on embryo mortality, timing of hatching, and hatchling growth in *Coregonus albellus*. Embryos were either treated with 100 ng/L ('Fluc100') or with 10 ng/L fluconazole ('Fluc10') to reduce microbial stress, sham treated, or exposed to various concentrations of estrogens. (A) Embryo mortality, (B) timing of hatching of the survivors (in degree days), (C) hatchling length one day and 10 days after hatching, (D) yolk sac volume one day and 10 days after hatching. All panels show means and the 95% confidence intervals based on family means. See text for statistics.

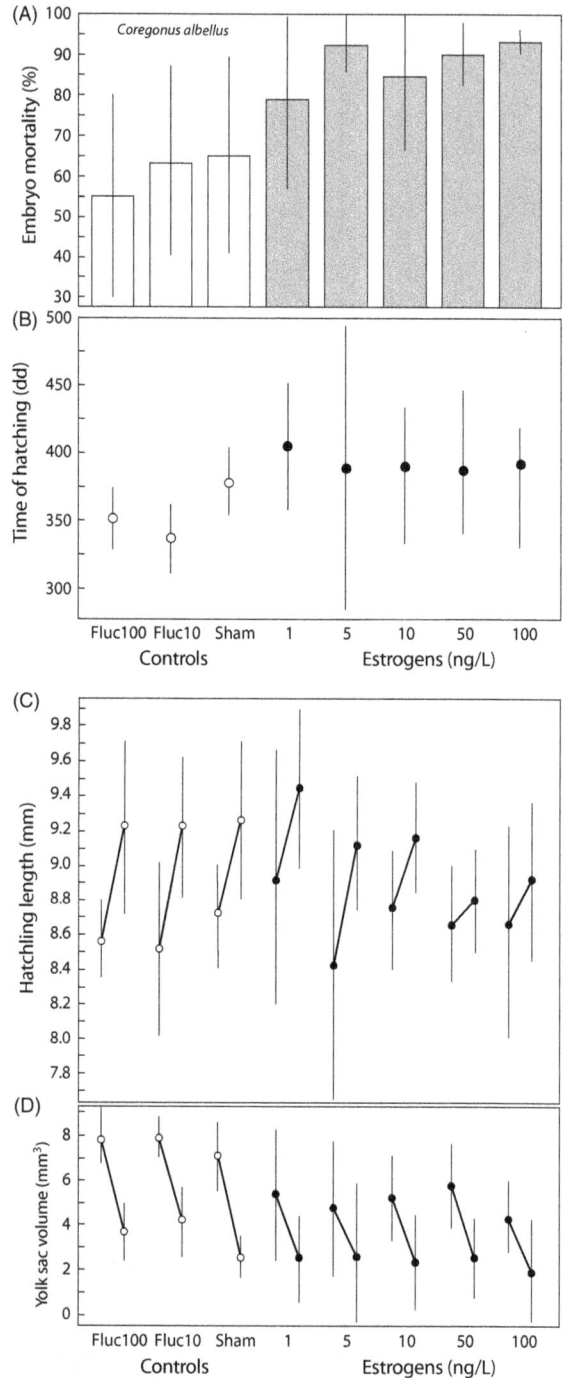

Figure 1 Experiments on *Coregonus palaea*: effects of exposure to the estrogen EE2 on (A) embryo mortality and (B) average timing of hatching (in degree days). The panels show means and the 95% confidence intervals based on family means. See text for statistics.

include unfertilized eggs revealed no significant treatment effects (C. palaea, day 13: χ^2 = 3.5, r_sP_c = 0.14, $P > 0.05$; C. albellus, day 16: χ^2 = 5.1, r_sP_c = 0.43, $P > 0.05$). Models that include EE2 treatment, dam, and sire effects revealed additive genetic variance for tolerance to EE2 in both whitefish species (the significant treatment × sire effects in Table 1), additionally to the overall additive genetic variance in viability that we found in both species (the significant sire main effects in Table 1), and the nonadditive genetic variance in viability that we found in C. palaea (the significant dam × sire effect in Table 1a).

Timing of hatching

We found significant dam and sire effects on the timing of hatching in both species (the main effects in Table 2). Estrogen treatment had a delaying effect on the timing of hatching in C. palaea ($F_{3,2054}$ = 167.7, r_sP_c = 0.80, $P < 0.01$; Fig. 1B). This could be confirmed in a three-way ANOVA that included the parental effects (treatment main effect in Table 2a). This ANOVA revealed additive genetic variance for the timing of hatching in response to the estrogen exposure (the significant treatment × sire effect in Table 2a). We also found significant a treatment × dam effect (Table 2a) and significant nonadditive genetic variance in response to the estrogen treatment (the dam × sire effect in Table 2a). None of these effects of EE2 treatment on the timing of hatching could be confirmed in C. albellus: Neither was the timing of hatching increasingly delayed with increasing estrogen concentration ($F_{5,110}$ = 1.3, r_sP_c = 0.19, $P > 0.05$; Fig. 2B), nor was there any significant parental effect in reaction to the treatment (Table 2b).

Table 1. Effect likelihood ratio tests on embryo mortality until hatching in (a) *Coregonus palaea* from Lake Geneva and (b) *Coregonus albellus* from Lake Brienz treated with various concentrations of the synthetic estrogens EE2.

Factor	χ^2	df	P
(a) *C. palaea* (N_{total} = 2304)			
Treatment (I)	3.9	3	0.27
Dam (D)	6.9	3	0.08
Sire (S)	28.8	5	<0.0001
T × D	13.3	9	**0.15**
T × S	25.6	15	**0.04**
D × S	25.1	15	0.05
(b) *C. albellus* excluding extra controls (N_{total} = 719)			
Treatment	37.9	5	<0.0001
Dam	7.3	1	0.007
Sire	19.5	2	<0.0001
T × D	12.6	5	**0.03**
T × S	21.0	10	**0.02**
D × S	0.05	2	0.98

P-values linked to parent × treatment effects are emphasized in bold.

Table 2. ANOVA on the timing of hatching (a) in *Coregonus palaea* and (b) in *Coregonus albellus* (notation as in Table 1). In (b), some degrees of freedom were lost because of high mortality in some experimental cells.

Factor	F	df	P
(a) *C. palaea* (N_{total} = 2055)			
Treatment	239.0	3	<0.0001
Dam	64.8	3	<0.0001
Sire	38.7	5	<0.0001
T × D	11.6	9	**<0.0001**
T × S	6.2	15	**<0.0001**
D × S	4.2	15	<0.0001
(b) *C. albellus* excluding extra controls (N_{total} = 115)			
Treatment	0.4	4	0.80
Dam	19.3	1	<0.0001
Sire	4.7	1	0.03
T × D	1.9	5	**0.11**
T × S	0.7	9	**0.67**
D × S	2.3	2	0.11

P-values linked to parent × treatment effects are emphasized in bold.

However, if the two additional controls that were treated with antimicrobials were included into the models, hatching was delayed with increased stress level ($F_{5,110}$ = 1.3, r_sP_c = 0.19, $P < 0.05$; Fig. 2B; the treatment effect in a three-way ANOVA analogous to the one in Table 2b would be: F = 7.2, df = 6, $P < 0.0001$).

Alevin size and growth

The body length of freshly hatched C. albellus alevins did not seem to be affected by the estrogen treatment ($F_{5,109}$ = 0.88, r_sP_c = 0.16, $P > 0.05$; Fig. 2C). However, yolk sac volume at the time of hatching was reduced ($F_{5,109}$ = 2.0, r_sP_c = 0.71, $P < 0.01$; Fig. 2D). Accordingly, negative effects of estrogen on hatchling length could be recorded 10 days after hatching ($F_{5,109}$ = 4.4, r_sP_c = 0.88, $P < 0.001$; Fig. 2C) and at the expected final alevin size ($F_{5,109}$ = 4.3, r_sP_c = 0.89, $P < 0.001$). Hatchling growth was not only reduced during the first 10 days after hatching ($F_{5,109}$ = 3.0, r_sP_c = 0.59, $P < 0.05$; Fig. 2C), but also the potential for further growth as yolk sac volume after 10 days was smaller with increasing exposure to estrogens ($F_{5,109}$ = 1.0, r_sP_c = 0.49, $P < 0.05$; Fig. 2D).

In C. albellus, alevin size at hatching was dependent on maternal effects (the significant dam effects in Table 3a), and even if no sire effects could be found on alevin size on hatching or later (Table 3), there was still significant additive genetic variance on growth because the time at which final alevin size was reached depended not only on dam but also on sire effects (Table 4). Estrogen treatment affects alevin growth differently for different dams (the treatment × dam effects in Table 3b,c).

Table 3. ANOVA on *Coregonus albellus* alevin size measured (a) 1 day after hatching, (b) 10 days after hatching, and (c) expected size at the time the yolk sac would be used up (extrapolated from loss of yolk sac volume and increase of alevin length during the first 10 days). Only estrogen- and sham-treated groups are included here. N_{total} = 114 for each statistical model. Including the two extra controls (the antimicrobial treatments) would not change any conclusions except that the main dam effects would always be significant at $P < 0.001$.

Factor	F	df	P
(a) *C. albellus* 1 day posthatching			
Treatment	0.06	4	0.99
Dam	14.2	1	0.0003
Sire	0.05	1	0.82
T × D	0.1	5	**0.99**
T × S	1.3	9	**0.23**
D × S	1.2	2	0.30
(b) *C. albellus* 10 day posthatching			
Treatment	2.1	4	0.08
Dam	2.7	1	0.10
Sire	1.7	1	0.19
T × D	2.6	5	**0.03**
T × S	0.7	9	**0.73**
D × S	0.8	2	0.43
(c) *C. albellus* expected final alevin size			
Treatment	2.1	4	0.09
Dam	3.0	1	0.09
Sire	1.9	1	0.18
T × D	2.5	5	**0.04**
T × S	0.7	9	**0.71**
D × S	0.9	2	0.43

P-values linked to parent × treatment effects are emphasized in bold.

Table 4. ANOVA on the total duration of embryo and larval development in *Coregonus albellus*, extrapolated from the yolk sac volume and its reduction during the first 10 days. Only estrogen- and sham-treated groups are included here (N_{total} = 114). Including the two extra controls would lead to very similar values and would not change the conclusions.

Factor	F	df	P
Treatment	0.1	4	0.96
Dam	10.4	1	0.002
Sire	3.9	1	0.05
T × D	1.3	5	**0.25**
T × S	0.7	9	**0.73**
D × S	1.2	2	0.32

P-values linked to parent × treatment effects are emphasized in bold.

Discussion

Estrogen pollution is a threat to the aquatic environments that has raised much concern (Sumpter 2005; Sumpter and Jobling 2013). Estrogens have repeatedly been demonstrated to induce negative effects on viability and development of fish within various orders (reviews in Scholz and Kluver 2009; Leet et al. 2011; Senior and Nakagawa 2013) and at concentrations that are often found in surface waters (Sumpter and Jobling 2013). We exposed singly kept whitefish embryos to the synthetic estrogen EE2 and found that EE2 significantly reduced viability and growth during embryogenesis even at the lowest concentration of 1 ng/L. Whitefish can therefore be added to the list of ray-finned fish that are very susceptible to estrogen pollution (Scholz and Kluver 2009; Senior and Nakagawa 2013). Increased concentrations of EE2 generally increased embryo mortality in both whitefish species we tested. However, there was much variation in general viability and the susceptibility to EE2 among the different sib groups in our study.

We found significant paternal effects on embryo survival in both species. Paternal origin also had significant effects on development rate in *C. albellus* where we determined embryo growth. Because whitefish are external fertilizers with no parental care, fathers only contribute genes to their offspring, and paternal effects on embryo survival and development rate therefore directly reveal additive genetic variance for general viability within the both populations that we sampled. It turned out that some males were of higher overall genetic quality than others, as previously observed in other samples of Alpine whitefish (Wedekind et al. 2001, 2007a, 2008a; Clark et al. 2014) and other salmonid populations (Jacob et al. 2007, 2010; Pitcher and Neff 2007; Wedekind et al. 2008b; Evans et al. 2010; Clark et al. 2013). Importantly, we also found significant interactions between paternal origin and the EE2 treatment on embryo viability in both species. Such interaction terms demonstrate that some genotypes are more tolerant to EE2 than others, even after controlling for the variation in overall genetic quality within the populations. We conclude that there is, in both study populations, significant genetic variation that would be required for rapid evolutionary responses to EE2 pollution.

When we tested for possible effects of EE2 on embryo growth and development, we found that hatching time was significantly affected in *C. palaea* but not in *C. albellus*. The apparent difference between the two species could be due to differences in sample sizes and the associated statistical power (these differences in sample sizes were partly due to higher embryo mortalities in *C. albellus* than in *C. palaea*; see Methods). In *C. palaea*, we also found hatching time to be generally determined by dam, sire, and dam × sire effects, that is, offspring of half-sib families hatched at different times even if each embryo was raised in isolation. With regard to hatching time, different maternal and paternal half-sib groups also reacted differently to the EE2 treatment. The significant sire × EE2 treatment effect demonstrates again a genetic variation in response to EE2.

Variation in hatching time may either reveal variation in developmental rate (if, at the conditions of our study, hatching is directly linked to a developmental stage) or could reveal a behavioral response to acute stress. Stress-induced precocious hatching is common in amphibians (Warkentin 2011) and has been demonstrated in whitefish in response to waterborne cues of infections or other threats (Wedekind 2002a; Wedekind and Müller 2005). However, in our samples, hatching was generally delayed in EE2-treated embryos. This suggests that the variation in hatching time that we observed revealed variation in developmental rates (as in Clark et al. 2014). The late hatching in EE2-treated embryos therefore suggests that EE2 reduces developmental rates in *C. palaea* and differently so for different genotypes, that is, some genotypes seemed again more susceptible than others to EE2 pollution.

When we analyzed body length and yolk sac volume in *C. albellus* hatchlings, we found not only significant dam effects (some females produced offspring that generally developed faster than those of other females) but also a significant interaction between dam effects and EE2 concentration on embryo growth and expected final size, that is, the progeny of some mothers were more susceptible to EE2 pollution than the progeny of others. Dam effects are expected to be a combination of maternal environmental effects (egg content and egg size) and additive genetic effects. The relative role of the latter remains unclear in this case, because the respective interaction between paternal effects and EE2 concentrations was not significant. However, individual growth rates can be fitness relevant in salmonids (e.g., Skoglund et al. 2012). Therefore, the reduction of embryo growth within some maternal sib groups let us to conclude that there are nonlethal toxic effects of EE2 that may affect fitness among the surviving embryos.

There are a number of differences between the controlled laboratory conditions and natural situations that could potentially affect the toxicity of EE2 and its congeners. Among the micro-ecological factors that could play a role are the composition and density of microbial symbiont communities associated to the embryos (L. G. E. Wilkins, A. Rogivue, L. Fumagalli and C. Wedekind, unpublished data). Very little is currently known about the importance of degradation of estrogenic chemicals in different aquatic environments, that is, it is still difficult to predict environmental concentrations of estrogenic compounds at different times and locations (Sumpter and Jobling 2013). Moreover, it remains to be shown how the effects that different hormone-active chemicals can have on fish development interact, for example, whether and to what degree their toxicity is additive (Sumpter and Jobling 2013). While laboratory studies like ours allow for qualitative conclusions about the existence of genetic and maternal environmental effects (Lynch and Walsh 1998), the relevant quantitative effects of EE2 on embryo growth and development remain to be confirmed under more natural conditions. Basing experiments like ours on larger number of breeders cannot solve this problem, even if larger samples would allow for better estimates of the variance components under our laboratory conditions (as, for example, in Clark et al. 2014).

Since the discovery of Purdom et al. (1994) that estrogenic chemicals in effluents of sewage treatment plants can cause significant alterations in fish, the industry and policy organizations of many countries have significantly invested into the treatment of wastewater to better remove estrogenic chemicals (e.g., Burkhardt-Holm et al. 2008; Sumpter and Jobling 2013). However, while the use of nonylphenol and related chemicals (a group of estrogenic pollutants) could be regulated via legislation in some parts of the world (Sumpter and Jobling 2013), EE2 may be more difficult to ban because it is an active ingredient of most hormonal contraceptives. To the best of our knowledge, no environmental quality standard has yet been defined by any legislative authority. Sumpter and Jobling (2013) suggested that an environmental quality standard of around 0.02 ng/L may be possible, but the authors stressed that the risks of potent chemicals like EE2 should never be fully dismissed even at very low concentrations.

Some whitefish populations in pre-Alpine lakes showed extraordinary high prevalences of gonadal deformations during recent years (Bernet et al. 2009). Potential pollution by endocrine disruptors has been a focus of various studies (e.g., Liedtke et al. 2009; Bogdal et al. 2010). Even if no suspicious contamination levels could be demonstrated so far, all pre-Alpine lakes receive effluents from sewage plants, that is, pollution by EE2 and other estrogens is an environmental risk also in low populated areas. We found that even low concentrations of EE2 would create strong selection pressures on two whitefish species that differ in many respects. Whitefish females produce large numbers of offspring (up to several thousands per year in the case of *C. palaea* and up to several hundreds per year in the case of *C. albellus*). These high reproductive rates in combination with the strong effects EE2 has on embryo survival and growth and the fact that both populations show additive genetic variation in the tolerance to EE2 suggest that rapid evolution in response to endocrine pollution is possible in Alpine whitefish. Our findings further illustrate the importance of genetic variation for natural populations that need to adapt to anthropogenic threats.

Acknowledgements

We thank the Fischereiinspektorat Bern (Bern canton) and the inspection de la pêche (Vaud canton) for permissions, B. Abegglen and A. Schmid for catching the fish, A. Babin, S. Büchel, P. Buri, E. Clark, D. Fell, F. Glauser, A. Jacob, K. Hine, N. Kunz, R. Nicolet, S. Nusslé, M. Pompini, B. Rieder, F. Russier, M. dos Santos, B. von Siebenthal, R. Stelkens, L. Wilkins and F. Witsenburg for assistance in the field or in the laboratory, M. Flück, F. Hofmann, C. Küng, and A. Roulin for discussion, C. Eizaguirre and two reviewers for helpful comments on the manuscript, and the Swiss National Science Foundation for funding.

Literature cited

Baroiller, J.-F., H. D'Cotta, and E. Saillant 2009. Environmental effects on fish sex determination and differentiation. Sexual Development 3:118–135.

Beck, I. C., R. Bruhn, J. Gandrass, and W. Ruck 2005. Liquid chromatography-tandem mass spectrometry analysis of estrogenic compounds in coastal surface water of the Baltic Sea. Journal of Chromatography A 1090:98–106.

Bernet, D., T. Wahli, S. Kipfer, and H. Segner 2009. Macroscopic gonadal deviations and intersex in developing whitefish *Coregonus lavaretus*. Aquatic Biology 6:1–13.

Bogdal, C., M. Scheringer, P. Schmid, M. Blauenstein, M. Kohler, and K. Hungerbuhler 2010. Levels, fluxes and time trends of persistent organic pollutants in Lake Thun, Switzerland: combining trace analysis and multimedia modeling. Science of the Total Environment 408:3654–3663.

Bonduriansky, R., and T. Day 2009. Nongenetic inheritance and its evolutionary implications. Annual Review of Ecology Evolution and Systematics 40:103–125.

Burkhardt-Holm, P., H. Segner, R. Burki, A. Peter, S. Schubert, M. J. F. Suter, and M. E. Borsuk 2008. Estrogenic endocrine disruption in Switzerland: assessment of fish exposure and effects. Chimia 62:376–382.

Caldwell, D. J., F. Mastrocco, P. D. Anderson, R. Lange, and J. P. Sumpter 2012. Predicted-no-effect concentrations for the steroid estrogens estrone, 17 beta-estradiol, estriol, and 17 alpha-ethinylestradiol. Environmental Toxicology and Chemistry 31:1396–1406.

Clark, E. S., R. B. Stelkens, and C. Wedekind 2013. Parental influences on pathogen resistance in brown trout embryos and effects of outcrossing within a river network. PLoS ONE 8:e57832.

Clark, E. S., M. Pompini, L. Marques da Cunha, and C. Wedekind 2014. Maternal and paternal contributions to pathogen resistance dependent on development stage in a whitefish (Salmonidae). Functional Ecology 28:714–723.

Cotton, S., and C. Wedekind 2009. Population consequences of environmental sex reversal. Conservation Biology 23:196–206.

Devlin, R. H., and Y. Nagahama 2002. Sex determination and sex differentiation in fish: an overview of genetic, physiological, and environmental influences. Aquaculture 208:191–364.

Dzieweczynski, T. L. 2011. Short-term exposure to an endocrine disruptor affects behavioural consistency in male threespine stickleback. Aquatic Toxicology 105:681–687.

Evans, M. L., B. D. Neff, and D. D. Heath 2010. Quantitative genetic and translocation experiments reveal genotype-by-environment effects on juvenile life-history traits in two populations of Chinook salmon (*Oncorhynchus tshawytscha*). Journal of Evolutionary Biology 23:687–698.

Fenske, M., G. Maack, C. Schafers, and H. Segner 2005. An environmentally relevant concentration of estrogen induces arrest of male gonad development in zebrafish, *Danio rerio*. Environmental Toxicology and Chemistry 24:1088–1098.

Ferrière, R., U. Dieckmann, and D. Couvet (eds) 2004. Evolutionary Conservation Biology. Cambridge University Press, Cambridge.

Gutendorf, B., and J. Westendorf 2001. Comparison of an array of *in vitro* assays for the assessment of the estrogenic potential of natural and synthetic estrogens, phytoestrogens and xenoestrogens. Toxicology 166:79–89.

Gutierrez, J. B., and J. L. Teem 2006. A model describing the effect of sex-reversed YY fish in an established wild population: the use of a Trojan Y chromosome to cause extinction of an introduced exotic species. Journal of Theoretical Biology 241:333–341.

Hamilton, P. B., E. Nicol, E. S. De-Bastos, R. J. Williams, J. P. Sumpter, S. Jobling, J. R. Stevens et al. 2014. Populations of a cyprinid fish are self-sustaining despite widespread feminization of males. BMC Biology 12:1.

Harris, C. A., P. B. Hamilton, T. J. Runnalls, V. Vinciotti, A. Henshaw, D. Hodgson, T. S. Coe et al. 2011. The consequences of feminization in breeding groups of wild fish. Environmental Health Perspectives 119:306–311.

Hendry, A. P., M. T. Kinnison, M. Heino, T. Day, T. B. Smith, G. Fitt, C. T. Bergstrom et al. 2011. Evolutionary principles and their practical application. Evolutionary Applications 4:159–183.

Jacob, A., S. Nusslé, A. Britschgi, G. Evanno, R. Müller, and C. Wedekind 2007. Male dominance linked to size and age, but not to 'good genes' in brown trout (*Salmo trutta*). BMC Evolutionary Biology 7:207.

Jacob, A., G. Evanno, B. A. von Siebenthal, C. Grossen, and C. Wedekind 2010. Effects of different mating scenarios on embryo viability in brown trout. Molecular Ecology 19:5296–5307.

Jobling, S., R. Williams, A. Johnson, A. Taylor, M. Gross-Sorokin, M. Nolan, C. R. Tyler et al. 2006. Predicted exposures to steroid estrogens in UK rivers correlate with widespread sexual disruption in wild fish populations. Environmental Health Perspectives 114:32–39.

Kime, D. H. 1998. Endocrine Disruption in Fish. Kluwer Academic Publishers, Norwell, MA.

Kipfer, S., H. Segner, M. Wenger, T. Wahli, and D. Bernet 2009. Long-term estrogen exposure of whitefish *Coregonus lavaretus* induces intersex but not Lake Thun-typical gonad malformations. Diseases of Aquatic Organisms 84:43–56.

Kirchhofer, A., and T. J. Lindt-Kirchhofer 1998. Growth and development during early life stages of *Coregonus lavaretus* from three lakes in Switzerland. Advances in Limnology 50:49–59.

Kolpin, D. W., E. T. Furlong, M. T. Meyer, E. M. Thurman, S. D. Zaugg, L. B. Barber, and H. T. Buxton 2002. Response to Comment on "Pharmaceuticals, hormones, and other organic wastewater contaminants in US streams, 1999–2000: a national reconnaissance". Environmental Science & Technology 36:4007–4008.

Lange, R., T. H. Hutchinson, C. P. Croudace, F. Siegmund, H. Schweinfurth, P. Hampe, G. H. Panter et al. 2001. Effects of the synthetic estrogen 17 alpha-ethinylestradiol on the life-cycle of the fathead minnow (*Pimephales promelas*). Environmental Toxicology and Chemistry 20:1216–1227.

Lange, A., G. C. Paull, P. B. Hamilton, T. Iguchi, and C. R. Tyler 2011. Implications of persistent exposure to treated wastewater effluent for breeding in wild roach (*Rutilus rutilus*) populations. Environmental Science & Technology 45:1673–1679.

Larsen, M. G., K. B. Hansen, P. G. Henriksen, and E. Baatrup 2008. Male zebrafish (*Danio rerio*) courtship behaviour resists the feminising effects of 17 alpha-ethinyloestradiol – morphological sexual characteristics do not. Aquatic Toxicology 87:234–244.

Larsson, D. G. J., M. Adolfsson-Erici, J. Parkkonen, M. Pettersson, A. H. Berg, P. E. Olsson, and L. Forlin 1999. Ethinyloestradiol – an undesired fish contraceptive? Aquatic Toxicology 45:91–97.

Leet, J. K., H. E. Gall, and M. S. Sepulveda 2011. A review of studies on androgen and estrogen exposure in fish early life stages: effects on gene and hormonal control of sexual differentiation. Journal of Applied Toxicology 31:379–398.

Leitritz, E., and R. C. Lewis 1976. Trout and salmon culture (hatchery methods). State of California Department of Fish and Game – Fish Bulletin 164:1–197.

Liedtke, A., R. Schonenberger, R. I. L. Eggen, and M. J. F. Suter 2009. Internal exposure of whitefish (*Coregonus lavaretus*) to estrogens. Aquatic Toxicology 93:158–165.

Lynch, M., and B. Walsh 1998. Genetics and Analysis of Quantitative Traits. Sinauer Associates Inc, Sunderland, MA.

Müller, R., M. Breitenstein, M. M. Bia, C. Rellstab, and A. Kirchhofer 2007. Bottom-up control of whitefish populations in ultra-oligotrophic Lake Brienz. Aquatic Sciences 69:271–288.

OECD 1992. OECD Guideline for the Testing of Chemicals. Organisation for Economic Cooperation and Development, Paris, France.

Pitcher, T. E., and B. D. Neff 2007. Genetic quality and offspring performance in Chinook salmon: implications for supportive breeding. Conservation Genetics 8:607–616.

Purdom, C. E., P. A. Hardiman, V. V. J. Bye, N. C. Eno, C. R. Tyler, and J. P. Sumpter 1994. Estrogenic effects of effluents from sewage treatment works. Chemistry and Ecology 8:275–285.

R Development Core Team 2011. R: A Language and Environment for Statistical Computing. R Foundation for Statistical Computing, Vienna, Austria. http://www.R-project.org (accessed on 15 January 2013).

Rice, W. R., and S. D. Gaines 1994. Extending nondirectional heterogeneity tests to evaluate simply ordered alternative hypotheses. Proceedings of the National Academy of Sciences, USA 91:225–226.

Rose, J., H. Holbech, C. Lindholst, U. Norum, A. Povlsen, B. Korsgaard, and P. Bjerregaard 2002. Vitellogenin induction by 17 beta-estradiol and 17 alpha-ethinylestradiol in male zebrafish (*Danio rerio*). Comparative Biochemistry and Physiology C-Toxicology & Pharmacology 131:531 539.

Scholz, S., and N. Kluver 2009. Effects of endocrine disrupters on sexual, gonadal development in fish. Sexual Development 3:136–151.

Segner, H., K. Caroll, M. Fenske, C. R. Janssen, G. Maack, D. Pascoe, C. Schafers et al. 2003a. Identification of endocrine-disrupting effects in aquatic vertebrates and invertebrates: report from the European IDEA project. Ecotoxicology and Environmental Safety 54:302–314.

Segner, H., J. M. Navas, C. Schafers, and A. Wenzel 2003b. Potencies of estrogenic compounds in in vitro screening assays and in life cycle tests with zebrafish in vivo. Ecotoxicology and Environmental Safety 54:315–322.

Senior, A. M., and S. Nakagawa 2013. A comparative analysis of chemically induced sex reversal in teleosts: challenging conventional suppositions. Fish and Fisheries 14:60–76.

Shore, L. S., M. Gurevitz, and M. Shemesh 1993. Estrogen as an environmental pollutant. Bulletin of Environmental Contamination and Toxicology 51:361–366.

von Siebenthal, B. A., A. Jacob, and C. Wedekind 2009. Tolerance of whitefish embryos to *Pseudomonas fluorescens* linked to genetic and maternal effects, and reduced by previous exposure. Fish and Shellfish Immunology 26:531–535.

Skoglund, H., S. Einum, T. Forseth, and B. T. Barlaup 2012. The penalty for arriving late in emerging salmonid juveniles: differences between species correspond to their interspecific competitive ability. Functional Ecology 26:104–111.

Soares, J., A. M. Coimbra, M. A. Reis-Henriques, N. M. Monteiro, M. N. Vieira, J. M. A. Oliveira, P. Guedes-Dias et al. 2009. Disruption of zebrafish (*Danio rerio*) embryonic development after full life-cycle parental exposure to low levels of ethinylestradiol. Aquatic Toxicology 95:330–338.

Stelkens, R. B., and C. Wedekind 2010. Environmental sex reversal, Trojan sex genes, and sex ratio adjustment: conditions and population consequences. Molecular Ecology 19:627–646.

Sumpter, J. P. 2005. Endocrine disrupters in the aquatic environment: an overview. Acta Hydrochimica et Hydrobiologica 33:9–16.

Sumpter, J. P., and S. Jobling 2013. The occurrence, causes, and consequences of estrogens in the aquatic environment. Environmental Toxicology and Chemistry 32:249–251.

Ternes, T. A., M. Stumpf, J. Mueller, K. Haberer, R. D. Wilken, and M. Servos 1999. Behavior and occurrence of estrogens in municipal sewage treatment plants – I. Investigations in Germany, Canada and Brazil. Science of the Total Environment 225:81–90.

Van den Belt, K., R. Verheyen, and H. Witters 2003. Effects of 17a-ethynylestradiol in a partial life-cycle test with zebrafish (*Danio rerio*): effects on growth, gonads and female reproductive success. Science of the Total Environment 309:127–137.

Vandegehuchte, M. B., and C. R. Janssen 2011. Epigenetics and its implication for ecotoxicology. Ecotoxicology 20:607–624.

Vonlanthen, P., D. Bittner, A. G. Hudson, K. A. Young, R. Müller, B. Lundsgaard-Hansen, D. Roy et al. 2012. Eutrophication causes speciation reversal in whitefish adaptive radiations. Nature 482:357–362.

Vulliet, E., and C. Cren-Olive 2011. Screening of pharmaceuticals and hormones at the regional scale, in surface and groundwaters intended to human consumption. Environmental Pollution 159:2929–2934.

Warkentin, K. M. 2011. Plasticity of hatching in amphibians: evolution, trade-offs, cues and mechanisms. Integrative and Comparative Biology 51:111–127.

Wedekind, C. 2002a. Induced hatching to avoid infectious egg disease in whitefish. Current Biology 12:69–71.

Wedekind, C. 2002b. Manipulating sex ratios for conservation: short-term risks and long-term benefits. Animal Conservation 5:13–20.

Wedekind, C. 2012. Managing population sex ratios in conservation practice: how and why? In: T. Povilitis, ed. Topics in Conservation Biology, pp. 81–96. InTech, Rijeka.

Wedekind, C. 2014. Fish populations surviving estrogen pollution. BMC Biology 12:10.

Wedekind, C., and R. Müller 2005. Risk-induced early hatching in salmonids. Ecology 86:2525–2529.

Wedekind, C., R. Müller, and H. Spicher 2001. Potential genetic benefits of mate selection in whitefish. Journal of Evolutionary Biology 14:980–986.

Wedekind, C., G. Rudolfsen, A. Jacob, D. Urbach, and R. Müller 2007a. The genetic consequences of hatchery-induced sperm competition in a salmonid. Biological Conservation 137:180–188.

Wedekind, C., B. A. von Siebenthal, and R. Gingold 2007b. The weaker points of fish acute toxicity tests and how tests on embryos can solve some issues. Environmental Pollution **148**:385–389.

Wedekind, C., G. Evanno, D. Urbach, A. Jacob, and R. Müller 2008a. 'Good-genes' and 'compatible-genes' effects in an Alpine whitefish and the information content of breeding tubercles over the course of the spawning season. Genetica **134**:21–30.

Wedekind, C., A. Jacob, G. Evanno, S. Nusslé, and R. Müller 2008b. Viability of brown trout embryos positively linked to melanin-based but negatively to carotenoid-based colours of their fathers. Proceedings of the Royal Society B-Biological Sciences **275**:1737–1744.

Wedekind, C., M. O. Gessner, F. Vazquez, M. Maerki, and D. Steiner 2010. Elevated resource availability sufficient to turn opportunistic into virulent fish pathogens. Ecology **91**:1251–1256.

Zhang, X., Y. Gao, Q. Li, G. Li, Q. Guo, and C. Yan 2011. Estrogenic compounds and estrogenicity in surface water, sediments, and organisms from Yundang Lagoon in Xiamen, China. Archives of Environmental Contamination and Toxicology **61**:93–100.

5

Harvest-induced evolution and effective population size

Anna Kuparinen,[1] Jeffrey A. Hutchings[2,3,4] and Robin S. Waples[5]

1 Department of Environmental Sciences, University of Helsinki, Helsinki, Finland
2 Department of Biology, Dalhousie University, Halifax, NS, Canada
3 Department of Biosciences, Centre For Ecological and Evolutionary Synthesis, University of Oslo, Oslo, Norway
4 Department of Natural Sciences, University of Agder, Kristiansand, Norway
5 National Marine Fisheries Service, National Oceanic and Atmospheric Administration, Northwest Fisheries Science Center, Seattle, WA, USA

Keywords

contemporary evolution, fisheries management, life history evolution, population genetics – empirical, wildlife management.

Correspondence

Robin S. Waples, Northwest Fisheries Science Center, 2725 Montlake Blvd. East, Seattle, WA, USA.

e-mail: robin.waples@noaa.gov

Abstract

Much has been written about fishery-induced evolution (FIE) in exploited species, but relatively little attention has been paid to the consequences for one of the most important parameters in evolutionary biology—effective population size (N_e). We use a combination of simulations of Atlantic cod populations experiencing harvest, artificial manipulation of cod life tables, and analytical methods to explore how adding harvest to natural mortality affects N_e, census size (N), and the ratio N_e/N. We show that harvest-mediated reductions in N_e are due entirely to reductions in recruitment, because increasing adult mortality actually increases the N_e/N ratio. This means that proportional reductions in abundance caused by harvest represent an upper limit to the proportional reductions in N_e, and that in some cases N_e can even increase with increased harvest. This result is a quite general consequence of increased adult mortality and does not depend on harvest selectivity or FIE, although both of these influence the results in a quantitative way. In scenarios that allowed evolution, N_e recovered quickly after harvest ended and remained higher than in the preharvest population for well over a century, which indicates that evolution can help provide a long-term buffer against loss of genetic variability.

Introduction

Increasingly in recent decades, humans have created a global experiment by subjecting natural populations to harvest at rates that equal or exceed the rate of natural mortality (Darimont et al. 2009). Some short-term consequences of harvest can be deduced from first principles. The additional harvest-induced mortality will truncate the age structure of the population because fewer individuals live to old age. Moreover, this additional mortality is often positively correlated with size, due to harvesting regulations and trophy hunting (Coltman et al. 2003; Allendorf and Hard 2009). Size in turn is correlated with age in species with indeterminate growth, such that the effect of age-structure truncation will be exacerbated.

These short-term demographic consequences can be expected to elicit evolutionary responses in species with the genetic capability to do so. Species with low rates of natural mortality as adults generally mature at older ages, because

investing limited energy into growth rather than early maturity means that they will be larger when they reach maturity (and hence have higher fecundity and potentially higher mating success), and they can expect to reap the benefits of higher fecundity for many years because mortality is low. If adult mortality is sharply increased, perhaps by a factor of 2 or more (Mertz and Myers 1998; Law 2007), individuals that delay reproduction no longer can expect to enjoy many seasons of high reproductive success, so relative fitness of that phenotype declines. The result is evolutionary pressure to mature at an earlier age and smaller size, to ensure at least some opportunities for reproduction before death. Precisely predicting evolutionary responses to harvest is difficult because changes in a population's vital rates can affect density dependence, particularly at juvenile life stages, as well as biotic interactions with other species (Polacheck et al. 2004; Howell et al. 2013; Kuparinen et al. 2014a). Nevertheless, numerous studies have estimated empirical rates of phenotypic change in harvested species that are in line with expectations from fisheries-induced evolution (FIE) (Hutchings and Baum 2005; Sharpe and Hendry 2009; Devine et al. 2012; Audzijonyte et al. 2013; Kendall et al. 2014).

Over the past decades, numerous studies have focused on FIE, to understand its mechanisms and to project its ecological consequences. However, this literature has largely ignored influences of FIE on effective population size (N_e). This is an important gap because N_e can influence virtually all evolutionary processes. Effective size determines not only the rates of inbreeding, allele frequency change, and loss of genetic variability in a population, but also the efficiency of natural selection (and hence the balance between random and directed evolutionary processes; see Edeline et al. 2007; Lanfear et al. 2014). N_e and the ratio of N_e to census size (N) are sensitive to population demography (Felsenstein 1971; Nunney 1993), so direct, short-term effects of harvest and longer-term evolutionary changes to a population's vital rates can both be expected to change N_e and N_e/N.

One notable exception to the above gap regarding effective population size is the study by Marty et al. (2015), who showed that considering random effects associated with FIE is important, particularly when evaluating potential for evolutionary recovery after fishing is relaxed. They showed that, in many circumstances, random factors related to N_e can be more important than FIE in eroding additive genetic variance, which provides evolutionary resilience to a population. Marty et al. (2015) simulated both neutral and adaptive genes and estimated N_e from neutral genes by tracking the rate of change in allele frequency over time (the temporal method; Waples 1989). They took samples every 20 years and converted this time interval into elapsed generations based on calculations of generation length (T) from the simulated demographies. This approach should be sufficient to provide rough estimates of N_e. However, the standard temporal method they used assumes discrete generations and is not ideally suited for iteroparous species with overlapping generations—exactly the type of species most likely to experience FIE (Hutchings and Fraser 2008). Based on the range of generation lengths in their modeled populations ($T = 7.5$–12.1 years; Marty et al. 2015), each 20-year period for estimating N_e encompassed only 1.7–2.7 generations, which is not enough to eliminate age-structure bias in $\hat{N}_e$ in the temporal method (Waples and Yokota 2007). Furthermore, the resulting estimates apply to a harmonic mean N_e over the period between samples and hence are difficult to relate to specific points in time.

Here, we take a different approach and calculate N_e directly from vital rates for simulated populations of Atlantic cod that experience various harvest scenarios previously modeled, for example, by Kuparinen et al. (2014a). We use a method for calculating N_e (AgeNe; Waples et al. 2011) that is designed for use with iteroparous, age-structured species and which can estimate effective size for individual cohorts. We consider both N_e and the ratio N_e/N (with N defined as the number of mature adults) because the latter allows us to disentangle the effects of changes in vital rates that affect the N_e/N ratio from effects on abundance, which can reduce N_e even if N_e/N is not reduced. To explore generality of our results, we supplement the simulations with analytical results and artificial manipulation of another life table for Atlantic cod.

Methods

Table 1 lists notation used in this study. Our analyses used two different life tables for Newfoundland's Northern cod, which we refer to as cod life table #1 and cod life table #2. These life tables are both based on empirical data, but for different areas and time periods with different histories of exploitation. Life table #1 was used to parameterize the simulations that evaluated demographic and evolutionary responses to fishing. These simulations included density dependence, again based on empirical data. To explore generality of our simulation results, we artificially manipulated cod life table #2 by increasing adult mortality. These analyses were purely demographic and did not consider evolution or density dependence. More details about each type of analysis are provided below.

Table 1. Notation used in this study.

N_T	Total population size, including juveniles
N	Adult population size (all mature individuals)
N_e	Effective population size per generation
α	Youngest age at which reproduction can occur
ω	Maximum age
N_1	Number of newborn offspring produced each year.
N_α	Number of offspring produced each year that survive to age at first reproduction, at which point they are known as *recruits*
b_x	Mean number of offspring per year produced by an individual of age x that survive to age of recruitment
s_x	Probability of survival from age x to age $x + 1$
d_x	$= 1 - s_x$ = probability of dying between age x and age $x + 1$
l_x	Cumulative survival through age x
T	Generation length = average age of parents of a newborn cohort
$V_{k\bullet}$	Lifetime variance in reproductive success among individuals in a single cohort
V_x	Variance in number of offspring produced by same-age, same-sex individuals in one time period
ϕ_x	V_x/b_x = ratio of the variance to mean number of offspring produced in one time period by individuals of age x
F	Instantaneous rate of fishing mortality (annual mortality $= 1 - e^{(-F)}$)
$L(t)$	Length at age t
L_∞	Asymptotic length
k	von Bertalanffy intrinsic growth coefficient

Simulation of cod dynamics and construction of life tables

To investigate the impacts of fishing and FIE on N_e, we constructed cod life table #1 at different phases of exploitation and fisheries-induced life-history evolution. To this end, we simulated cod dynamics using an individual-based modeling approach that integrates quantitative genetics, life-history evolution, and ecological dynamics of the population. Individual life histories are described through von Bertalanffy growth trajectories (von Bertalanffy 1938), $L(t) = L_\infty - (L_\infty - L_0)e^{-kt}$, where L_0 and $L(t)$ are length at ages 0 and t, L_∞ is asymptotic body length, and k is the intrinsic growth coefficient describing the speed at which L_∞ is reached.

Genetic contributions to life histories were described through additive effects of 10 diploid loci (coded 0 or 1), to mimic the fact that quantitative traits are typically coded by many loci with small additive effects (Roff 2002). The sum of allelic values (ranging between 0 and 20) was coupled with a small amount of environmental variation (drawn from a normal distribution with mean = 0, SD = 3.5) to yield realistic heritabilities of ~0.2–0.3 for life-history traits (Mousseau and Roff 1987; Carlson and Seamons 2008; but see also Postma 2014) and translated linearly into the value of L_∞. The correlations between k and L_∞ and between L_∞ and the length at maturation are well-established life-history relationships (Charnov 1993; Charnov et al. 2013), so the value of k and the length at maturation could be estimated based on L_∞. Empirical bases for the growth parameters and their relationships were obtained from growth trajectories estimated from otoliths collected in a landlocked cod population in Baffin Island, northern Canada. Cod life histories in this population are similar to marine cod populations in northern latitudes, and the population is unexploited and, therefore, reflects natural phenotypic diversity of cod life histories. The empirically observed range of L_∞ was 30–130 cm, and k could be estimated through regression as $\log(k) = 0.609 - 0.0139 \times L_\infty$ (with residual standard error of 0.305) (Kuparinen et al. 2012). L_0 was set to 4 cm for each growth trajectory. The age–length relationship was estimated from the same cod data as $weight = 3.52 \times 10^6 \times length^{3.19}$.

Population dynamics were simulated through time such that at each time step (year) the processes of natural mortality, growth, maturation, and reproduction were modeled on an individual basis. Demographic stochasticity was accounted for by drawing appropriate random numbers to describe the outcome of each process. Baseline instantaneous natural mortality was assumed to be 0.12, to which a survival cost of reproduction of 0.1 was added for mature individuals; these values provide the closest match between the empirically observed cod growth trajectories and those

predicted by the model (Kuparinen et al. 2012). Growth occurred such that at each time step an individual progressed along its von Bertalanffy growth trajectory according to a time increment $\Delta t = e^{15-17.69c} (1 + e^{15-17.69c})^{-1}$, where c is the ratio of population biomass to carrying capacity (K). In a sparse population, Δt was approximately 1, corresponding to 1 year increment in simulation time, whereas in a dense population the progress is slower. Maturation was assumed to occur at a body length 66% of L_∞ (Jensen 1997), and maximum age was set to 25 years.

At each time step, all mature individuals reproduced, such that for each mature female a mature male was assigned randomly (no sexual selection was assumed). Alleles were passed from parents to juveniles stochastically through Mendelian inheritance. Egg production was predicted through $eggs = \{0.48 \times [(female weight + 0.37]/ 1.45) + 0.12\} \times 10^6$, as estimated for Northern cod in the 1960s (Hutchings 2005). At that time, abundance of the Northern cod stock was assumed to be at about 40% of its carrying capacity. Density dependence of juvenile production was assumed to be compensatory, such that the above egg production was scaled up or down according to the abundance-specific relative fecundity estimates reported in Kuparinen et al. (2014b). Survival from egg to a 3-year-old recruit was set to 1.13×10^{-6} (Hutchings 2005). For further details of the model and its parameterization, see Kuparinen et al. (2012, 2014b).

Dynamics of preadapted cod populations were simulated first for 100 years in equilibrium conditions, followed by a 50-year period of fishing and a 150-year period of recovery in the absence of fishing. Simulations were repeated with and without life-history evolution. In nonevolving simulations, juvenile alleles were drawn from a parental pool recorded during equilibrium conditions. We considered three alternative fishing pressures ($F = 0.15$, $F = 0.20$, and $F = 0.25$, where F is instantaneous fishing mortality expressed as a fraction of total biomass) and two fishing selectivity scenarios (logistic typical for trawl, and no size selectivity). These fishing intensities are well within the range of population-specific target fishing mortality levels for Atlantic cod (F_{MSY}: 0.18–0.40; www.ices.dk). However, we needed to model levels that were sustainable over five decades and left a large enough population to allow calculation of age-specific vital rates. At each time step throughout the simulations, we recorded age-specific survival (s_x), fecundity (b_x), and the proportion of mature individuals, as well as total annual recruit production. Life tables for the simulated populations were then compiled by averaging across replicates at specific years representing the period of equilibrium (year 100); early fishing (years 110, 130); late fishing (year 150), by which point fisheries-induced evolution had occurred in evolving populations; initial recovery following the end of fishing (years 160, 180); mid recovery (year 220); and late recovery, by which time biomass had rebuilt back to equilibrium levels (year 300).

Census size and effective population size

Census size

In a stable, age-structured population, total population size (N_T) depends on two parameters: the number of newborns each year (N_1) and cumulative survivorship over time (l_x), calculated through the maximum age (ω). Adult population size (N) can be obtained by replacing newborns with recruits (N_α = the number of offspring that survive to age at maturity, α), defining l_α to be 1, and taking the sum across the years of the adult life span (α to ω):

$$N = N_\alpha \sum_{x=\alpha}^{\omega} l_x \qquad (1)$$

Because age at maturity varies in cod (Table 2), in calculating adult N from eqn (1) we used $\alpha = 3$ (the minimum age any individuals matured in our study) and adjusted Σl_x to account for the fraction mature at each age.

If adult mortality is constant at the rate d per year, then it can be shown that $\Sigma l_x = 1/d$ and

$$N = N_\alpha/d \qquad (2)$$

This result is exact for a species with an arbitrarily long life span (Waples, in review) and is a good approximation for a long-lived species like Atlantic cod.

Effective population size

We used the software AgeNe (Waples et al. 2011) to calculate N_e and N_e/N at specific time steps, based on population vital rates calculated as described above. AgeNe uses Hill's (1972) general formula for calculating N_e for species with overlapping generations but retains the direct link to population vital rates provided by the method of Felsenstein (1971):

$$N_e = \frac{4N_\alpha T}{V_{k\bullet} + 2}, \qquad (3)$$

where (in our notation) N_α is the number of offspring produced each time period that survive to become recruits, $V_{k\bullet}$ is lifetime variance in reproductive success of the N_α recruits in a cohort, and T is generation length. AgeNe calculates lifetime $V_{k\bullet}$ from a population's vital rates by grouping individuals by age at death (see Waples et al. 2011). N_α is a scaling parameter; N and N_e both increase linearly with N_α, but the ratio N_e/N does not depend on N_α. Similarly, mortality that occurs before maturity affects both N and N_e in the same way but not the ratio N_e/N. AgeNe automatically rescales relative age-specific fecundities to produce a stable population, and it also follows the Felsenstein and Hill models in assuming stable age

Table 2. Fraction of individuals that survive to age 3–10 that are sexually mature, at three time points in simulations with (E) and without (NE) evolution.

Age	Year 100 Equilibrium	Year 150 NE	Year 150 E	Year 300 NE	Year 300 E	150E/ Eq	150NE/ Eq
1	0	0	0	0	0	–	–
2	0	0	0	0	0	–	–
3	0.005	0.047	0.068	0.006	0.017	14.0	9.7
4	0.039	0.180	0.247	0.043	0.088	6.4	4.7
5	0.123	0.379	0.425	0.132	0.227	3.5	3.1
6	0.249	0.492	0.624	0.264	0.397	2.5	2.0
7	0.396	0.662	0.732	0.413	0.555	1.8	1.7
8	0.531	0.775	0.837	0.553	0.683	1.6	1.5
9	0.650	0.812	0.858	0.666	0.784	1.3	1.2
10	0.739	0.839	0.930	0.750	0.850	1.3	1.1

Year 100 is the end of the equilibrium period before fishing; year 150 is the end of fishing and beginning of recovery, and year 300 is late recovery. The last two columns on the right show the ratio of results for year 150 with (and without) evolution to year 100 equilibrium. These data are for selective fishing with $F = 0.2$ and are based on cod life table #1.

structure and independence of survival and reproduction across time periods.

One final piece of information is required to calculate N_e: $\phi_x = V_x/b_x$ = the ratio of variance to mean reproductive success in one season for individuals of age x. If reproductive success of same-age, same-sex individuals is random, then each age and sex behaves like a mini Wright–Fisher ideal population, and $\phi \approx 1$. Values of $\phi > 1$ therefore represent overdispersed variance in reproductive success. To parameterize this part of the model, we drew on experimental data for three captive populations in which parentage analysis was used to assign offspring (fertilized eggs) to potential parents (see Supporting Information for details). Table S1 shows an example of age-specific vital rates for the simulated population at equilibrium before harvest (year 100), after harvest (year 150), and late recovery (year 300).

AgeNe is based on discrete-time life tables and requires the user to specify a maximum age, ω. In each scenario, we chose ω as the oldest age (≤ 25) for which both age-specific survival and fecundity data were available; this was limited by low numbers of individuals that survived to advanced age, particularly in populations whose abundance declined sharply due to harvest. Resulting life tables for representative scenarios can be found in Table S1. At each time period in each scenario, the mean number of recruits produced per year was used as the value for N_α in the AgeNe calculations. Because vital rates in the simulations were only tracked for females, we used the same estimates for males in the AgeNe analyses.

Artificial manipulation of a life table for cod

Finally, to further explore generality of the above results, we artificially manipulated another life table for Atlantic cod (cod life table #2), based on data from Hutchings (2011) as modified by Waples et al. (2013). In this population, cod do not mature until age 7 and have maximum age $\omega = 20$, constant annual adult survival at $s_x = 0.82$, and fecundity that increases with age (Table S2). We created variations of this life table by allowing annual adult survival to drop to 0.72, 0.62, and 0.52 to reflect an increasing but uniform harvest rate that changed annual adult mortality to $d = 0.28, 0.38$, and 0.48, respectively. In the original population, the fraction of adults reaching age 7 that were still alive at age 20 was $0.82^{13} = 0.076$. Therefore, in the three artificial populations we truncated the life table at $\omega =$ the first age when cumulative survival from age 7 dropped below 0.076. In the variations with $s_x = 0.72, 0.62$, and 0.52, this resulted in $\omega = 16, 14$, and 12, respectively. We considered three general scenarios, each with variable adult survival: (i) fecundity is constant and ϕ is fixed at 3, which is roughly the value we estimated for age 15 in a pristine population; (ii) relative fecundity increases with age in the same relative proportions as in the original life table, and ϕ is fixed at 3; and (iii) fecundity and ϕ both increase with age, with the increase in ϕ following the same schedule we used for the simulated populations, except we started with $\phi = 1$ at age 7 rather than age 3. These scenarios did not consider either evolution of earlier age at maturity or potential density-dependent effects of increasing adult mortality on population dynamics, so N_1 was assumed to remain constant. Nonetheless, they provide insights into consequences for age structure and N_e/N associated with changes in adult mortality.

Results

Simulations of harvest and recovery

Fishing led to steep declines of cod population biomass, such that by the end of the fishing period the biomass had dropped below 20% of population carrying capacity (Fig. 1A, with selective harvest). Owing to selective removal of large individuals, fisheries-induced evolution caused asymptotic body length to decline across the fishing period by about 7 cm (Fig. 1B). Similar declines were also seen in the age and size at maturation, but the difference between evolving and nonevolving scenarios was less pronounced, as relaxed density-dependent competition accelerated growth and allowed fish to reach maturity earlier (Fig. 1C,D). After fishing ceased, biomass recovered rapidly to the prefishing level, but evolutionary recovery of the life-history traits was much slower, and clear differences in asymptotic length and age and size at maturity could still be seen at the end of the simulations.

Changes in demographic parameters

Figure 2 shows how key demographic parameters changed over the course of a typical simulation (selective harvest at $F = 0.2$, with evolution). Adding harvest on top of natural mortality roughly doubled the total adult mortality experienced by the population. As a consequence, adult N declined sharply during harvest before rapidly returning to its original status after harvest finished. The number of recruits (N_α) also declined sharply during harvest, but not as much as did N. Changes in annual survival between the equilibrium population and the end of fishing (year 150) are shown for several scenarios in Figure S1.

The purely demographic consequences for age at maturity of harvesting at this level can be seen by focusing on results where evolution was not allowed (Table 2). By the end of fishing at year 150, the fraction that were mature at young ages (3–5) was 3–10 times higher than in the equilibrium population before fishing (year 100), and the first age at which 50% of the population was mature had been reduced from 8 to 6. This occurred because increased adult mortality reduced overall abundance, and juvenile growth was enhanced owing to reduced density-dependent competition, allowing fish to reach body size at which they matured (66% of $L\infty$) at a younger age. By year 300 (late recovery), age at maturity in scenarios without evolution had largely returned to the preharvest equilibrium pattern (Table 2).

Patterns of change in N, N_e and N_e/N

N_e always declined sharply (by 50% or more) during fishing, while the ratio N_e/N always increased over the same time period (Fig. 3). This figure shows results for selective and nonselective harvest at $F = 0.2$ with and without evolution, but this same general pattern was found in every scenario we examined, including those in which the initial population size was doubled or halved (Fig. 4). During recovery, N_e and N_e/N both approached their original values fairly quickly, and this pattern was also consistent across scenarios.

The increases in N_e/N during fishing have a simple explanation: declines in N_e almost exactly mirrored declines in the number of recruits (N_1), while N declined at a faster rate (Fig. 2). As discussed later, the more rapid declines in N can be attributed to the fact that, whereas declines in recruitment affect N and N_e to the same extent, truncation of age structure caused by increased adult mortality also reduces adult N but by itself does not directly change N_e.

Effects of evolution

In our model, evolution could increase the probability of maturing at an earlier age through its effect on von Bertalanffy parameters, but in our simulations no individuals matured before age $\alpha = 3$. By the end of fishing at year 150

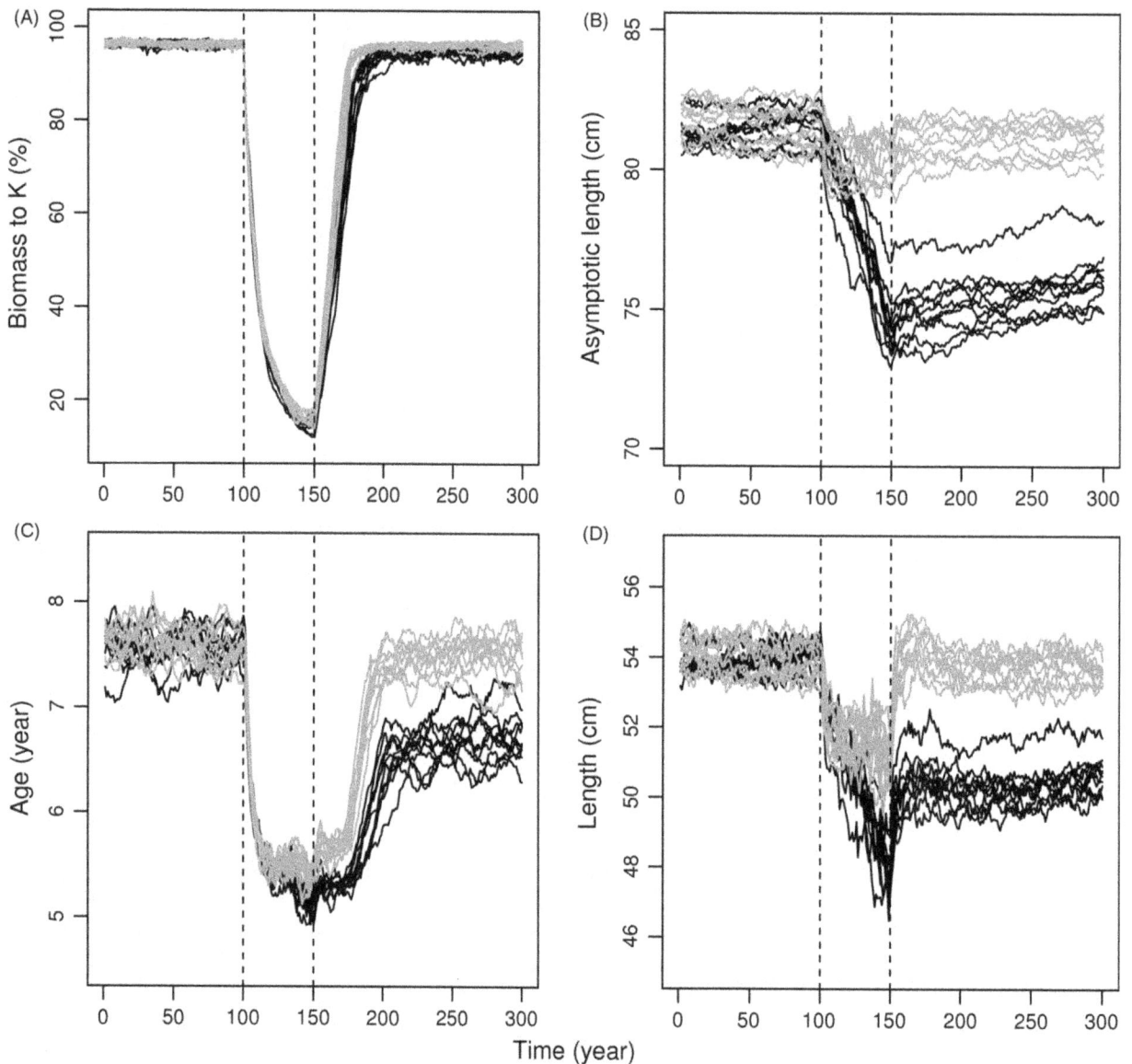

Figure 1 The temporal development of cod population biomass (A), asymptotic body length (B), age at maturity (C), and size at maturity (D) in ten replicated simulation runs, each described by a solid line. Evolving simulations are drawn with black and nonevolving simulations with gray. The beginning and the end of fishing period are denoted with vertical dashed lines.

in scenarios that allowed evolution, the fraction that were mature at young ages (3–5) was 3–14 × higher than in the equilibrium population before fishing, compared to 3–10 × higher for scenarios that did not allow evolution (Table 2). Thus, most of the age-structure changes by year 150 can be attributed directly to demographic consequences of increased adult mortality, although evolution enhanced this effect in scenarios where it was allowed. By year 300 (late recovery), the fraction mature at earlier ages was still elevated in scenarios that allowed evolution.

These demographic patterns were reflected in patterns of change in effective population size. Whether or not

evolution was allowed had little effect on N_e and relatively minor effect (±about 15%) on N_e/N during harvest. During recovery, however, N_e and N_e/N were both slower to return to their prefishery equilibrium values in scenarios involving evolution, and even at year 300 they had not fully recovered.

Selective versus nonselective harvest

Nonselective harvest resulted in more dramatic reductions in overall population size and hence N_e. For example, by the end of fishing (year 150) with $F = 0.2$ and evolution, size-selective harvest had reduced N_e from 1969 to 683, a

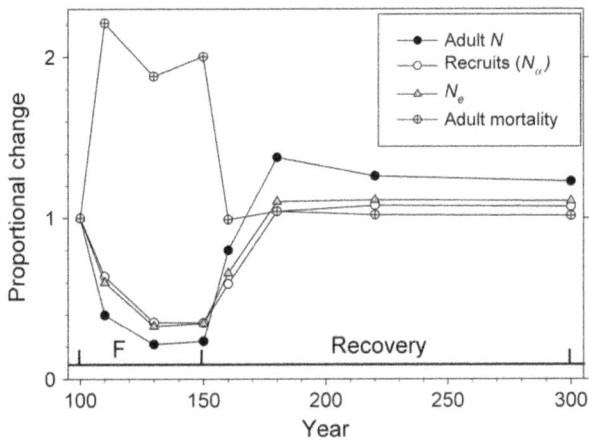

Figure 2 Proportional change in key demographic parameters over the course of the simulations. Results are for selective fishing at $F = 0.2$, with evolution. Time periods indicate the end of equilibrium and start of fishing (year 100), end of fishing and beginning of recovery (year 150), and late recovery (year 300).

decline of 65%, while nonselective harvest reduced N_e from 1959 to 131, a decline of 93% (Fig. 3). These stronger declines in N occurred because selective harvest could remove 20% of the biomass by harvesting a relatively small number of larger, older fish, while nonselective harvest that included many smaller fish would have to remove more individuals to take the same biomass. Whether harvest was selective or not had only modest effects on N_e/N because additional reductions associated with nonselective harvest were similar for N_e and N (Fig. 3).

Different levels of harvest

Allowing different levels of F had predictable consequences for population size and N_e but did not change the basic patterns described above. Harvesting at a level of $F = 0.25$ led to greater reductions in N_e, while reducing F to 0.15 produced a smaller reduction (Fig. 5). By the end of fishing (year 150), selective harvest at $F = 0.25$ with evolution had increased N_e/N to 1.67, compared to 1.52 and 1.35 for

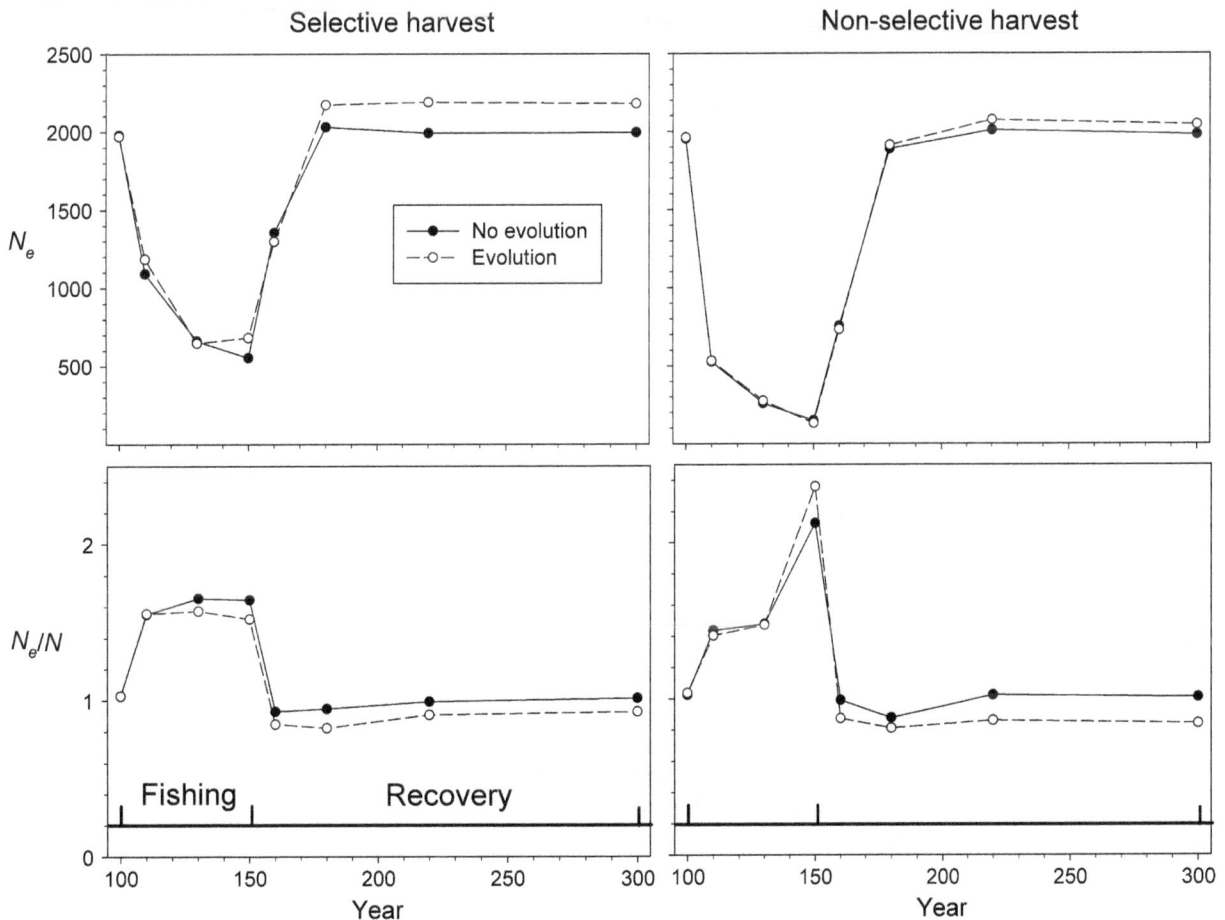

Figure 3 Changes in N_e and N_e/N over the course of simulations with $F = 0.2$. Results are shown for scenarios with selective harvest (left panels) and nonselective harvest (right panels), and that do (open circles) and do not (filled circles) allow evolution of life-history traits.

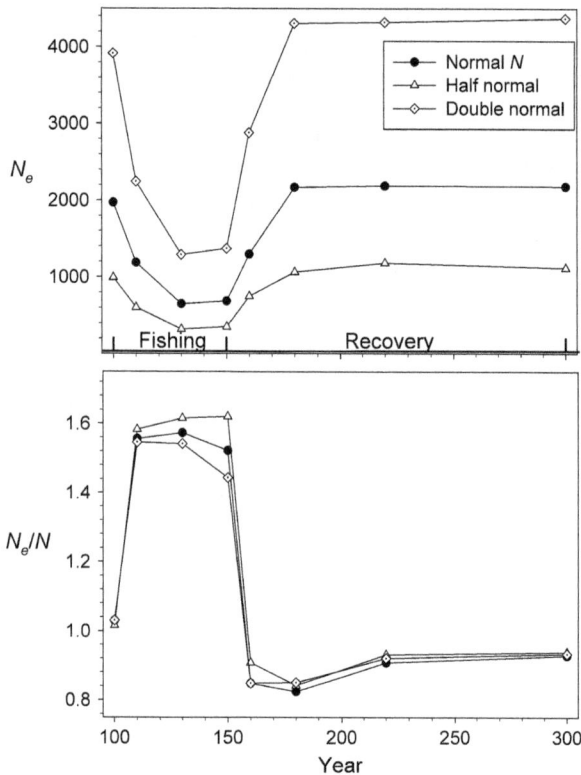

Figure 4 Effects of varying initial population size for simulated cod populations. Results are for selective fishing at $F = 0.2$ with evolution. The filled circles (Normal N) reproduce results for $F = 0.2$ shown in Fig. 3; the other lines and symbols show results for scenarios in which initial size was half or double the 'Normal' level.

$F = 0.2$ and 0.15, respectively. All of these patterns related to varying levels of F were qualitatively similar under scenarios without evolution (Figure S2).

Changes in T *and* $V_{k\bullet}$
Additional mortality associated with harvest sharply reduced both generation length and lifetime variance in reproductive success, but $V_{k\bullet}$ declined more rapidly so the ratio $T/V_{k\bullet}$ increased (Fig. 6). When fishing stopped, both T and $V_{k\bullet}$ ncreased again and approached their preharvest equilibrium values, with a predictable lag for scenarios involving evolution. Immediately after fishing stopped, T increased more rapidly than $V_{k\bullet}$, leading to the spike in $T/V_{k\bullet}$ at year 160. Figure 6 shows results for selective fishing with $F = 0.2$ and allowing evolution, but again this general pattern was evident in all scenarios.

Analysis of alternative life table

Artificially reducing adult survival from 0.82/year to 0.62/year in cod life table #2 dramatically reduced (from 26% to 4%) the fraction of the adult population made up of

individuals age 13 or older, and the population became increasingly dominated by younger individuals (49% of the adult population was age-7 individuals with annual survival = 0.52, compared to 19% in the real population with natural survival = 0.82; Table 3). Truncating the age structure as adult mortality increased from $d = 0.18$ to 0.48 reduced the adult population size by 60.8% (Table 4). This is close to the value predicted from eqn (2) ($N_2/N_1 = (1/0.48)/(1/0.18) = 0.375$, a decline of 62.5%), which would apply to a population with arbitrarily long life span.

In the base population (Scenario II in Table 4), in which fecundity increased with age and ϕ was constant, generation length also decreased but by a smaller amount (35.8% for $d = 0.48$). Although both N and T are inversely related to adult mortality (Figures S3 and S4; see also Nunney 1991), T cannot be lower than the age at maturity ($\alpha = 7$ in this population), and this constrained the rate at which (and amount by which) T could be reduced as d increased. Furthermore, increasing mortality also reduced $V_{k\bullet}$ (by 26%), and this largely offset reductions in N_e caused by lower T. As a consequence, N_e only declined by 17.4% when d increased to 0.48. Because this was much less than the reduction in N, the ratio N_e/N more than doubled, from 0.70 to 1.48.

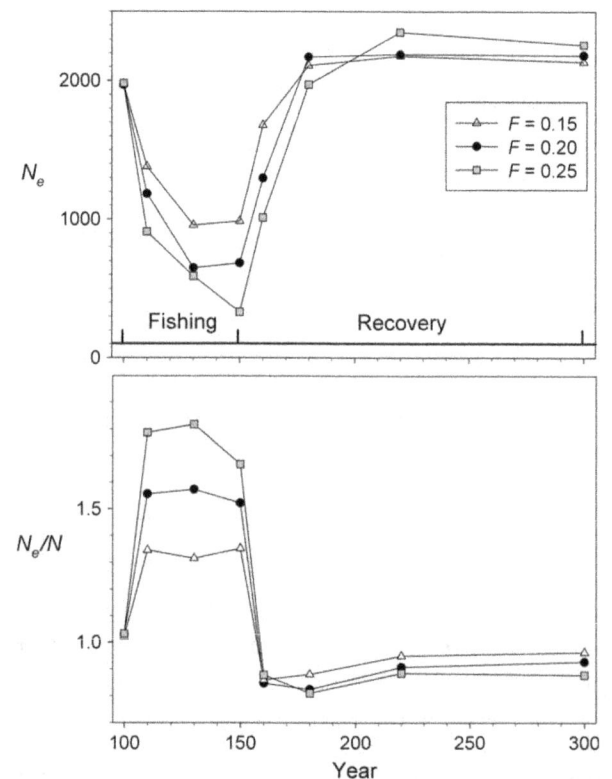

Figure 5 Effects of varying levels of fishing intensity. Results are for simulations with selective fishing with evolution.

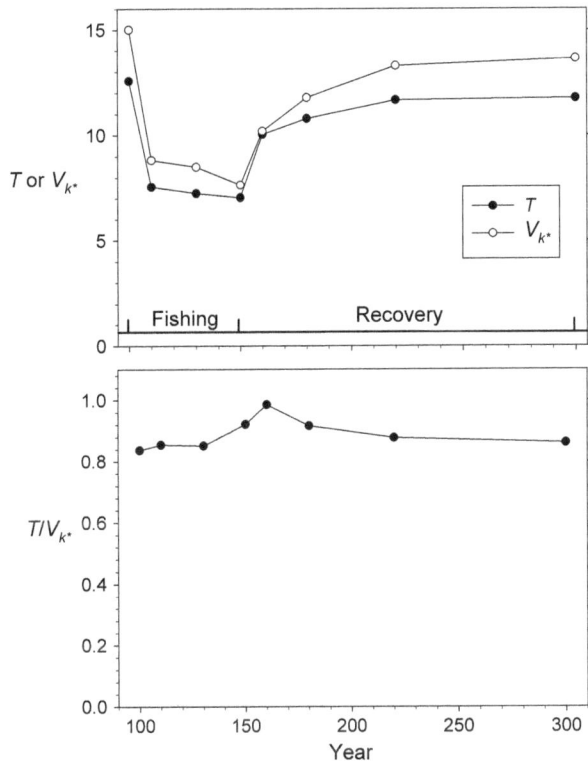

Figure 6 Changes over the course of the simulations in generation length (T), lifetime variance in reproductive success ($V_{k\bullet}$), and their ratio. Results are for selective fishing at $F = 0.2$ with evolution.

Table 3. Fraction of adult population in each age class for a Northern cod population experiencing various hypothetical levels of annual adult mortalit.

Age class	Adult survival			
	0.82	0.72	0.62	0.52
7	0.192	0.291	0.388	0.490
8	0.157	0.209	0.241	0.255
9	0.129	0.151	0.149	0.132
10	0.106	0.109	0.093	0.069
11	0.087	0.078	0.057	0.036
12	0.071	0.056	0.036	0.019
13+	0.258	0.106	0.036	—*

The first column shows data for the reference population (cod life table #2, for which annual adult survival = 0.82) from Hutchings (2011), as modified by Waples et al. (2013). The other columns depict results for hypothetical populations with the same age-specific fecundity relationship but different rates of adult survival that reflect natural mortality + fishing mortality.

*In this scenario, maximum age was truncated to $\omega = 12$ based on the rules described in the text.

In Scenario I in Table 4 with constant fecundity (as might be applicable for some harvested species, such as birds), the reduction in N_e was slightly greater (20.4%). This occurred because under constant fecundity, and starting from relatively high survival, T declines more rapidly with increasing mortality than does $V_{k\bullet}$. (Figure S4).

A different pattern was seen in Scenario III, in which both fecundity and ϕ were proportional to age. In this case, increasing mortality had a stronger effect on reducing $V_{k\bullet}$, such that the ratio $T/V_{k\bullet}$ increased by 23.7% as d increased to 0.48; as a consequence, N_e actually was 7.5% higher with $d = 0.48$ than with $d = 0.18$ (again, under the assumption that N_α remained constant).

Results in Table 4 help to illustrate how changes in recruitment and adult mortality interact to determine adult census size [eqn (2)]. For cod life table #2, we lacked empirical data regarding density dependence, so we adopted a simple assumption of no change in recruitment, which would occur only under full productivity compensation (i.e., if the reduced number of adults still produced the same number of offspring per time period). Therefore, this table probably underestimates the reduction in effective size, because N_e is also linearly related to the number of recruits [eqn (1)]. If instead we had assumed that per capita production of recruits remained constant when mortality increased (i.e., no productivity compensation), then N_e would have been reduced by an additional 60.8% for $d = 0.48$. In that event, however, N also would have experienced the same additional reduction, so assumptions about density dependence and recruitment had no effect on the N_e/N ratio.

In the simulated populations using cod life table #1 (which included density dependence), recruitment dropped substantially with harvest, but not as much as did adult abundance (Fig. 2). This shows at least partial productivity compensation at low density, even if it was not sufficient to fully offset the reduction in adult numbers. It is important to note here that the recruitment and mortality terms in eqns (1) and (2) can interact over time in a feedback loop that can produce cumulative changes over time much larger than predicted from a single iteration. For example, if increased mortality in time period 1 reduces adult N and this reduces recruitment, adult N will be reduced further in the next time period, and, in the absence of strong productivity compensation, this process can continue until the population collapses. Given our initial conditions, the duration of fishing, and the empirically based form of density dependence we modeled, we found that was the case for simulated populations with F greater than about 0.25.

Discussion

The major results from our study can be summarized as follows:

Table 4. Results of artificial manipulation of cod life table #2.

Adult Mortality (d)	Adult N	%	N_e	%	N_e/N	%	T	%	$V_{k\bullet}$	%	$T/V_{k\bullet}$	%
Scenario I: constant b_x; ϕ fixed at 3												
0.18	10421	–	8224	–	0.789	–	10.63	–	8.34	–	1.275	–
0.28	6875	−34.0	7172	−12.8	1.043	32.2	9.18	−13.6	8.24	−1.2	1.114	−12.6
0.38	5148	−50.6	6716	−18.3	1.305	65.3	8.45	−20.5	8.07	−3.2	1.047	−17.8
0.48	4084	−60.8	6548	−20.4	1.603	103.2	7.96	−25.1	7.73	−7.3	1.030	−19.2
Scenario II: b_x increases with age; ϕ fixed at 3												
0.18	10421	–	7310	–	0.701	–	12.87	–	12.08	–	1.065	–
0.28	6875	−34.0	6361	−13.0	0.925	31.9	10.22	−20.6	10.85	−10.2	0.942	−11.6
0.38	5148	−50.6	6027	−17.6	1.171	66.9	9.02	−29.9	9.97	−17.5	0.905	−15.1
0.48	4084	−60.8	6040	−17.4	1.479	110.8	8.26	−35.8	8.94	−26.0	0.924	−13.3
Scenario III: b_x and ϕ increases with age												
0.18	10421	–	8453	–	0.811	–	12.87	–	10.18	–	1.264	–
0.28	6875	−34.0	8252	−2.4	1.200	48.0	10.22	−20.6	7.91	−22.3	1.292	2.2
0.38	5148	−50.6	8431	−0.3	1.638	101.9	9.02	−29.9	6.55	−35.7	1.377	8.9
0.48	4084	−60.8	9083	7.5	2.224	174.2	8.26	−35.8	5.28	−48.1	1.564	23.7

The original life table (Scenario II) had constant adult mortality of $d = 0.18$ (see Table 3) and fecundity (b_x) that increases with age; we also assumed that $\phi = 3$ for all ages. We considered how increases in adult mortality in this life table would affect key demographic parameters. We also considered two other hypothetical scenarios: one with constant fecundity and ϕ fixed at 3 (Scenario I), and one in which fecundity and ϕ both increase with age (Scenario III). Results were calculated using AgeNe assuming that the number of recruits produced per year was constant at $N_\alpha = 2000$. Within each scenario, '%' indicates the percent change from the value when $d = 0.18$.

1 Increasing adult mortality through harvest reduces both census and effective size, but the ratio N_e/N increases because N is reduced more than N_e.

2 This general result occurs regardless whether harvest is size-selective or not, and regardless whether evolution of life-history traits is allowed or not—that is, those other factors affect the outcome in a quantitative way but do not change the qualitative patterns.

3 The intensity of fishing affects the magnitude of change in a predictable way but also does not change these general patterns.

4 The effects of evolution were more pronounced late in the recovery period than they were during harvest. In scenarios without evolution, population parameters rapidly returned to near their equilibrium values after harvest ended, but in the scenarios with evolution the population never achieved its original status by year 300. This was true of biomass, size, age at maturity (Fig. 1), census size and effective size (Fig. 2), and generation length and variance in reproductive success (Fig. 6). Although both N and N_e were higher at year 300 than they were at equilibrium in scenarios with evolution, the proportional increase in N was larger, so the N_e/N ratio was lower (Figs 2 and 3).

Below we discuss these points and explain why we believe they are not specific to our study system but instead represent quite general expectations for the consequences of increased adult mortality.

The N_e/N ratio

The increase in N_e/N during fishing while N_e went down can be easily understood based on two key insights from inspection of eqns (1)–(3). First, both N and N_e are linear functions of the number of recruits that reach age at maturity (N_α). This means that any changes in recruitment have proportional changes in N_e and N that are exactly the same, so the ratio N_e/N is not affected by recruitment. Therefore, changes in the N_e/N ratio are entirely determined by differences in the way N and N_e respond to changes in adult mortality (d). The effects of changes in d on N are again straightforward: increased mortality truncates the age structure and reduces the number of adults as a simple function of the mortality profile as described in eqns (1) and (2). In contrast, N_e is not directly affected by changes in mortality; it is only indirectly influenced by the effects of changes in mortality on generation length and lifetime variance in reproductive success [eqn (3)].

As discussed above and illustrated in Figure S3, the exact patterns of change in T and $V_{k\bullet}$ associated with a change in adult mortality are complex and depend on age-specific vital rates and age-specific ϕ. However, because a) the direction of change in T and $V_{k\bullet}$ with increasing mortality is the same (Figure S3 and S4), and b) T occurs in the numerator of eqn (3) while $V_{k\bullet}$ occurs in the denominator, mortality-mediated changes in T and $V_{k\bullet}$ largely cancel each other (Fig. 6), which greatly constrains the degree to which changes in adult mortality directly affect N_e. To a first approximation, therefore, change in N_e associated with fishing can be explained solely by changes in recruitment, while changes in N depend on both recruitment and mortality. The net result is that increases in adult mortality reduce N more than N_e, so the ratio N_e/N goes up, even though N_e will generally decline (absent complete productivity compensation).

Another type of compensation, sometimes termed 'genetic compensation', can affect both N_e and N_e/N; this occurs when variance in reproductive success declines at low density, presumably because reduced competition for mates allows otherwise inferior individuals to successfully reproduce. As a consequence of reduced $V_{k\bullet}$, the ratio N_e/N is often higher when population abundance is reduced. Empirical studies that have reported this type of result include Palstra and Ruzzante (2008), Beebee (2009), and Saarinen et al. (2010). Although this could potentially be an important phenomenon in populations subjected to higher adult mortality through harvest, we did not have any empirical information to parameterize this effect with cod. To the extent that it does occur, it would reinforce the pattern we observed in which N_e/N increases with fishing intensity.

The N_e/N ratios shown in Figs 3–5, especially those during harvest, are higher than most reported in the literature (e.g., Frankham 1995, Palstra and Fraser 2012). In general, it has been thought that N_e must be $<N$ in natural populations, but recently it has been shown that this is not necessarily the case for species with overlapping generations, particularly those (like cod) with delayed age at maturity (Waples et al. 2013). However, N_e/N in iteroparous species is very sensitive to the variance in reproductive success among individuals of the same age and sex (ϕ_x), and high N_e/N ratios are only possible if ϕ_x is relatively low. In this study, we used empirical data for a captive population to parameterize ϕ_x, and as a result it increased from about one at age at maturity to over four by age 25. Values of ϕ_x in wild populations could potentially be much higher, especially in species that experience 'sweepstakes' reproductive success as proposed by Hedgecock (1994). Unfortunately, however, very few estimates of ϕ_x are available for wild populations of any species. Nevertheless, it is easy to evaluate how hypothetical values would affect the N_e/N ratio for the simulated populations. For example, in our simulated populations based on cod life table #1, initial N_e/N would be reduced from nearly 1.0 to below 0.1 if the age-specific ϕ_x values we used were all multiplied by a fixed factor 50 (Figure S4A). Such a population would have much lower N_e and N_e/N, but the pattern of change over time in these parameters (Figure S4B) would be similar to that shown in Figs 3–5.

Finally, because changes in adult mortality can have large effects on the N_e/N ratio (as demonstrated here), and because anthropogenic changes to all of earth's ecosystems have dramatically changed mortality profiles for many species, it is risky to assume that the N_e/N ratio is constant, absent a good reason to believe that is the case.

Effects of evolution

The typical evolutionary response to increased adult mortality is to evolve mechanisms that allow earlier maturation, which increases the chances of having at least one opportunity to reproduce before being harvested. What are the likely consequences for N_e? If increased adult mortality causes an evolutionary response toward earlier maturation, that would reduce generation length and, all else being equal, that would reduce N_e [eqn (1)]. However, earlier maturation could also mean that more total individuals survive to maturity, which would increase the number of recruits (N_α) and, all else being equal, increase N_e. Therefore, the net effects of evolution on N_e and N_e/N are expected to depend on the relative importance of these two factors. The effects on generation length are easier to predict, while those on recruitment depend on assumptions about ecological processes such as competition and density dependence.

In the simulated populations, reductions in N caused by higher harvest rates enhanced juvenile growth and survival through relaxation of density dependence, and as a consequence, a larger fraction of individuals matured at earlier ages (Table 2). This was a purely ecological phenomenon that also caused age-structure shifts in populations without evolution. Allowing evolution of age at maturity, therefore, only added a relatively small component to a fundamentally ecological process (compare last two columns in Table 2). This tended to blur the distinction between results for scenarios that did and did not allow evolution, at least during the period of harvest.

The major (and quite consistent) difference between the evolution and nonevolution scenarios can be found at the end of the long recovery period (year 300), by which time the vital rates of all populations simulated without evolution had returned to essentially the same place they were before harvest commenced. In contrast, at year 300 in scenarios that involved evolution, N_e was always slightly higher and N_e/N slightly lower than it was in the equilibrium preharvest population. This result is consistent with empirical observations from other studies (e.g., Pigeon et al. 2016) that document rapid evolution of life history under strong selection, but slower evolution toward initial phenotypes once selection is relaxed, presumably because selection in the wild is seldom as strong as selection humans impose through harvest (Allendorf and Hard 2009).

Two factors combined to produce the higher N_e at year 300: higher N (Fig. 2) and higher $T/V_{k\bullet}$ (Fig. 6) compared to their values at year 100. However, the ratio N_e/N was lower at year 300 than at year 100. This occurred because evolution of the age-at-maturity reaction norm toward earlier maturity meant that a larger fraction of the population was mature at an earlier age, and this increased adult N faster than it did N_e. The net effect was a reduction in N_e/N, even though N_e was slightly higher in late-recovery populations that allowed evolution than it was at preharvest equilibrium.

Model assumptions

The Felsenstein–Hill models that AgeNe is based upon assume constant population size and stable age structure. These assumptions were met in the preharvest equilibrium population (year 100) and nearly met in the late recovery phases (after about year 200), but harvest led to rapid changes in population demography that affected data collected in years 110–180. Therefore, because AgeNe calculates N_e for individual cohorts based on vital rates calculated at specific points in time, our results are best interpreted as estimates of instantaneous N_e that would apply to a population that remained stable with those mean vital rates. Nevertheless, several lines of evidence suggest that our results should be fairly robust to these demographic changes. Felsenstein (1971) showed that his model accurately estimates N_e for populations that are increasing or declining at a constant rate, and this was approximately met during the decline due to fishing and the resulting rebound after fishing stopped. Waples et al. (2011, 2014) showed that eqn (1) provides robust results in simulated populations that incorporate random demographic stochasticity and with N_e as low as 200 (lower than the levels reached in any of our scenarios except those with nonselective fishing). Furthermore, substantial generational overlap and long adult life span (as are found in cod populations) help to buffer a population against cyclical environmental fluctuations (Gaggiotti and Vetter 1999). Finally, although Hill (1972; 289) did not formally evaluate the assumptions of constant population size and random mating, he did provide arguments why he believed that 'neither effect has much influence on effective population size'.

We did not simulate very small effective sizes (N_e <100) because that is difficult to do in a long-lived species with many age classes. If effective size is that small, random changes in allele frequency can overwhelm the effects of selection, which would make predictions regarding FIE less reliable. However, because most of the changes we reported were dominated by demographic changes related to increases in adult mortality rather than evolutionary changes, we believe our results would also be qualitatively true for smaller N_e values than we modeled.

The AgeNe model also assumes that probabilities of survival and reproduction are independent across time. That will not always be the case. If, for example, individuals (especially females) who reproduce in one time period have a reduced probability of reproducing for one or more subsequent time periods, N_e will be slightly higher than calculated under AgeNe because skip breeding tends to reduce extreme variation in lifetime reproductive success (Waples and Antao 2014). Conversely, if certain individuals are consistently above or below average in their reproductive output, N_e will be reduced (Lee et al. 2011). Although these phenomena can influence effective population size, they should not affect the general patterns of change in N_e and N_e/N in response to increases in adult mortality.

Implications for conservation and management

We demonstrated that N_e is likely to decline, perhaps substantially, in response to elevated adult mortality associated with harvest. Our results thus support the conclusion by Marty et al. (2015) that failure to account for stochastic processes associated with reduced N_e can lead to incorrect conclusions about eco-evolutionary dynamics associated with fishery-induced evolution. However, these results also add some important nuances to our understanding of this complex topic.

First, the good news is that increasing harvest rates can be expected to increase the N_e/N ratio. This means that the proportional reductions in N_e will be smaller than the effects of harvest on total abundance. As the latter are easier to predict, the expected reduction in N can be used as an upper limit to the expected reduction in N_e, with the expectation that increases in the N_e/N ratio will at least partially buffer the overall reduction in effective size.

The second important point is that although adding anthropogenic harvest to natural mortality can promote fishery-induced evolution, direct demographic consequences of elevated adult mortality explain most of the reductions in effective size that we observed in the modeled populations. Reductions in N_e are caused primarily by reductions in recruitment, as the effects of elevated harvest on T and V_k tend to cancel each other [eqn (3)]. We did, however, find that long after harvest stopped, N_e was higher in the scenarios that involved evolution, which indicates a potentially important role for evolution in maintaining genetic diversity in populations recovering from periods of elevated harvest-related mortality.

Although this does not directly relate to effective size, it is worth noting that, because substantial generational overlap and the storage effect (Warner and Chesson 1985) help buffer a long-lived species against environmental fluctuations, truncation of age structure resulting from increased adult mortality will reduce this buffering capacity, leaving the population more vulnerable to random events.

The eco-evolutionary patterns described here are quite general and should be applicable to a wide range of species that experience increased mortality from anthropogenic factors, including but not limited to harvest. In a recent study, Dowling et al. (2014) monitored effective size over 15 years in a species (razorback sucker, *Xyrauchen texanus*) experiencing reduced survival in altered habitat and found that effective size was stable or increased while N declined, so N_e/N increased. These were genetically based estimates and did not consider demography, but the authors also used AgeNe to evaluate the consequences of truncating the life span from 44 to 20 years. Dowling et al. (2014) found this truncation caused little change in N_e/N, so they concluded that the increase in the effective: census size ratio was due to reduced variance in reproductive success. However, simply truncating a life table at a certain age does not properly mimic a scenario with increasing adult mortality, as the latter will reduce abundance in all ages from age at maturity onwards. We altered the life table for razorback sucker (published in Waples et al. 2013) by reducing adult survival from 0.8 to 0.6 and truncating at 20 years, and this raised N_e/N from about 1.0 to 1.6, comparable to changes we report here. Thus, although it is certainly possible that variance in reproductive success has been reduced in this species, it is not necessary to postulate that to explain the empirical pattern in the estimates of N_e/N.

One important factor that applies to species subject to trophy hunting is that harvest that targets males can skew the sex ratio and hence reduce N_e (Coltman et al. 2003; Hard et al. 2006). Although the AgeNe model can easily incorporate sex-specific vital rates to fully account for sex-ratio effects on N_e, harvest of cod is thought to be sex-neutral and we do not have evidence for sex-based differences in survivorship. We can, however, predict the general consequences of male-targeting trophy hunting on N_e and N_e/N using the framework developed here. When males and females have different vital rates, the simple formula developed by Wright (1938) can be used to calculate overall N_e as a function of the effective numbers of females and males. Sharply increasing mortality of adult males will reduce male N but at the same time will increase male N_e/N, for reasons described above. As a consequence, male N_e will not decline as fast as male N, so the effects on overall N_e will be less than would be predicted simply from the reduction in the number of adult males. The net results for overall N_e will depend on population-specific patterns in vital rates that determine how the ratio $T/V_{k\bullet}$ changes with increasing adult mortality.

Acknowledgements

The Academy of Finland provided funding to AK and a Natural Sciences and Engineering Research Council of Canada Discovery Grant and Loblaw Companies Ltd provided funding to JAH to support this research. The comments of the associate editor and two anonymous reviewers considerably improved the manuscript.

Literature cited

Allendorf, F. W., and J. J. Hard 2009. Human-induced evolution caused by unnatural selection through harvest of wild animals. In J. C. Avise, and F. J. Ayala, eds. In the Light Of Evolution, vol. **III**: Two Centuries of Darwin, pp. 129–148. The National Academies Press, Washington, DC.

Audzijonyte, A., A. Kuparinen, and E. A. Fulton 2013. How fast is fisheries-induced evolution? Quantitative analysis of modelling and empirical studies. Evolutionary Applications **6**:585–595.

Beebee, T. J. C. 2009. A comparison of single-sample effective size estimators using empirical toad (Bufo calamita) population data: genetic compensation and population size-genetic diversity correlations. Molecular Ecology **18**:4790–4797.

von Bertalanffy, L. 1938. A quantitative theory of organic growth (inquiries on growth laws II). Human Biology **10**:181–213.

Carlson, S. M., and T. R. Seamons 2008. A review of quantitative genetic components of fitness in salmonids: implications for adaptation to future change. Evolutionary Applications **1**:222–238.

Charnov, E. 1993. Life History Invariants: Some Explorations of Symmetry in Evolutionary Ecology. Oxford University Press, Oxford.

Charnov, E. L., H. Gislason, and J. G. Pope 2013. Evolutionary assembly rules for fish life histories. Fish and Fisheries **14**:213–224.

Coltman, D. W., P. O'Donoghue, J. T. Jorgenson, J. T. Hogg, C. Strobeck, and M. Festa-Bianchet 2003. Undesirable evolutionary consequences of trophy hunting. Nature **426**:655–658.

Darimont, C. T., S. M. Carlson, M. T. Kinnison, P. C. Paquet, T. E. Reimchen, and C. C. Wilmers. 2009. Human predators outpace other agents of trait change in the wild. Proceedings of the National Academy of Sciences of the United States of America **106**:952–954.

Devine, J. A., P. J. Wright, H. M. Pardoe, and M. Heino 2012. Comparing rates of contemporary evolution in life-history traits for exploited fish stocks. Canadian Journal of Fisheries and Aquatic Sciences **69**:1105–1120.

Dowling, T. E., T. F. Turner, E. W. Carson, M. J. Saltzgiver, D. Adams, B. Kesner, and P. C. Marsh 2014. Time-series analysis reveals genetic responses to intensive management of razorback sucker (Xyrauchen texanus). Evolutionary Applications **7**:339–354.

Edeline, E., S. M. Carlson, L. C. Stige, I. J. Winfield, J. M. Fletcher, J. B. James, T. O. Haughen et al. 2007. Trait changes in a harvested population are driven by a dynamic tug-of-war between natural and harvest selection. Proceedings of the National Academy of Sciences of the United States of America **104**:15799–15804.

Felsenstein, J. 1971. Inbreeding and variance effective numbers in populations with overlapping generations. Genetics **68**:581–597.

Gaggiotti, O. E., and R. D. Vetter 1999. Effect of life history strategy, environmental variability, and overexploitation on the genetic diversity of pelagic fish populations. Canadian Journal of Fisheries and Aquatic Sciences **56**:1376–1388.

Hedgecock, D. 1994. Does variance in reproductive success limit effective population size of marine organisms? In A. Beaumont, ed. Genetics and Evolution of Aquatic Organisms, pp. 122–134. Chapman & Hall, London.

Hill, W. G. 1972. Effective size of population with overlapping generations. Theoretical Population Biology 3:278–289.

Howell, D., A. A. Filin, B. Bogstad, and J. E. Stiansen 2013. Unquantifiable uncertainty in projecting stock response to climate change: example from North East Arctic cod. Marine Biology Research 9:920–931.

Hutchings, J. A. 2005. Life-history consequences of overexploitation to population recovery in Northwest Atlantic cod (Gadus morhua). Canadian Journal of Fisheries and Aquatic Sciences 62:824–832.

Hutchings, J.A. 2011. Chapter 5, Population Ecology. In B. Freedman et al. , eds. Ecology: a Canadian Context. Nelson Education, Toronto.

Hutchings, J. A., and J. K. Baum 2005. Measuring marine fish biodiversity: temporal changes in abundance, life history, and demography. Philosophical Transactions of the Royal Society of London 360:315–338.

Hutchings, J. A., and D. J. Fraser 2008. The nature of fishing- and farming-induced evolution. Molecular Ecology 17:294–313.

Jensen, A. L. 1997. Origin of the relation between K and Linf and synthesis of relations among life history parameters. Canadian Journal of Fisheries and Aquatic Sciences 54:987–989.

Kendall, N. W., U. Dieckmann, M. Heino, A. E. Punt, and T. P. Quinn 2014. Evolution of age and length at maturation of Alaskan salmon under size-selective harvest. Evolutionary Applications 7:313–322.

Kuparinen, A., D. C. Hardie, and J. A. Hutchings 2012. Evolutionary and ecological feedbacks of the survival cost of reproduction. Evolutionary Applications 5:245–255.

Kuparinen, A., N. C. Stenseth, and J. A. Hutchings 2014a. Fundamental population–productivity relationships can be modified through density-dependent feedbacks of life-history evolution. Evolutionary Applications 7:1218–1225.

Kuparinen, A., D. M. Keith, and J. A. Hutchings 2014b. Allee effect and the uncertainty of population recovery. Conservation Biology 3:790–798.

Lanfear, R., H. Kokko, and A. Eyre-Walker 2014. Population size and the rate of evolution. Trends in Ecology and Evolution 29:33–41.

Law, R. 2007. Fisheries-induced evolution: present status and future directions. Marine Ecology Progress Series 335:271–277.

Lee, A. M., S. Engen, and B.-E. Sæther 2011. The influence of persistent individual differences and age at maturity on effective population size. Proceedings of the Royal Society of London B: Biological Sciences 278:3303–3312.

Marty, L., U. Dieckmann, and B. Ernande 2015. Fisheries-induced neutral and adaptive evolution in exploited fish populations and consequences for their adaptive potential. Evolutionary Applications 8:47–63.

Mertz, G., and R. A. Myers 1998. A simplified formulation for fish production. Canadian Journal of Fisheries and Aquatic Sciences 55:478–484.

Mousseau, T. A., and D. A. Roff 1987. Natural selection and the heritability of fitness components. Heredity 59:181–197.

Nunney, L. 1991. The influence of age structure and fecundity on effective population size. Proceedings of the Royal Society of London B: Biological Sciences 246:71–76.

Nunney, L. 1993. The influence of mating system and overlapping generations on effective population size. Evolution 47:1329–1341.

Palstra, F. P., and D. J. Fraser 2012. Effective/census population size ratio estimation: a compendium and appraisal. Ecology and Evolution 2:2357–2365.

Palstra, F. P., and D. E. Ruzzante 2008. Genetic estimates of contemporary effective population size: what can they tell us about the importance of genetic stochasticity for wild population persistence? Molecular Ecology 17:3428–3447.

Pigeon, G., Festa-Bianchet M., Coltman D.W., and F. Pelletier 2016. Intense selective hunting leads to artificial evolution in horn size. Evolutionary Applications (published online 26 January 2016; doi:10.1111/eva.12358).

Polacheck, T., J. P. Eveson, and G. M. Laslett 2004. Increase in growth rates of southern bluefin tuna (Thunnus maccoyii) over four decades: 1960 to 2000. Canadian Journal of Fisheries and Aquatic Sciences 61:307–322.

Postma, E. 2014. Four decades of estimating heritabilities in wild vertebrate populations: improved methods, more data, better estimates? In: A. Charmantier, D. Garant, and L. E. B. Kruuk, eds. Quantitative Genetics in the Wild, pp. 16–33. Oxford University Press, Oxford, UK.

Roff, D. A. 2002. Life History Evolution. Sinauer, Sunderland, MA.

Saarinen, E. V., J. D. Austin, and J. C. Daniels 2010. Genetic estimates of contemporary effective population size in an endangered butterfly indicate a possible role for genetic compensation. Evolutionary applications 3:28–39.

Sharpe, D., and A. Hendry 2009. Life history change in commercially exploited fish stocks: n analysis of trends across studies. Evolutionary Applications 2:260–275.

Waples, R. S. 1989. A generalized approach for estimating effective population size from temporal changes in allele frequency. Genetics 121:379–391.

Waples, R. S., and T. Antao 2014. Intermittent breeding and constraints on litter size: consequences for effective population size per generation (N_e) and per reproductive cycle (N_b). Evolution 68:1722–1734.

Waples, R. S., and M. Yokota 2007. Temporal estimates of effective population size in species with overlapping generations. Genetics 175:219–233.

Waples, R. S., C. Do, and J. Chopelet 2011. Calculating N_e and N_e/N in age-structured populations: a hybrid Felsenstein-Hill approach. Ecology 92:1513–1522.

Waples, R. S., G. Luikart, J. R. Faulkner, and D. A. Tallmon 2013. Simple life history traits explain key effective population size ratios across diverse taxa. Proceedings of the Royal Society of London B: Biological Sciences 280:20131339.

Waples, R. S., T. Antao, and G. Luikart 2014. Effects of overlapping generations on linkage disequilibrium estimates of effective population size. Genetics 197:769–780.

Warner, R. R., and P. L. Chesson 1985. Coexistence mediated by recruitment fluctuations: a field guide to the storage effect. American Naturalist 125:769–787.

Wright, S. 1938. Size of population and breeding structure in relation to evolution. Science 87:430–431.

6

A quantitative genetic approach to assess the evolutionary potential of a coastal marine fish to ocean acidification

Alex J. Malvezzi,[1] Christopher S. Murray,[2] Kevin A. Feldheim,[3] Joseph D. DiBattista,[4] Dany Garant,[5] Christopher J. Gobler,[1] Demian D. Chapman[1] and Hannes Baumann[2]

1 School of Marine and Atmospheric Sciences, Stony Brook University, Stony Brook, NY, USA
2 Department of Marine Sciences, University of Connecticut, Groton, CT, USA
3 Pritzker Laboratory for Molecular Systematics and Evolution, Field Museum of Natural History, Chicago, IL, USA
4 Red Sea Research Center, King Abdullah University of Science and Technology, Thuwal, Saudi Arabia
5 Département de Biologie, Université de Sherbrooke, Sherbrooke, QC, Canada

Keywords

animal model, ASReml, Atlantic Silverside *Menidia menidia*, genotyping, heritability, microsatellites, pedigree analysis, survival.

Correspondence

Hannes Baumann, Department of Marine Sciences, University of Connecticut, Groton, CT 06340-6048, USA.

e-mail: hannes.baumann@uconn.edu

Abstract

Assessing the potential of marine organisms to adapt genetically to increasing oceanic CO_2 levels requires proxies such as heritability of fitness-related traits under ocean acidification (OA). We applied a quantitative genetic method to derive the first heritability estimate of survival under elevated CO_2 conditions in a metazoan. Specifically, we reared offspring, selected from a wild coastal fish population (Atlantic silverside, *Menidia menidia*), at high CO_2 conditions (~2300 µatm) from fertilization to 15 days posthatch, which significantly reduced survival compared to controls. Perished and surviving offspring were quantitatively sampled and genotyped along with their parents, using eight polymorphic microsatellite loci, to reconstruct a parent–offspring pedigree and estimate variance components. Genetically related individuals were phenotypically more similar (i.e., survived similarly long at elevated CO_2 conditions) than unrelated individuals, which translated into a significantly nonzero heritability (0.20 ± 0.07). The contribution of maternal effects was surprisingly small (0.05 ± 0.04) and nonsignificant. Survival among replicates was positively correlated with genetic diversity, particularly with observed heterozygosity. We conclude that early life survival of *M. menidia* under high CO_2 levels has a significant additive genetic component that could elicit an evolutionary response to OA, depending on the strength and direction of future selection.

Introduction

Ocean acidification (OA) has been recognized among the key anthropogenic processes threatening humanity and its environment (Rockstrom et al. 2009). Concerns about OA stem from (i) the certainty of continued oceanic CO_2 uptake with projected further atmospheric CO_2 increases and hence further declines in ocean pH and calcium carbonate saturation (Caldeira and Wickett 2003; Sabine et al. 2004; Bates et al. 2012) and (ii) from fast accumulating evidence that such predicted conditions affect many contemporary marine organisms in complex and often negative ways (Hendriks et al. 2010; Doney et al. 2012). High OA sensitivities, as inferred from reductions in survival, growth, and calcification rates under experimentally

elevated CO_2 conditions, have been observed during the early life stages of calcifying marine invertebrates (Orr et al. 2005; Kleypas et al. 2006; Doney et al. 2009; Talmage and Gobler 2010), but also in offspring of some marine fishes. In the latter, documented adverse effects of high CO_2 exposure range from behavioral abnormalities (Munday et al. 2009, 2014; Dixson et al. 2010, 2014; Chivers et al. 2014) and tissue damage (Frommel et al. 2012b) to directly reduced growth and survival rates (Baumann et al. 2012; Chambers et al. 2014). Cumulatively, these findings imply that OA could profoundly alter marine ecosystems, likely to the detriment of humans (Branch et al. 2013; Bednaršek et al. 2014).

Detecting widespread CO_2 sensitivities in marine organisms, however, is only a first step in assessing OA's long-term ecological consequences (Pfister et al. 2014; Sunday et al. 2014). Although anthropogenic OA is likely unprecedented in both pace and magnitude (Caldeira and Wickett 2003), the predicted environmental shifts will still happen gradually over the next few hundred years, that is, within a few (e.g., whales, turtles, many sharks) or hundreds (e.g., many fish, mollusks) or hundred thousands of generations (single cell plankton). Hence, an equally important but much less understood question is to what extent organisms will adapt to acidifying oceans through natural selection of either extant or randomly arising genotypes of greater fitness. This knowledge gap is being increasingly recognized, and several approaches have so far been used to address it (Munday et al. 2013; Reusch 2013; Sunday et al. 2014).

One approach has been to demonstrate standing genetic variation in CO_2 reaction norms within marine species or populations. For example, different strains of cyanobacteria from different oceanic regions show large variations in CO_2-dependent nitrogen fixation rates, suggesting that some strains benefit from increasing CO_2 more than others (Hutchins et al. 2013). Differentially adverse CO_2 effects are also evident, for example, between strains of the coccolithophore *Emiliania huxleyi* (Langer et al. 2009), or genetically different oyster *Saccostrea glomerata* lines (Parker et al. 2011), sea urchin populations (Kelly et al. 2013), and Atlantic cod *Gadus morhua* populations (coastal Norway versus Baltic Sea, Frommel et al. 2012a,b). Such observations suggest that rising CO_2 levels are gradually shifting the fitness landscape in the ocean, thus triggering changes in the genotypic composition of marine species, even in cases where little or no phenotypic change will be observed (Pespeni et al. 2013; Sunday et al. 2014).

A second type of approach is to assess evolutionary responses to OA *in vitro*. Lohbeck et al. (2012) found that adverse growth and calcification effects of high CO_2 (1000 µatm) in single clones of *E. huxleyi* partially disappeared after rearing 500 asexual generations at high CO_2 levels in the laboratory, suggesting a *de novo* evolutionary response. Unfortunately, such *in vitro* approaches are largely unfeasible in metazoans with substantially longer generation times. *In vitro* selection for OA-relevant genes was demonstrated via analysis of single nucleotide polymorphisms (SNPs) in purple sea urchin *Strongylocentrotus purpuratus* larvae cultured under different CO_2 levels (Pespeni et al. 2013). This revealed consistent shifts in allele frequencies predominantly within functional genomic groups associated with calcification and osmoregulatory control.

A third type of approach has been to infer the evolutionary potential of a trait by estimating its heritability h^2 (Charmantier and Garant 2005; Garant and Kruuk 2005; Sunday et al. 2014). Heritability is the proportion of a traits interindividual phenotypic variation that is not environmentally but genetically determined. Knowing a traits heritability is valuable, because together with the strength of selection (selection coefficient, singe-trait S, multiple traits β), the rate of evolutionary change (ΔZ) can be estimated with the breeders equation ($\Delta Z = h^2 S$, Falconer and Mackay 1996). Thereby, highly heritable traits that come under strong selection will evolve rapidly within few generations, whereas low heritability combined with small selection differentials produce small or negligible evolutionary change (but see Merilä et al. 2001; Morrissey et al. 2010).

To experimentally estimate a traits heritability, mean phenotypic change (ΔZ) could be measured following selection over multiple generations, but as with experimental evolution, this is practical mostly for organisms with short generation times (days to weeks, Lohbeck et al. 2012). Alternatively, trait heritability can be quantified in single-generation experiments, provided the design allows the different components of phenotypic variance (V_P) to be distinguished (Lynch and Walsh 1998; Sunday et al. 2014). This can be achieved by measuring V_P in offspring of single mother × father (=dam × sire) crosses that are separately reared under standardized laboratory conditions. Unfortunately, this can necessitate an impractically large number of experimental vessels (e.g., Sunday et al. 2011: >1000), particularly if model species or later developmental stages demand larger rearing volumes (e.g., fish larvae) or if the genetic diversity of a wild population is to be represented (20+ spawners per sex, Conover and Present 1990; Baumann and Conover 2011a). Finally, heritability estimates associated with OA are most needed for life history traits (survival, growth, fecundity), which are most directly related to fitness (Mousseau and Roff 1987; Sunday et al. 2014).

Here we evaluate a quantitative genetic approach to estimate trait heritability during single-generation experiments that involve large numbers of spawners but without the onerous requirement of separate experimental units for each sire × dam combination. Instead, the approach relies on resolving genetic relationships between offspring and parents *post-mortem* by genotyping all individuals using microsatellite markers. Microsatellites are routinely used to assign parentage with high confidence, for example, in human paternity tests (Marshall et al. 1998) or studies of wild population genetics (Chapman et al. 2009; Feldheim et al. 2014). Here, the approach is applied to derive the first heritability estimate for larval survival under elevated CO_2 conditions in a marine fish. We chose the Atlantic silverside (*Menidia menidia*), because (i) it is an ecologically important, annual forage fish along most of the western Atlantic coast, (ii) elevated CO_2 levels have previously been shown to reduce early life survival in this and related species (Baumann et al. 2012; Murray et al. 2014), and (iii) the required species-specific primers for 10 microsatellite loci were already available (Sbrocco and Barber 2011).

To explore the approach, we reared newly fertilized offspring from wild-caught parents at CO_2 conditions projected for the average open ocean within the next 300 years (~2000 μatm, Caldeira and Wickett 2003), while quantitatively sampling all perished and surviving larvae for downstream genetic analyses. Thus, we determined 'Days Survived at high CO_2' (hereafter: DS) and the genotype of each larva, which were used to construct and analyze a parent–offspring pedigree. We then tested the null hypothesis that genetically related offspring were no more phenotypically similar (i.e., survived similarly long) than unrelated offspring. Other genetic parameters such as observed heterozygosity or allelic richness were quantified for each replicate to examine the relationship between genetic diversity and offspring survival in high CO_2 environments.

Materials and methods

Parental collection and laboratory rearing
We collected ripe, adult M. menidia at the beginning of the spawning season (25 April 2013) from a tidal salt marsh on the North shore of Long Island (Poquot, 40°58.12′N, 73°5.28′W). Males and females were caught with a 30 × 2 m beach seine, then transported to our laboratory facility (Flax Pond Marine Laboratory) and held overnight in separate temperature-controlled baths (200 L, 21°C). The next morning, 29 females were strip-spawned by gently squeezing their hydrated eggs into a large shallow container containing seawater-activated sperm of 42 males and a large sheet of window screen (1 mm mesh). Within 15 min, fertilized eggs attached to the screen via uncoiled chorionic filaments, in contrast to unfertilized eggs that could later be gently rinsed off. The window screen was randomly cut into many small pieces, and within 2 h after fertilization, 3–6 randomly selected pieces totaling 100 embryos were suspended into each of 10 replicate rearing containers (20 L) preset for the high CO_2 treatment (~2300 μatm, Table 1). Another 100 embryos were placed into each of five control replicates (~460 μatm). The high CO_2 treatment represents the maximum level predicted for the open ocean within the next 300 years (Caldeira and Wickett 2003), but also a seasonal condition occurring within many productive coastal habitats already today (Wallace et al. 2014). Control replicates were used to

validate the assumption that high CO_2 conditions indeed reduced offspring survival, as has been documented in previous studies on this and closely related species (Baumann et al. 2012; Murray et al. 2014). Throughout the experiment, temperature, salinity, and photoperiod were held constant at 24°C, 25 psu, and 15L:9D, respectively. Larvae hatched ~120 h postfertilization and were immediately provided with daily ad libitum rations of newly hatched brine shrimp nauplii, Artemia salina, (San Francisco strain; Brine Shrimp Direct, Inc. Ogden, UT, USA) until the end of the experiment.

Seawater chemistry manipulation

Seawater chemistry was modified and monitored throughout the experiment according to recommended best practices for OA research (Riebesell et al. 2010). We used a gas proportioning system (Cole Parmer® flowmeters on a multitube frame, Cole Parmer, Vernon Hills, IL, USA) to deliver precise mixes of 5% ultra pure CO_2 and air (air only for controls) directly to the bottom of each rearing container via air stones. Target pH for high CO_2 (~2300 μatm, pH = 7.45) and control levels (~460 μatm, pH = 8.10) were monitored daily with an Orion ROSS Ultra pH/ATC Triode and Orion Star A121 pH portable meter, which were calibrated regularly using three-point National Institute of Standards and Technology (NIST) traceable pH references. In addition, discrete water samples were taken once from each replicate/treatment (borosilicate bottles, preserved with 200 μL HgCl$_2$) and analyzed for dissolved inorganic carbon (DIC, mol kg seawater^{-1}) with an EGM-4 Environmental Gas Analyzer (for detailed descriptions see Talmage and Gobler 2010; Gobler et al. 2014). Actual levels of CO_2 and total alkalinity were then calculated based on measured DIC, pH (NIST), temperature, salinity, and first and second dissociation constants of carbonic acid in seawater (Roy et al. 1993) using CO2SYS (http://cdiac.ornl.gov/ftp/co2sys/, Table 1).

Estimation of 'Days survived at high CO_2' (DS)

To quantify DS for each hatched larva, the bottom of every high CO_2 rearing container was gently siphoned twice daily, and all dead individuals were removed, recorded, and individually transferred to a tissue lysis solution (100 μL tissue lysis buffer + 12 μL proteinase K) for subsequent

Table 1. Carbon chemistry parameters during the experiment as determined from discrete water samples.

Treatment	Replicates	pH (NIST)	pCO$_2$	Total dissolved inorganic carbon	Total alkalinity
Control	5	8.11 ± 0.07	459 ± 73	2175 ± 126	2262 ± 123
High CO$_2$	10	7.45 ± 0.04	2294 ± 134	2324 ± 166	2235 ± 179

genomic DNA extraction. Frequent siphoning was critical, because fish larvae decompose beyond recognition within hours after death (at 24°C). The pattern of daily posthatch mortality was typical for early larval fish, with mortality peaking soon after hatch (1–4 dph), but then declining exponentially over the remaining days of the experiment (Fig. 1). After two consecutive days without any mortality (days 14 and 15 posthatch), the experiment was terminated and all surviving larvae (i.e., $DS = 15$) were individually transferred to tissue lysis solution for DNA extraction. To specifically evaluate the resistance of the population to elevated CO_2, only hatched larvae of the 10 high CO_2 replicates were sampled for genetic analyses. Unhatched embryos were recorded but not genotyped. To validate the assumption that high CO_2 was a significant source of mortality in the experimental replicates, live larvae were counted on day 10 posthatch by gently scooping small groups into replacement containers.

DNA extraction and amplification

Adult spawner DNA was extracted from tail clips (15–35 mg) following the animal tissue protocol of Qiagen DNeasy® kits. Larval DNA (DS ranging from 1 to 15 dph) was extracted using a more cost-effective 'salting out' protocol (Sunnucks et al. 1996) that yielded useable DNA from larval silversides. Ten polymorphic microsatellite loci for *M. menidia* (Sbrocco and Barber 2011) were amplified for all parents and offspring in a 10 µL reaction containing

$1\times$ PCR buffer, $10\times$ bovine serum albumin, 1.5–3.5 mM $MgCl_2$ (Table 2), 0.12 mM dNTPs, 0.16 µM of the reverse primer and fluorescently labeled M13 primer (Schuelke 2000), 0.04 µM of the species-specific forward primer and 1 unit Taq polymerase, and 10–40 ng of genomic DNA. Thermal cycling consisted of 5 min at 94°C followed by 35 cycles of 94°C for 30 s, primer specific annealing temperature (T_a, Table 2) for 30 s and 72°C for 60 s with a final extension at 72°C for 10 min. Fluorescently labeled PCR products were electrophoresed on an ABI 3730 DNA analyzer along with an internal fluorescent ladder (LIZ-500; Applied Biosystems). Alleles were scored by a single analyst (AJM) using the software Peakscanner 1.0 (Applied Biosystems®, Life Technologies, Grand Island, NY, USA). A subset of approximately 10% of genotypes was verified by a second analyst (KAF) using GENEMAPPER v4.0 (Applied Biosystems). One locus (Mm09) did not amplify and another one did not yield easily scored peaks (Mm119), hence only 8 of 10 loci were eventually used for parentage assignment and genetic analyses.

Genetic analyses

We used CERVUS 3.0 (www.fieldgenetics.com) to calculate summary statistics (e.g., number of occurrences of each allele at each locus) and assess the suitability of loci for parentage analysis (Table 2). CERVUS was then used to calculate the following four measures for each of the 10 high CO_2 replicates: relative allelic richness (i.e., $N_{Alleles, replicate}/N_{Alleles, total}$), observed heterozygosity, number of dams, and number of sires. Linear regression was used to relate the four proxies to survival. Subsequently, CERVUS was used to assign individual offspring to candidate mothers and fathers. For each offspring tested, candidate parents were assigned with at least 95% confidence using exclusion probability first and the maximum likelihood score second (Marshall et al. 1998). Comparing each offspring to all parent genotypes, exclusion of candidate parents followed if

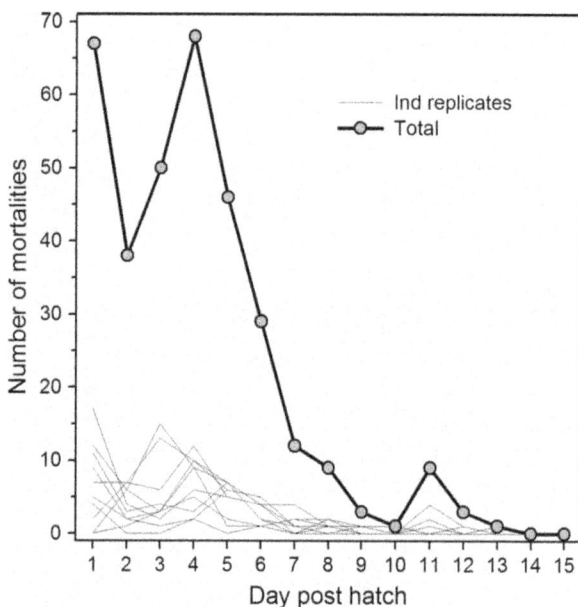

Figure 1 *Menidia menidia.* Daily posthatch mortality at high pCO_2 levels (2300 µatm) in each of 10 replicates (thin gray lines). The bold, black line depicts the total mortality across all replicates.

Table 2. Allele frequency summary statistics for eight microsatellite loci analyzed for all parent and offspring *Menidia menidia*: number of alleles (K), annealing temperature (T_a), salt concentration used ($MgCl_2$), number of individuals typed (N), observed heterozygosity (H_{Obs}), expected heterozygosity (H_{Exp}), and polymorphic information content (PIC).

Locus	T_a	$MgCl_2$	K	N	H_{Obs}	H_{Exp}	PIC
002	56	2.5	18	636	0.700	0.898	0.889
108	47	1.5	8	800	0.668	0.755	0.716
202	45	3.5	12	759	0.768	0.783	0.753
204	42	3.5	16	807	0.664	0.850	0.833
240	42	2	12	789	0.791	0.811	0.787
248	51	1.5	10	718	0.758	0.787	0.752
251	47	3	21	757	0.682	0.858	0.844
272	44	3.5	19	802	0.643	0.86	0.845

mismatches occurred at more than one locus. In total, 704 of 772 individuals (91%) were successfully assigned to parents, a success rate comparable to other studies on fish (75–95%; Estoup et al. 1998; Eldridge et al. 2002; Vandeputte et al. 2004). Unassigned offspring, resulting from poor amplification at some loci, were excluded from further analyses. All females, but only 35 of 42 males, were assigned to offspring, likely because the sperm from seven males was not activated or the fertilized eggs were not selected for in the experiment.

Last, we ran a hierarchical set of univariate 'animal models' for the trait DS, using the restricted maximum likelihood (REML) software ASReml v3.0.5 (©VSN International Ltd, Wood Lane, Hemel Hempstead, UK). An 'animal model' is a mixed model (i.e., a form of linear regression with 'fixed' and 'random' effects as explanatory variables) that has been found advantageous for estimating phenotypic variance components in wild populations (Lynch and Walsh 1998; Kruuk 2004; Garant and Kruuk 2005; Thériault et al. 2007; Dibattista et al. 2009). For fixed effects, we only considered replicate ID (1–10), which corresponds to the experimental vessel, given that age (all individuals had the same day of fertilization) and sex (individuals sampled in their entirety) were not relevant factors. Replicate ID significantly influenced DS (df = 9, $P < 0.001$, univariate general linear model SPSS; IBM) therefore all subsequent analyses included replicate ID as a fixed effect to remove its effect prior to the estimation of genetic parameters. We then estimated the heritability of DS by testing three distinct models: (i) the base model including only the fixed effect, (ii) a model including the fixed effect and an additive genetic random effect (α_i with variance V_A), and (iii) a model including the fixed effect, additive genetic random effect, and dam identity (m_i with variance V_M) as a random effect. The components of total phenotypic variance in DS ($V_{DS} = V_A + V_M + V_R$) were then used to calculate narrow sense heritability h^2 of DS as the ratio of additive genetic variance to total phenotypic variance ($h^2 = V_A/V_{DS}$) and maternal effects m^2 of DS as the ratio of maternal variance to total phenotypic variance ($m^2 = V_M/V_{DS}$). Statistical significance was assessed with likelihood-ratio tests [LRT = -2 * (likelihood base model–likelihood model with random effects)] tested against the chi-square distribution (df = 1).

Results

High CO_2 conditions in the experimental replicates significantly lowered average offspring survival by 34% compared to control replicates (10 dph survival ± SD: High CO_2 = 46.1 ± 15.5%, Control: 69.6 ± 17.5%, t-test, df = 13, t = 2.63, P = 0.02, Fig. 2). In the experimental replicates, 843 of 1000 fertilized embryos hatched, and of these

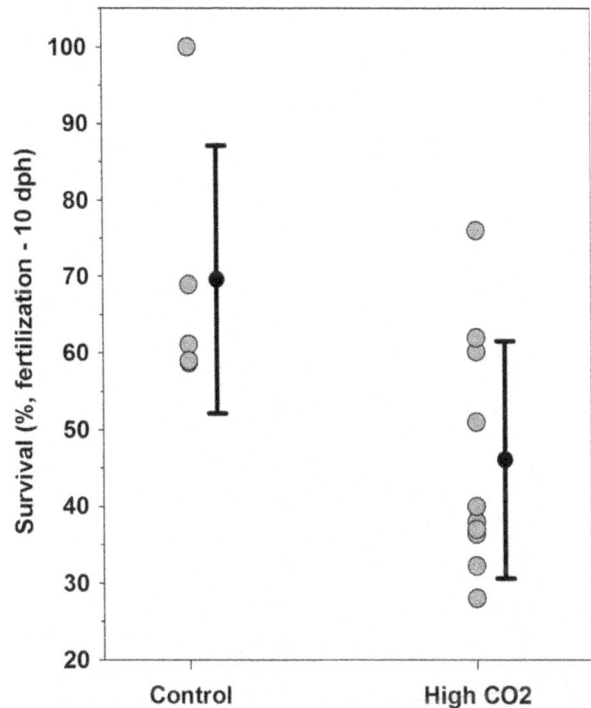

Figure 2 *Menidia menidia.* Survival from fertilization to 10 dph in offspring exposed to high (2300 µatm) versus ambient (control: 460 µatm) pCO_2 levels.

336 dead and 436 surviving individuals were recovered throughout the experiment. The remaining 7.1% were unaccounted for, likely due to rapid decomposition between siphoning. We assumed that decomposition rates were independent of genotype and therefore did not bias our results.

Pedigree analysis revealed that the minimum number of sires per dam was one, resulting in one offspring (dead on day 7) produced from that pair. The maximum number of sires to fertilize a single dam's eggs was 31, resulting in 115 offspring, of which 83 survived and 32 died (Fig. 3). Dams with the most offspring were also sired by the most males, and this increase in offspring number with sire number was exponential and greater for survivors than for perished offspring (Fig. 3). Survival among the 10 high CO_2 replicates ranged from 28% to 76%, with a mean (SD) survival of 46% (15%). Survival tended to increase with all four measures of genetic diversity (relative allelic richness, observed heterozygosity, number of dams, number of sires); however, the linear relationship was only significant for observed heterozygosity (P = 0.025, Fig. 4).

The estimated genetic variance components for DS are presented in Table 3. Including genotype as an additive genetic random effect to the base model improved the model significantly (LRT = 22.26, $P < 0.001$). Adding maternal identity as a second random effect resulted in a small and nonsignificant model improvement (LRT = 2.20,

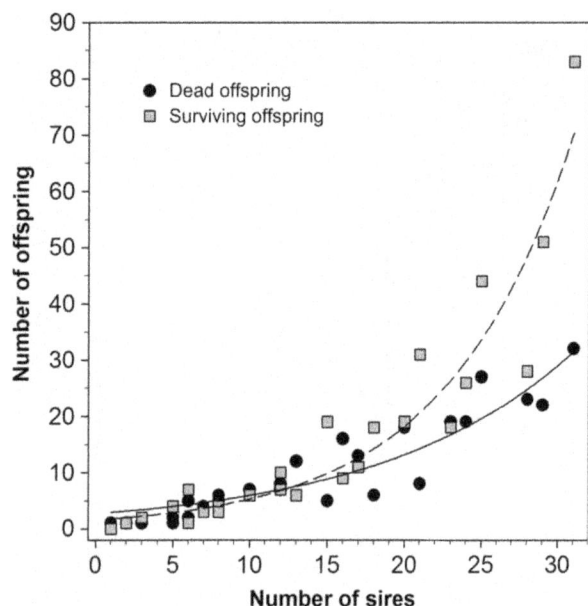

Figure 3 *Menidia menidia*. Relationship between the number of offspring from each female and the number of sires detected to fertilize offspring from each female. Black circles denote offspring that perished during the experiment, whereas gray squares depict survivors (i.e., the data are paired, such that each female is represented once along the x-axis). Both perished and surviving offspring were fitted with an exponential curve (dashed and solid line, respectively). $N_{Dead} = 2.7 \times e^{0.08 \times Sires}$, $R^2 = 0.85$, $P < 0.0001$; $N_{Surv} = 1.56 \times e^{0.12 \times Sires}$, $R^2 = 0.89$, $P < 0.0001$.

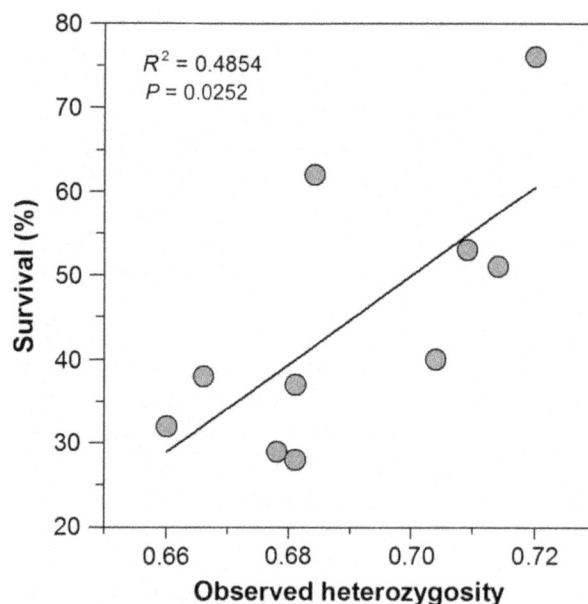

Figure 4 *Menidia menidia*. 15 dph offspring survival at high CO_2 conditions in relation to observed heterozygosity at 10 replicate rearing containers.

$P = 0.14$). Hence, in the final model without maternal identity, the estimated total phenotypic variance ($\pm$SE) V_{DS} was 29.15 $\pm$ 1.69, with an additive genetic variance V_A of 5.72 $\pm$ 2.10 and thus a heritability of larval survival at high CO_2 of 0.20 $\pm$ 0.07.

Discussion

We used a quantitative genetic approach to estimate the heritability of fish survival during the early developmental stages at elevated CO_2 conditions. Our approach had three priorities, that is, using a (i) single-generation experiment with (ii) enough parents to approximate a wild population's genetic diversity, while (iii) allowing all sire × dam crosses to be reared in a (replicated) common garden environment. We found that genetically related larvae were indeed phenotypically more similar (i.e., survived similarly long at elevated CO_2 conditions) than unrelated larvae, which translated into a significant component of additive genetic variance and a small but nonzero heritability of 0.20 $\pm$ 0.07. Compared to average larval survival at the controls (70%), elevated CO_2 conditions significantly reduced survival (to 46%, a 34% reduction) in addition to CO_2-unrelated mortality agents. Hence, our heritability

estimate for survival under elevated CO_2 conditions should be considered a maximum, because it assumed that all mortality resulted from these conditions.

The small contribution of maternal effects was unexpected, given that early life history traits in marine organisms often reflect maternal traits and investment (Green 2008; Sunday et al. 2011; Gao and Munch 2013). A low likelihood of detecting maternal effects in this case might have been due to collecting all adults from the same site and rearing offspring in common garden environments at unrestricted food levels (Marshall and Uller 2007; Allan et al. 2008). We also detected a positive relationship between observed heterozygosity and early life survival at high CO_2 levels. Assuming that neutral genetic variation is a proxy for adaptive genetic variation, this finding highlights the importance of standing genetic variation for a population's capacity to withstand environmental stress.

When evaluating the usefulness of our approach, its advantages (i.e., single-generation, genetic representation, common garden rearing) should be weighed against feasibility, cost, and time considerations. In most OA-sensitive organisms, CO_2-related mortality occurs during the earliest and thus smallest life stages (Dupont et al. 2008; Gobler et al. 2014), which rapidly decompose after death in experimental vessels. Hence, the requirement of individually sampling and genotyping all dying offspring may pose a challenge in species with smaller and more fragile offspring (e.g., mollusks, crustaceans) than newly hatched silverside larvae (~5 mm SL, ~2 mg wet weight; Baumann and Conover 2011b). In those cases, a potential solution might

Table 3. Variance component estimates for two univariate animal models predicting 'days survived at high CO_2' (DS) as a function of replicate ID (fixed effect) and genotype α_i.

Model	Variance component	Estimate	Standard error
$DS_i = ID + \alpha_i + \varepsilon_i$	V_{DS}	29.145	1.685
	V_A	5.716	2.102
	V_R	23.428	1.925
Heritability	$h^2 = V_A/V_{DS}$	0.196	0.067

V_{DS}, total phenotypic variance of DS; V_A, additive genetic variance (genotype); V_R, residual variance.

be to perform full factorial crosses and meticulously standardize the contribution of each cross to common rearing containers, which may then allow deducing the 'dead' from parental and survivor genotypes. In our case, meticulous siphoning resulted in near complete recovery (93%) of hatched individuals, and prior trials had ascertained that single larvae yield sufficient genomic material for amplification and genotyping (Malvezzi 2013). Whether this is feasible in other species has yet to be determined. We also assumed that a robust heritability estimate required approximately 1000 offspring and a good mixture of full siblings, half siblings, and unrelated individuals, which was feasible in this case by strip-spawning a large number of adults and using a large number of replicates (10 × 100 embryos). However, the costs associated with DNA extraction, amplification, and automated analysis of PCR products were approximately $10/individual (in 2013). Our approach was also time intensive, not the least because all experimental individuals (offspring + parents) had to be manually genotyped for each of nine polymorphic microsatellite loci via ABI-Analyzer scans [(total of 772 offspring + 71 parents)*9 loci = 7587 scans]. Adding to the time budget is the interpretation of genotypes and then evaluating intra- and inter-reader consistency.

In our final model (fixed + additive genetic random effect), the heritability estimate (±SE) for early life survival at high CO_2 levels was 0.20 ± 0.07, which is well within the range of narrow sense heritability reported for survival in fish in the literature (Mousseau and Roff 1987; Law 2000). For example, heritability of early life survival in Atlantic salmon under aquaculture conditions ranged between 0 and 0.34 (Standal and Gjerde 1987; Jonasson 1993), while estimates for juvenile brook trout were 0.16–0.51 (Robison and Luempert 1984). In a meta-analysis of over 1100 narrow sense heritability estimates across taxa (Mousseau and Roff 1987) reported that 50% of the estimates for life history traits (e.g., survival, growth, fecundity) ranged between 0.14 and 0.34 (mean = 0.27 ± 0.03, $n = 79$), in contrast to generally higher heritability values associated with morphological traits like length or weight (mean = 0.51 ± 0.02, $n = 140$). The current understand-

ing of the many sources of genetic variation in natural populations predicts this difference, because life history traits are more closely associated with fitness than more distantly related morphological traits (Falconer 1981; Mousseau and Roff 1987). Given the novelty of the approach and lack of heritability estimates for early life survival of metazoans under OA, the values are difficult to relate to previous works. However, heritability of early larval size under OA have been derived for a mussel and a sea urchin species (Sunday et al. 2011), where estimates at ambient CO_2 levels (0.12 ± 0.09 and 0.14 ± 0.16, respectively) were higher than at elevated CO_2 (0.00 ± 0.00 and 0.09 ± 0.10, respectively). Assuming that $h^2_{size} > h^2_{survival}$, our study's heritability estimates for early survival in a fish (0.10–0.20) were notably above the two invertebrates studied thus far.

Does this mean that *M. menidia* will adapt relatively rapidly to OA? We believe that this conclusion is not warranted yet, given the current lack of understanding regarding the strength and form of OA-induced selection in the wild (but see Bednaršek et al. 2014). In addition to heritability, rapid evolutionary change requires strong, uni-

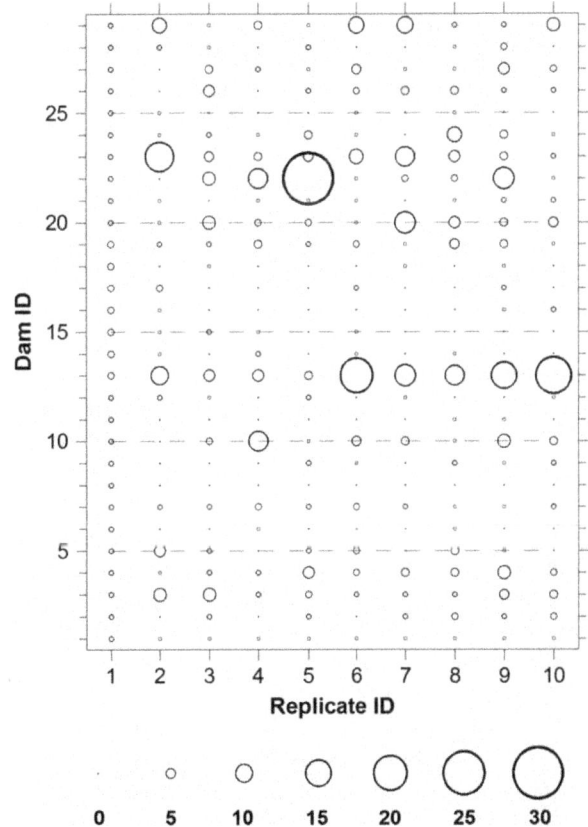

Figure 5 *Menidia menidia*. Distribution of offspring from 29 dams among 10 replicates based on the microsatellite based parent–offspring pedigree.

directional selection. Conover and Munch (2002) found that strong artificial selection on *M. menidia* body size (only 10% of the largest/smallest adults were allowed to reproduce) resulted in a significant divergence between up- and down-selected lines within only four generations. Incidentally, the heritability of growth rate in this experiment was also 0.20. More recently, Brown et al. (2008) used this heritability estimate to model the rate of evolutionary change under more realistic selection differentials typical for commercial fisheries, reporting a comparable evolutionary change within 30 generations. OA-induced selection is surely weaker and not necessarily unidirectional. For example, exposure of newly fertilized embryos to 2300 µatm CO_2 in the current experiment reduced average survival by 34%. In reality, however, high CO_2 levels will not suddenly but gradually occur in the average open ocean over the next three centuries. In addition, indirect effects or other environmental changes (e.g., temperature, oxygen, plankton production) might mitigate or even reverse OA-induced selection (Pfister et al. 2014), which is inherently difficult to predict. Lastly, recent findings suggest that *M. menidia* offspring are CO_2-sensitive only during some (early) parts of the spawning season, while becoming CO_2 tolerant later in the season, when elevated respiration rates increase CO_2 levels in tidal marsh spawning habitats (Murray et al. 2014; Baumann et al. 2015). Seasonally varying CO_2 sensitivities are likely a result of parent–offspring linkages known as transgenerational plasticity, and its probable ubiquity adds yet another dimension to the question of species' evolutionary potential to OA (Salinas et al. 2013; Allan et al. 2014). One hypothesis is that coastal marine species such as *M. menidia* are adapted to highly variable CO_2 conditions through intra- and transgenerational plasticity (Reusch 2013; Crozier and Hutchings 2014), which could effectively stall selection and evolutionary adaptation until perhaps reaching a threshold with adverse consequences (a form of evolutionary trap, Schlaepfer et al. 2002). In summary, we found that larval *M. menidia* are negatively affected by high CO_2 levels and that early life survival has a significant additive genetic component that could elicit an evolutionary response if selection pressures in the wild resemble – at least in direction – the artificial conditions in the laboratory. While challenging, this constitutes one of the most important issues for future OA research to address (Merilä and Hendry 2014; Pfister et al. 2014; Sunday et al. 2014).

In addition to heritability, our quantitative genetic approach revealed links between phenotypic and genotypic differences among the 10 'replicates' of this study. The strong statistical effect of replicate ID was most likely due to a nonrandom distribution of genotypes among experimental vessels, despite our meticulous efforts to randomize offspring at the beginning of the trial (Fig. 5). Small biotic (e.g., larval density) and/or abiotic differences between replicates might have been contributing factors, even though daily temperature and pH monitoring suggested no systematic differences in physiochemical conditions. However, between-replicate survival varied threefold (28–76%, mean = 46%), while genotyping revealed that the number of dams (16–23, $N_{Parents}$: 44–54), relative allelic richness (0.56–0.72), and observed heterozygosity (0.66–0.72) all varied substantially between replicates. The latter was significantly positively correlated to survival and explained almost 50% of the variation. All proxies for genetic variability were calculated across all individuals per replicate (regardless of dead and surviving), hence assuming that neutral (microsatellites) and adaptive genetic variability are correlated, our finding corroborates the notion that more genetic variability and a higher overall allelic richness among progeny positively affect survival (Ruzzante et al. 1996). This is consistent with studies showing diminished genetic variability in natural populations resulting in increased genetic bottlenecks, inbreeding, reductions in effective population size and the potential for reduced adaptability and productivity (Hauser et al. 2002; O'Leary et al. 2013).

Our work has emphasized the suitability of using a quantitative genetic approach to evaluate species evolutionary potential in the face of continuing OA. We believe that such estimates are of increasing interest to a research field that needs to address the questions of short- and long-term adaptability in marine organisms to allow a more realistic assessment of the OA threat.

Acknowledgements

We kindly acknowledge the assistance of Shannon O'Leary, Andrew Fields, and the staff of the Field Museum's Pritzker Lab. M. Pespeni and one anonymous reviewer provided helpful comments to an earlier version of this manuscript. The National Science Foundation provided full funding for A.J.M. and C.S.M. and partial funding for H.B. and C.J.G. (NSF #1129622). C.J.G. was also partially funded by NOAA's Ocean Acidification Program through award #NA12NOS4780148 from the National Centers for Coastal Ocean Science and the Chicago Community Trust.

Literature cited

Allan, R. M., Y. M. Buckley, and D. J. Marshall 2008. Offspring size plasticity in response to intraspecific competition: an adaptive maternal effect across life-history stages. The American Naturalist **171**:13.

Allan, B. J. M., G. M. Miller, M. I. Mccormick, P. Domenici, and P. L. Munday 2014. Parental effects improve escape performance of juvenile reef fish in a high-CO_2 world. Proceedings of the Royal Society B-Biological Sciences **281**:20132179.

Bates, N. R., M. H. P. Best, K. Neely, R. Garley, A. G. Dickson, and R. J. Johnson 2012. Detecting anthropogenic carbon dioxide uptake and ocean acidification in the North Atlantic Ocean. Biogeosciences 9:2509–2522.

Baumann, H., and D. O. Conover 2011a. Adaptation to climate change: contrasting patterns of thermal-reaction-norm evolution in Pacific versus Atlantic silversides. Proceedings of the Royal Society B-Biological Sciences 278:2265–2273.

Baumann, H., and D. O. Conover 2011b. Thermal reaction norms of growth in Atlantic and Pacific silverside fishes. doi: 10.1594/PANGAEA.773233. Pangaea Database.

Baumann, H., S. C. Talmage, and C. J. Gobler 2012. Reduced early life growth and survival in a fish in direct response to increased carbon dioxide. Nature Climate Change 2:38–41.

Baumann, H., R. Wallace, T. Tagliaferri, and C. J. Gobler 2015. Large natural pH, CO_2 and O_2 fluctuations in a temperate tidal salt marsh on diel, seasonal and interannual time scales. Estuaries and Coasts 38:220–231.

Bednaršek, N., R. A. Feely, J. C. P. Reum, B. Peterson, J. Menkel, S. R. Alin, and B. Hales 2014. *Limacina helicina* shell dissolution as an indicator of declining habitat suitability owing to ocean acidification in the California Current Ecosystem. Proceedings of the Royal Society B-Biological Sciences 281:20140123.

Branch, T. A., B. M. Dejoseph, L. J. Ray, and C. A. Wagner 2013. Impacts of ocean acidification on marine seafood. Trends in Ecology & Evolution 28:178–186.

Brown, C. J., A. J. Hobday, P. E. Ziegler, and D. C. Welsford 2008. Darwinian fisheries science needs to consider realistic fishing pressures over evolutionary time scales. Marine Ecology Progress Series 369:257–266.

Caldeira, K., and M. E. Wickett 2003. Anthropogenic carbon and ocean pH. Nature 425:365–365.

Chambers, R. C., A. C. Candelmo, E. A. Habeck, M. E. Poach, D. Wieczorek, K. R. Cooper, C. E. Greenfield et al. 2014. Ocean acidification effects in the early life-stages of summer flounder, *Paralichthys dentatus*. Biogeosciences 11:1613–1626.

Chapman, D. D., E. A. Babcock, S. H. Gruber, J. D. Dibattista, B. R. Franks, S. A. Kessel, T. Guttridge et al. 2009. Long-term natal site-fidelity by immature lemon sharks (*Negaprion brevirostris*) at a subtropical island. Molecular Ecology 18:3500–3507.

Charmantier, A., and D. Garant 2005. Environmental quality and evolutionary potential: lessons from wild populations. Proceedings of the Royal Society B-Biological Sciences 272:1415–1425.

Chivers, D. P., M. I. Mccormick, G. E. Nilsson, P. L. Munday, S.-A. Watson, M. G. Meekan, M. D. Mitchell et al. 2014. Impaired learning of predators and lower prey survival under elevated CO_2: a consequence of neurotransmitter interference. Global Change Biology 20:515–522.

Conover, D. O., and S. B. Munch 2002. Sustaining fisheries yields over evolutionary time scales. Science 297:94–96.

Conover, D. O., and T. M. C. Present 1990. Countergradient variation in growth rate: compensation for length of the growing season among Atlantic silversides from different latitudes. Oecologia 83:316–324.

Crozier, L. G., and J. A. Hutchings 2014. Plastic and evolutionary responses to climate change in fish. Evolutionary Applications 7:68–87.

Dibattista, J. D., K. A. Feldheim, D. Garant, S. H. Gruber, and A. P. Hendry 2009. Evolutionary potential of a large marine vertebrate: quantitative genetic parameters in a wild population. Evolution 63:1051–1067.

Dixson, D. L., P. L. Munday, and G. P. Jones 2010. Ocean acidification disrupts the innate ability of fish to detect predator olfactory cues. Ecology Letters 13:68–75.

Dixson, D. L., A. R. Jennings, J. Atema, and P. L. Munday 2014. Odor

tracking in sharks is reduced under future ocean acidification conditions. Global Change Biology. doi:10.1111/gcb.12678.

Doney, S. C., V. J. Fabry, R. A. Feely, and J. A. Kleypas 2009. Ocean acidification: the other CO_2 problem. Annual Review of Marine Science 1:169–192.

Doney, S. C., M. Ruckelshaus, J. Emmett Duffy, J. P. Barry, F. Chan, C. A. English, H. M. Galindo et al. 2012. Climate change impacts on marine ecosystems. Annual Review of Marine Science 4:11–37.

Dupont, S., J. Havenhand, W. Thorndyke, L. Peck, and M. C. Thorndyke 2008. Near-future level of CO_2-driven ocean acidification radically affects larval survival and development in the brittlestar *Ophiothrix fragilis*. Marine Ecology Progress Series 373:285–294.

Eldridge, W. H., M. D. Bacigalupi, I. R. Adelman, L. M. Miller, and A. R. Kapuscinski 2002. Determination of relative survival of two stocked walleye populations and resident natural-origin fish by microsatellite DNA parentage assignment. Canadian Journal of Fisheries and Aquatic Sciences 59:282–290.

Estoup, A., K. Gharbi, M. Sancristobal, C. Chevalet, P. Haffray, and R. Guyomard 1998. Parentage assignment using microsatellites in turbot (*Scophthalmus maximus*) and rainbow trout (*Oncorhynchus mykiss*) hatchery populations. Canadian Journal of Fisheries and Aquatic Sciences 55:715–723.

Falconer, D. S. 1981. Introduction to Quantitative Genetics, 2nd edn. Longman, London, UK.

Falconer, D., and T. Mackay. eds. 1996. Heritability. In Introduction to Quantitative Genetics, pp. 160–183. Longman, London.

Feldheim, K. A., S. H. Gruber, J. D. Dibattista, E. A. Babcock, S. A. Kessel, A. P. Hendry, E. K. Pikitch et al. 2014. Two decades of genetic profiling yields first evidence of natal philopatry and long-term fidelity to parturition sites in sharks. Molecular Ecology 23:110–117.

Frommel, A., A. Schubert, U. Piatkowski, and C. Clemmesen 2012a. Egg and early larval stages of Baltic cod, *Gadus morhua*, are robust to high levels of ocean acidification. Marine Biology 160:1825–1834.

Frommel, A. Y., R. Maneja, D. Lowe, A. M. Malzahn, A. J. Geffen, A. Folkvord, U. Piatkowski et al. 2012b. Severe tissue damage in Atlantic cod larvae under increasing ocean acidification. Nature Climate Change 2:42–46.

Gao, J., and S. B. Munch 2013. Genetic and maternal variation in early growth in the Atlantic silverside *Menidia menidia*. Marine Ecology Progress Series 485:211–222.

Garant, D., and L. E. Kruuk 2005. How to use molecular marker data to measure evolutionary parameters in wild populations. Molecular Ecology 14:1843–1859.

Gobler, C. J., E. Depasquale, A. Griffith, and H. Baumann 2014. Hypoxia and acidification have additive and synergistic negative effects on the growth, survival, and metamorphosis of early life stage bivalves. PLoS One 9:e83648.

Green, B. S. 2008. Maternal effects in fish populations. Advances in Marine Biology 54:1–105.

Hauser, L., G. J. Adcock, P. J. Smith, J. H. B. Ramírez, and G. R. Carvalho 2002. Loss of microsatellite diversity and low effective population size in an overexploited population of New Zealand snapper (*Pagrus auratus*). Proceedings of the National Academy of Sciences of the United States of America 99:11742–11747.

Hendriks, I. E., C. M. Duarte, and M. Álvarez 2010. Vulnerability of marine biodiversity to ocean acidification: a meta-analysis. Estuarine, Coastal and Shelf Science 86:157–164.

Hutchins, D. A., F.-X. Fu, E. A. Webb, N. Walworth, and A. Tagliabue 2013. Taxon-specific response of marine nitrogen fixers to elevated carbon dioxide concentrations. Nature Geoscience 6:790–795.

Jonasson, J. 1993. Selection experiments in salmon ranching. I. Genetic and environmental sources of variation in survival and growth in freshwater. Aquaculture 109:225–236.

Kelly, M. W., J. L. Padilla-Gamiño, and G. E. Hofmann 2013. Natural variation and the capacity to adapt to ocean acidification in the keystone sea urchin Strongylocentrotus purpuratus. Global Change Biology 19:2536–2546.

Kleypas, J. A., R. A. Feely, V. J. Fabry, C. Langdon, C. L. Sabine, and L. L. Robbins 2006. Impacts of ocean acidification on coral reefs and other marine calcifiers: a guide for future research, 88 pp. report of a workshop held 18–20 April 2005, St. Petersburg, FL, sponsored by NSF, NOAA, and the U.S. Geological Survey.

Kruuk, L. E. B. 2004. Estimating genetic parameters in natural populations using the 'animal model'. Philosophical Transactions of the Royal Society of London. Series B: Biological Sciences 359:873–890.

Langer, G., G. Nehrke, I. Probert, J. Ly, and P. Ziveri 2009. Strain-specific responses of Emiliania huxleyi to changing seawater carbonate chemistry. Biogeosciences 6:2637–2646.

Law, R. 2000. Fishing, selection, and phenotypic evolution. ICES Journal of Marine Science 57:659–668.

Lohbeck, K. T., U. Riebesell, and T. B. H. Reusch 2012. Adaptive evolution of a key phytoplankton species to ocean acidification. Nature Geoscience 5:346–351.

Lynch, M., and B. Walsh 1998. Genetics and Analysis of Quantitative Traits. Sinauer Associates, Inc., Sunderland, MA, 980 pp.

Malvezzi, A. 2013. Using Quantitative Genetics Methods to Estimate Heritability of Larval CO_2-Resistance in a Coastal Marine Fish. School of Marine and Atmospheric Sciences, Stony Brook University, Stony Brook.

Marshall, D. J., and T. Uller 2007. When is a maternal effect adaptive? Oikos 116:1957–1963.

Marshall, T. C., J. Slate, L. E. B. Kruuk, and J. M. Pemberton 1998. Statistical confidence for likelihood-based paternity inference in natural populations. Molecular Ecology 7:639–655.

Merilä, J., and A. P. Hendry 2014. Climate change, adaptation, and phenotypic plasticity: the problem and the evidence. Evolutionary Applications 7:1–14.

Merilä, J., B. C. Sheldon, and L. E. B. Kruuk 2001. Explaining stasis: microevolutionary studies in natural populations. Genetica 112–113:199–222.

Morrissey, M. B., L. E. B. Kruuk, and A. J. Wilson 2010. The danger of applying the breeder's equation in observational studies of natural populations. Journal of Evolutionary Biology 23:2277–2288.

Mousseau, T. A., and D. A. Roff 1987. Natural selection and the heritability of fitness components. Heredity 59:181–197.

Munday, P. L., D. L. Dixson, J. M. Donelson, G. P. Jones, M. S. Pratchett, G. V. Devitsina, and K. B. Døving 2009. Ocean acidification impairs olfactory discrimination and homing ability of a marine fish. Proceedings of the National Academy of Sciences of the United States of America 106:1848–1852.

Munday, P. L., R. R. Warner, K. Monro, J. M. Pandolfi, and D. J. Marshall 2013. Predicting evolutionary responses to climate change in the sea. Ecology Letters 16:1488–1500.

Munday, P. L., A. J. Cheal, D. L. Dixson, J. L. Rummer, and K. E. Fabricius 2014. Behavioural impairment in reef fishes caused by ocean acidification at CO_2 seeps. Nature Climate Change 4:487–492.

Murray, C. S., A. J. Malvezzi, C. J. Gobler, and H. Baumann 2014. Offspring sensitivity to ocean acidification changes seasonally in a coastal marine fish. Marine Ecology Progress Series 504:1–11.

O'Leary, S. J., L. A. Hice, K. A. Feldheim, M. G. Frisk, A. E. Mcelroy, M. D. Fast, and D. D. Chapman 2013. Severe inbreeding and small effective number of breeders in a formerly abundant marine fish. PLoS One 8:e66126.

Orr, J. C., V. J. Fabry, O. Aumont, L. Bopp, S. C. Doney, R. A. Feely, A. Gnanadesikan et al. 2005. Anthropogenic ocean acidification over the twenty-first century and its impact on calcifying organisms. Nature 437:681–686.

Parker, L., P. Ross, and W. O'Connor 2011. Populations of the Sydney rock oyster, Saccostrea glomerata, vary in response to ocean acidification. Marine Biology 158:689–697.

Pespeni, M. H., E. Sanford, B. Gaylord, T. M. Hill, J. D. Hosfelt, H. K. Jaris, M. Lavigne et al. 2013. Evolutionary change during experimental ocean acidification. Proceedings of the National Academy of Sciences 110:6937–6942.

Pfister, C. A., A. J. Esbaugh, C. A. Frieder, H. Baumann, E. E. Bockmon, M. M. White, B. R. Carter et al. 2014. Detecting the unexpected: a research framework for ocean acidification. Environmental Science & Technology 48:9982–9994.

Reusch, T. B. H. 2013. Climate change in the oceans: evolutionary versus phenotypically plastic responses of marine animals and plants. Evolutionary Applications 7:104–122.

Riebesell, U., V. J. Fabry, L. Hansson, and J. P. Gattuso. eds. 2010. Guide to Best Practices for Ocean Acidification Research and Data Reporting. Publications Office of the European Union, Luxembourg, 260 pp.

Robison, O. W., and L. G. Luempert 1984. Genetic variation in weight and survival of brook trout (Salvelinus fontinalis). Aquaculture 38:155–170.

Rockstrom, J., W. Steffen, K. Noone, A. Persson, F. S. Chapin, E. F. Lambin, T. M. Lenton et al. 2009. A safe operating space for humanity. Nature 461:472–475.

Roy, R. N., L. N. Roy, K. M. Vogel, C. Porter-Moore, T. Pearson, C. E. Good, F. J. Millero et al. 1993. The dissociation constants of carbonic acid in seawater at salinities 5 to 45 and temperatures 0 to 45°C. Marine Chemistry 44:249–267.

Ruzzante, D. E., C. T. Taggart, and D. Cook 1996. Spatial and temporal variation in the genetic composition of a larval cod (Gadus morhua) aggregation: cohort contribution and genetic stability. Canadian Journal of Fisheries and Aquatic Sciences 53:2695–2705.

Sabine, C. L., R. A. Feely, N. Gruber, R. M. Key, K. Lee, J. L. Bullister, R. Wanninkhof et al. 2004. The oceanic sink for anthropogenic CO_2. Science 305:367–371.

Salinas, S., S. C. Brown, M. Mangel, and S. B. Munch 2013. Non-genetic inheritance and changing environments. Non-Genetic Inheritance 1:38–50.

Sbrocco, E. J., and P. H. Barber 2011. Ten polymorphic microsatellite loci for the Atlantic Silverside, Menidia menidia. Conservation Genetics Resources 3:585–587.

Schlaepfer, M. A., M. C. Runge, and P. W. Sherman 2002. Ecological and evolutionary traps. Trends in Ecology & Evolution 17:474–480.

Schuelke, M. 2000. An economic method for the fluorescent labeling of PCR fragments. Nature Biotechnology 18:233–234.

Standal, M., and B. Gjerde 1987. Genetic variation in survival of Atlantic salmon during the sea-rearing period. Aquaculture 66:197–207.

Sunday, J. M., R. N. Crim, C. D. G. Harley, and M. W. Hart 2011. Quantifying rates of evolutionary adaptation in response to ocean acidification. PLoS One 6:e22881.

Sunday, J. M., P. Calosi, S. Dupont, P. L. Munday, J. H. Stillman, and T. B. H. Reusch 2014. Evolution in an acidifying ocean. Trends in Ecology & Evolution 29:117–125.

Sunnucks, P., P. R. England, A. C. Taylor, and D. F. Hales 1996. Micro-satellite and chromosome evolution of parthenogenetic Sitobion aphids in Australia. Genetics **144**:747–756.

Talmage, S. C., and C. J. Gobler 2010. Effects of past, present, and future ocean carbon dioxide concentrations on the growth and survival of larval shellfish. Proceedings of the National Academy of Sciences of the United States of America **107**:17246–17251.

Thériault, V., D. Garant, L. Bernatchez, and J. J. Dodson 2007. Heritability of life-history tactics and genetic correlation with body size in a natural population of brook charr (*Salvelinus fontinalis*). Journal of Evolutionary Biology **20**:2266–2277.

Vandeputte, M., M. Kocour, S. Mauger, M. Dupont-Nivet, D. De Guerry, M. Rodina, D. Gela et al. 2004. Heritability estimates for growth-related traits using microsatellite parentage assignment in juvenile common carp (*Cyprinus carpio* L.). Aquaculture **235**:223–236.

Wallace, R. B., H. Baumann, J. S. Grear, R. C. Aller, and C. J. Gobler 2014. Coastal ocean acidification: the other eutrophication problem. Estuarine, Coastal and Shelf Science **148**:1–13.

7

Fisheries-induced neutral and adaptive evolution in exploited fish populations and consequences for their adaptive potential

Lise Marty,[1] Ulf Dieckmann[2] and Bruno Ernande[1,2]

1 IFREMER, Laboratoire Ressources Halieutiques, Unité Halieutique Manche-Mer du Nord, Boulogne-sur-mer, France
2 IIASA, Evolution and Ecology Program, Laxenburg, Austria

Keywords
eco-genetic model, effective population size, fisheries-induced evolution, genetic drift, genetic erosion, genetic markers, life-history traits, natural selection and contemporary evolution.

Correspondence
Lise Marty, Center for Ocean Life, DTU-Aqua, Jægersborg Allé 1, 2920 Charlottenlund, Denmark.

e-mail: lisma@aqua.dtu.dk

Abstract

Fishing may induce neutral and adaptive evolution affecting life-history traits, and molecular evidence has shown that neutral genetic diversity has declined in some exploited populations. Here, we theoretically study the interplay between neutral and adaptive evolution caused by fishing. An individual-based eco-genetic model is devised that includes neutral and functional loci in a realistic ecological setting. In line with theoretical expectations, we find that fishing induces evolution towards slow growth, early maturation at small size and higher reproductive investment. We show, first, that the choice of genetic model (based on either quantitative genetics or gametic inheritance) influences the evolutionary recovery of traits after fishing ceases. Second, we analyse the influence of three factors possibly involved in the lack of evolutionary recovery: the strength of selection, the effect of genetic drift and the loss of adaptive potential. We find that evolutionary recovery is hampered by an association of weak selection differentials with reduced additive genetic variances. Third, the contribution of fisheries-induced selection to the erosion of functional genetic diversity clearly dominates that of genetic drift only for the traits related to maturation. Together, our results highlight the importance of taking into account population genetic variability in predictions of eco-evolutionary dynamics.

Introduction

Anthropogenic activities do not only affect population dynamics in the wild, but can also have evolutionary consequences. Harvesting, habitat fragmentation, pollution and many other pressures affect the genetic composition of populations through selection for certain trait values or through altered rates of genetic drift (Allendorf et al. 2008; Hendry et al. 2011). Selection and genetic drift interact via their effects on standing genetic variation, and together determine evolutionary dynamics.

Harvesting may increase genetic drift, because it reduces population size and alters population structure in age, size and maturity status; it may also modify sex ratio, as in trophy hunting in which males are selectively targeted (e.g. in ungulates; Martínez et al. 2002). All these effects may reduce the effective population size N_e (Wright 1938), which, in the absence of sources of new alleles (migration

or mutation), may deplete genetic variability, a process known as genetic erosion. Unlike most genes coding for quantitative traits, which are still in the process of being identified by modern genomic methods, neutral molecular markers have been increasingly used over the past decades to aid the conservation of natural populations (e.g. Palsbøll et al. 2007). Regarding fish stocks, they have allowed investigating the historical influences of past fishing pressures on neutral genetic diversity, using DNA extracted from archived otoliths or scales. Some of these studies, mostly based on microsatellites and/or mitochondrial DNA, found a loss of neutral genetic diversity (Hauser et al. 2002; Hutchinson et al. 2003; Hoarau et al. 2005), while other did not (Ruzzante et al. 2001; Therkildsen et al. 2010). This disparity could originate in the low numbers of individuals and/or loci sampled, resulting in inaccurate N_e estimates due to sampling error in measuring population allele frequencies (Waples 1998). In marine species, bias can also be caused by high gene flow (Wang and Whitlock 2003). However, a recent cross-species analysis comparing neutral genetic diversity in harvested and nonharvest fish stocks has found it to be lower in the former case than in the latter (Pinsky and Palumbi 2014).

Also functional genetic diversity is theoretically susceptible to erosion through genetic drift. Without selection, and when N_e becomes very small (a few tens), additive genetic variance is, along with neutral genetic variance, expected to increase logistically with effective population size, because of weaker drift and an increased number of mutations (Willi et al. 2006). However, a loss of additive genetic variation due to genetic drift in harvested fish stocks has not been empirically documented, and is not expected according to current knowledge on effective population sizes in marine fish, which are several orders higher than values required for this to happen.

In contrast, harvest-induced genetic changes through the adaptive evolution of life-history traits have been reported in a number terrestrial species (e.g. Jachmann et al. 1995; Coltman et al. 2003), as well as for harvested fish (see reviews by Jørgensen et al. 2007; Kuparinen and Merilä 2007; Hutchings and Fraser 2008). According to life-history theory, high levels of fishing mortality and size-selectivity favour individuals with slow growth, early maturation at small size and high reproductive investment – predictions that have been empirically corroborated (reviewed in Jørgensen et al. 2007). These phenotypic changes can undermine a stock's renewal capacity if the reduction in fecundity due to smaller adult body size is higher than the gain in fecundity due to higher reproductive investment and may become maladaptive once fishing pressure is released. Besides affecting average phenotypes, fisheries-induced selection may also affect genetic diversity in quantitative traits. After a directional episode, selection can turn stabilizing and, if not counterbalanced by gene flow or high mutation rates, may reduce additive genetic variance. The latter determines not only the rate at which characters respond to selection, but also the scope of this response in the short term.

Eco-genetic models have been used to explore the eco-evolutionary consequences of harvesting on fish populations (e.g. Baskett et al. 2005; Dunlop et al. 2007, 2009a,b; Thériault et al. 2008; Enberg et al. 2009). Previous studies found, in particular, that fisheries-induced genetic adaptations were reversed once fishing was stopped but that the evolutionary reversal rate was much slower than the fisheries-induced evolutionary rate, because natural selection pressures were weaker than fisheries-induced ones. However, the influence of fishing on the amount of additive genetic variance and its consequences for the reversal of fisheries-induced adaptations have not yet been investigated. This is because previous studies investigated trait

evolution using the infinitesimal quantitative genetic model that describes traits as affected by an infinite number of loci, which precludes any significant loss of additive genetic variation (but see Wang and Höök 2009; Kuparinen and Hutchings 2012). In addition, the response to selection also depends on effective population size, because, in small populations, genetic drift is stronger and can counteract selection.

Here, we develop an individual-based eco-genetic model (Dunlop et al. 2009b) with gametic inheritance of traits coded by finite numbers of loci and alleles per locus, so that some alleles can be lost due to selection and/or drift. We address the evolutionary dynamics at neutral loci, only affected by genetic drift, and at functional loci coding for several life-history traits, on which both genetic drift and selection act. Resulting population genetic and demographic properties are examined. We specifically tackle three questions: (i) Are the fisheries-induced evolutionary dynamics of life-history traits and their reversal during a subsequent moratorium different when traits are explicitly described by finite numbers of loci and alleles?; (ii) What is the relative importance of reduced selection strength, reduced additive genetic variance, and reduced effective population size on the timescale of the evolutionary recovery of life-history traits?; and (iii) What are the relative contributions of selection and genetic drift on fisheries-induced changes in additive genetic variance?

Model description

We model individuals that are diploid hermaphrodites (i.e. we do not distinguish between males and females, albeit we base our model on female life history). Their genotypes comprise a set of neutral loci, to study neutral evolution, and a set of functional loci, to investigate the combination of neutral and adaptive evolution of life-history traits. Functional loci code for the life-history traits of individuals and thus, together with plastic responses of traits to environmental variation, affect the life-history processes of growth, maturation, reproduction and mortality, which jointly determine population dynamics. Ultimately, individual fitness, controlling the production of offspring to which genetic material is transmitted, is determined by an individual's functional loci and environment. The latter is altered by fishing, the emerging population dynamics through density dependence, and environmental stochasticity affecting resources are available for growth and recruitment.

Below we describe in turn the genetics of neutral markers and life-history traits, phenotypic expression, the associated individual-level life-history processes from which population dynamics emerge, and the measures of population genetic diversity used to monitor evolutionary dynamics (see Table 1 for variable definitions and Table 2 for parameter values).

Individual genotypes and phenotypes

Genetic inheritance

For both neutral and functional loci, genetic inheritance is modelled according to Mendelian laws under sexual reproduction: haploid gametes are formed for mature individuals by randomly drawing one of the two alleles at each locus, representing allelic segregation during meiosis. This is carried out independently between loci, so alleles can recombine freely, that is linkage is not included. Reproduction occurs between pairs of mature individuals (see section 'Reproduction' for details) and

the fusion of two randomly picked gametes creates a diploid offspring.

Neutral markers

To assess genetic drift, each individual carries 30 neutral loci. Genetic diversity at each neutral locus is represented by 10 possible allelic states distributed in the population, which mimics allelic diversity at microsatellite markers (Poulsen et al. 2006, for instance, reported a mean allelic diversity across loci and populations of 9.4 for cod).

Functional loci and genotypic values of life-history traits

Individuals have five evolving life-history traits, namely the juvenile growth rate g; two traits that specify the maturation schedule, that is the slope s and the intercept y of a

Table 1. Model variables.

	Variable	Symbol	Unit	Equations
Genotypic values	Alleles ($j \in \{1, 2\}$) at locus k coding for trait x	$z_{x,k,j}(i)$	–	1a
	Allelic value of allele $z_{x,k,j}(i)$	$A_{x,k,j}(i)$	See traits	1a,b
	Genotypic value of trait x	$A_x(i)$	See traits	1b,2a,b
Phenotypic traits	Growth coefficient	$g(i,t)$	cm year^{-1}	2a,b, 3b,c, 6b
	Growth investment at maturation onset	$\alpha(i)$	–	2a, 3a
	Annual ratio of decay in postmaturation growth investment	$\chi(i)$	–	2a, 3a
	PMRN intercept	$y(i)$	cm	2a
	PMRN slope	$s(i)$	cm year^{-1}	2a
Emerging traits and individual state	Age	$a(i,t)$	Year	3a
	Age at maturation	$a_m(i)$	Year	3a
	Fraction of productive season allocated to growth	$p(i,t)$	–	3a,b,c
	Somatic length	$\ell(i,t)$	cm	3b,c, 4a, 6a, 7a
	Somatic weight	$w(i,t)$	g	
	Gonadic weight	$G(i,t)$	g	3c
	GSI	$\Gamma(i,t)$	–	
	Fecundity	$Q(i,t)$	–	5
	Maturation probability	$m(i,t)$	–	4a
	Length at 50% maturation probability	$\ell_{p50}(i,t)$	cm	4a
Population	Population biomass	$B(t)$	g	2b
	Number of recruits	$N_0(t)$	–	5
	Mean offspring number per year at age a	$b_a(t)$	–	
Mortality	Instantaneous size-dependent predation mortality rate	$d_s(i,t)$	–	6a, 8
	Instantaneous growth-dependent mortality rate	$d_g(i,t)$	–	6b, 8
	Size-selectivity function of fishery	$f_{trawl}(i,t)$	–	7a,b
	Instantaneous harvest mortality rate	$d_F(i,t)$	–	7b,8
	Total instantaneous mortality rate	$Z(i,t)$		8
	Death probability	$D(i,t)$		
	Survival probability until age a	λ_a	–	
Population genetic diversity	Frequency of allele l at locus k	$\Pi_{k,l}(t)$	–	9a,b, 11a,b
	Standardized variance in allele frequency change	F	–	9a,b
	Effective population size	N_e	–	9b
	Generation time	T	Year	9b
	Additive genetic variance of trait x	$V_A(x)$	trait unit^2	10
	Phenotypic variance of trait x	$V_P(x)$	trait unit^2	10
	Heritability of trait x	$h^2(x)$	–	10
	Expected heterozygosity	$H_e(t)$	–	11a,b

–, dimensionless variable.

Table 2. Model parameters.

	Parameter	Symbol	Value	Unit	Equations	Source
Genome structure	Number of neutral loci	K_n	30	–	9a, 11a	
	Number of alleles per neutral locus	L_n	10	–	9a, 11a	(1)
	Number of functional loci	K_f	8	–	1a, 11b	
	Number of alleles per functional locus	L_f	10	–	1a, 11b	(2)
Initial ranges of genotypic values	Growth coefficient	$[A_{g,min}, A_{g,max}]$	[6.0, 22.0]	cm year^{-1}	1a	(3)
	Growth investment at maturation onset	$[A_{\alpha,min}, A_{\alpha,max}]$	[0.4, 1.0]	–	1a	(3)
	Rate of decay in postmaturation growth investment	$[A_{\chi,min}, A_{\chi,max}]$	[0.1, 0.5]	–	1a	(3)
	PMRN intercept	$[A_{y,min}, A_{y,max}]$	[40.0, 90.0]	cm	1a	(3)
	PMRN slope	$[A_{s,min}, A_{s,max}]$	[−1.0, 1.0]	cm year^{-1}	1a	(3)
Expression noise	Noise coefficient of growth coefficient	$\varepsilon_g(i)$	1 (1.19)*	–	2a,b	(4)
	Noise coefficient of growth investment at maturation onset	$\varepsilon_\alpha(i)$	1 (0.05)*	–	2a	(4)
	Noise coefficient of annual ratio of decay in postmaturation growth investment	$\varepsilon_\chi(i)$	1 (0.05)*	–	2a	(4)
	Noise coefficient of PMRN slope	$\varepsilon_y(i)$	1 (0.24)*	–	2a	(4)
	Noise coefficient of PMRN intercept	$\varepsilon_s(i)$	1 (6.7)*	–	2a	(4)
Growth	Strength of density dependence in growth	ρ	3×10^{-9}	g^{-1}	2b	(3)
	Production exponent	β	2/3	–		(3)
	Constant in allometric weight–length relationship	Ω	0.01	g cm^{-3}	3c	(5)
	Initial length	ℓ_0	10	cm		(3)
Maturation	PMRN envelope width	ω	20	cm	4b	(3)
	Lower bound of PMRN envelope	p_{low}	0.25	–	4b	(6)
	Upper bound of PMRN envelope	p_{up}	0.75	–	4b	(6)
Reproduction	Ratio of somatic to gonadic wet-weight energy densities	γ	0.62	–	3c	(3)
	Weight of an egg	w_{egg}	4×10^{-4}	g		(7)
	Maximum survival probability of recruits	η_1	22×10^{-7}	–	5	(7)
	Strength of density dependence in recruitment	η_2	23×10^{-12}	–	5	(7)
	Noise coefficient of recruitment	$\varepsilon_R(t)$	1 (0.1)*	–	5	(7)
Natural mortality	Size-independent instantaneous natural mortality rate	d_0	0.2	year^{-1}	8	(3)
	Maximum instantaneous predation mortality rate	c_s	0.6	year^{-1}	6a	(3)
	Scaling factor of predation mortality rate	ℓ_s	14	cm	6a	(3)
	Minimum instantaneous growth-dependent mortality rate	c_g	0.02	year^{-1}	6b	(3)
	Scaling factor of growth-dependent mortality rate	g_0	6	cm year^{-1}	6b	(3)
Fishing mortality	Steepness of the fishery's size-selectivity curve	θ	0.2	cm^{-1}	7a	(7)
	Length at 50% selectivity	ℓ_{50}	60	cm	7a	(8)
	Maximum instantaneous harvest rate	H	[0.2, 1.0]	year^{-1}	7b	

–, dimensionless parameters.
(1) Poulsen et al. (2006); (2) by analogy with (1); (3) values chosen such that the life-history characteristics resemble those of North Sea cod (e.g. Marty et al. 2014); (4) standard deviation for each trait is determined such that the total expressed variance σ_E^2 is related to: (i) an assumed initial additive genetic variance σ_A^2 determined by an assumed initial genetic coefficient of variation CV_g of 6% and the initial mean trait values, and (ii) an assumed initial heritability h^2 of 0.2, as $\sigma_E^2 = \sigma_A^2 (h^{-2} - 1)$; (5) values obtained from http://www.fishbase.org; (6) definition of PMRN width based on quartiles; (7) values taken from Enberg et al. (2009) and slightly modified when necessary; (8) between EU minimum landing size (35 cm) and asymptotic body size.
*Mean (standard deviation).

linear probabilistic maturation reaction norm (PMRN; Heino et al. 2002a); and two parameters related to energy allocation between growth and reproduction after maturation, that is the proportion α of energy devoted to somatic growth in the first adult year and the annual ratio χ at which this proportion decreases throughout adult life (Table 1; see section 'Life-History Processes' for more details).

For each individual i, the genotypic value of each trait results from K_f diploid functional loci. The two alleles at each locus can take L_f different possible allelic states. For a given trait, functional alleles act additively at and between loci. Dominance between alleles and epistasis between loci is modelled as a stochastic process through an expression noise (see next section). We denote by $z_{x,k,1}(i)$ and $z_{x,k,2}(i)$ the two alleles at a given locus $k \in \{1, 2, \ldots, K_f\}$ coding for trait $x \in \{g, \alpha, \chi, y, s\}$ and define them as two integers lying between 1 and L_f. For simplicity, loci coding for the same trait has the same number of possible alleles. To translate these integers into two allelic values $A_{x,k,1}(i)$ and $A_{x,k,2}(i)$, we assume an initial minimum and maximum genotypic value, $A_{x,\min}$ and $A_{x,\max}$, respectively, for each trait x in the population (Table 2). Allelic values are then defined as

$$A_{x,k,j}(i) = \frac{A_{x,\min}}{2K_f} + (z_{x,k,j}(i) - 1)\frac{A_{x,\max} - A_{x,\min}}{2K_f L_f}, \quad (1a)$$

for $j \in \{1, 2\}$. The genotypic value $A_x(i)$ of trait x is then given by the sum of allelic values across loci

$$A_x(i) = \sum_{k=1}^{K_f} (A_{x,k,1}(i) + A_{x,k,2}(i)). \quad (1b)$$

Phenotypic expression of life-history traits

At birth, an individual's expressed traits (g, α, χ, y, and s) deviate from their genotypic values, reflecting micro-environmental variation and nonadditive genetic effects (dominance and epistasis). This variation is described by an expression noise ε_x with $x \in \{g, \alpha, \chi, y, s\}$ that acts multiplicatively on the genotypic value. Any trait value x for individual i is then obtained from the following generic equation,

$$x(i) = \varepsilon_x(i)A_x(i), \quad (2a)$$

where $\varepsilon_x(i)$ is randomly drawn, once per lifetime, from a normal distribution with mean 1 and a standard deviation specific to each trait x (Table 2).

Besides expression noise, the juvenile growth rate $g(i,t)$ is also affected by population-level density dependence,

$$g(i, t) = \frac{g(i)}{1 + \rho B(t)} = \frac{\varepsilon_g(i)A_g(i)}{1 + \rho B(t)}, \quad (2b)$$

where $B(t)$ is the population biomass in year t and $1/\rho$ is the total population biomass at which density dependence halves the juvenile growth rate.

Life-history processes

An individual i in year t is characterized by its five life-history traits ($g(i,t)$, $\alpha(i)$, $\chi(i)$, $y(i)$ and $s(i)$); its age $a(i,t)$, its length $\ell(i,t)$ and its age at maturation $a_m(i)$. Together, these determine the four annual life-history processes: somatic and gonadic growth, maturation, reproduction and mortality.

Energy allocation to somatic and gonadic growth

Energy allocation between growth and reproduction is described following Quince et al.'s (2008) biphasic seasonal growth model. An individual's net energy acquisition rate,

that is the energy surplus after accounting for maintenance, is assumed to scale with its somatic weight $w(i,t)$ as $w^\beta(i,t)$, where β denotes the production exponent. Juveniles allocate all energy available to somatic growth, whereas adults start by allocating all energy to somatic growth and switch to allocating all energy to gonadic growth after a fraction $p(i,t)$ of the productive season. $p(i,t)$ is given by

$$p(i, t) = \begin{cases} 1 & \text{for } a(i,t) < a_m(i) \\ \alpha(i)\chi(i)^{a(i,t)-(a_m(i)+1)} & \text{for } a(i,t) \geq a_m(i) \end{cases}, \quad (3a)$$

where $\alpha(i)$ and $\chi(i)$ lie between 0 and 1 (Quince et al. 2008). A newly matured individual allocates a proportion $\alpha(i)$ of the productive season to somatic growth; this proportion then decreases geometrically with an annual ratio $\chi(i)$ as the individual ages. This is consistent with von Bertalanffy adult growth.

Assuming a production exponent β of 2/3 (Kozlowski and Wiegert 1986; Kozłowski and Wiegert 1987; Lester et al. 2004) and that somatic weight scales with body length according to $w = \Omega \ell^3$, an individual's length-at-age trajectory is given by

$$\ell(i, t) = \ell(i, t-1) + p(i, t)g(i, t). \quad (3b)$$

Before maturation, the gonad weight $G(i,t)$ equals 0; after maturation, it is a function of body length at the end of each productive season,

$$G(i, t) = 3\gamma\Omega \ell^2(i, t)(1 - p(i, t))g(i, t), \quad (3c)$$

where γ is the ratio of the wet-weight energy density of somatic tissue to gonad tissue. The gonado-somatic index (GSI) is defined as the ratio between gonadic weight and somatic weight, $\Gamma(i,t) = G(i,t)/w(i,t)$.

Maturation

We model sexual maturation using probabilistic maturation reaction norms (PMRNs; Heino et al. 2002b). PMRNs describe the probability that an immature individual matures in dependence on its age and size. We assume a linear PMRN characterized by its intercept y and slope s. The maturation probability is then given by the logistic equation

$$m(i, t) = \frac{1}{1 + \exp(-(\ell(i, t) - \ell_{p50}(i, t))/\delta)}, \quad (4a)$$

where $\ell_{p50}(i,t) = y(i) + s(i)a(i,t)$ is individual i's length at 50% maturation probability. The parameter δ is determined by the PMRN envelope width ω,

$$\delta = \frac{\omega}{\text{logit}(p_{up}) - \text{logit}(p_{low})}, \quad (4b)$$

where $\text{logit}(p) = \ln(p/(1-p))$, and p_{up} and p_{low} are the upper and lower probability bounds, respectively, for which the PMRN envelope width is defined.

Maturation is modelled as a stochastic process of Bernoulli trials, taking place if a number randomly drawn from a uniform distribution between 0 and 1 is smaller than $m(i,t)$.

Reproduction

An individual's fecundity is defined by dividing its gonad weight by an egg weight, $Q(i,t) = G(i,t)/w_{egg}$. From this, the total number of newborns in each year, N_0, is determined by a Beverton–Holt recruitment function (Beverton and Holt 1994),

$$N_0(t) = \frac{\eta_1 \sum_i Q(i,t)}{1 + \eta_2 \sum_i Q(i,t)} e^{-\varepsilon_R(t)}, \qquad (5)$$

which depends on the population fecundity $\sum_i Q(i,t)$ and a lognormally distributed population-level interannual noise factor $e^{-\varepsilon_R(t)}$, where $\varepsilon_R(t)$ is normally distributed (Table 2), which represents the influence of environmental fluctuations (e.g. temperature or larval food supply) on recruitment. The parameter η_1 specifies the survival probability of recruits when population fecundity is low and $1/\eta_2$ is the population fecundity at which density dependence halves recruitment, resulting in a maximum asymptotic number of η_1/η_2 recruits.

So far, we have only defined the number of newborns in each year. Each newborn is assigned two parents, selected in proportion to their gonad size through a stochastic process of Bernoulli trials. The adult having the largest gonad weight is determined in each year. Then, an individual is randomly drawn (with replacement) and a number is randomly drawn from a uniform distribution between 0 and this maximal gonad weight. If this number is less than the drawn individual's gonad weight, this individual will be selected as a parent. Hence, the higher an individual's fecundity, the more likely it is to become a parent. The random draws continue until enough parents have been selected. Individuals selected as parents are then randomly combined into pairs for each recruit. Therefore, on average, each individual i has a number of offspring in year t that is a fraction of the population's recruitment $N_0(t)$, equalling the proportion $Q(i,t)/\sum_j Q(j,t)$ of population fecundity contributed by that individual. After two parents are selected for a newborn, this offspring receives, for each diploid locus, one allele from each parent, carried by two randomly selected parental gametes.

Natural and fishing mortality

Three different sources of natural mortality are considered. First, a constant mortality d_0 originates from diseases, senescence, or any stochastic source of mortality unrelated to body size. Second, size-dependent predation generates mortality decreasing with body size according to

$$d_s(i,t) = c_s \exp(-\ell(i,t)/\ell_s), \qquad (6a)$$

where c_s is the maximum instantaneous predation mortality rate and ℓ_s is the size at which the predation mortality is decreased by the factor $1/e$. Third, a growth-dependent mortality,

$$d_g(i,t) = c_g \exp(g(i,t)/g_0), \qquad (6b)$$

with c_g denoting the minimum instantaneous growth-dependent mortality rate and g_0 the growth rate at which the growth-dependent mortality is increased by the factor e, accounts for the trade-off between growth and survival. Such a trade-off can originate physiologically when faster growth is achieved at the expense of less energy being available for maintenance (Ernande et al. 2003) or ecologically when the higher energy intake required for faster growth is achieved through more active foraging and thus at the expense of a higher exposure to predation (Abrams 1991; Werner and Anholt 1993; Walters and Korman 1999).

Individuals may also experience size-dependent fishing mortality. The fishery considered is composed of trawlers, characterized by a sigmoid size-selectivity curve,

$$f_{trawl}(i,t) = \frac{1}{1 + \exp(-\theta(\ell(i,t) - \ell_{50}))}, \qquad (7a)$$

where θ specifies the steepness of the sigmoid and ℓ_{50} is the body size at which the harvest rate equals 50% of its maximum. The instantaneous fishing mortality rate depends on this selectivity curve and on the maximum instantaneous harvest rate H,

$$d_F(i,t) = f_{trawl}(i,t)H. \qquad (7b)$$

The total instantaneous mortality rate is then given by

$$Z(i,t) = d_0 + d_s(i,t) + d_g(i,t) + d_F(i,t), \qquad (8)$$

so the probability that individual i dies during $T = (t + 1 \text{ year}) - t = 1$ year equals $D(i,t) = 1 - \exp(-Z(i,t)T)$. As for maturation and reproduction, individual mortality is modelled as a stochastic process of Bernoulli trials and occurs if a number randomly drawn from a uniform distribution between 0 and 1 is lower than $D(i,t)$.

Measures of population genetic diversity

We follow the effect of fishing on neutral and functional genetic diversity using three genetic diversity indices.

Effective population size

We borrow empirical methods of population genetics and estimate effective population size N_e using the temporal method (Waples 1989; Waples and Yokota 2007).

Considering the standardized measure of variance in allele frequency change F between two samples at different time points,

$$F = \frac{1}{L_{\mathrm{n}}} \sum_{l=1}^{L_{\mathrm{n}}} \frac{1}{K_{\mathrm{n}}} \sum_{k=1}^{K_{\mathrm{n}}} \frac{(\Pi_{k,l}(t_1) - \Pi_{k,l}(t_2))^2}{(\Pi_{k,l}(t_1) + \Pi_{k,l}(t_2))/2 - \Pi_{k,l}(t_1)\Pi_{k,l}(t_2)}$$

$$(9a)$$

where $\Pi_{k,l}(t)$ is the frequency of allelic state l at locus k at time t, and t_1 and t_2 are two sampling years, the effective population size is then obtained as

$$N_e = \frac{t_2 - t_1 + 1}{2T(F - 1/(2N_1) - 1/(2N_2) + 1/N_1)}$$
$$= \frac{t_2 - t_1 + 1}{2T(F + 1/(2N_1) - 1/(2N_2))},$$

$$(9b)$$

where T is the generation time in years, and N_1 and N_2 are the population sizes at times t_1 and t_2, respectively (Nei and Tajima 1981; Waples 1989). Denoting by λ_a the fraction of a cohort surviving to age a and by b_a the mean number of offspring produced in one time interval by individuals aged a, T is given by $T = \sum_a a\lambda_a b_a$. We calculate F using neutral markers of the whole population every 20 years and then estimate effective population size N_e for each time interval. We chose an estimation interval of 20 years, because the bias in N_e estimates due to overlapping generations decreases with the number of generations in the estimation interval (Waples and Yokota 2007).

Additive genetic variance and heritability of quantitative traits

Knowing the genotypic value $A_x(i)$ of any trait $x(i)$ for each individual i, the population's mean genotypic value $\bar{A}_x$ and additive genetic variance $V_A(x)$ are directly calculated from the population composition in each year, and the heritability $h^2(x)$ is obtained as

$$h^2(x) = V_A(x)/V_P(x) \qquad (10)$$

(see, e.g. Lynch and Walsh 1998).

Expected heterozygosity

To assess the relative contributions of genetic drift and selection imposed by fishing on the evolution of life-history traits, we estimate the expected heterozygosity over neutral loci on the one hand (eqn 11a) and over functional loci coding for each life-history trait on the other hand (eqn 11b) (Nei and Roychoudhury 1974),

$$H_e(t) = \frac{1}{K_{\mathrm{n}}} \sum_{k=1}^{K_{\mathrm{n}}} \left(1 - \sum_{l=1}^{L_{\mathrm{n}}} \Pi_{k,l}^2(t) \right) \qquad (11a)$$

$$H_e(t) = \frac{1}{K_{\mathrm{f}}} \sum_{k=1}^{K_{\mathrm{f}}} \left(1 - \sum_{l=1}^{L_{\mathrm{f}}} \Pi_{k,l}^2(t) \right) \qquad (11b)$$

where $\Pi_{k,l}(t)$ is the frequency of allelic state l at locus k at time t.

Expected heterozygosity will decrease as variability in allele frequencies, and thus genetic diversity, decreases owing to genetic drift and/or selection. Both evolutionary forces affect functional diversity, while neutral heterozygosity depends only on genetic drift, and thus constitutes a baseline allowing detecting the effects of selection whenever functional heterozygosity differs from this baseline.

Selection differentials

To assess the strength of selection pressures, selection differentials $S(x)$ are estimated in each year by inverting the breeder's equation $R(x) = h^2(x)S(x)$, where heritability $h^2(x)$ is estimated as described above and the selection response $R(x)$ is computed as the difference in mean trait value between the new cohort and the population that produced it. Mean-standardized selection differentials are also calculated, by multiplying selection differentials with the trait's mean and dividing by its variance (Matsumura et al. 2012).

Model parameterization, initial values and runs

We model a population resembling North Sea cod (*Gadus morhua*) with parameters taken from the literature or, when not available, fixed to values yielding plausible emergent properties and patterns; for more details, see Table 2.

The initial population is comprised of 220 000 juveniles. For each individual, alleles at each neutral and functional locus are randomly drawn from a uniform distribution between 1 and 10 (i.e. the number of allelic states per neutral and functional locus). Juveniles are initially given a random age between 2 and 6 years old, to avoid the high mortality at younger ages (which, without reproduction, would steeply reduce population size). For each juvenile i, initial body size is determined by its genotypic juvenile growth rate, i.e. $g(i,0) = A_g(i)$, and its randomly attributed age $a(i,0)$, while neglecting density dependence and expression noise, so that $\ell(i,0) = \ell_0 + A_g(i)a(i,0)$.

We first let the population reach a demographic and evolutionary equilibrium during 17 000 years without fishing. We display all results from this time onward. Model runs start without fishing for 100 extra years. Harvesting then begins and lasts another 100 years. The maximum instantaneous fishing rate H takes values from 0.2 to 1 (Table 2). Finally, we stop harvesting and explore the genetic trait dynamics after fishing, for 200 more years. All results pre-

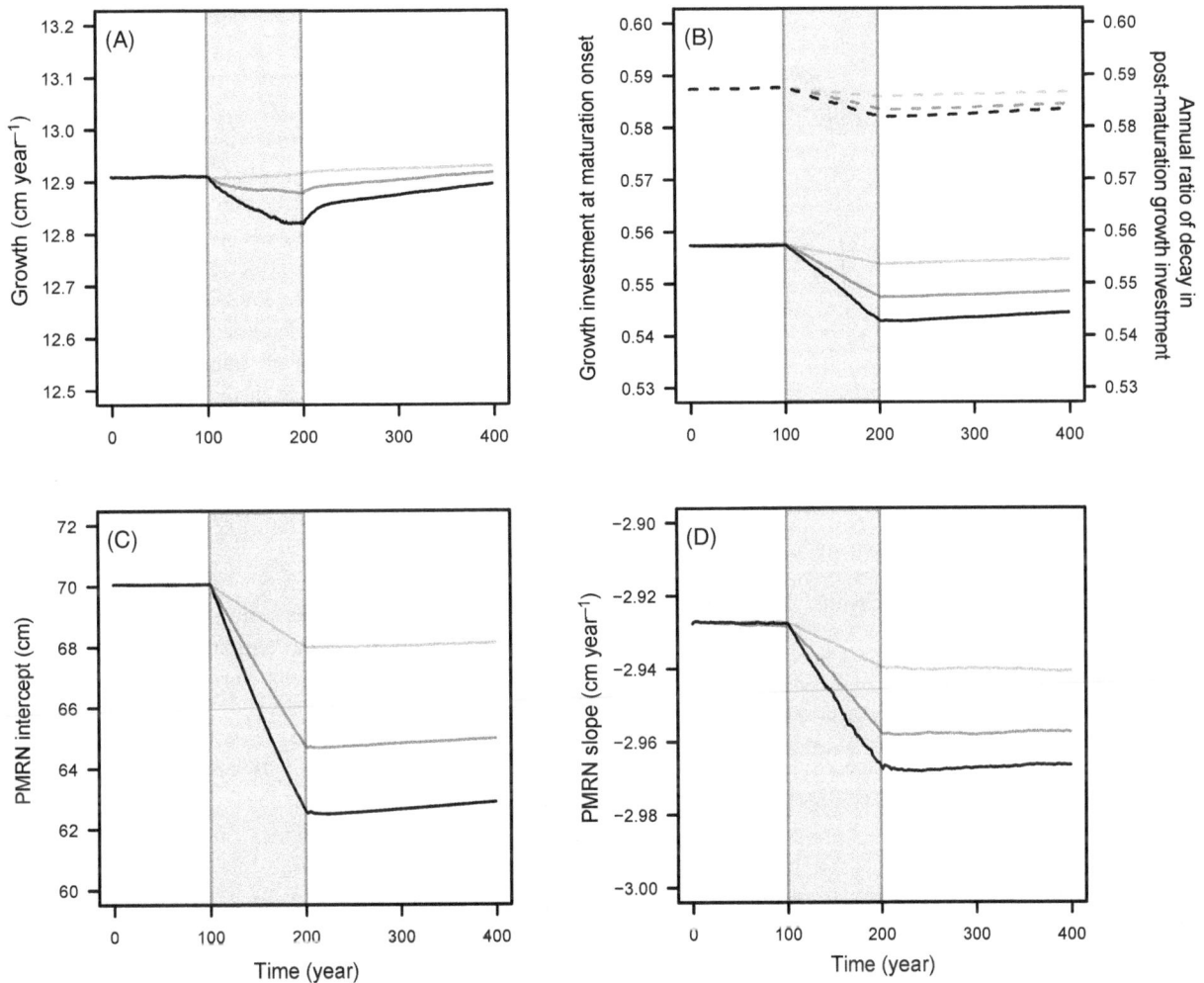

Figure 1 Dynamics of the mean genotypic values of life-history traits before, during and after harvesting. Harvesting (grey shading) starts at $t = 100$ year and stops at $t = 200$ year. Dynamics are shown for three different maximum instantaneous harvest rates: $H = 0.2$ year^{-1} (light grey curves), $H = 0.6$ year^{-1} (dark grey curves) and $H = 1$ year^{-1} (black curves). (A) Juvenile growth rate g. (B) Energy allocation to growth after maturation: growth investment α at maturation onset (continuous curve) and annual ratio χ of decay in postmaturation growth investment (dashed curve). (C) PMRN intercept y. (D) PMRN slope s.

sented are averages of 25 replicate model runs, carried out with different random seeds.

Results

Effects of fishing on life-history trait means
As harvesting occurs, the mean genotypic values of all five life-history traits decrease – the higher the harvest intensity, the stronger these effects (Fig. 1A–D). When fishing stops, the mean genotypic value of juvenile growth rate almost recovers, for all considered fishing intensities, within the next 200 years, while all other genetic values remain low, despite some very shallow upward trends mostly noticeable for the highest fishing intensity. The phenotypic dynamics of emergent life-history traits (age and size at maturation,

length-at-age, and GSI-at-age) follow from the dynamics of the five genetically coded traits in combination with the demographic effects of fishing (Figs S1 and S2).

Irreversibility of life-history trait evolution after fishing: possible causes
Genetic drift due to demographic effects
Fishing modifies effective population size and thus changes the rate of genetic drift affecting allelic frequencies (Fig. 2). As harvesting starts, effective population size first increases and then steeply drops, both occurring with larger amplitudes when fishing intensity is high. After fishing is stopped, effective population size recovers to levels equal or slightly higher than the initial ones, but with a time lag that

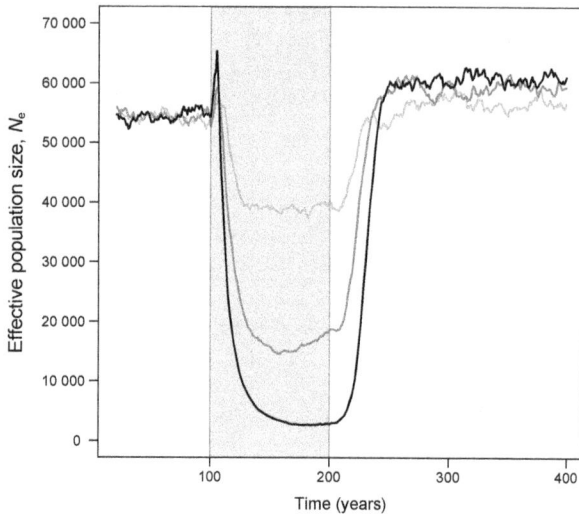

Figure 2 Dynamics of effective population size N_e, before, during and after harvesting. Harvesting (grey shading) starts at $t = 100$ year and stops at $t = 200$ year. Dynamics are shown for three different maximum instantaneous harvest rates: $H = 0.2$ year^{-1} (light grey curves), $H = 0.6$ year^{-1} (dark grey curves) and $H = 1$ year^{-1} (black curves). The effective population size N_e is shown as a function of time measured in years, and not in generations, because the population's generation time changes throughout the modelled period, equalling, on average, 12.1 year before fishing, 9.9, 8.0 and 7.5 year during fishing, and 11.9, 11.6 and 11.4 year after fishing (at $H = 0.2$ year^{-1}, $H = 0.6$ year^{-1}, and $H = 1$ year^{-1}, respectively).

increases when fishing intensity is stronger (around 30 years at $H = 0.2$ year^{-1}, 40 years at $H = 0.6$ year^{-1} and 50 years at $H = 1$ year^{-1}). Effective population sizes after recovery are larger when fishing intensities are higher (Fig. 2), mirroring trends in population size (Figs S2 and S3). This is because genetic adaptations during fishing drive individuals towards higher fecundity and earlier maturation, which raises the number and lifetime fecundity of mature individuals, and thus increases recruitment and population size.

Selection differentials

Before fishing, the population is at an evolutionary equilibrium and no selection differential is statistically different from 0 at the 1% risk level (α: $t(98) = -0.32$, $P = 0.74$; χ: $t(98) = 0.83$, $P = 0.41$; y: $t(98) = -2.21$, $P = 0.03$; s: $t(98) = -1.5$, $P = 0.14$), except for growth selection differentials (g: $t(98) = 98.1$, $P < 2.2 \times 10^{-16}$), which are positive with a mean of 0.1 (Fig. 3). The latter, however, do not reflect selection, but just the trade-off between growth and survival: because of this trade-off, slow-growing phenotypes are overrepresented in older age classes (similar to 'Lee's phenomenon'; Lee 1912), so that the difference in mean growth between a new cohort and the whole population is positive.

During harvesting, selection differentials of all traits decrease, with the differences to preharvest values being statistically significant at the 1% risk level for all traits (g: $t(99) = -3.46$, $P < 0.001$; α: $t(99) = -13.41$, $P < 2.2 \times 10^{-16}$; χ: $t(99) = -5.3$, $P < 10^{-6}$; y: $t(99) = -38.9$, $P < 2.2 \times 10^{-16}$; s: $t(99) = -8.1$, $P < 10^{-11}$). The initial decrease is followed by an increase for the PRMN intercept and slope (red lines in Fig. 3D,E), and by an initial increase and a subsequent drop for the other traits (red lines in Fig. 3A–C).

After fishing, selection differentials increase and slightly exceed preharvest values for all traits (Fig. 3A–D) except the PMRN slope (Fig. 3E). Even though they are statistically higher than their preharvest values at the 1% risk level for juvenile growth rate (g: $t(99) = 3.04$, $P = 0.003$; Fig. 3A) and growth investment at maturation onset (α: $t(99) = 2.6$, $P = 0.01$; Fig. 3B), selection differentials remain very low.

Mean-standardized selection differentials (Fig. 3F), which allow comparing selection strength across traits, undergo reductions of similar amplitude across all traits – except for the PMRN intercept, for which negative selection is much stronger.

Effects of fishing on genetic and phenotypic variances of life-history traits

Before harvesting, the dynamics of additive genetic and phenotypic variances of all traits are steady, indicating the absence of selection (Fig. 4A–E). Heritability equals 0.11 for growth, 0.22 for growth investment at maturation onset and its subsequent annual ratio of decay, 0.18 for the PMRN intercept, and 0.25 for the PMRN slope.

As fishing starts, the dynamics of both functional genetic and phenotypic diversity are affected. First, fluctuations in genetic and/or phenotypic variances are amplified, with larger amplitudes of change in the latter. Second, high fishing pressure induces a reduction in the genetic and phenotypic variances of several traits. Most noticeable is the PMRN intercept, with a decrease of roughly 3 cm^2 in genetic variance for the highest fishing mortality (Fig. 4D). Genetic variance decreases to a lesser degree for growth (Fig. 4A) and the PMRN slope (Fig. 4E) and slightly increases for the annual ratio of decay in postmaturation growth investment (Fig. 4C). Most variances that are reduced by fishing do not recover to previous levels after fishing, although shallow upward trends can be noticed for the PMRN intercept.

Relative contributions of genetic drift and selection to losses in functional genetic variability

At the beginning of model runs, neutral heterozygosity is slightly higher than functional heterozygosity (Fig. 5), because stabilizing selection acts on functional loci during

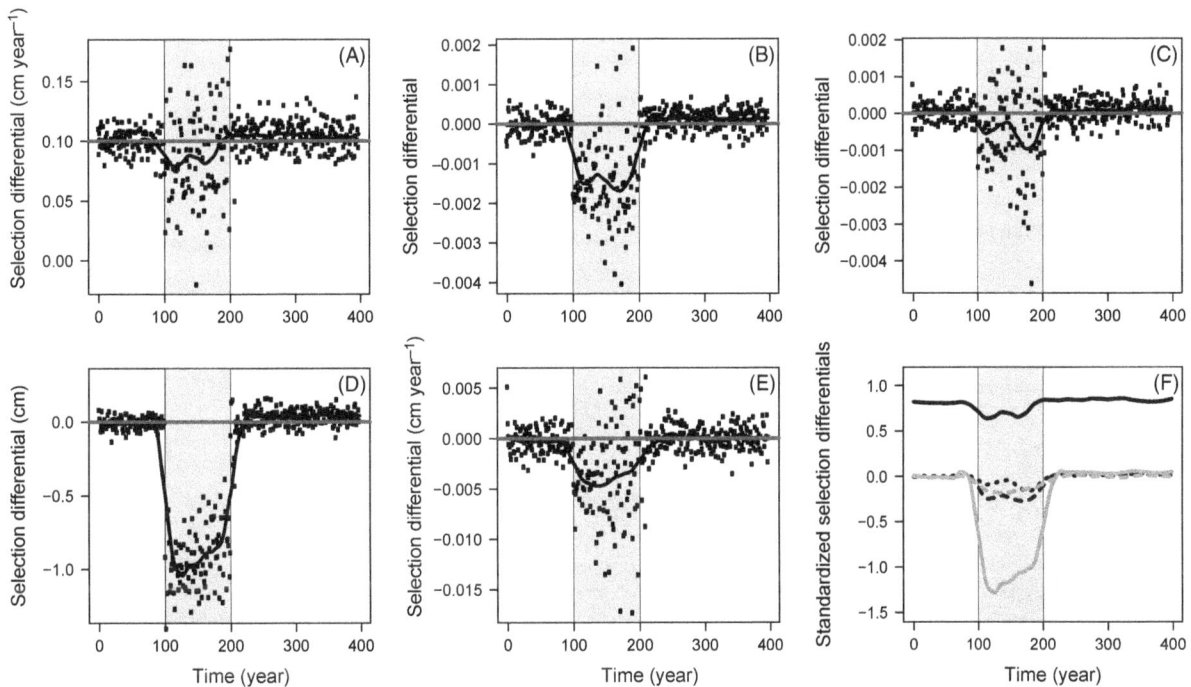

Figure 3 Selection differentials (A–E) and mean-standardized selection differentials (F) of life-history traits before, during, and after harvesting. Harvesting (grey shading) starts at $t = 100$ year and stops at $t = 200$ year. Selection differentials are shown for a maximum instantaneous harvest rate of $H = 1$ $year^{-1}$. In (A–E), the red line is a smoothing function (loess with a span of 0.2) and horizontal grey lines give the baseline of 0 selection differential for all traits except for juvenile growth rate (baseline of 0.1). (A) Juvenile growth rate g (continuous black line in F). (B) Growth investment α at maturation onset (dashed black line in F). (C) Annual ratio χ of decay in postmaturation growth investment (dotted black line in F). (D) PMRN intercept y (continuous grey line in F). (E) PMRN slope s (dashed grey line in F). (F) Mean-standardized selection differentials of the five life-history traits, shown by smoothing functions (loess with a span of 0.2).

the initialization period meant to reach evolutionary equilibrium (17 000 years), eroding part of the functional genetic diversity. As a consequence, the amplitudes of changes in neutral and functional heterozygosity are to be compared, rather than their absolute values, to distinguish the impacts of neutral and adaptive evolution on functional genetic diversity.

As expected from the results for effective population size, neutral heterozygosity is constant under moderate fishing mortality (continuous grey lines in Fig. 5), suggesting weak fisheries-induced genetic drift, whereas it diminishes under high fishing mortality (dashed grey lines in Fig. 5) due to increased genetic drift. Notice, however, that the decrease amounts to <1% of the initial value, which is expected given the range of effective populations sizes (3000–40 000, Fig. 2), and the fact that heterozygosity decreases by the ratio $1 - 1/(2N_e)$ per generation under genetic drift (Wright 1931). At high fishing intensity, a much more pronounced decrease in functional heterozygosity is observed for the PMRN intercept ($\approx 6\%$ Fig. 5D) and PMRN slope ($\approx 2\%$, Fig. 5E). In contrast, functional heterozygosity decreases at a rate similar to neutral heterozygosity for juvenile

growth (Fig. 5A), decreases at a lower rate for growth investment at maturation onset (Fig. 5B) or stays approximately constant for the annual ratio of decay in postmaturation growth investment (Fig. 5C).

However, as the rate of change in genetic diversity is expected to be influenced by its own initial value, whether evolution is neutral (Nei et al. 1975) or adaptive (Lande 1980; Lande and Arnold 1983), the differences between neutral and functional heterozygosity levels at the beginning of model runs preclude any conclusive interpretation of slight differences in changes when fishing occurs, especially when changes occur at a scale of only a few percentage or less. This holds particularly for the traits with the strongest initial differences, that is for juvenile growth rate, the annual ratio of decay in growth investment and the PMRN slope (Fig 5A,C,E). Therefore, one can firmly conclude that both genetic drift and selection are contributing to diminishing functional genetic diversity only for the PMRN intercept (Fig. 5D), with their respective contributions being unclear for growth (Fig. 5A) and the PMRN slope (Fig. 5E). For the two traits involved in growth investment after maturation (Fig. 5B,C), selection seems to counteract the effect of genetic drift, but the amount of

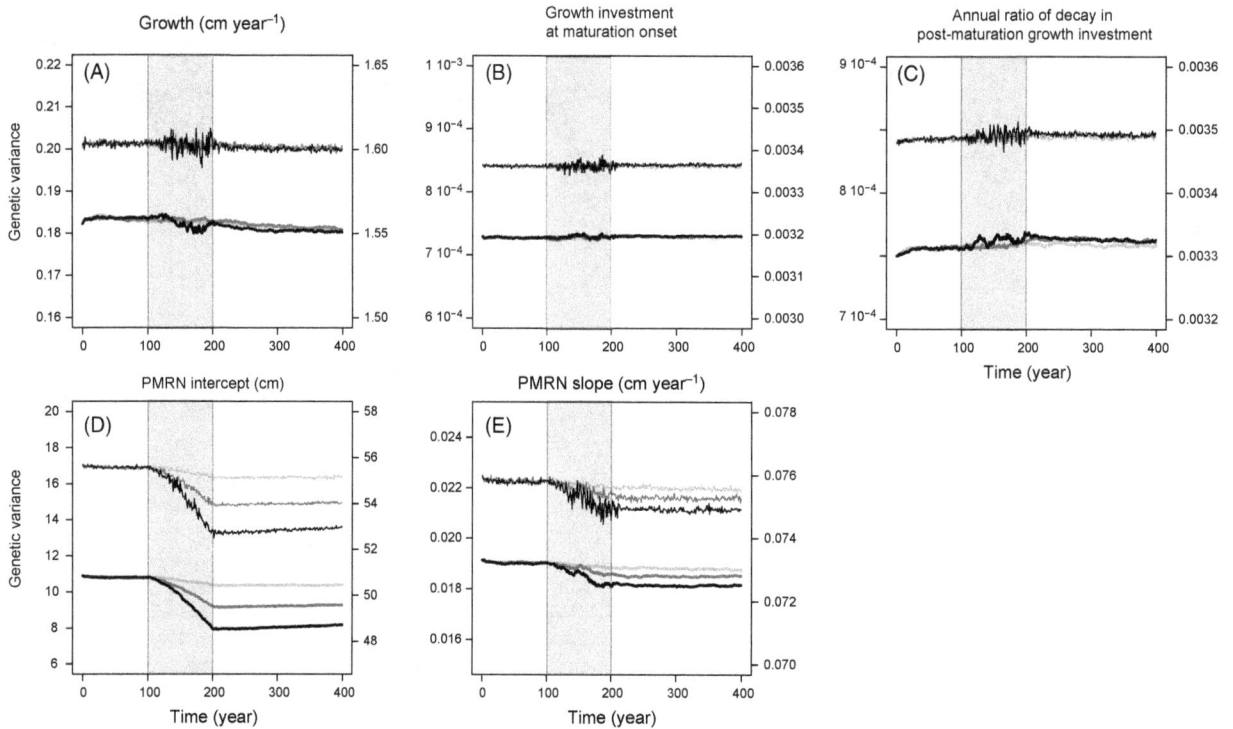

Figure 4 Evolutionary dynamics of genetic variances (thick lines, left vertical axes) and phenotypic variances (thin lines, right vertical axes) of life-history traits before, during and after harvesting. Harvesting (grey shading) starts at t = 100 year and stops at t = 200 year. Results are shown for three different maximum instantaneous harvest rates: H = 0.2 year^{-1} (light grey curves), H = 0.6 year^{-1} (dark grey curves) and H = 1 year^{-1} (black curves). (A) Juvenile growth rate g. (B) Growth investment α at maturation onset. (C) Annual ratio χ of decay in postmaturation growth investment. (D) PMRN intercept y. (E) PMRN slope s.

change in neutral and functional heterozygosity (<1%) can be deemed negligible. Most importantly, whatever the source of loss in genetic diversity and its amplitude, there is little recovery, if any, after fishing stops, except for the PMRN intercept under the strongest fishing intensity (black-dashed line in Fig. 5D).

Discussion

Fisheries-induced adaptive evolution and its reversal: comparison with previous models

Eco-genetic individual-based models have been recently used to explore the eco-evolutionary dynamics of harvested fish populations (e.g. Baskett et al. 2005; Dunlop et al. 2007, 2009a,b; Thériault et al. 2008; Enberg et al. 2009). Our model belongs to this modelling framework with the main novelty lying in an explicit description of the population genetics of life-history traits, using a finite number of loci and alleles coding for each trait, together with the gametic transmission of alleles (see Wang and Höök 2009; Kuparinen and Hutchings 2012 for earlier approaches of this kind). In contrast, most previous models relied on a quantitative genetic modelling approach, which assumes that life-history traits are influenced by an

infinite number of loci, each of small effect (Huisman and Tufto 2012).

Three observations are noteworthy in comparison with those previous studies. First, our model makes consistent predictions for the evolution of the mean genotypic values of life-history traits under fishing pressure, namely reduced juvenile growth, increased reproductive investment and earlier maturation through a reduced PMRN intercept. Second, the reversal of genetic adaptations during a fishing moratorium was faster in previous eco-genetic studies than in our model. Therefore, weak selection differentials cannot be the only cause of the slow recovery of trait genotypic values to their initial levels. Instead, our model has revealed a fisheries-induced erosion of adaptive potential that is hampering this recovery. This erosion went unnoticed in eco-genetic models based on quantitative genetic principles, in which the loss of genetic variation is not observed. Most probably, this is because, unlike multilocus model such as ours, those models assumed a constant genetic variance under linkage equilibrium (Huisman and Tufto 2012). Third, we find that phenotypic values of emergent life-history traits – age and size at maturation, length-at-age and GSI-at-age – partly recover thanks to phenotypic compensation (Fig. S1). Such partial phenotypic recovery

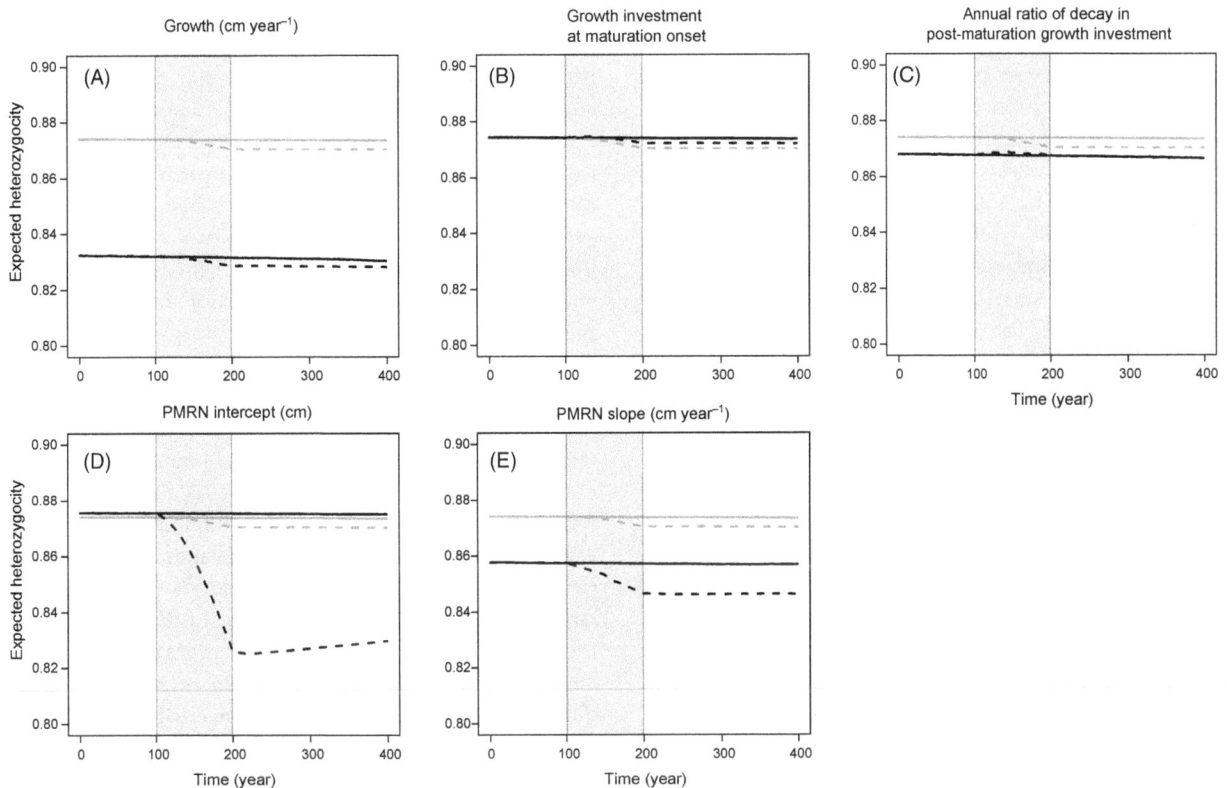

Figure 5 Comparison of temporal trends in neutral and functional genetic diversity before, during and after harvesting. Harvesting (grey shading) starts at $t = 100$ year and stops at $t = 200$ year. Expected heterozygosity H_e of neutral loci (grey lines) and functional loci (black lines) are compared for the five life-history traits: (A) juvenile growth rate g, (B) growth investment α at maturation onset, (C) annual ratio χ of decay in postmaturation growth investment, (D) PMRN intercept y and (E) PMRN slope s. Results are shown for maximum instantaneous harvest rates that are moderate ($H = 0.4$ year^{-1}; continuous lines) or strong ($H = 1$ year^{-1}; dashed lines).

also explains, at least partly, the lack of genetic recovery: it reduces the gap between the expressed phenotype and the new phenotypic trait values favoured when fishing stops and hence lowers the selection pressures towards the initial genotypic trait values.

Processes affecting evolutionary recovery

We have examined the influence of three nonmutually exclusive processes that can hamper the reversal of fisheries-induced evolution: increased rate of genetic drift, low strength of selection and reduced additive genetic variance.

Effective population size
Our results indicate that fishing may decrease effective population size, thus increasing the rate of neutral evolution due to genetic drift. This decrease is in agreement with some empirical genetic studies on exploited fish population (Smith et al. 1991; Hauser et al. 2002; Turner et al. 2002; Hutchinson et al. 2003; Hoarau et al. 2005; Pinsky and Palumbi 2014; but see Ruzzante et al. 2001; Therkildsen et al. 2010). Such a decrease could limit a population's adaptive

response to selection, because genetic drift in small populations decreases the chance of fixation of beneficial alleles, which can counteract the effects of selection. This, however, may only happen when effective population sizes reach values of a few tens, which are not observed in our model (Robertson 1960). The balancing of selection by genetic drift therefore mostly applies to small populations, for example of marine coral reef fish or freshwater species. In addition, our model shows that when fishing ceases, effective population size bounces back and surpasses its prefishing level. This suggests that the effect of fishing on the rate of genetic drift does not last long during a subsequent moratorium. We therefore expect that for most large marine fish populations, as in our model, the rate of genetic drift will not counteract the reversal of fisheries-induced evolution.

A strong decrease in effective population size due to fishing raises concerns about potential losses of genetic variability due to genetic drift. Such losses imply a risk of inbreeding, which may in turn increase extinction risks (reviewed in Frankham 2005). However, our results suggest that, whatever the fishing intensity considered, genetic diversity (expected heterozygosity) remains almost unaltered by genetic drift in large exploited marine fish populations, as expected from the range of N_e values observed in our model (Wright 1931). Although the loss of genetic variability may thus appear to be a secondary issue, empirical studies of N_e dynamics, using DNA from archived otoliths or scales (Poulsen et al. 2006; Nielsen and Hansen 2008), could still enable a 'retrospective monitoring' of other aspects of conservation and management interest, such as inferring the historical demography of exploited fish stocks from the link between N_e and population abundance (Fig. S2). Reconstructing demographic history would enable integrating baseline estimates, in terms of preharvest parameters, as reference points in fisheries management.

In our model, the ratio N_e/N of effective population size to population census size equals approximately 0.1 on average (Fig. S3), which agrees with empirical evidence in general (Frankham 1995; Palstra and Ruzzante 2008), although empirical estimates of N_e/N suffer from uncertainty (Palstra and Fraser 2012). However, studies on marine fishes have also documented extremely low ratios N_e/N, around 10^{-2} to 10^{-6} (Smith et al. 1991; Hauser et al. 2002; Turner et al. 2002; Hoarau et al. 2005). Such large discrepancies between N_e and N in marine fish species have been attributed mainly to high interindividual variability in reproductive success. This reproductive skew can arise from (i) the influence of environmental variability on recruitment (which, combined with large fecundity, may lead to 'sweepstake recruitment' events; Hedgecock 1994) and/or from (ii) productivity differences among isolated subpopulations (Turner et al. 2002). Both mechanisms are not accounted for in our model, which could explain why we observe a larger N_e/N ratio than some empirical studies. How this affects the interpretation of our results is an interesting question: for a much lower N_e/N ratio, it is theoretically possible that fishing only hits the 'noneffective' part of a population, while sparing its effective part. In practice, however, it is well known that, in most fishes, the most successful spawners are the older and larger ones, which typically also are most vulnerable to fishing. We therefore think that all ecological mechanisms potentially further reducing N_e would rather increase the risk of inbreeding due to the fisheries-induced decrease in effective population size observed in our model.

Selection differentials and additive genetic variance
Our study highlights that the pace of reversal of fisheries-induced life-history evolution after the cessation of fishing is hampered by two processes: small natural selection differentials compared to those imposed by fishing and a reduction in the genetic variability of traits, as shown by the decrease in their additive genetic variance. While the

former effect had already been pointed out in previous eco-genetic studies analysing fisheries-induced evolution (Enberg et al. 2009; Kuparinen and Hutchings 2012), the latter effect so far has not received any attention within this framework. At the timescale considered in our study, the creation of new functional alleles through mutation is a rare event, and thus neglected, so standing quantitative genetic variation is the main determinant of the modelled population's ability to evolve. This genetic variation can be eliminated at a fast rate through selection and genetic drift, raising concerns for a population's adaptability (Ryman et al. 1995).

While the considered multilocus genetic architecture of life-history traits takes our model an important step closer towards the genetic complexity of real populations, leaving out the effects of dominance and epistasis remains a simplifying assumption. Overcoming this simplification in future research would be desirable, as the nonadditive genetic variance enabled by dominance and epistasis can represent a large proportion of a population's total genetic variance in traits associated with reproductive fitness (Heath et al. 1994; Pante et al. 2002; but see Nilsson 1992). Also, nonadditive genetic variance may be converted into additive genetic variance during processes of severe population reduction, thus increasing additive genetic variance instead of decreasing it (e.g. Bryant et al. 1986; Fernández et al. 1995). However, this effect is expected to be a short-termed (Frankham 2005), and it is unknown whether it is widespread or not.

Fishing reduces functional genetic variation

In our results, losses of genetic variability due to pure genetic drift (neutral loci) are extremely weak, which is as expected at the observed range of effective population sizes: given that heterozygosity decreases by the factor $1-1/2N_e$ per generation under pure genetic drift, 100 generations (i.e. more than 700 years in our model) are required to reduce it by 2% when $N_e = 3000$, which is the lowest N_e value observed in our model under strong fishing intensity. It is therefore unlikely that in large marine fish populations, fisheries-induced genetic drift significantly reduces functional genetic variation.

We also find that differences between the temporal trends of neutral and functional expected heterozygosity increase when fisheries-induced changes in the mean genotypic values of life-history traits are larger: the loss in genetic variability at neutral and functional loci is almost equal for most traits, except for the PMRN intercept. Therefore, strong stabilizing selection obviously occurs for the latter, whereas for the other four traits, losses in functional genetic diversity remain very small and equivalent to losses in neutral genetic diversity.

It also interesting to consider whether the observed fisheries-induced evolutionary responses of mean genotypic life-history trait values are due to selective pressures or genetic drift or a mixture of both. As the observed directions of these evolutionary responses are adaptive (although they might become maladaptive after the cessation of fishing), they are most likely due to fisheries-induced selection. Moreover, genetic drift modifies allelic frequencies randomly, thus equally leading to beneficial or detrimental changes in individual fitness, which would tend to counteract the effects of selection. This further supports the interpretation that the evolutionary dynamics of at least the mean genotypic values of life-history traits are due to fisheries-induced selection, in agreement with new experimental results using guppies as model species (Van Wijk et al. 2013).

Besides these considerations, the co-variation observed between heterozygosity of neutral and functional loci suggest that the temporal dynamics of neutral genetic diversity might possibly be a useful surrogate of the temporal dynamics of functional genetic diversity (Merilä and Crnokrak 2001; Leinonen et al. 2007). As yet, however, this conjecture needs to be treated with some caution, considering the current lack of empirical evidence and the fact that most changes in genetic diversity observed in our model were relatively weak, except for the PMRN intercept.

Two main results of our study are that the observed losses in additive genetic variance and functional allelic diversity are permanent, prevailing even after fishing is stopped, and that the resulting loss in evolutionary potential impedes the genetic recovery of life-history traits. Long-term selection has been shown to deplete additive genetic variance in animal models such as *Drosophila melanogaster* (e.g. Robertson 1955), but not always, in particular when population size was reasonably large (e.g. Yoo 1980). We neglected mutations and nonadditive genetic effects in our model, which could replenish additive genetic variance and allow initial mean genotypic values to be restored. However, mutation rates of functional DNA are so low (10^{-9} per locus per generation; Li 1997) that they are irrelevant at the timescale considered (although this rate could in theory increase under environmental stress; Lamb et al. 1998).

Management implications

Our results show the importance of managing genetic impacts of fishing on exploited populations. First, fisheries-induced evolution may be detrimental for a stock's productivity, as population biomass does not fully recover to its prefishing level after fishing is stopped (Fig. S2). This is a concern for present and future profits of the fishing industry, as our model suggests that genetic adaptations induced by fishing do not recover even within 200 years without fishing.

Mitigating fisheries-induced genetic drift is important for reducing the potential risk of inbreeding and its impact on evolutionary potential. Although conflicting empirical results emerge from different estimates of effective population size (but see meta-analysis by Pinsky and Palumbi 2014), maintaining its level appears critical. To do so, genetic monitoring of neutral molecular markers, or alternatively, under certain circumstances, population demographic properties can be used to infer effective population size (Waples and Yokota 2007). In large populations, estimates of effective population size should be treated with caution, as the genetic-drift signal becomes small, and this may lead to low estimation precision (Hare et al. 2011). In our analyses, we did not test whether the decrease in effective population size due to fishing occurs only due to demographic effects or also due to the selective effects of fishing (e.g. through an increase in the variance of reproductive fitness among individuals). Our model could be used to carry out such a test as part of future research, by comparing estimates of effective population size when the modelled population is evolving to estimates when evolution is artificially precluded.

Regarding the reduction of functional diversity due to fisheries-induced adaptive evolution, our findings for the PMRN intercept highlight the importance of mitigating selection on maturation imposed by fishing gears, both to limit changes in mean genotypic and phenotypic values of life-history traits, and to avoid losses of evolutionary potential. A theoretical study on Arctic cod has shown that fishing with a dome-shaped size-selectivity curve could mitigate the evolution of mean age and size at maturation (Jørgensen et al. 2009), but whether this could also avoid significant losses of functional genetic variance remains to be examined.

Acknowledgements

This study has been carried out with financial support from the European Commission, as part of the Specific Targeted Research Project on 'Fisheries-induced evolution' (FinE, contract number SSP-2006-044276) under the Scientific Support to Policies cross-cutting activities of the European Community's Sixth Framework Programme. It does not necessarily reflect the views of the European Commission and does not anticipate the Commission's future policy in this area.

Literature cited

Abrams, P. A. 1991. Life history and the relationship between food availability and foraging effort. Ecology **72**:1242–1252.

Allendorf, F. W., P. R. England, G. Luikart, P. A. Ritchie, and N. Ryman 2008. Genetic effects of harvest on wild animal populations. Trends in Ecology & Evolution 23:327–337.

Baskett, M. L., S. A. Levin, S. D. Gaines, and J. Dushoff 2005. Marine reserve design and the evolution of size at maturation in harvested fish. Ecological Applications 15:882–901.

Beverton, R. J. H., and S. J. Holt 1994. On the dynamics of exploited fish populations. Reviews in Fish Biology and Fisheries 4:259–260.

Bryant, E. H., S. A. McCommas, and L. M. Combs 1986. The effect of an experimental bottleneck upon quantitative genetic variation in the housefly. Genetics 114:1191–1211.

Coltman, D. W., P. O'Donoghue, J. T. Jorgenson, J. T. Hogg, C. Strobeck, and M. Festa-Bianchet 2003. Undesirable evolutionary consequences of trophy hunting. Nature 426:655–658.

Dunlop, E. S., B. J. Shuter, M. Heino, and U. Dieckmann 2007. Demographic and evolutionary consequences of selective mortality: predictions from an eco-genetic model for smallmouth bass. Transactions of the American Fisheries Society 136:749–765.

Dunlop, E. S., M. L. Baskett, M. Heino, and U. Dieckmann 2009a. Propensity of marine reserves to reduce the evolutionary effects of fishing in a migratory species. Evolutionary Applications 2:371–393.

Dunlop, E. S., M. Heino, and U. Dieckmann 2009b. Eco-genetic modeling of contemporary life-history evolution. Ecological Applications 19:1815–1834.

Enberg, K., C. Jørgensen, E. S. Dunlop, M. Heino, and U. Dieckmann 2009. Implications of fisheries-induced evolution for stock rebuilding and recovery. Evolutionary Applications 2:394–414.

Ernande, B., P. Boudry, and J. Clobert 2003. Plasticity in resource allocation based life history traits in the Pacific oyster, Crassostrea gigas. I. Spatial variation in food abundance. Journal of Evolutionary Biology 17:342–356.

Fernández, A., M. A. Toro, and C. López-Fanjul 1995. The effect of inbreeding on the redistribution of genetic variance of fecundity and viability in Tribolium castaneum. Heredity 75:376–381.

Frankham, R. 1995. Effective population size/adult population size ratios in wildlife: a review. Genetics Research 66:95–107.

Frankham, R. 2005. Genetics and extinction. Biological Conservation 126:131–140.

Hare, M. P., L. Nunney, M. K. Schwartz, D. E. Ruzzante, M. Burford, R. S. Waples, K. Ruegg et al. 2011. Understanding and estimating effective population size for practical application in marine species management: applying effective population size estimates to marine species management. Conservation Biology 25:438–449.

Hauser, L., G. J. Adcock, P. J. Smith, J. H. Bernal Ramírez, and G. R. Carvalho 2002. Loss of microsatellite diversity and low effective population size in an overexploited population of New Zealand snapper (Pagrus auratus). Proceedings of the National Academy of Sciences of the United States of America 99:11742–11747.

Heath, D. D., R. H. Devlin, J. W. Heath, and G. K. Iwama 1994. Genetic, environmental and interaction effects on the incidence of jacking in Oncorhynchus tshawytscha (chinook salmon). Heredity 72:146–154.

Hedgecock, D. 1994. Does variance in reproductive success limit effective population size of marine organisms. In A. Beaumont, ed. Genetics and Evolution of Aquatic Organisms, pp. 122–134. Chapman and Hall, London.

Heino, M., U. Dieckmann, and O. R. Godø 2002a. Estimating reaction norms for age and size at maturation with reconstructed immature size distributions: a new technique illustrated by application to Northeast Arctic cod. ICES Journal of Marine Science 59:562–575.

Heino, M., U. Dieckmann, and O. R. Godø 2002b. Measuring probabilistic reaction norms for age and size at maturation. Evolution 56:669–678.

Hendry, A. P., M. T. Kinnison, M. Heino, D. Troy, T. B. Smith, G. Fitt, C. T. Bergstrom et al. 2011. Evolutionary principles and their practical application. Evolutionary Applications 4:159–183.

Hoarau, G., E. Boon, D. N. Jongma, S. Ferber, J. Palsson, H. W. V. der Veer, A. D. Rijnsdorp et al. 2005. Low effective population size and evidence for inbreeding in an overexploited flatfish, plaice (Pleuronectes platessa L.). Proceedings of the Royal Society of London, Series B: Biological Sciences 272:497–503.

Huisman, J., and J. Tufto 2012. Comparison of non-Gaussian quantitative genetic models for migration and stabilizing selection. Evolution 66:3444–3461.

Hutchings, J. A., and D. J. Fraser 2008. The nature of fisheries- and farming-induced evolution. Molecular Ecology 17:294–313.

Hutchinson, W. F., C. van Oosterhout, S. I. Rogers, and G. R. Carvalho 2003. Temporal analysis of archived samples indicates marked genetic changes in declining North Sea Cod (Gadus morhua). Proceedings of the Royal Society of London, Series B: Biological Sciences 270:2125–2132.

Jachmann, H., P. S. M. Berry, and H. Imae 1995. Tusklessness in African elephants: a future trend. African Journal of Ecology 33:230–235.

Jørgensen, C., K. Enberg, E. S. Dunlop, R. Arlinghaus, D. S. Boukal, K. Brander, B. Ernande et al. 2007. Managing evolving fish stocks. Science 318:1247–1248.

Jørgensen, C., B. Ernande, and Ø. Fiksen 2009. Size-selective fishing gear and life history evolution in the Northeast Arctic cod. Evolutionary Applications 2:356–370.

Kozlowski, J., and R. G. Wiegert 1986. Optimal allocation of energy to growth and reproduction. Theoretical Population Biology 29:16–37.

Kozłowski, J., and R. G. Wiegert 1987. Optimal age and size at maturity in annuals and perennials with determinate growth. Evolutionary Ecology 1:231–244.

Kuparinen, A., and J. A. Hutchings 2012. Consequences of fisheries-induced evolution for population productivity and recovery potential. Proceedings of the Royal Society of London, Series B: Biological Sciences 279:2571–2579.

Kuparinen, A., and J. Merilä 2007. Detecting and managing fisheries-induced evolution. Trends in Ecology & Evolution 22:652–659.

Lamb, B. C., M. Saleem, W. Scott, N. Thapa, and E. Nevo 1998. Inherited and environmentally induced differences in mutation frequencies between wild strains of Sordaria fimicola from 'Evolution Canyon'. Genetics 149:87–99.

Lande, R. 1980. The genetic covariance between characters maintained by pleiotropic mutations. Genetics 94:203–215.

Lande, R., and S. J. Arnold 1983. The measurement of selection on correlated characters. Evolution 37:1210–1226.

Lee, R. M. 1912. An investigation into the methods of growth determination in fishes by means of scales. Publications de Circonstance Conseil Permanent International pour l'Exploration de la Mer 63:1–34.

Leinonen, T., R. B. O'Hara, and J. Merilä 2007. Comparative studies of quantitative trait and neutral marker divergence: a meta-analysis. Journal of Evolutionary Biology 21:1–17.

Lester, N. P., B. J. Shuter, and P. A. Abrams 2004. Interpreting the von Bertalanffy model of somatic growth in fishes: the cost of reproduction. Proceedings of the Royal Society of London. Series B: Biological Sciences 271:1625–1631.

Li, W. H. 1997. Molecular Evolution. Sinauer Associates Inc., Sunderland.

Lynch, M., and B. Walsh 1998. Genetics and Analysis of Quantitative Traits. Sinauer Associates Inc., Sunderland.

Martínez, J. G., J. Carranza, J. L. Fernández-García, and C. B. Sánchez-Prieto 2002. Genetic variation of red deer populations under hunting exploitation in Southwestern Spain. Journal of Wildlife Management 66:1273–1282.

Marty, L., M. Rochet, and B. Ernande 2014. Temporal trends in age and size at maturation of four North Sea gadid species: cod, haddock, whiting and Norway pout. Marine Ecology Progress Series 497:179–197.

Matsumura, S., R. Arlinghaus, and U. Dieckmann 2012. Standardizing selection strengths to study selection in the wild: a critical comparison and suggestions for the future. BioScience 62:1039–1054.

Merilä, J., and P. Crnokrak 2001. Comparison of genetic differentiation at marker loci and quantitative traits. Journal of Evolutionary Biology 14:892–903.

Nei, M., and A. K. Roychoudhury 1974. Sampling variances of heterozygosity and genetic distance. Genetics 76:379–390.

Nei, M., and F. Tajima 1981. Genetic drift and estimation of effective population size. Genetics 98:625–640.

Nei, M., T. Maruyama, and R. Chakraborty 1975. The bottleneck effect and genetic variability in populations. Evolution 29:1–10.

Nielsen, E. E., and M. M. Hansen 2008. Waking the dead: the value of population genetic analyses of historical samples. Fish and Fisheries 9:450–461.

Nilsson, J. 1992. Genetic parameters of growth and sexual maturity in Arctic char (Salvelinus alpinus). Aquaculture 106:9–19.

Palsbøll, P. J., M. Bérubé, and F. W. Allendorf 2007. Identification of management units using population genetic data. Trends in Ecology & Evolution 22:11–16.

Palstra, F. P., and D. J. Fraser 2012. Effective/census population size estimation: a compendium and appraisal. Ecology and Evolution 2:2357–2365.

Palstra, F. P., and D. E. Ruzzante 2008. Genetic estimates of contemporary effective population size: what can they tell us about the importance of genetic stochasticity for wild population persistence? Molecular Ecology 17:3428–3447.

Pante, M. J. R., B. Gjerde, I. McMillan, and I. Misztal 2002. Estimation of additive and dominance genetic variances for body weight at harvest in rainbow trout, Oncorhynchus mykiss. Aquaculture 204:383–392.

Pinsky, M. L., and S. R. Palumbi 2014. Meta-analysis reveals lower genetic diversity in overfished populations. Molecular Ecology 23:29–39.

Poulsen, N. A., E. E. Nielsen, M. H. Schierup, V. Loeschcke, and P. Grønkjaer 2006. Long-term stability and effective population size in North Sea and Baltic Sea cod (Gadus morhua). Molecular Ecology 15:321–331.

Quince, C., P. A. Abrams, B. J. Shuter, and N. P. Lester 2008. Biphasic growth in fish I: theoretical foundations. Journal of Theoretical Biology 254:197–206.

Robertson, F. W. 1955. Selection response and the properties of genetic variation. Cold Spring Harbor Symposium on Quantitative Biology 20:166–167.

Robertson, A. 1960. A theory of limits in artificial selection. Proceedings of the Royal Society of London, Series B: Biological Sciences 153:234–249.

Ruzzante, D. E., C. T. Taggart, R. W. Doyle, and D. Cook 2001. Stability in the historical pattern of genetic structure of Newfoundland cod (Gadus morhua) despite the catastrophic decline in population size from 1964 to 1994. Conservation Genetics 2:257–269.

Ryman, N., F. Utter, and L. Laikre 1995. Protection of intraspecific biodiversity of exploited fishes. Reviews in Fish Biology and Fisheries 5:417–446.

Smith, P. J., R. I. C. C. Francis, and M. McVeagh 1991. Loss of genetic diversity due to fishing pressure. Fisheries Research 10:309–316.

Thériault, V., E. S. Dunlop, U. Dieckmann, L. Bernatchez, and J. J. Dodson 2008. The impact of fishing-induced mortality on the evolution of alternative life-history tactics in brook charr. Evolutionary Applications 1:409–423.

Therkildsen, N. O., E. E. Nielsen, D. P. Swain, and J. S. Pedersen 2010. Large effective population size and temporal genetic stability in Atlantic cod (Gadus morhua) in the southern Gulf of St. Lawrence. Canadian Journal of Fisheries and Aquatic Sciences 67:1585–1595.

Turner, T. F., J. P. Wares, and J. R. Gold 2002. Genetic effective size is three orders of magnitude smaller than adult census size in an abundant, estuarine-dependent marine fish (Sciaenops ocellatus). Genetics 162:1329–1339.

Van Wijk, S. J., M. I. Taylor, S. Creer, C. Dreyer, F. M. Rodrigues, I. W. Ramnarine, C. van Oosterhout et al. 2013. Experimental harvesting of fish populations drives genetically based shifts in body size and maturation. Frontiers in Ecology and the Environment 11:181–187.

Walters, C., and J. Korman 1999. Linking recruitment to trophic factors: revisiting the Beverton-Holt recruitment model from a life history and multispecies perspective. Reviews in Fish Biology and Fisheries 9:187–202.

Wang, H.-Y., and T. O. Höök 2009. Eco-genetic model to explore fishing-induced ecological and evolutionary effects on growth and maturation schedules. Evolutionary Applications 2:438–455.

Wang, J., and M. C. Whitlock 2003. Estimating effective population size and migration rates from genetic samples over space and time. Genetics 163:429–446.

Waples, R. S. 1989. A generalized approach for estimating effective population size from temporal changes in allele frequency. Genetics 121:379–391.

Waples, R. S. 1998. Separating the wheat from the chaff: patterns of genetic differentiation in high gene flow species. Journal of Heredity 89:438–450.

Waples, R. S., and M. Yokota 2007. Temporal estimates of effective population size in species with overlapping generations. Genetics 175:219–233.

Werner, E. E., and B. R. Anholt 1993. Ecological consequences of the trade-off between growth and mortality rates mediated by foraging activity. American Naturalist 142:242–272.

Willi, Y., J. van Buskirk, and A. A. Hoffmann 2006. Limits to the adaptive potential of small populations. Annual Review of Ecology, Evolution, and Systematics 37:433–458.

Wright, S. 1931. Evolution in mendelian populations. Genetics 16:97–159.

Wright, S. 1938. Size of population and breeding structure in relation to evolution. Science 87:430–431.

Yoo, B. H. 1980. Long-term selection for a quantitative character in large replicate populations of Drosophila melanogaster. Part 3: The nature of residual genetic variability. Theoretical and Applied Genetics 57:25–32.

8

Intense selective hunting leads to artificial evolution in horn size

Gabriel Pigeon,[1,2] Marco Festa-Bianchet,[1] David W. Coltman[3] and Fanie Pelletier[1,2]

1 Département de Biologie and Centre d'Études Nordiques, Université de Sherbrooke, Sherbrooke, QC, Canada
2 Département de Biologie, Canada Research Chair in Evolutionary Demography and Conservation, Université de Sherbrooke, Sherbrooke, QC, Canada
3 Department of Biological Sciences, University of Alberta, Edmonton, AB, Canada

Keywords
conservation biology, contemporary evolution, quantitative genetics.

Correspondence
Gabriel Pigeon, Département de Biologie, Université de Sherbrooke, 2500 boulevard de l'Université, Sherbrooke J1K 2R1, QC, Canada.

e-mail: Gabriel.Pigeon@USherbrooke.ca

Abstract

The potential for selective harvests to induce rapid evolutionary change is an important question for conservation and evolutionary biology, with numerous biological, social and economic implications. We analyze 39 years of phenotypic data on horn size in bighorn sheep (*Ovis canadensis*) subject to intense trophy hunting for 23 years, after which harvests nearly ceased. Our analyses revealed a significant decline in genetic value for horn length of rams, consistent with an evolutionary response to artificial selection on this trait. The probability that the observed change in male horn length was due solely to drift is 9.9%. Female horn length and male horn base, traits genetically correlated to the trait under selection, showed weak declining trends. There was no temporal trend in genetic value for female horn base circumference, a trait not directly targeted by selective hunting and not genetically correlated with male horn length. The decline in genetic value for male horn length stopped, but was not reversed, when hunting pressure was drastically reduced. Our analysis provides support for the contention that selective hunting led to a reduction in horn length through evolutionary change. It also confirms that after artificial selection stops, recovery through natural selection is slow.

Introduction

Human activities such as habitat modifications, expanding road networks, overexploitation and climate change affect animal populations. While the demographic impacts of humans on wild species are clear, their evolutionary impacts are debated (Loehr et al. 2007; Hard et al. 2008). Intense exploitation by humans may outpace (Darimont et al. 2009) or oppose (Carlson et al. 2007) the selective effects of natural predators, potentially leading to evolutionary changes in behaviour, phenotype or life history (Hard et al. 2008; Devine et al. 2012). van Wijk et al. (2013) showed that selective harvesting of guppies (*Poecilia reticulata*) led to changes in size and in the frequency of alleles associated with size in just two generation. Human-induced evolution may also impair population persistence or prevent recovery (Swain et al. 2007; Uusi-Heikkilä et al. 2015). While numerous studies of fishes report evidence of evolution induced by intense harvest (reviewed in

Hutchings and Fraser 2008), evidence for evolution through selective harvest in terrestrial species remains scarce and controversial (Coltman et al. 2003; Garel et al. 2007; Mysterud 2011; Traill et al. 2014), partly because the statistical techniques used to quantify evolutionary changes using pedigrees in earlier studies have been questioned (Postma 2006; Hadfield et al. 2010).

Trophy hunting can be an important component of many conservation programs (Leader-Williams et al. 2001), and its economic revenues are partly driven by expectation of large trophy size (Festa-Bianchet and Lee 2009; Crosmary et al. 2013). In most of Canada, sport harvest of mountain sheep (*Ovis canadensis* and *O. dalli*) rams is based on a phenotypic definition of minimum horn curl that establishes whether or not a ram can be shot, with an unlimited number of permits available to resident hunters (Festa-Bianchet et al. 2014). In wild sheep, horn size is a key determinant of success in male-male competition over breeding opportunities (Coltman et al. 2002). Artificial selection favoring shorter horns through hunting mortality, however, sets in 2–3 years before natural selection favoring longer horns through reproductive success (Coltman et al. 2002). Multiple studies report that males with fast-growing horns, that would enjoy high mating success at 8–10 years of age, are harvested at 4–7 years, conferring a reproductive advantage to small-horned males that, in the absence of size-selective harvests, would normally be outcompeted (Festa-Bianchet et al. 2004, 2014; Loehr et al. 2007; Hengeveld and Festa-Bianchet 2011; Douhard et al. In Press).

One approach to study evolution in nature, often referred to as the animal model, involves mixed models combining a pedigree with data on phenotype and environmental conditions to estimate genetic parameters (Kruuk 2004). Using this approach, Coltman et al. (2003) used a pedigree up to six generation deep to report a decline in estimated breeding values (EBV) of horn length and body mass in bighorn rams over 30 years, suggesting an evolutionary response to size-selective harvests. Their analyses, however, were criticised for not adequately accounting for environmental effects on phenotype, for the error in estimation of breeding values and for the effect of drift; possibly leading to exaggerated estimates of evolutionary change (Postma 2006; Hadfield et al. 2010). Hence, the importance of evolution in the observed change in phenotype following selective harvesting is still debated.

A recent paper used data from the individually monitored population of bighorn sheep of Ram Mountain to parameterise an Integral Projection Model and show a decline in body mass, but argued that the phenotypic response to harvest was only demographic (Traill et al. 2014). The statistical criticisms and alternative analyses listed above cast doubt on the conclusion that selective hunting could lead to evolutionary changes. Coltman et al. (2003) drew that conclusion after analysing data for the only sport-hunted population of ungulates for which a pedigree and horn measurements are available (Pelletier et al. 2012). By extension, these criticisms also question phenotype-based studies that reported long-term trends consistent with an evolutionary impact of selective hunting (Garel et al. 2007; Hengeveld and Festa-Bianchet 2011). A clear understanding of the importance of evolutionary change due to selective harvesting is of critical importance to those responsible for managing harvested wild populations (Allendorf and Hard 2009). A reanalysis of the Ram Mountain data is therefore warranted, particularly because the 10-fold decline in harvests after 1996 provides an opportunity to test the impacts of changes in harvest pressure on trait evolution (Douhard et al. In Press).

Here, we use a Bayesian animal model to analyse an expanded database on bighorn sheep from Ram Mountain, adding 9 years of data to those available to Coltman et al. (2003) and taking into account subsequent

statistical criticisms (Postma 2006; Hadfield et al. 2010). We also compare a period of intense harvest with a period when harvest was first dramatically reduced, then stopped. This allowed us to compare temporal trends in genetic values under heavy and very light artificial selection. To maximise the use of phenotypic information, we considered data on male and female traits using a multivariate model. Genetic correlations have already been established among some of these traits (Poissant et al. 2012), and proper estimation of breeding values must account for genetic covariance (Wolak et al. 2015). By including phenotypic data on females we could also compare temporal changes in traits that are (male horn base and female horn length) and are not (female horn base) genetically correlated to male horn length (Poissant et al. 2012). We expected to see temporal changes in EBV in male horn length only under heavy harvest. Male horn base circumference is particularly interesting because it is correlated with horn length and likely affects male-male competition by contributing to horn mass, but is not a direct target of selective hunting. We expected strong selective effects on male horn length, the trait most directly related to the legal definition of harvestable ram (Festa-Bianchet et al. 2014). We expected a response similar to male horn length for male horn base and female horn length given their strong genetic correlations with male horn length, and no response in female horn base, which has a weak genetic correlation with male horn length (genetic correlations of 0.72, 1 and −0.28 respectively; Poissant et al. 2012). To compare our results with previous studies on this population, we also built animal models using univariate one-sex (with phenotypic data on males only) and two-sex (phenotypic data from both sexes) approach. Univariate models are also less prone to problems when fitted with limited data given their simpler structure (Wilson et al. 2010).

Material and methods

Study population and phenotypic data

Bighorn sheep at Ram Mountain, Alberta, Canada are intensively monitored. The study area is 30 km east of the Rockies (52°8′N, 115°8′W, elevation 1082–2173 m), on a mountainous outcrop dominated by cliffs, rock scree and alpine meadows. Since 1972, sheep have been marked with ear tags and collars. Each year, between May and September, sheep were repeatedly captured in a corral trap baited with salt. Rams were captured on average 2.6 times per year. At each capture, horn length in cm was measured along the outside curvature with a flexible tape. To reduce the potential measurement error caused by horn wear or breakage, we used the longest horn in analyses. Horn base circumference was also measured in cm, and we analysed the mean of the left and right measurements. Nearly all individuals (95%) were first captured as lambs or yearlings, so their exact age was known. For the others, age was determined using horn annuli.

The study population was hunted until 2011 based on a morphological definition of 'legal' ram. From late August to October, rams were at a risk of being shot only if they met that definition, which specified a minimum degree of horn curl and was correlated with horn length (Festa-Bianchet et al. 2014). Artificial selection through hunting, however, changed over time. In 1996, the minimum horn curl of a 'legal' ram was increased from 4/5 to full curl (Fig. S1). This change, implemented at a time when horn size had declined (Coltman et al. 2003), drastically decreased the harvest, with only four rams shot in the following 15 years. Mean harvest was 2.26 rams/year in 1973–1995 and only 0.27 rams/year in 1996–2010. Hunting was closed in 2011. Therefore, we compared trends in the EBV of morphological traits for cohorts of 1973–1996 (referred to as the hunted period) and 1996–2011 (non-hunted period; see below). Based on the average age of fathers at Ram Mountain (7.3 years), we monitored 3.3 generations under strong artificial selection followed by 2.2 generations under natural selection.

We first adjusted all traits to September 15 using a mixed model approach (Martin and Pelletier 2011). As adult females and adult males display different growth curves, we used sex-specific linear models to account for capture date and fitted one model per year to allow for environmental variability. Trait was fitted as a function of the square root of Julian date, considering May 25th as day 1. With this modelling approach, individual identity can be used to estimate an individual intercept and slope, providing a more accurate standardization than classical least square regression (Martin and Pelletier 2011). The procedure was used for horn length and horn base. A total of 2295 adjusted phenotypic measurements where obtained from 510 females and 497 males.

Pedigree reconstruction

Since 1972, maternities were assigned from observation of suckling behaviour. Since 1988, DNA samples have allowed the assignation of paternities based on 26 microsatellite loci with a confidence threshold of 95% using CERVUS (Coltman et al. 2005). The pedigree in 2014 contained 864 maternal links involving 254 dams and 528 paternal links involving 79 sampled and 37 unsampled sires, the latter identified using COLONY (Jones and Wang 2010). Unsampled sires include rams that died before we began sampling for DNA and immigrants that are on Ram Mountain only for the rut.

Quantitative genetic analyses

Analyses of horn base include phenotypic data of individuals aged 2–10 years between 1975 and 2013. For horn length, however, we only included data for sheep aged 2–4 years. Horns frequently break, and the chance of horn damage increases with age. Many old males have broken horns, missing up to the first 2 years of growth. Our data suggest that by 4 years of age ewes have reached 97% of the horn length they will have at age 8 (including the effect of breakage), and rams 73%. In addition, after age 4 the sample of rams is biased because those with longer horns are removed by hunters (Coltman et al. 2002). To reduce the importance of maternal effects on phenotypes (Wilson et al. 2005), analyses excluded phenotypic data of lambs and yearling (Réale et al. 1999; Wilson, Kruuk, and Coltman 2005).

The multivariate animal model was fitted using four traits: male horn length, female horn length, male horn base and female horn base. Phenotypic variance was then partitioned into its components, including additive genetic variance. The model also included sheep identity, year of measurement and year of birth as random effects to assess the amount of variance due to permanent, yearly and cohort environmental effects respectively. Including year of measurement and year of birth as random effects accounts for both short- and long-term environmental effects, including changes in density, weather and forage quality. The year effect is necessary to obtain unbiased estimates of breeding value but it may also partly absorb temporal genetic trends, making this analysis conservative. Maternal identity was not included since the exclusion of lambs and yearlings minimized maternal effects. Age was included as a categorical fixed effect. To compare our results with previously published studies on this population we also examined univariate animal models (see supplementary material). We tested univariate models using male phenotype only (SI 2) to obtain results comparable to Coltman et al. (2003). We also fitted univariate models including both male and female phenotype to increase power (SI 3). These models are further described and their results presented in the supplementary material (SI 2–3).

The animal model estimates the breeding value of each individual. To correctly estimate breeding values and their associated error (Hadfield et al. 2010), the model was fitted using a Bayesian method with MCMCglmm version 2.21. We used a multivariate inverse-Wishart prior to obtain the most objective results possible. Models were run using two chains for 8 500 000 iterations, with a thinning of 75 000 and a burn-in of 1 000 000 iterations. A sensitivity analysis evaluated the robustness of the model to different prior specifications (Fig. S2).

Temporal change in estimated breeding value

We compared temporal trends in EBV to those obtained based on different models of evolutionary change. For each realization of the MCMC chain of the animal model, we calculated mean EBV by cohort and the slope in mean EBV as a function of cohort (ße) for both the hunted and not-hunted periods to obtain a posterior distribution of slopes. We compared this distribution to the posterior distribution of slopes of alternative models, which included no change, drift, stasis and expected evolutionary response. To compare the posterior distributions of slopes, we subtracted each realization of the posterior distribution of the alternative model to that of the distribution ße, obtaining a distribution of differences. From this distribution, we can obtain the mean difference between the expected and observed distributions as well as the confidence interval of the difference.

First, as done previously by Coltman et al. (2003), we compared slopes in EBV to 0. Second, following Hadfield et al. (2010), we compared the slopes in EBV to those obtained from simulated drift. To do so, we simulated random breeding values down the pedigree for each of the 1000 posterior samples of the animal model based on the estimated additive genetic variance. We then fitted a linear regression to the cohort mean of these random breeding values to obtain the slopes due to drift for each posterior sample. Third, we compared observed change in estimated breeding value to stasis (Hunt 2007), a pattern likely to occur under stabilizing selection. To simulate stasis, mean cohort breeding values were randomly drawn from a normal distribution with a mean of 0 and a variance equal to the observed variance in mean cohort EBV.

We also compared observed change in EBV to the response to selection predicted by the secondary theorem of selection (Morrissey et al. 2012). This theorem states that change should be equal to the additive genetic covariance between the trait of interest and relative fitness. We used longevity as a fitness measure, which we divided by mean cohort longevity to obtain relative fitness. We used longevity rather than reproductive success because molecular assignments of paternities only began in 1988. We then fitted separate bivariate animal models of trait and fitness for each of the studied traits. The predicted response to selection was then extracted from the G matrix and divided by the mean generation time (7.3 years, the average age of fathers in our population) to obtain a predicted change per year (Table S1). Predicted response to selection could only be estimated for the hunted period due to the limited number of individuals of known longevity born after hunting pressure was reduced. The proportion of iterations for which the slope for the estimated breeding value (ße) is lower than that of the random breeding value (ßr) was also

calculated to estimate the probability that the trend was not caused solely by drift.

Results

Animal model analyses

Estimates of variance components and heritability for horn length and base (Table 1) showed that heritability was >0 for all traits. The trait with the highest posterior mode for heritability was male horn length, followed by male horn base, female horn length and female horn base. Permanent environmental effects explained much of the variance in female but not in male traits. Cohort always explained a significant part of phenotypic variance, while the effects of year and permanent environment varied among traits (Table 1). Confirming previous analyses (Poissant et al. 2012), genetic correlations between male horn length, female horn length and male horn base were high while female horn base had low genetic correlation with other traits (Table 2).

Temporal changes in EBV

Temporal changes in mean phenotypic values over 39 years differed between traits (Fig. 1). A temporal change in EBV was also observed (Fig. 2). During the hunted period from 1973 to 1996, the EBV of male horn length declined significantly (ß = −0.119; CI = −0.248, −0.006). Similarly, genetically correlated traits also appeared to decline. Female horn length breeding value declined with a slope of −0.027 (CI = −0.063, 0.013), while EBV for male horn base had a slope of −0.030 (CI = −0.076, 0.019). Unlike male horn length, the breeding value of female horn base appeared to increase, with a slope of 0.005 (CI = −0.008, 0.016). We then compared observed changes in EBV for male horn length to those expected under various models of evolutionary change (Table S2; Fig. 3). The observed temporal change in estimated breeding value differed significantly from 0 (Pr[ße < 0] = 0.974). While observed EBV did not differ significantly from that predicted by other models, the probability of declining more than expected by drift alone (Pr[ße < ßr]) was 0.901. The observed temporal change in EBV was most similar to that predicted by the secondary theorem of selection (expected change per generation of −0.76, Fig. 3) with a posterior difference of 0.016, while the posterior differences of other models of evolution ranged from 0.117 to 0.120 (Table S2). Similarly, for female horn length, observed trends were most similar to those predicted by the secondary theorem of selection (expected change per generation of −0.10) with a posterior difference of 0.013. Other models of evolution all had similar differences of 0.027. The probability of declining more than expected by drift alone (Pr[ße < ßr])

Table 1. Variance components and heritability of horn length and horn base in bighorn sheep at Ram Mountain, Canada, according to multivariate animal models. The posterior mode of the proportion of phenotypic variance explained by each component is followed by the 95% Bayesian posterior interval of highest density in parentheses.

	Horn length male	Horn length female	Horn base male	Horn base female
h^2	0.397 (0.203–0.534)	0.223 (0.090–0.446)	0.250 (0.119–0.413)	0.265 (0.148–0.335)
ID	0.025 (0.003–0.211)	0.376 (0.203–0.540)	0.098 (0.016–0.268)	0.171 (0.110–0.265)
yr	0.110 (0.039–0.168)	0.022 (0.010–0.052)	0.193 (0.109–0.289)	0.161 (0.112–0.268)
Cohort	0.363 (0.211–0.528)	0.149 (0.071–0.286)	0.203 (0.097–0.354)	0.212 (0.107–0.291)

h^2 refers to the narrow-sense heritability, ID refers to the proportion of phenotypic variance explained by permanent environment (identity of the sheep), yr refers to the proportion of phenotypic variance explained by year of measurement and cohort refers to the proportion of phenotypic variance explained by year of birth.

Table 2. Genetic correlations and covariance matrix for horn size in bighorn sheep. Values on the diagonal (grey shading) are posterior modes of genetic additive variance.

	Hl-M	Hl-F	Hb-M	Hb-F
Hl-M	17.884 (9.82–25.881)	0.921 (0.557–0.981)	0.878 (0.729–0.959)	0.189 (−0.285–0.538)
Hl-F	5.345 (1.928–8.144)	1.622 (0.748–3.963)	0.799 (0.275–0.939)	0.368 (−0.063–0.610)
Hb-M	5.435 (2.797–9.666)	1.274 (0.318–2.881)	2.915 (1.124–4.485)	0.286 (−0.203–0.656)
Hb-F	0.070 (−0.542–1.059)	0.187 (−0.062–0.481)	0.182 (−0.164–0.508)	0.183 (0.119–0.270)

Values below the diagonal are the posterior modes of genetic covariance between traits: male horn length (HL-M), female horn length (HL-F), male horn base (HB-M) and female horn base (HB-F). Values above the diagonal are the posterior modes of genetic correlations. Values in parentheses represent the 95% Bayesian posterior interval of highest density.

was 0.816. For male horn base, observed trends were also most similar to those predicted by the secondary theorem of selection (expected change per generation of −0.13) with a posterior difference of 0.013. Other models of evolution all had similar differences of 0.030. The probability of declining more than expected by drift alone (Pr[ße < ßr]) was 0.796. Finally, for female horn base, all models were similar. The predicted response according to the secondary theorem of selection (expected change per generation of 0.005) had a difference of −0.004. Other models had differences of −0.005.

After the near-cessation of hunting in 1996, average EBV remained stable or showed a weak tendency to increase, with slopes of 0.053 (CI: −0.174, 0.282) and 0.021 (CI: −0.054, 0.104) for horn length of males and females respectively. For horn base, EBVs after the change in regulations had slopes of 0.032 (CI: −0.056, 0.137) and −0.006 (CI: −0.029, 0.018) for males and females respectively. The probabilities that the slope in EBV of male and female horn length increased after the change in hunting regulations were 0.894 and 0.847 respectively. Similarly, the probabilities that the slope in EBVs of male and female horn base increased after the change in regulations were 0.866 and 0.226. Unfortunately, we could not compare observed changes in EBVs for male horn length to those expected under various models of evolutionary change for the not-hunted period. Because of the shorter period and a smaller

population size, we did not have adequate statistical power to estimate predicted responses to selection. A comparison of observed temporal trends in EBV to alternative models of evolutions such as drift or stasis suggested that all models were quite similar (Table S2).

Results for the univariate animal models were qualitatively similar to the multivariate model presented here. For the male-only model, the posterior probabilities of declining more than expected by drift (Pr[ße < ßr]) were 0.874 and 0.629 for horn length and horn base. Posterior probabilities of declining more than expected by drift changed to 0.560, and 0.637 respectively after the change in hunting regulations (for complete results, see Table S3). For the two-sex model, the posterior probabilities that breeding values declined more than expected by drift (Pr[ße < ßr]) were 0.985 and 0.503 for horn length of males and females, respectively. These probabilities were 0.745 and 0.582 for horn base of males and females (Table S4). Changes in breeding value had a low probability of being steeper than expected from drift after 1996 (0.560 and 0.637 for horn length, and horn base of males; 0.543 and 0.555 for the same traits in females).

Discussion

We assessed whether temporal genetic trends in wild bighorn sheep were consistent with evolutionary changes

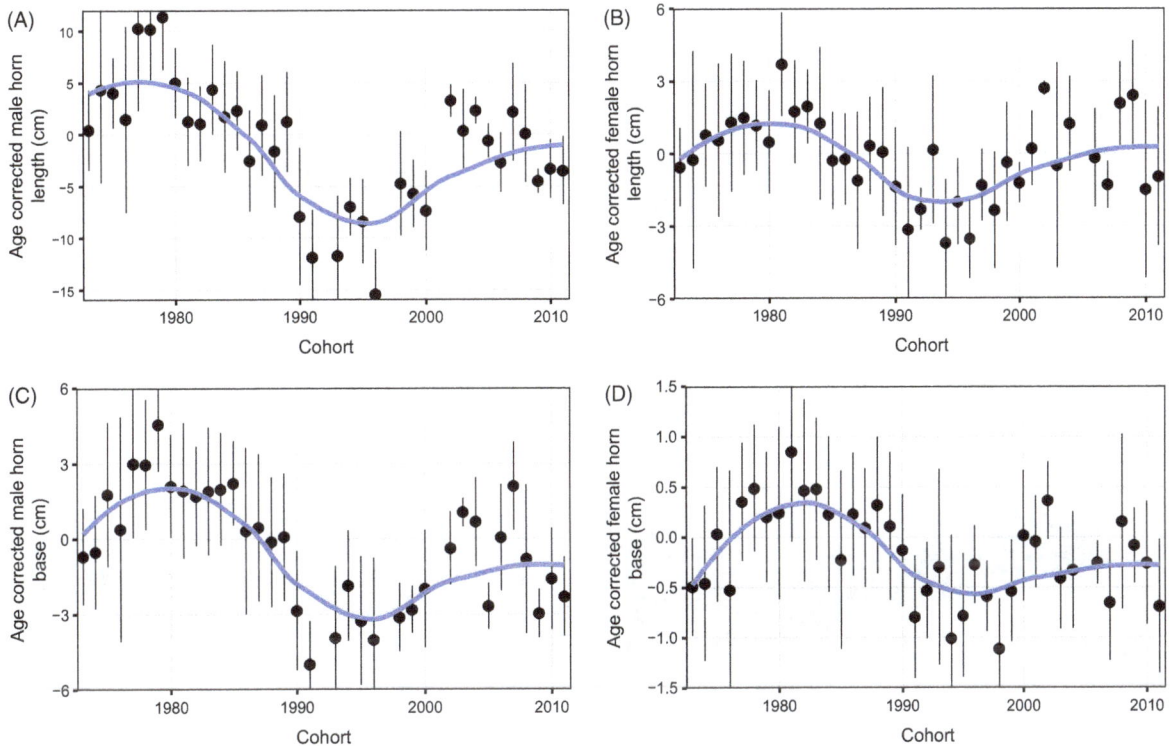

Figure 1 Temporal trends in age-corrected phenotypic traits for bighorn sheep cohorts born at Ram Mountain, Canada, between 1973 and 2011. Panels show mean (A, B) horn length and (C, D) horn base in cm. Black dots and error bars represent the cohort average (±1 SD) phenotype after correcting for age. Smooths (blue line) were fitted using loess.

expected from selective pressures acting on traits targeted or not targeted by trophy hunting. Using a 39-year dataset, we expand upon previous results (Coltman et al. 2003), using a statistical approach (Hadfield et al. 2010) that is robust to biases likely affecting earlier estimates of breeding values. A model including a term for the random effect of year, as suggested by Postma (2006) confirms a statistically significant negative trend in EBV for male horn length during a period of intense harvest. Hadfield et al. (2010) suggested an even more conservative test, comparing the observed change in breeding value to simulated changes that may occur through genetic drift. The observed decline in male horn length breeding value had a probability of 90.1% of being greater than expected from drift alone, although this probability varied when using simpler univariate models (87.4% and 98.5% depending on the univariate animal model used; Tables S1–S2). The decline in breeding value had a very high probability of being greater than that expected under genetic drift in a univariate model that included information on phenotype and pedigree from both sexes but did not include genetic covariance with other traits (Table S2).

The decline in male horn length breeding values appeared to stop when hunting pressure was greatly reduced. While horn length declined during the hunting period, female horn base, a trait not subjected to trophy hunting and with low genetic correlation (0.189) to male horn length (Table 2), did not decline, supporting the contention that the decline in horn length was partly due to artificial selection. Further, female horn length and male horn base, traits genetically correlated to male horn length but not under selection, showed responses similar to male horn length. Overall, these results provide compelling evidence of a response to artificial selection while refuting the hypothesis that the observed changes were entirely caused by changes in environment. Our study population is small (average of 28.5 adult rams, yearly range 8–61) and after the hunting regulations were changed it declined partly through cougar (*Puma concolor*) predation (Festa-Bianchet et al. 2006), averaging 17 rams. Therefore, drift may play a substantial role in changes in allele frequencies and fluctuations in breeding values over time.

Traill et al. (2014) suggested that all phenotypic changes in mass observed at Ram Mountain were due to demographic changes in response to hunting. Our analyses of horn length, however, support the result of Coltman et al. (2003) and suggest that observed changes in horn length were due to an evolutionary response to artificial selection.

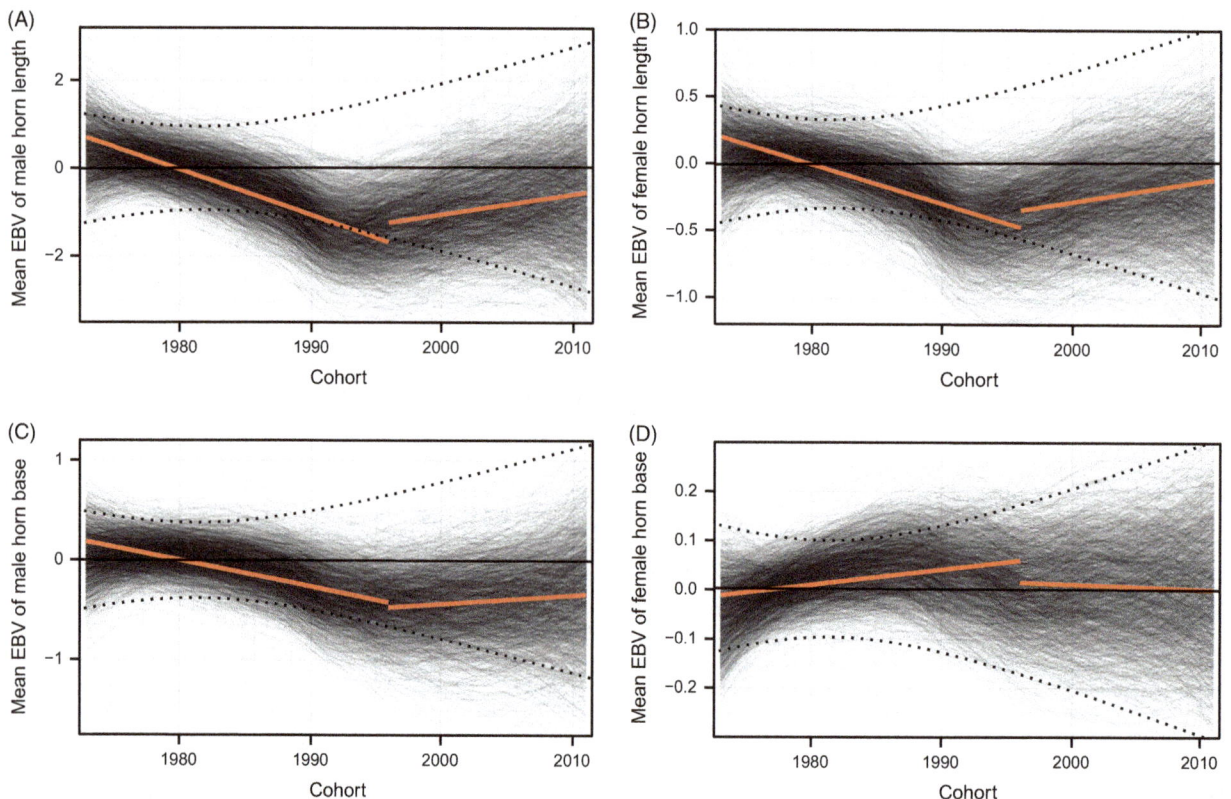

Figure 2 Changes in mean estimated breeding values (EBV) for bighorn sheep cohorts born at Ram Mountain between 1973 and 2011, according to a multivariate model. Panels present the EBV of (A, B) horn length and (C, D) horn base in cm. The left column shows results for males and the right column for females. Each grey line represents the average estimated breeding value through time for one iteration of the MCMC chain of the animal model using loess. Red lines represent the posterior mean trend using linear regression for the hunted and non-hunted period. The blue line represents the average response expected by drift alone, with 95% confidence interval in dashed blue lines.

The difference between these studies can be explained in two ways. First, the simulations presented by Traill et al. (2014) were based on body mass. Although horn length and body mass have a moderate genetic correlation (0.48, Poissant et al. 2012), mass is not a direct target of trophy hunting. More importantly, the inheritance function in Traill et al. (2014) links parent and offspring phenotype solely upon the relationship between parental mass at conception and offspring mass at weaning: it does not allow large fathers to produce offspring that grow to become large adults (Hedrick et al. 2014; Chevin 2015) despite strong heritability of adult mass in this population (Poissant et al. 2012). The 'inheritance' function is nearly zero for father-offspring, while the mother-offspring function explains only about 5% of the variance in weaning mass (Festa-Bianchet and Jorgenson 1998; Réale et al. 1999).

Between 1973 and 1996, the horn length of bighorn rams on Ram Mountain declined by nearly 30% (Coltman et al. 2003). It has since recovered by about 13%. When the artificial selection stopped, EBV did not increase, but there was a phenotypic increase in horn length. The very low

population density in the last 15 years may have contributed to the non-genetic increase in mean age-corrected horn length, which remains smaller than 30–40 years ago (Fig. 1). Environmental factors such as population density and weather play important roles in horn growth (Jorgenson et al. 1993; Festa-Bianchet et al. 2014). For example, a doubling of population size at Ram Mountain contributed to a decline in ram horn length, which, however, remained stable during an earlier period of experimental population control through ewe removals (Jorgenson et al. 1998). Therefore, it is important to adequately partition environmental and genetic phenotypic changes. Including both cohort and year in the animal model should control for both long- and short-term effects of these variables on the phenotype (Wilson et al. 2010).

It seems reasonable to expect that strong artificial selection on heritable traits may lead to evolutionary changes (Garland and Rose 2009). A study on 74 domestic sheep (*Ovis aries*) breeds found strong genetic signals of selection for the absence of horns and for other traits such as body size, reproduction and pigmentation (Kijas et al. 2012).

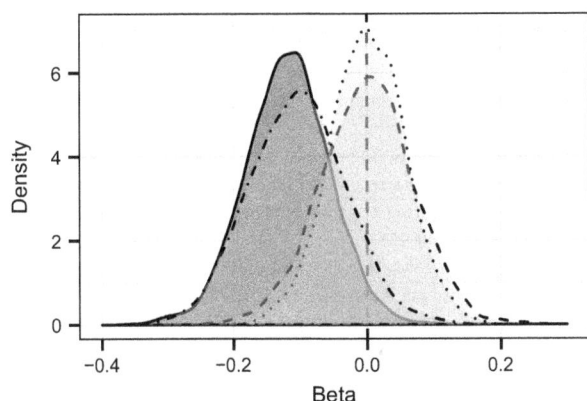

Figure 3 Posterior density plots for the slopes in mean estimated breeding values of male horn length and predicted change in estimated breeding value according to different models of evolutionary change for bighorn sheep cohorts born at Ram Mountain from 1973 to 1996. The dark filled distribution with solid line represents the posterior distribution of slopes in mean cohort breeding values (ße) for male horn length. The distribution with dot-dashed line represents predicted annual evolutionary response according to the secondary theorem of selection. The distribution with dotted line represents predicted change according to simulation of stasis. The distribution with dashed line represents predicted changes due to drift according to simulation of random breeding values.

Evidence for the evolutionary effects of selective hunting in wild terrestrial species, however, remains scarce and controversial (Mysterud 2011). We suggest that evidence is scarce partly because it requires detailed long-term data on genotypes, phenotypes, vital rates, population fluctuations, harvest pressure and environmental changes in harvested populations. Most longitudinal studies of wild vertebrates that have collected these data have been conducted on unharvested populations. There is abundant support for artificial selection in commercially-exploited fish and recent studies provide evidence of a genetic response to that selection over a few generations (Swain et al. 2007; van Wijk et al. 2013). Therefore, it should not be surprising to find an effect of artificial selection over about 3–4 generations of bighorn sheep, given that rams with 4/5-curl horn faced a 40% yearly probability of being shot and that the negative selective pressure through hunting started 2–3 years before large-horned rams could achieve high reproductive success (Coltman et al. 2002). Long-term phenotypic data from harvested rams support this contention by showing temporal declines in horn length in populations subject to high harvest pressure. Age-specific horn size of Rocky Mountain bighorn rams declined in Alberta (Festa-Bianchet et al. 2014) but not in the neighbouring province of British Columbia, where a more conservative definition of 'legal' ram reduces harvest pressure (Hengeveld and Festa-Bianchet 2011). Similarly, in Stone's rams (*Ovis dalli*), early horn growth declined under intense

selective harvest, but not under lower hunting pressure (Douhard et al. In Press).

Using detailed monitoring of a harvested population, we provide evidence that horn length – a trait directly targeted by trophy hunting – declined in response to intense artificial selection. The lack of evolutionary recovery in mean horn length breeding values after harvest stopped supports the hypothesis that recovery from potentially maladaptive human-induced evolution is slow, likely because natural selective pressures are weaker than artificial ones (Swain et al. 2007; Allendorf and Hard 2009). Given the substantial economic importance of trophy hunting (Foote and Wenzel 2009) and its potential role in conservation (Leader-Williams et al. 2001), it is critical to assess what levels of selective harvest can drive evolution in game species.

Acknowledgements

We are grateful to Bill Wishart, Anne Hubbs, Chiarastella Feder, and Jon Jorgenson for their support of the Ram Mountain research program, to Jack Hogg for initiating tissue sample collections and analyses, to Dany Garant and Alastair Wilson for insightful discussions, and to all assistants and students who worked on this program over decades. M.F.B., D.W.C., and F.P. are funded by NSERC Discovery Grants. F.P. holds the Canada Research Chair in Evolutionary Demography and Conservation. Our research was also supported by the Government of Alberta, the Université de Sherbrooke and an Alberta Conservation Association Challenge Grant in Biodiversity. The authors declare no conflicts of interest.

Literature cited

Allendorf, F. W., and J. J. Hard 2009. Human-induced evolution caused by unnatural selection through harvest of wild animals. Proceedings of the National Academy of Sciences of the United States of America **106**(Suppl 1):9987–9994.

Carlson, S. M., E. Edeline, L. Asbjørn Vøllestad, T. O. Haugen, I. J. Winfield, J. M. Fletcher, J. Ben James et al. 2007. Four decades of opposing natural and human-induced artificial selection acting on Windermere pike (*Esox lucius*). Ecology Letters **10**:512–521.

Chevin, L.-M. 2015. Evolution of adult size depends on genetic variance in growth trajectories: a comment on analyses of evolutionary dynamics using integral projection. Methods in Ecology and Evolution **6**:981–986.

Coltman, D. W., M. Festa-Bianchet, J. T. Jorgenson, and C. Strobeck 2002. Age-dependent sexual selection in bighorn rams. Proceedings of the Royal Society B: Biological Sciences **269**:165–172.

Coltman, D. W., P. O'Donoghue, J. T. Jorgenson, J. T. Hogg, C. Strobeck, and M. Festa-Bianchet 2003. Undesirable evolutionary consequences of trophy hunting. Nature **426**:655–658.

Coltman, D. W., P. O'Donoghue, J. T. Hogg, and M. Festa-Bianchet 2005. Selection and genetic (co)variance in bighorn sheep. Evolution **59**:1372–1382.

Crosmary, W. G., A. J. Loveridge, H. Ndaimani, S. Lebel, V. Booth, S. D.

Côté, and H. Fritz 2013. Trophy hunting in Africa: long-term trends in antelope horn size. Animal Conservation 16:648–660.

Darimont, C. T., S. M. Carlson, M. T. Kinnison, P. C. Paquet, T. E. Reimchen, and C. C. Wilmers. 2009. Human predators outpace other agents of trait change in the wild. Proceedings of the National Academy of Sciences United States of America 106:952–954.

Devine, J. A., P. J. Wright, H. E. Pardoe, M. Heino, and D. J. Fraser 2012. Comparing rates of contemporary evolution in life-history traits for exploited fish stocks. Canadian Journal of Fisheries and Aquatic Sciences 69:1105–1120.

Douhard, M., M. Festa-Bianchet, F. Pelletier, J.-M. Gaillard, and C. Bonenfant In Press. Changes in horn size of Stone's sheep over four decades correlate with trophy hunting pressure. Ecological Applications.

Festa-Bianchet, M., and J. T. Jorgenson 1998. Selfish mothers: reproductive expenditure and resource availability in bighorn ewes. Behavioral Ecology 9:144–150.

Festa-Bianchet, M., and R. Lee 2009. Guns, sheep, and genes: when and why trophy hunting may be a selective pressure. In B. Dickson, J. Hutton, and W. M. Adams, eds. Recreational Hunting. Conservation and Rural Livelihoods, pp. 94–107. Wiley-Blackwell, Oxford.

Festa-Bianchet, M., D. W. Coltman, L. Turelli, and J. T. Jorgenson 2004. Relative allocation to horn and body growth in bighorn rams varies with resource availability. Behavioral Ecology 15:305–312.

Festa-Bianchet, M., T. Coulson, J. M. Gaillard, J. T. Hogg, and F. Pelletier 2006. Stochastic predation events and population persistence in bighorn sheep. Proceedings of the Royal Society B: Biological Sciences 273:1537–1543.

Festa-Bianchet, M., F. Pelletier, J. T. Jorgenson, C. Feder, and A. Hubbs 2014. Decrease in horn size and increase in age of trophy sheep in Alberta over 37 years. The Journal of Wildlife Management 78:133–141.

Foote, L., and G. W. Wenzel 2009. Polar bear conservation hunting in Canada: economics, culture and unintended consequences. In M. M. R. Freeman, and L. Foote, eds. Inuit, Polar Bear and Sustainable Use, pp. 13–24. CCI Press, Edmonton.

Garel, M., J. M. Cugnasse, D. Maillard, J. M. Gaillard, A. J. Hewison, and D. Dubray 2007. Selective harvesting and habitat loss produce long-term life history changes in a mouflon population. Ecological Applications 17:1607–1618.

Garland, T., and M. R. Rose 2009. Experimental Evolution: concepts, Methods, and Applications of Selection Experiments. University of California Press, Berkeley, 752 pp.

Hadfield, J. D., A. J. Wilson, D. Garant, B. C. Sheldon, and L. E. Kruuk 2010. The misuse of BLUP in ecology and evolution. The American Naturalist 175:116–125.

Hard, J. J., M. R. Gross, M. Heino, R. Hilborn, R. G. Kope, R. Law, and J. D. Reynolds 2008. SYNTHESIS: Evolutionary consequences of fishing and their implications for salmon. Evolutionary Applications 1:388–408.

Hedrick, P. W., D. W. Coltman, M. Festa-Bianchet, and F. Pelletier. 2014. Not surprisingly, no inheritance of a trait results in no evolution. Proceedings of the National Academy of Sciences United States of America 111:E4810.

Hengeveld, P. E., and M. Festa-Bianchet 2011. Harvest regulations and artificial selection on horn size in male bighorn sheep. The Journal of Wildlife Management 75:189–197.

Hunt, G. 2007. The relative importance of directional change, random walks, and stasis in the evolution of fossil lineages. Proceedings of the National Academy of Sciences U.S.A. 104:18404–18408.

Hutchings, J. A., and D. J. Fraser 2008. The nature of fisheries- and farming-induced evolution. Molecular Ecology 17:294–313.

Jones, O. R., and J. Wang 2010. COLONY: a program for parentage and sibship inference from multilocus genotype data. Molecular Ecology Resources 10:551–555.

Jorgenson, J. T., M. Festa-Bianchet, M. Lucherini, and W. D. Wishart 1993. Effects of body size, population density, and maternal characteristics on age at first reproduction in bighorn ewes. Canadian Journal of Zoology 71:2509–2517.

Jorgenson, J. T., M. Festa-Bianchet, and W. D. Wishart 1998. Effects of population density on horn development of bighorn rams. Journal of Wildlife Management 62:1011–1020.

Kijas, J. W., J. A. Lenstra, B. Hayes, S. Boitard, L. R. Porto Neto, M. San Cristobal, B. Servin et al. 2012. Genome-wide analysis of the world's sheep breeds reveals high levels of historic mixture and strong recent selection. PLoS Biology 10:e1001258.

Kruuk, L. E. 2004. Estimating genetic parameters in natural populations using the "animal model". Philosophical Transactions of the Royal Society of London: Series B, Biological Sciences 359:873–890.

Leader-Williams, N., R. J. Smith, and M. J. Walpole 2001. Elephant hunting and conservation. Science 293:2203–2204.

Loehr, J., J. Carey, M. Hoefs, J. Suhonen, and H. Ylonen 2007. Horn growth rate and longevity: implications for natural and artificial selection in thinhorn sheep (Ovis dalli). Journal of Evolutionary Biology 20:818–828.

Martin, J. G. A., and F. Pelletier 2011. Measuring growth patterns in the field: effects of sampling regime and methods on standardized estimates. Canadian Journal of Zoology 89:529–537.

Morrissey, M. L. B., D. J. Parker, P. R. Korsten, J. M. Pemberton, L. E. B. Kruuk, and A. J. Wilson 2012. The prediction of adaptive evolution: empirical application of the secondary theorem of selection and comparison to the Breeder's equation. Evolution 66:2399–2410.

Mysterud, A. 2011. Selective harvesting of large mammals: how often does it result in directional selection? Journal of Applied Ecology 48:827–834.

Pelletier, F., M. Festa-Bianchet, and J. T. Jorgenson 2012. Data from selective harvests underestimate temporal trends in quantitative traits. Biology Letters 8:878–881.

Poissant, J., C. S. Davis, R. M. Malenfant, J. T. Hogg, and D. W. Coltman 2012. QTL mapping for sexually dimorphic fitness-related traits in wild bighorn sheep. Heredity 108:256–263.

Postma, E. 2006. Implications of the difference between true and predicted breeding values for the study of natural selection and microevolution. Journal of Evolutionary Biology 19:309–320.

Réale, D., M. Festa-Bianchet, and J. T. Jorgenson 1999. Heritability of body mass varies with age and season in wild bighorn sheep. Heredity 83:526–532.

Swain, D. P., A. F. Sinclair, and J. Mark Hanson 2007. Evolutionary response to size-selective mortality in an exploited fish population. Proceedings of the Royal Society B: Biological Sciences 274:1015–1022.

Traill, L. W., S. Schindler, and T. Coulson 2014. Demography, not inheritance, drives phenotypic change in hunted bighorn sheep. Proceedings of the National Academy of Sciences United States of America 111:13223–13228.

Uusi-Heikkilä, S., A. R. Whiteley, A. Kuparinen, S. Matsumura, P. A. Venturelli, C. Wolter, J. Slate et al. 2015. The evolutionary legacy of size-selective harvesting extends from genes to populations. Evolutionary Applications 8:597–620.

van Wijk, S. J., M. I. Taylor, S. Creer, C. Dreyer, F. M. Rodrigues, I. W.

Ramnarine, C. van Oosterhout et al. 2013. Experimental harvesting of fish populations drives genetically based shifts in body size and maturation. Frontiers in Ecology and the Environment 11:181–187.

Wilson, A. J., D. W. Coltman, J. M. Pemberton, A. D. Overall, K. A. Byrne, and L. E. Kruuk 2005. Maternal genetic effects set the potential for evolution in a free-living vertebrate population. Journal of Evolutionary Biology 18:405–414.

Wilson, A. J., L. E. B. Kruuk, and D. W. Coltman 2005. Ontogenetic patterns in heritable variation for body size: using random regression models in a wild ungulate population. The American naturalist 166: E177–E192.

Wilson, A. J., D. Réale, M. N. Clements, M. M. Morrissey, E. Postma, C. A. Walling, L. E. Kruuk et al. 2010. An ecologist's guide to the animal model. The Journal of Animal Ecology 79:13–26.

Wolak, M. E., D. A. Roff, and D. J. Fairbairn 2015. Are we underestimating the genetic variances of dimorphic traits? Ecology and Evolution 5:590–597.

9

Evolution of invasive traits in nonindigenous species: increased survival and faster growth in invasive populations of rusty crayfish (*Orconectes rusticus*)

Lindsey W. Sargent[1] and David M. Lodge[2]

1 Department of Biological Sciences, University of Notre Dame, Notre Dame, IN, USA
2 Department of Biological Sciences and Environmental Change Initiative, University of Notre Dame, Notre Dame, IN, USA

Keywords
adaptation, antipredator behavior, fish, food quality, growth rate, invasive species, mortality, predator.

Correspondence
Lindsey W. Sargent, Department of Biological Sciences, University of Notre Dame, Notre Dame, IN 46556, USA.

e-mail: lsargen1@nd.edu

Abstract

The importance of evolution in enhancing the invasiveness of species is not well understood, especially in animals. To evaluate evolution in crayfish invasions, we tested for differences in growth rate, survival, and response to predators between native and invaded range populations of rusty crayfish (*Orconectes rusticus*). We hypothesized that low conspecific densities during introductions into lakes would select for increased investment in growth and reproduction in invasive populations. We reared crayfish from both ranges in common garden experiments in lakes and mesocosms, the latter in which we also included treatments of predatory fish presence and food quality. In both lake and mesocosm experiments, *O. rusticus* from invasive populations had significantly faster growth rates and higher survival than individuals from the native range, especially in mesocosms where fish were present. There was no influence of within-range collection location on growth rate. Egg size was similar between ranges and did not affect crayfish growth. Our results, therefore, suggest that growth rate, which previous work has shown contributes to strong community-level impacts of this invasive species, has diverged since *O. rusticus* was introduced to the invaded range. This result highlights the need to consider evolutionary dynamics in invasive species mitigation strategies.

Introduction

Evolution in nonindigenous populations contributes to the success and harmful impacts of some invasive species (Siemann and Rogers 2001; Handley et al. 2011), but how often this phenomenon occurs, especially for animals, is inadequately understood. Nonindigenous species can dramatically change biotic communities and ecosystem processes, sometimes causing extensive ecological or economic harm (Sala et al. 2000; Keller et al. 2009; Butchart et al. 2010). However, only a small percentage of nonindigenous species have strong community level impacts (Ricciardi and Kipp 2008). *r*-selected life history traits such as rapid growth and high fecundity are common among many of those nonindigenous species that have strong impacts (Sakai et al. 2001; van Kleunen et al. 2010; Lamarque et al. 2011), and characteristics of the environment and biotic community within the introduced range are also often important for invasion success (Catford et al. 2009). A subset of introduced species already possess *r*-selected life history traits upon arrival in a novel habitat, but other species may evolve toward these traits in response to selection in the invaded range.

Theory suggests that populations with lower conspecific densities should have greater *r*-selection (Lewontin 1965), and recent empirical data support this theory. For example, cane toads (*Bufo marinus* Linnaeus) from populations on an expanding range edge grow more rapidly than those from longer established populations when raised in common conditions and, therefore, reach reproductive maturity more quickly (Phillips 2009). In addition, a recent review of trait evolution in nonindigenous populations reveals that some invasive populations have evolved faster growth, wide environmental tolerance, shorter generation time, and increased reproductive capability in the invaded range (Whitney and Gabler 2008). However, few studies have tested for differences in invasive traits between populations using reciprocal transplant or common garden experiments (as opposed to comparative field observations), and almost all of these studies focus on introduced plants (but see Koskinen et al. 2002; Lee et al. 2003; Phillips 2009). Therefore, it remains unclear how often invasive traits evolve in nonindigenous animal populations. Though evolution within the invasive range may alter the impact of nonindigenous species, evolutionary potential is rarely included in risk assessments and policy decisions involving species introductions (Whitney and Gabler 2008).

To evaluate the likelihood of evolution influencing crayfish invasiveness, we conducted a series of common garden experiments to test whether differences existed in growth rate, survival, and response to predators in young of year (YOY) rusty crayfish (*Orconectes rusticus* Girard) from native and invasive populations. *O. rusticus* is one of many species of crayfish that have been introduced globally (Lodge et al. 2012). *O. rusticus*, in particular, cause major community level impacts in their invaded range. *O. rusticus* is native to streams in the Ohio River drainage in Ohio, Indiana, and Kentucky and was introduced by anglers to northern Wisconsin and Michigan lakes in the mid 1960s as well as to Illinois, Minnesota, Ontario (Canada), the Laurentian Great Lakes, and portions of 11 other states (Olden et al. 2006; Peters et al. 2014).

For this study, we focused on comparing well-studied invasive *O. rusticus* populations from northern Wisconsin to lesser studied native populations from the Ohio River drainage. Where *O. rusticus* has become abundant in Wisconsin and Michigan lakes, it has displaced resident crayfishes, reduced the abundance and richness of macrophytes and macroinvertebrates, and caused declines in the abundance of panfish (*Lepomis* spp.) (Capelli 1982; Lodge et al. 1994; Wilson et al. 2004; Olden et al. 2006). Faster growth of *O. rusticus* also contributes to the displacement of resident crayfishes (Hill et al. 1993; Garvey et al. 1994). In addition, *O. rusticus* reaches higher densities than other crayfishes in this region, which contributes to its strong impacts (Wilson et al. 2004). To our knowledge, the community level impacts of *O. rusticus* in the native range have not been investigated. Pintor and Sih (2009) found that *O. rusticus* from an invasive population grew more rapidly than *O. rusticus* from a native population when competing with congeners in mesocosms. However, because adult crayfish collected from the field were used in this study, it is unclear whether this result was due to evolution or to environmental differences between the two collection locations. Here, we use experiments to test for divergence in *r*-selected traits, specifically YOY growth rate and survival, in invasive *O. rusticus* populations in Wisconsin.

To determine whether there are widespread growth rate and survival differences between *O. rusticus* from native

and invasive populations, we first reared crayfish from both ranges in enclosures in three lakes within the invaded range in summer 2011. We selected lakes with different abundances of predatory fish and macroinvertebrate prey to determine whether differences existed in growth rate or survival among different invasive range environments. Then in summer 2012, to provide evidence for the hypothesis that there is a genetic basis for the differences we observed, we reared crayfish in mesocosms where we controlled temperature and varied the presence of predatory fish and food quality to determine which factors were important in controlling *O. rusticus* growth rate and survival. Previous research indicates that predatory fish can reduce crayfish feeding activity (Stein and Magnuson 1976; Hill and Lodge 1995) and growth (Hill and Lodge 1999). We hypothesized that crayfish from the invaded range would respond less (i.e. smaller reduction in feeding activity) to predatory fish than those from the native range because there is likely to be a greater fitness benefit to allocating time to feeding (growth) within the invaded range. In addition, food quality (Hill et al. 1993) and temperature (Mundahl and Benton 1990) are important for crayfish growth. It is possible that rapid growth of invaded range crayfish can only be achieved in locations with abundant, high quality food resources. Our study is the first to test whether food quality and predator abundance have different effects on *O. rusticus* from native and invasive populations. Finally, we investigated the potential influence of maternal effects on results by testing for the effects of clutch and initial egg weight. Our study examines whether there are growth differences between replicated populations of rusty crayfish from the invaded and native range, and whether the observed differences have an environmental or genetic basis.

Methods

Lake common garden experiment

To test whether differences exist in growth rate and survival between native and invasive *O. rusticus*, we reared YOY crayfish from native and invasive populations in enclosures in invaded range lakes in summer 2011. We hand collected berried females (females with eggs attached to their abdomen) from the Great Miami (39°56'N, 83°44'W and 39°56'N, 83°42'W) and Little Miami (38°54'N, 83°34'W) river drainages in the native range in Ohio, USA and from High Lake (46°08'N, 89°32'W), Big Lake (46°11'N, 89°26'W), and Papoose Lake (46°10'N, 89°48'W) in the invaded range in Wisconsin, USA. Because temperatures are warmer in the native range than in the invaded range, *O. rusticus* females extrude eggs earlier in the native range. We collected females from streams in the native range in late April and from lakes in the invaded range in late May.

Each female was placed in an individual container (18 × 18 cm) in the laboratory with constantly aerated well water, a shelter constructed from a polyvinyl chloride pipe, and gravel substrate. Eggs hatched, and young became independent from females 3–4 weeks after collection. Females were removed from containers once young were independent. YOY were fed a combination of spirulina disks and shrimp pellets *ad libitum* while in the laboratory. YOY were placed in lakes 1–2 weeks after they became independent from females (in late May for native range YOY and early July for invaded range YOY). All YOY were removed from lakes on September 9th after native range crayfish were in lakes for 15 weeks and invaded range crayfish were in lakes for 10 weeks.

Within lakes, crayfish were each housed in an individual clear plastic container (18 × 18 × 12.7 cm) with large rectangular holes (14 × 8 cm) cut into each side and replaced with window screen. Screened sides prevented crayfish escape, but allowed crayfish to be in contact with the physical and chemical lake environment and to receive visual cues from predators. Four to six grams of natural woody debris (sticks) were added to each container and four stones were glued to the bottom to increase container weight and provide shelter. Containers were placed between 0.25 and 0.5 m depth at one of two sites (sites were 50–100 m apart) in each of three lakes, Big Lake, High Lake, and Papoose Lake in Wisconsin, USA. Two sites were used in each lake to hedge against total loss of crayfish in the case of disturbance by humans, other animals, or storms. Each site contained 1 YOY from each brood (13 invasive females and 13 native females), so that there were a total of 26 YOY housed at each site and 52 YOY housed in each lake.

Lakes were chosen based on different invasion histories. Papoose Lake had high densities of *O. rusticus* (35 *O. rusticus* per trap in 2011), and therefore we expected this lake to have reduced macrophyte, macroinvertebrate, and panfish populations resulting from *O. rusticus* impacts (Wilson et al. 2004). High Lake had low densities of *O. rusticus* (4 *O. rusticus* per trap in 2011), and Big Lake had moderate densities of *O. rusticus* (19 *O. rusticus* per trap in 2010).

Total length of YOY was measured every 6–8 days. Growth rate was calculated as the difference between initial and final length divided by days in the experiment. Mortalities that occurred within the first 3 weeks of the experiment were replaced with individuals from the same brood that were housed in the laboratory with the same husbandry and conditions as provided after hatching.

Data from a previous preliminary experiment in Big Lake indicated that Big Lake YOY grown in containers (and fed only though natural colonization of the containers by macroinvertebrates) were equal in length to those YOY growing outside of containers at the end of the summer.

Therefore, we did not add food to the experimental containers. Containers in all three lakes were quickly colonized by macroinvertebrates which provided food for crayfish. To provide an index of any differences in food availability among lakes, six containers without crayfish were placed in each lake in June. Macroinvertebrates were removed from these containers in August and preserved in 70% ethanol. We compared the ash free dry mass of these macroinvertebrates among lakes.

We also assessed temperature and the abundance of predatory fishes in each lake because they might affect crayfish growth rates. Hourly temperature was recorded at each site using temperature loggers (Onset Computer Corporation). Predatory fish abundance was assessed in each lake using three fyke nets set for one night at the end of June and one night at the end of July. Fyke nets were set within 50 m of crayfish sites, and thus were intended to measure fish activity at those specific locations.

Mesocosm common garden experiment

To examine whether differences in growth rate and survival observed between YOY crayfish from native and invasive populations could be genetically based, and to identify important factors influencing growth rate and survival, we raised crayfish in common conditions in mesocosms in summer 2012. We used a 2 × 2 × 2 factorial design to examine the effects of range, predators, and food quality on the growth rate, survival, and behavior of rusty crayfish. As in the 2010 lake experiments, berried females were collected earlier from native range locations due to differences in reproductive timing between the two ranges. We hand collected berried females in early April from the Little Miami (38°54′N, 83°34′W and 39°47′N, 83°51′W) and Scioto River (40°00′N, 83°23′W) drainages in Ohio, USA. In early May, we collected berried females from High Lake (46°08′N, 89°32′W), Big Lake (46°11′N, 89°26′W), and Papoose Lake (46°10′N, 89°48′W) in Wisconsin, USA. Housing of females in the laboratory and all other husbandry practices were the same as for lake experiment unless specified below. YOY from the native range were placed in experimental mesocosms in late May, and YOY from the invaded range were placed in experimental mesocosms in late June. While native and invasive YOY were placed in mesocosms at different times during the summer, we controlled temperature and food availability so that all crayfish experienced the same environmental conditions throughout the experiment (as described below).

Within each mesocosm, ten invaded range and ten native range YOY crayfish were each housed individually in a clear plastic container with screened sides (identical to those used in lake experiments), so that the growth of each crayfish was independent and was not affected by the other crayfish in the mesocosm. Two stones were glued to one side of the bottom of the container to provide shelter for crayfish. On the opposite side, we attached a small nylon nut and bolt which held disks of prepared food (described below) securely in place. Crayfish, therefore, had to choose between feeding and hiding. Total length of crayfish was measured once every 7 days, and crayfish were removed from the experiment after 7 weeks. We replaced mortalities that occurred within the first 2 weeks of the experiment with crayfish from the same range and, if possible, the same brood. Replacement crayfish were housed in the laboratory with the same husbandry and conditions as provided after hatching.

Mesocosms consisted of 416 L plastic tanks with flow-through, aerated well-water, and were located in a wooded area on the shore of Trout Lake (Wisconsin, USA), under a suspended tarp to reduce light, falling debris, and heat load. There were 12 mesocosms in total, with 20 YOY *O. rusticus* in individual containers (10 invasive and 10 native) reared in each mesocosm. Temperature was maintained in each mesocosm by a 300 W heater, and each mesocosm was aerated to maintain high dissolved oxygen (8–10 mg/L) and uniform temperature. Hourly temperature was recorded in each mesocosm using temperature loggers (Onset Computer Corporation; mean temperature = 18.1°C, summertime range = 10–25°C). These temperatures are cooler than summer invaded range lake temperatures (mean temperature was 24.3°C in lake epilimnia during the first 7 weeks of crayfish growth in 2011); however, we were only able to heat well water to this extent in early summer, and needed to keep temperatures consistent later in the summer so that invaded range crayfish experienced the same conditions as native range crayfish. We tested whether temperature was different during native and invaded range crayfish growth periods using ANOVA of average weekly temperature in each mesocosm. To examine whether predators had an effect on growth, six of the twelve mesocosms contained predatory fish (three bluegill, *Lepomis macrochirus* Rafinesque, and three smallmouth bass, *Micropterus dolomieu* Lacépède). Bluegill ranged in size from 9.5 to 13 cm total length during the experiment, and smallmouth bass ranged in size from 10.5 to 14.5 cm total length. Fish were fed *O. rusticus* (three per mesocosm) once per week and earthworms (*Lumbricus terrestris* Linnaeus) twice per week for the duration of the experiment. Fish readily consumed both food types. Bluegill and smallmouth bass are common in both the native and invaded range of *O. rusticus* (Boschung et al. 1983).

To examine the effect of food quality of crayfish growth, half of the crayfish in each mesocosm were fed high quality food and half were fed low quality food, which we created by mixing 500 mL of plant and animal matter with 20 g of sodium alginate and 750 mL of water. Using methods similar to those used by Cronin et al. (2002), we solidified food by pouring dissolved calcium chloride (14 g calcium chloride in 500 mL water) over a thin sheet of this food mixture. Food was cut into 2.5–5 g squares and secured in each container weekly with the nylon nut and bolt. Except for a few of the largest crayfish at the end of the experiment, some food remained in each container at the end of each week, so the amount of food was *ad libitum*. High quality food consisted of 40% macrophytes (*Potamogeton amplifolius* Tuck, *Potamogeton richardsonii* (Bennett) Rydberg, and *Sagittaria graminea* Michaux) and 60% animal matter (earthworms and bluegill filets). Low quality food was made from the same organisms, but contained 80% macrophytes and 20% animal matter. All food was frozen after it was made, and thawed within 1 day of placing it in the experiment. Native and invasive crayfish were fed from the same batch of food during the same week of growth.

Statistical analyses

For the lake common garden experiment, we used ANOVA to examine the effects of range (native or invasive) and lake on growth rate (mm/day). We also included initial length of YOY and maternal identity (clutch) as covariates to account for potential effects of these variables. Only crayfish that survived for the entirety of the experiment were used in the growth analysis. We also used ANOVA to examine the effect of the collection location within each range on growth. For this analysis, we ran one ANOVA for native range crayfish and one ANOVA for invaded range crayfish, and also included the effect of lake (where YOY were housed) in each model. Initial length and clutch were not included in this second analysis because they were found to be unimportant in the first growth model. Because native and invaded range crayfish were placed in lakes at different times, we also conducted an analysis of weekly growth to better account for the varying effects of temperature and crayfish length throughout the summer. We used a linear mixed effects model to examine the effects of range, lake, length of crayfish (at the start of the week), and average temperature (during the one week growth period) on weekly crayfish growth. We included crayfish identity in this model as a random effect.

For the mesocosm common garden experiment, we used a linear mixed effects model to examine the effects on crayfish growth rate of range, fish, food quality, and all interactions between these variables. In addition, mesocosm nested within fish treatment was included as a random effect. We also included the effects of average temperature and initial length in the model to account for potential effects of these variables. Although initial length was found to be unimportant in the lake common garden analysis, we included it in this analysis because YOY in the mesocosm common garden were housed in the laboratory slightly longer; therefore, variance in initial size was greater and could have had a more substantial effect on growth. Because each crayfish was reared in a separate, individual container within the 12 mesocosms and was provided with its own food, the growth of each crayfish was independent and was not affected by the growth of other crayfish in the same mesocosm. Therefore, we analyzed the effects of range, food quality, temperature, and initial length at the individual crayfish level, and controlled for the influence of the mesocosm by including it as a random effect. On the other hand, fish treatment was applied at the mesocosm level, and thus we nested the effect of fish within mesocosm in the analysis, so that we conducted this analysis at the mesocosm level. We did not test for the effect of clutch in this analysis because YOY from females were not evenly divided between treatments as in the lake experiment. YOY used in this analysis came from 53 different clutches, with an average ± standard error (SE) of 3.75 ± 0.26 crayfish per clutch. It is, therefore, unlikely that the genotype of any one parent would drive growth rate trends. We included all crayfish that survived for at least 30 days in this analysis, so that we could increase our sample size while allowing sufficient time for crayfish to grow. There was no significant effect of survival time (30–50 days) on growth rate (invasive: $P = 0.84$, $r^2 = 0.0005$; native: $P = 0.57$, $r^2 = 0.006$). As for the lake experiment, we tested the effect of collection location within each range on growth rate. For this analysis, we used separate linear mixed models for crayfish from the native and invaded range and included the effects of fish treatment, average temperature, and initial length as fixed effects as well as mesocosm nested within fish treatment as a random effect. Food quality was not included because it was found to be unimportant in the first growth model.

For both the lake and mesocosm experiments, we used Cox Proportional Hazards Models to test the effect of range on YOY survival. We included lake as a fixed effect in the lake experiment model and predatory fish treatment and food quality as fixed effects in the mesocosm experiment model. We also included mesocosm nested within fish treatment as a random effect in the mesocosm experiment model.

To examine whether maternal investment was important for growth rate differences between native and invaded range crayfish, we compared egg mass between native and invaded range females in spring 2012. We obtained blotted wet weight for five to nine eggs from each of six females from the native range and nine females from the invaded range. Because female size may also affect egg mass, we analyzed these data using ANCOVA to test the effects of range (native or invasive) and maternal carapace length on egg weight. We found that maternal carapace length was an important predictor of egg mass, and therefore could use

maternal carapace length as an index of egg size for those crayfish grown in the common garden. We could not directly measure egg size for crayfish used in the experiment because removal of eggs from females causes egg mortality. We used ANCOVA to test the effects of range, maternal carapace length (as a index of egg size), and their interaction on the initial and final length of YOY, and a linear model to examine how range, maternal carapace length, and their interaction influenced YOY growth rate. We also included fish treatment and average temperature as fixed effects in the linear model because these were important factors controlling growth rate. To examine how maternal carapace length (as a proxy for egg size) influenced YOY survival, we added maternal carapace length to the Cox Proportional Hazards Model for the mesocosm experiment.

In the mesocosm experiment, we also tested whether crayfish behavior differed between fish treatments by recording the location of each YOY when containers were opened once a week to measure crayfish and replace food. Starting in the fourth week of crayfish growth, we recorded the crayfish as 'in shelter' if it was under or motionless next to the rocks or screened sides of the container and 'out of shelter' if it was found away from the rocks and screen. We quantified the percentage of observations that were classified as 'out of shelter' for all native and invaded range crayfish in each mesocosm. We tested the effects of range and fish treatment on the percent of observations out of shelter in a mixed effects model with mesocosm included as a random effect.

Results

Lake common garden experiment

Over the course of the summer, *O. rusticus* from invasive populations grew more rapidly than *O. rusticus* from native populations ($F_{1,83} = 22.13$, $P = 0.0033$); lake ($F_{2,82} = 73.87$ $P < 0.0001$) and the interaction between lake and range also significantly affected growth rate ($F_{2,82} = 8.56$, $P = 0.0175$; Fig. 1A). Crayfish from invasive populations grew about 20% faster than crayfish from native populations in Big Lake and High Lake, but growth rates were similar between native and invaded range crayfish and about 30% slower in Papoose Lake (Fig. 1A). There was no significant effect of clutch or initial length on growth rate, or any other significant interactions between range, lake, clutch, or initial length ($P > 0.4$; Table S1).

We tested for differences in temperature between lakes to determine if temperature could cause differences in YOY growth rate observed between lakes. Temperature did not differ greatly between lakes. Average temperature in Big Lake (24.5°C) was very similar to that recorded in Papoose Lake (24.4°C) and High Lake (24.1°C) (see details in Fig. S1).

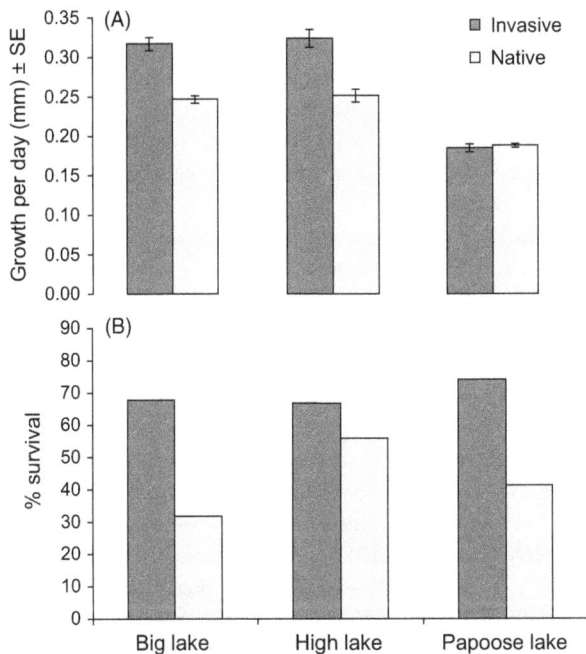

Figure 1 (A) Growth rate of *O. rusticus* from native and invasive range populations in lake common gardens. (B) Percent survival of native and invasive range crayfish over the course of the lake common garden experiment.

In contrast to temperature, we found substantial differences in fish and invertebrate abundance between lakes. Predatory fish species collected in fyke nets included bluegill, pumpkinseed (*L. gibbosus* Linnaeus), smallmouth bass, largemouth bass (*M. salmoides* Lacépède), rock bass (*Ambloplites rupestris* Rafinesque), and yellow perch (*Perca flavescens* Mitchill). Predatory fish were most abundant in High Lake (397 fish per trap night) followed by Big Lake (33 fish per trap night) and then Papoose Lake (19 fish per trap night) (see details in Table S2). Thus, growth rate differences were not consistent with inhibition of feeding in the presence of predatory fishes. Differences in invertebrate colonization among lakes were consistent with differences in growth rate of crayfish among lakes: invertebrates colonizing containers were most abundant in High Lake (0.066 ± 0.37 g ash free dry mass ± SE) followed by Big Lake (0.025 ± 0.006 g ash free dry mass ± SE), the two lakes where growth rates were highest, and then Papoose Lake (0.007 ± 0.001 g ash free dry mass ± SE) (see details in Fig. S2).

We also tested for within-range variation in growth rates to determine whether growth rate differences occur throughout the native and invaded range or were dependent on sampling location. While there was a significant impact of range on growth rate in the lake experiment, there was little within-range variation. For crayfish from the invaded range, lake of origin (population) was not a

significant predictor of growth rate ($F_{2,52} = 2.41$, $P = 0.1011$), and there was no significant interaction between population and the lake where YOY were grown ($P > 0.05$). Similarly, for crayfish from the native range, river of origin (population) was not a significant predictor of growth rate ($F_{1,29} = 0.57$, $P = 0.4589$) and there was no interaction between population and the lake where YOY were grown ($P > 0.05$).

Crayfish from the invaded range also grew more rapidly than those from the native range in the weekly growth analysis ($F_{1,695} = 7.74$, $P = 0.0069$), and there was still a significant effect of lake on growth rate ($F_{1,695} = 17.48$, $P < 0.0001$). Temperature was also important for weekly growth ($F_{1,695} = 17.48$, $P < 0.0001$), and there was an interaction between length and lake indicating that there was no effect of length on growth rate in some lakes, but larger crayfish grew more slowly in other lakes ($F_{1,695} = 8.53$, $P = 0.0002$). There was also an interaction between range and length whereby larger native range crayfish grew more slowly, but there was no effect of length on growth rate in invaded range crayfish ($F_{1,695} = 10.34$, $P = 0.0014$).

In addition to differences in growth rate, we also tested for differences in survival. Native range crayfish were about 12% less likely to survive than invaded range crayfish within invaded range lakes (Cox Proportional Hazards Model coefficient = 1.1206, $z_{1,201} = 3.056$, $P = 0.0022$; Fig. 1B). Neither lake nor the interaction between lake and range were significant predictors of survival ($P > 0.1$).

Mesocosm common garden experiment

Growth results in the mesocosm experiment were similar to those from the lake experiment. Crayfish from invasive populations grew about 50% to 120% faster (depending on fish treatment and food quality) than crayfish from native populations ($F_{1,145} = 21.41$, $P < 0.0001$; Fig. 2A). We also found effects of fish presence and temperature on growth. Growth rates were about 40% lower in mesocosms with fish present ($F_{1,11} = 5.48$, $P = 0.0412$; Fig. 2A) and increased with temperature ($F_{1,145} = 6.00$, $P < 0.0001$). Despite our attempts to control temperature, crayfish from native populations experienced slightly warmer temperatures on average than crayfish from invasive populations ($F_{1,145} = 6.01$, $P = 0.0154$). Average temperature experienced by native range crayfish (±SE) was 18.4 ± 0.2°C, while average temperature experienced by invaded range crayfish (±SE) was 17.8 ± 0.2°C. The slower growing, native range crayfish experienced warmer temperatures; therefore, the positive effect of temperature on growth rate was weaker than the effect of range. There was no significant effect of food quality ($F_{1,145} = 0.39$, $P = 0.5328$) or initial length ($F_{1,145} = 0.55$, $P = 0.4586$) on crayfish

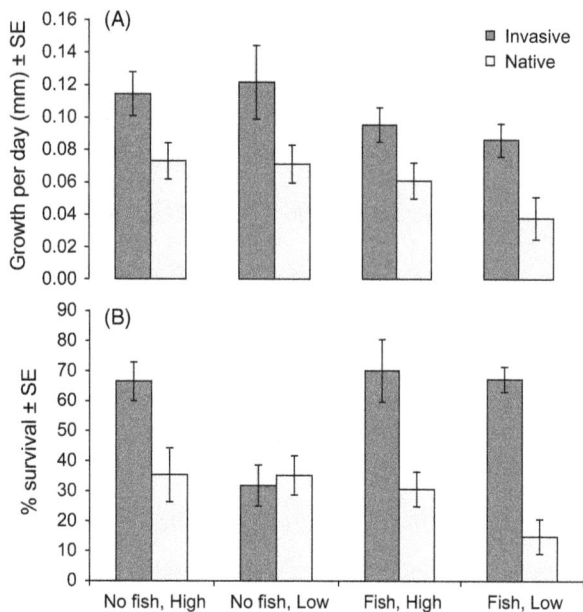

Figure 2 (A) Growth rate of *O. rusticus* from native and invasive range populations in mesocosm common gardens. (B) Percent survival of native and invasive range crayfish over the course of the mesocosm common garden experiment. Treatments include predatory fish absent or present × high or low quality food.

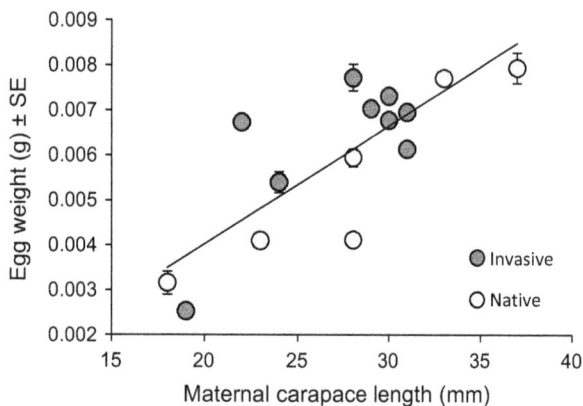

Figure 3 Relationship between maternal carapace length and egg weight in native and invasive range *O. rusticus*.

growth. Average initial length ±SE of YOY at the start of the experiment (when placed in the mesocosms) was 13.9 ± 0.2 mm for crayfish from invasive populations and 13.2 ± 0.2 mm for crayfish from native populations. There was a significant interaction between initial length and range ($F_{1,145} = 6.99$, $P = 0.0093$), indicating that growth rate decreased with initial size in invaded range crayfish ($r^2 = 0.03$), but increased with initial size in native range crayfish ($r^2 = 0.08$). All other interactions between range, fish treatment, food quality and initial length were non-significant ($P > 0.1$; Table S3).

Also similar to the lake experiment, no significant effect existed of within-range lake or river of origin on growth rate for either invaded range or native range crayfish ($F_{1,92} = 0.24$, $P = 0.7881$, and $F_{1,52} = 1.31$, $P = 0.2859$, respectively). In addition, there were no significant interactions between within-range population and any other variable in the models ($P > 0.1$).

As in the lake experiment, crayfish from native populations were about 12% less likely to survive during the experiment than crayfish from invasive populations across treatments (Table 1; Fig. 2B). In addition, crayfish that received low quality food were roughly 13% less likely to survive than crayfish that received high quality food (Table 1; Fig. 2B). Significant interactions existed between range, fish treatment, and food quality on crayfish survival (Table 1; Fig. 2B). Overall, within crayfish from the invaded range, individuals had the lowest survival when fish were absent and they received low quality food. Within crayfish from the native range, individuals had the lowest survival when fish were present and they received low quality food.

Maternal effects

Overall, there was little evidence for significant effects of egg weight on growth rate or survival. There was no significant difference in egg weight between crayfish from native and invasive populations ($F_{1,11} = 3.16$, $P = 0.1030$; Fig. 3), and no interaction between range and maternal carapace length on egg weight ($F_{1,11} = 0.01$, $P = 0.9250$; Fig. 3), indicating that native and invaded range females of the same size produced eggs of the same size. However, larger females from both ranges produced significantly larger eggs than small females ($F_{1,11} = 24.82$, $P = 0.0004$; Fig. 3).

Further, while larger females produced larger young, there was no effect of maternal size on growth rate or survival. At the beginning of the mesocosm experiment (when YOY were placed in mesocosms), there was a significant positive relationship between maternal carapace length (as a proxy for egg size) and carapace length of offspring

Table 1. Cox Proportional Hazards Model for crayfish survival in the mesocosm common garden experiment. A total of 333 crayfish were used in this analysis.

Factor	Coefficient	Z	P
Range (native)	1.122	3.16	0.0016*
Fish (present)	0.106	0.25	0.8000
Food Quality (low)	1.128	3.18	0.0015*
Range*Fish	0.042	0.09	0.9300
Range*Food Quality	−1.178	−2.67	0.0075*
Fish*Food Quality	−0.988	−1.88	0.0600
Range*Fish*Food Quality	1.254	1.99	0.0470*

*$P < 0.05$.

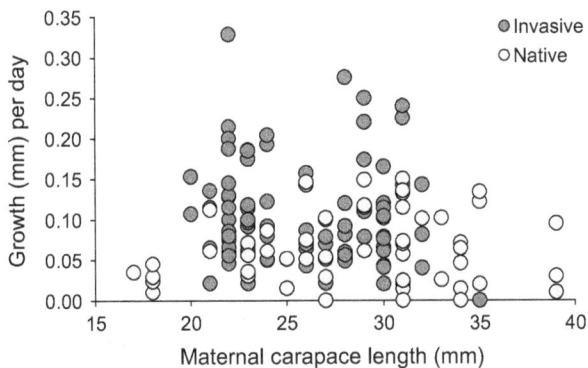

Figure 4 Relationship between maternal carapace length (as a proxy for egg size) and growth per day in native and invasive range *O. rusticus*.

$(F_{1,142} = 33.21, \ P < 0.0001)$. There was also a significant interaction between maternal carapace length and range on offspring size $(F_{1,142} = 5.26, \ P = 0.0233)$, indicating that maternal carapace length had a greater positive effect on invaded range offspring than native range offspring. At the end of the experiment, maternal carapace length still had a positive influence on offspring size $(F_{1,142} = 6.32, \ P = 0.0131)$, but the interaction between maternal carapace length and range on offspring size was non-significant $(F_{1,142} = 0.08, \ P = 0.7810)$. The relationship between maternal carapace length and offspring growth rate during the experiment was also non-significant $(F_{1,142} < 0.01, \ P = 0.9823;$ Fig. 4), and there were no significant interactions between maternal carapace length and any other variable in the growth model $(P > 0.1)$. Further, there was no significant effect of maternal carapace length on survival (coefficient = 0.0268, $z_{1,332} = 0.29, \ P = 0.77$) nor any significant interaction between maternal carapace length and any other variable in the Cox Proportional Hazards Model $(P > 0.4)$. Overall, these results indicate that egg weight did not drive the observed differences in growth rate and survival between native and invaded range crayfish.

Crayfish behavior

Crayfish behavior differed between fish treatments. Crayfish in mesocosms without fish were more likely to be found outside of shelter $(F_{1,23} = 27.11, \ P < 0.0001)$. In addition, there was a non-significant trend suggesting that crayfish from native populations spent more time outside of shelter than crayfish from invasive populations $(F_{1,23} = 3.24 \ P = 0.0878)$. Invaded range crayfish were found outside of shelter $(\pm SE)$ 75 ± 5% of the time in mesocosms without fish and 52 ± 2% of the time in mesocosms with fish, and native range crayfish were found outside of shelter 86 ± 1% of the time in mesocosms without

fish and 59 ± 7% of the time in mesocosms with fish. There was no interaction between range and fish treatment on behavior $(P = 0.65)$.

Discussion

Growth rate differences

In both lake and mesocosm common garden experiments, invasive crayfish had faster growth rates than native crayfish. Data indicate that these growth rate differences were not due to differences in egg weight between the two ranges. Overall, these findings are consistent with evolution of faster growth rates within the invaded range. While larger females initially produced larger young (presumably because of the positive relationship between maternal carapace length and egg weight), there was no significant effect of maternal length on growth. In addition, in both lake and mesocosm experiments we found that within each range young collected as eggs from different lakes or streams had similar growth rates. These results provide further evidence that the observed growth rate differences are due to differences that characterize the ranges (native vs. invaded) rather than sampling locations within each range.

While our data are consistent with evolution of faster growth in invasive populations of *O. rusticus*, we cannot completely rule out the influence of maternal effects. However, maternal effects are less likely to control the differences we observed in growth rates than genetic differences because eggs from the largest females in our study were roughly 3× larger than the eggs from the smallest females, and we did not detect a significant effect of this difference in egg size on YOY growth rate (Fig. 4). Other research has found little influence of maternal effects on offspring quality in other decapods (Tropea et al. 2012; Swiney et al. 2013), or that maternal effects scale with female size (Sato and Suzuki 2010). However, there could potentially be other maternal effects such as differences in hormones or specific nutrients within eggs that could affect growth rate. We intentionally collected females from lakes with variable invertebrate prey availability, and there was no effect of within-range lake or river of origin on growth rate in either common garden experiment. We therefore expect that the differences in growth rates we observed were most likely genetically based.

Within the lake common garden experiment, we placed crayfish from the native range in the lakes earlier than crayfish from the invaded range because of differences in timing of reproduction. Growth differences, therefore, could have been due to differences in temperature and/or food availability during the initial weeks of YOY growth. However, because we were able to control the external environment including temperature and food availability in mesocosms, our results are consistent with a genetic basis for the observed differences in growth rates between native and invaded range crayfish. Lake experiments suggest that this phenomenon occurs not only in the laboratory, but also in natural environments.

Although we were not completely successful controlling temperature in the mesocosm experiment, the differences in temperature experienced by invasive and native range crayfish are not consistent with the differences in growth rate between populations (i.e., if temperature had been the primary driver of growth rates, the differences in growth rate would have been in the opposite direction from those we observed). Still, because crayfish were reared at different times in the mesocosm experiment, it is possible that an unmeasured factor affecting growth rate could have influenced our results. However, this is unlikely because we observed consistent differences in growth rate between native and invaded range crayfish across mesocosms where we varied important factors such temperature, food availability and predator presence. Eggs were also exposed to the environment within their lake or river of origin for a few weeks before collection. We also think this is less likely than genetic differences to be responsible for the observed differences in growth rate because we collected eggs from diverse environments within each range and the majority of egg development occurred in identical conditions in the laboratory.

Our results suggest that food availability differences among lakes were important for differences in crayfish growth rate. In the lake common garden experiment, the positive relationship between prey abundance and crayfish growth rate suggests that food availability was an important driver of growth rate differences; however, we found no effect of food quality on growth rate in the mesocosm experiment. It is possible that less food was available in the lake with the lowest invertebrate biomass (Papoose Lake) than we provided in the low quality food treatment of the mesocosm experiment (which still contained 20% animal matter). We expect that providing less food in the mesocosm experiment would have made food quality an important predictor of growth rate in this experiment as well.

Predatory fish presence, in contrast, was a significant predictor of crayfish growth rate in the mesocosm experiment but not in the lake experiment (i.e., the lake with the slowest growth had the lowest abundance of predatory fish). We expect that the effect of fish was more pronounced in the mesocosm experiment because fish were completely absent from some mesocosms but were present at different densities in all lakes. Behavioral data suggest that reduced growth rates associated with fish presence are likely due to the behavioral response of crayfish to fish. Results indicated that crayfish spent more time hiding (and therefore not consuming food) when fish were present. Together these data suggest that nonconsumptive effects of

fish do reduce crayfish growth rates, but in natural systems, low densities of predators can have similar effects to high densities of predators.

The mesocosm experiment also revealed that initial length and temperature were important predictors of YOY growth rate. Invaded range crayfish were slightly larger on average than native range crayfish at the start of the mesocosm experiment (by an average of 0.7 mm carapace length), and there was a significant interaction between initial length and range on growth. Larger invaded range crayfish tended to grow slower over the course of the experiment, which is consistent with the well-documented pattern of declining growth rate with increasing size in many animals (Ricklefs 1967). Native range crayfish did not get as large as invaded range crayfish in the mesocosm study which may be why there was no decline in growth rate for these individuals and the largest individuals grew most rapidly. We also observed a negative relationship between native range crayfish length and growth rate in the lake study likely for the same reason. Native range crayfish were largest at the end of the lake study because they had a longer growing period. As observed in previous studies (e.g., Mundahl and Benton 1990), crayfish grew faster in warmer water, but this was clearly not sufficient to overshadow the differences between native and invasive populations.

While crayfish from northern Wisconsin grew more rapidly than those from the Ohio River drainage, it is unclear whether this would lead to larger young within the invaded range compared to those within the native range because of temperature differences between these two locations. Previous studies examining growth of YOY *O. rusticus* have found YOY carapace lengths ranging from 9 to 16.5 mm in September in northern Wisconsin (Lorman 1980) and YOY carapace lengths ranging from 8 to 17 mm in September in northern Kentucky (Prins 1968). In a preliminary study in 2010, we collected YOY crayfish from Big Lake in northern Wisconsin in August, which ranged from 9.5 to 15.5 mm carapace length. While these measurements are restricted to specific locations within each range, and not necessarily representative of growth rates throughout each range, they suggest that if there are differences in crayfish size between these two ranges, they are not large.

Survival differences

Not only did native range crayfish have reduced growth, they also had reduced survival in both the lake and mesocosm common garden experiments. This could be a result of local adaptation of invasive *O. rusticus* populations to environmental conditions in the invaded range, especially if some characteristics in the mesocosm experiment more closely resembled lakes in northern Wisconsin than streams in the Ohio River drainage. We expect calcium concentration was lower in lakes and mesocosms than it is in Ohio streams, and flow also differs between these environment types. Differences in growth rates could also be attributed to local adaptation of the invasive population to environmental characteristics such as these. However, if the differences in growth rates observed between the two ranges were due to local adaptation to calcium concentration or flow rate, we would expect to see larger YOY at the end of the summer in the native range where temperatures are warmer.

In the mesocosm experiment, food quality and predatory fish presence had similar effects on native and invaded range crayfish growth, but these factors differentially influenced native and invaded range crayfish survival. Invaded range crayfish were more likely to survive than native range crayfish in all treatments except when food quality was low and no fish were present. Higher mortality within this group was unexpected, but may be due to the combination of rapid growth and low quality food, which could potentially lead to higher mortality due to unavailability of essential nutrients. In addition, crayfish in this group had the fastest growth rates on the low quality diet, and thus likely ate a greater quantity of low quality food than other crayfish. Therefore, secondary metabolites from macrophytes in the low quality diet could also be responsible for the observed increase in crayfish mortality within this group. Crayfish from the native range had the lowest survival when food quality was low and fish were present. This finding suggests a strong behavioral response of native range crayfish to fish that results in reduced feeding or increased energy expenditure. This response was not observed in invaded range crayfish. Because of high mortality, those crayfish that had the greatest behavioral response to fish may not have survived long enough to be included in the growth results, which may be why there is no interaction between range and fish presence apparent from the growth data. In addition, higher mortality of native range crayfish may also explain the trend that surviving native range crayfish spent more time outside of shelter than surviving invaded range crayfish. If predation pressure is similar between the two ranges, we expect it will be more beneficial for invaded range crayfish to favor feeding over predator avoidance because there should be a greater fitness benefit associated with fast growth in this range.

Mechanisms leading to growth rate evolution in invasive populations

The finding that *O. rusticus* from invasive populations have faster growth rates than those from native populations was consistent with our expectations of how natural selection within the invaded range would alter this trait. Larger crayfish produce more eggs than small crayfish (Savolainen et al. 1997; Skurdal et al. 2011), so crayfish with faster growth have greater reproductive output. Life history theory predicts that optimal life-history strategies will differ between density-regulated and non-density-regulated populations, with higher fitness associated with high reproductive rates in non-density-regulated populations (Roughgarden 1971; Burton et al. 2010). Invasive populations are non-density-regulated in the early stages of an introduction. Some previous studies have found evidence for evolution of *r*-selected life history traits in invasive populations or during range expansion (Burton et al. 2010; Phillips et al. 2010; Flory et al. 2011); however, other studies have found no evidence for the evolution of these traits (e.g. Bossdorf et al. 2004; Cripps et al. 2009). We expect that there may be a more lasting effect of life history evolution in aquatic invasive species compared to most terrestrial species. Range edges, or locations with low conspecific densities, are scattered throughout the invaded range for most aquatic species. Lakes are insular environments and uncolonized lakes are spread throughout the invaded range; therefore, many aquatic invasive populations are serially introduced into locations with low conspecific densities. Thus, we hypothesize that compared to invaders in most terrestrial or marine environments, aquatic invaders will experience exponential growth more often, and there will be a stronger or longer lasting effect of *r*-selection in these populations.

Evolution of increased competitive ability (EICA) is another mechanism which could lead to evolution of rapid growth in invasive populations. EICA postulates that release from natural enemies such as predators and parasites allows nonindigenous species to allocate more resources toward growth (Keane and Crawley 2002; Inderjit and van der Putten 2010). However, because there are native congeners in northern Wisconsin lakes, there are many predators and parasites that readily consume or infect *O. rusticus*. Predatory fish are important in controlling *O. rusticus* populations (Roth et al. 2007) and high levels of parasitism by trematodes have been observed in some lakes (Roesler 2009). Therefore, we think this mechanism is less likely to be responsible for the higher growth rates observed in invasive *O. rusticus* than life history trait selection.

Hybridization may enhance the likelihood that nonindigenous populations will evolve invasive traits (Ellstrand and Schierenbeck 2000). Within the invaded range, *O. rusticus* hybridizes with a resident congener, *Orconectes propinquus* Girard (Perry et al. 2001). *O. rusticus* is competitively superior, and hybrids produce offspring that are most likely to backcross with *O. rusticus* (Perry et al. 2001). It is unclear whether *O. propinquus* alleles remain in invasive *O. rusticus* (hybrid) populations over time (Perry et al. 2001). Since the early 1980s, no *O. propinquus* have been detected in any of the lakes where we collected *O. rusticus*, and *O. propinquus* has never been detected in Big Lake (Lodge unpublished data); therefore, we were not examining populations that hybridized recently. *O. propinquus* grow more slowly than *O. rusticus* (Hill et al. 1993), so rapidly growing invasive *O. rusticus* represent novel genotypes that are dissimilar from both parental populations. It is, however, possible that earlier hybridization and introgression provided increased additive genetic variance or created novel epistatic interactions that allowed *O. rusticus* to evolve faster growth rates in the invaded range.

Community impacts of growth rate divergence

Rapid growth rates in invasive *O. rusticus* have had major community-level consequences. *Orconectes rusticus* has a greater impact on the ecological community than congeners, *O. virilis* Hagen and *O. propinquus*, and often causes declines in macrophyte and macroinvertebrate abundance and richness when replacing these species (Wilson et al. 2004). The ability of *O. rusticus* to replace *O. propinquus* has been attributed in part to its faster growth rate and ability to outcompete smaller individuals for shelter (Hill et al. 1993; Garvey et al. 1994; Hill and Lodge 1994). Faster growth also causes *O. rusticus* to escape predation from gape-limited fish more rapidly (Stein 1977). Understanding how often nonindigenous organisms evolve invasive traits is crucial for understanding the costs and consequences associated with introducing species to new locations.

Implications for management of invasions

We recommend that the potential for populations to evolve increasingly invasive traits be considered when moving species to new locations. Even though a species may not be problematic in its native range, or may be unproblematic initially in a new location, traits such as rapid growth and high reproductive output that may increase ecological impacts can evolve within the invaded range. This may be especially likely to occur in populations which are serially introduced to insular environments such as aquatic organisms in lakes.

Risk assessments that do not include evolutionary potential may underestimate the likelihood of a species to cause ecological and economic harm. Species that are likely to hybridize with native species may also be especially likely to evolve in response to selection within the invaded range because of increased additive genetic variance in hybrids. Crayfish in North America are a prime example of organisms that are likely to evolve invasive traits when intro-

duced to new locations because they live in patchy insular environments (lakes or stream drainages), and because they are likely to encounter native crayfishes with which hybridization may be possible. Especially when introduced to new locations within North America, crayfish are often exposed to closely-related, native species with which they are likely to hybridize (Perry et al. 2002). Seventy-five percent of the world's crayfish species are found within the United States (Lodge et al. 2000). Despite these problems, many states in the United States do not regulate the movement of crayfish or encourage voluntary practices to restrict moving crayfish, and many other states have legislation that only restricts moving certain species that are known to be problematic (Peters and Lodge 2009; Dresser and Swanson 2013). Our research suggests that invasive traits can evolve in nonindigenous crayfish populations, and this risk could be considered when weighing the costs and benefits of moving crayfish to new locations.

Acknowledgements

We thank Ashley Baldridge and Jill Deines for assistance with crayfish collection. We also greatly appreciate the efforts of Joshua Morse, Bradley Wells, June Shrestha, and Iris Petersen in assisting with field and mesocosm experiments. Andy Deines and Erin Grey provided helpful feedback on the manuscript and statistical approaches. We also appreciate helpful comments on the manuscript from two anonymous reviewers. We are grateful to the University of Wisconsin's Trout Lake Research Station for providing the location for mesocosm experiments and laboratory space for hatching crayfish. Funding for this research was provided by the University of Notre Dame Environmental Research Center and NSF IGERT grant award #0504495 to the GLOBES graduate training program at the University of Notre Dame. This is a publication of the Notre Dame Environmental Change Initiative.

Literature Cited

Boschung, H. T., J. D. Williams, D. W. Gotshall, D. K. Caldwell, M. C. Caldwell, C. Nehring, and J. Verner 1983. The Audubon Society field guide to North American Fishes, Whales, and Dolphins. Alfred A. Knopf, New York.

Bossdorf, O., D. Prati, H. Auge, and B. Schmid 2004. Reduced competitive ability in an invasive plant. Ecology Letters 7:346–353.

Burton, O. J., B. L. Phillips, and J. M. J. Travis 2010. Trade-offs and the evolution of life-histories during range expansion. Ecology Letters 13:1210–1220.

Butchart, S. H. M., M. Walpole, B. Collen, A. van Strien, J. P. W. Scharlemann, R. E. A. Almond, J. E. M. Baillie et al. 2010. Global biodiversity: indicators of recent declines. Science 328:1164–1168.

Capelli, G. M. 1982. Displacement of northern Wisconsin crayfish by *Orconectes rusticus* (Girard). Limnology and Oceanography 27:741–745.

Catford, J. A., R. Jansson, and C. Nilsson 2009. Reducing redundancy in invasion ecology by integrating hypotheses into a single theoretical framework. Diversity and Distributions 15:22–40.

Cripps, M. G., H. L. Hinz, J. L. McKenney, W. J. Price, and M. Schwarz-lander 2009. No evidence for an 'evolution of increased competitive ability' for the invasive *Lepidium draba*. Basic and Applied Ecology 10:103–112.

Cronin, G., D. M. Lodge, M. E. Hay, M. Miller, A. M. Hill, T. Horvath, R. C. Bolser et al. 2002. Crayfish feeding preferences for freshwater macrophytes: the influence of plant structure and chemistry. Journal of Crustacean Biology 22:708–718.

Dresser, C., and B. Swanson 2013. Preemptive legislation inhibits the anthropogenic spread of an aquatic invasive species, the rusty crayfish (*Orconectes rusticus*). Biological Invasions 15:1049–1056.

Ellstrand, N. C., and K. A. Schierenbeck 2000. Hybridization as a stimulus for the evolution of invasiveness in plants? Proceedings of the National Academy of Sciences of the United States of America 97:7043–7050.

Flory, S. L., F. R. Long, and K. Clay 2011. Invasive *Microstegium* populations consistently outperform native range populations across diverse environments. Ecology 92:2248–2257.

Garvey, J. E., R. A. Stein, and H. M. Thomas 1994. Assessing how fish predation and interspecific prey competition influence a crayfish assemblage. Ecology 75:532–547.

Handley, L. J. L., A. Estoup, D. M. Evans, C. E. Thomas, E. Lombaert, B. Facon, A. Aebi et al. 2011. Ecological genetics of invasive alien species. BioControl 56:409–428.

Hill, A. M., and D. M. Lodge 1994. Diel changes in resource demand: competition and predation in species replacement among crayfishes. Ecology 75:2118–2126.

Hill, A. M., and D. M. Lodge 1995. Multi-trophic-level impact of sublethal interactions between bass and omnivorous crayfish. Journal of the North American Benthological Society 14:306–314.

Hill, A. M., and D. M. Lodge 1999. Replacement of resident crayfishes by an exotic crayfish: The roles of competition and predation. Ecological Applications 9:678–690.

Hill, A. M., D. M. Sinars, and D. M. Lodge 1993. Invasion of an occupied niche by the crayfish *Orconectes rusticus*: potential importance of growth and mortality. Oecologia 94:303–306.

Inderjit, and W. H. van der Putten 2010. Impacts of soil microbial communities on exotic plant invasions. Trends in Ecology & Evolution 25:512–519.

Keane, R. M., and M. J. Crawley 2002. Exotic plant invasions and the enemy release hypothesis. Trends in Ecology & Evolution 17:164–170.

Keller, R. P., D. M. Lodge, M. A. Lewis, and J. F. Shogren 2009. Bioeconomics of Invasive Species: Integrating Ecology, Economics, Policy, and Management. Oxford University Press, New York.

van Kleunen, M., E. Weber, and M. Fischer 2010. A meta-analysis of trait differences between invasive and non-invasive plant species. Ecology Letters 13:235–245.

Koskinen, M. T., T. O. Haugen, and C. R. Primmer 2002. Contemporary fisherian life-history evolution in small salmonid populations. Nature 419:826–830.

Lamarque, L. J., S. Delzon, and C. J. Lortie 2011. Tree invasions: a comparative test of the dominant hypotheses and functional traits. Biological Invasions 13:1969–1989.

Lee, C. E., J. L. Remfert, and G. W. Gelembiuk 2003. Evolution of physiological tolerance and performance during freshwater invasions. Integrative and Comparative Biology 43:439–449.

Lewontin, R. C. 1965. Selection for colonizing ability. In: H. G. Baker, and G. L. Setebbins, eds. The Genetics of Colonizing Species, pp. 79–94. Academic Press, New York.

Lodge, D. M., M. W. Kershner, J. E. Aloi, and A. P. Covich 1994. Effects of an omnivorous crayfish (*Orconectes rusticus*) on a freshwater littoral food web. Ecology 75:1265–1281.

Lodge, D. M., C. A. Taylor, D. M. Holdich, and J. Skurdal 2000. Nonindigenous crayfishes threaten North American freshwater biodiversity: lessons from Europe. Fisheries 25:21–25.

Lodge, D. M., A. Deines, F. Gherardi, D. C. J. Yeo, T. Arcella, A. K. Baldridge, M. A. Barnes et al. 2012. Global introductions of crayfishes: evaluating the impact of species invasions on ecosystem services. Annual Review of Ecology, Evolution, and Systematics 43:449–472.

Lorman, J. G. 1980. Ecology of the crayfish *Orconectes rusticus* in northern Wisconsin. Dissertation. University of Wisconsin, Madison.

Mundahl, N. D., and M. J. Benton 1990. Aspects of the thermal ecology of the rusty crayfish *Orconectes rusticus* (Girard). Oecologia 82:210–216.

Olden, J. D., J. M. McCarthy, J. T. Maxted, W. W. Fetzer, and M. J. Vander Zanden 2006. The rapid spread of rusty crayfish (*Orconectes rusticus*) with observations on native crayfish declines in Wisconsin (U.S.A.) over the past 130 years. Biological Invasions 8:1621–1628.

Perry, W. L., J. L. Feder, G. Dwyer, and D. M. Lodge 2001. Hybrid zone dynamics and species replacement between *Orconectes* crayfishes in a northern Wisconsin lake. Evolution 55:1153–1166.

Perry, W. L., D. M. Lodge, and J. L. Feder 2002. Importance of hybridization between indigenous and nonindigenous freshwater species: an overlooked threat to North American biodiversity. Systematic Biology 51:255–275.

Peters, J. A., and D. M. Lodge 2009. Invasive species policy at the regional level: a multiple weak links problem. Fisheries 34:373–381.

Peters, J. A., M. J. Cooper, S. M. Creque, M. S. Kornis, J. T. Maxted, W. L. Perry, F. W. Schueler et al. 2014. Historical changes and current status of crayfish diversity and distribution in the Laurentian Great Lakes. Journal of Great Lakes Research. 40:35–46.

Phillips, B. L. 2009. The evolution of growth rates on an expanding range edge. Biology Letters 5:802–804.

Phillips, B. L., G. P. Brown, and R. Shine 2010. Life-history evolution in range-shifting populations. Ecology 91:1617–1627.

Pintor, L. M., and A. Sih 2009. Differences in growth and foraging behavior of native and introduced populations of an invasive crayfish. Biological Invasions 11:1895–1902.

Prins, R. 1968. Comparative ecology of the crayfishes *Orconectes rusticus rusticus* and *Cambarus tenebrosus* in Doe Run, Meade County, Kentucky. Internationale Revue der gesamten Hydrobiologie und Hydrographie 53:667–714.

Ricciardi, A., and R. Kipp 2008. Predicting the number of ecologically harmful exotic species in an aquatic system. Diversity and Distributions 14:374–380.

Ricklefs, R. E. 1967. A graphical method of fitting equations to growth curves. Ecology 48:978–983.

Roesler, C. 2009. Distribution of a crayfish parasite, *Microphallus* sp., in northern Wisconsin lakes and apparent impacts on rusty crayfish populations. Report for the Wisconsin Department of Natural Resources.

Roth, B. M., J. C. Tetzlaff, M. L. Alexander, and J. F. Kitchell 2007. Reciprocal relationships between exotic rusty crayfish, macrophytes, and *Lepomis* species in northern Wisconsin lakes. Ecosystems 10:74–85.

Roughgarden, J. 1971. Density-dependent natural selection. Ecology 52:453–468.

Sakai, A. K., F. W. Allendorf, J. S. Holt, D. M. Lodge, J. Molofsky, K. A.

With, S. Baughman et al. 2001. The population biology of invasive species. Annual Review of Ecology and Systematics **32**:305–332.

Sala, O. E., F. S. Chapin, J. J. Armesto, E. Berlow, J. Bloomfield, R. Dirzo, E. Huber-Sanwald et al. 2000. Biodiversity: global biodiversity scenarios for the year 2100. Science **287**:1770–1774.

Sato, T., and N. Suzuki 2010. Female size as a determinant of larval size, weight, and survival period in the coconut crab, *Birgus latro*. Journal of Crustacean Biology **30**:624–628.

Savolainen, R., K. Westman, and M. Pursiainen 1997. Fecundity of Finnish noble crayfish, *Astacus astacus*, and signal crayfish, *Pacifastacus leniusculus* (Dana), females in various natural habitats and in culture in Finland. Freshwater Crayfish **11**:319–338.

Siemann, E., and W. E. Rogers 2001. Genetic differences in growth of an invasive tree species. Ecology Letters **4**:514–518.

Skurdal, J., D. O. Hessen, E. Garnas, and L. A. Vollestad 2011. Fluctuating fecundity parameters and reproductive investment in crayfish: driven by climate or chaos? Freshwater Biology **56**:335–341.

Stein, R. A. 1977. Selective predation, optimal foraging, and the predator-prey interaction between fish and crayfish. Ecology **58**:1237–1253.

Stein, R. A., and J. J. Magnuson 1976. Behavioral response of crayfish to a fish predator. Ecology **57**:751–761.

Swiney, K. M., G. L. Eckert, and G. H. Kruse 2013. Does maternal size affect red king crab, *Paralithodes camtschaticus*, embryo and larval quality? Journal of Crustacean Biology **33**:470–480.

Tropea, C., M. Arias, N. S. Calvo, and L. S. L. Greco 2012. Influence of female size on offspring quality of the freshwater crayfish *Cherax quadricarinatus* (Parastacidae: Decapoda). Journal of Crustacean Biology **32**:883–890.

Whitney, K. D., and C. A. Gabler 2008. Rapid evolution in introduced species, 'invasive traits' and recipient communities: challenges for predicting invasive potential. Diversity and Distributions **14**:569–580.

Wilson, K. A., J. J. Magnuson, D. M. Lodge, A. M. Hill, T. K. Kratz, W. L. Perry, and T. V. Willis 2004. A long-term rusty crayfish (*Orconectes rusticus*) invasion: dispersal patterns and community change in a north temperate lake. Canadian Journal of Fisheries and Aquatic Sciences **61**:2255–2266.

10

Evolution determines how global warming and pesticide exposure will shape predator–prey interactions with vector mosquitoes

Tam T. Tran,[1,2] Lizanne Janssens,[2] Khuong V. Dinh,[1,3] Lin Op de Beeck[2] and Robby Stoks[2]

1 Institute of Aquaculture, Nha Trang University, Nha Trang, Vietnam
2 Laboratory of Aquatic Ecology, Evolution and Conservation, University of Leuven, Leuven, Belgium
3 National Institute of Aquatic Resources, Technical University of Denmark, Copenhagen, Denmark

Keywords
biological control, climate change, contaminants, *Ischnura elegans*, latitudinal gradient, life history evolution, range shifts, thermal evolution.

Correspondence
Tam Thanh Tran, Institute of Aquaculture, Nha Trang University, No 2., Nguyen Dinh Chieu Street, Nha Trang, Vietnam.

e-mail: thanhtam.ntu.edu@gmail.com

Abstract

How evolution may mitigate the effects of global warming and pesticide exposure on predator–prey interactions is directly relevant for vector control. Using a space-for-time substitution approach, we addressed how 4°C warming and exposure to the pesticide endosulfan shape the predation on *Culex pipiens* mosquitoes by damselfly predators from replicated low- and high-latitude populations. Although warming was only lethal for the mosquitoes, it reduced predation rates on these prey. Possibly, under warming escape speeds of the mosquitoes increased more than the attack efficiency of the predators. Endosulfan imposed mortality and induced behavioral changes (including increased filtering and thrashing and a positional shift away from the bottom) in mosquito larvae. Although the pesticide was only lethal for the mosquitoes, it reduced predation rates by the low-latitude predators. This can be explained by the combination of the evolution of a faster life history and associated higher vulnerabilities to the pesticide (in terms of growth rate and lowered foraging activity) in the low-latitude predators and pesticide-induced survival selection in the mosquitoes. Our results suggest that predation rates on mosquitoes at the high latitude will be reduced under warming unless predators evolve toward the current low-latitude phenotype or low-latitude predators move poleward.

Introduction

How global warming will affect vector species and associated diseases is one of the pressing questions with relevance for human health (Kovats et al. 2001; Ramasamy and Surendran 2012; Parham et al. 2015). While much attention is going to how vectorborne disease dynamics will change in a warmer world, much less attention is going to how warming will shape biotic interactions with vector species (Parham et al. 2015). Yet, biotic interactions such as predator–prey interactions may be an important factor controlling vector mosquitoes (Kamareddine 2012). Despite the general insight that predator–prey interactions are important for the local persistence of prey populations under global warming (Gilman et al. 2010; Zarnetske et al. 2012), few studies directly looked at how warming affects the outcome of

these interactions (but see, e.g., De Block et al. 2013; Hayden et al. 2015). Moreover, none of these studies considered vector prey species. Another challenge for understanding how predators may control vector populations is that in many areas, pest control provided by natural enemies has been lowered by the use of pesticides (MEA 2005). Moreover, pesticide use is expected to increase under global warming (Kattwinkel et al. 2011). Therefore, to assess the future potential of predation to play a role in vector control in a warming world, we need to study how predator–prey interactions are jointly shaped by warming and pesticides (Schmitz and Barton 2014).

Many species have the potential to evolve in response to warming (Merilä and Hendry 2014; Stoks et al. 2014). Therefore, a relevant applied question in this context is whether gradual thermal evolution of the predator may mitigate how warming and pesticide exposure shape predator–prey interactions with vector species, and with pest species in general (Roderick et al. 2012). Importantly, gradual evolution under global warming may thereby also shape the vulnerability to pesticides. Indeed, adaptation to a warmer climate may come at the cost of a reduced tolerance to contaminants (Moe et al. 2013). Besides direct effects of thermal evolution, also indirect effects mediated through evolved changes in life history, particularly in voltinism (number of generations per year), may affect the vulnerability to pesticides (e.g., Dinh Van et al. 2014a). Indeed, at warmer temperatures, invertebrates typically show more generations per year and in accordance evolve a faster growth and development as each generation will have less time to complete the larval stage (Seiter and Kingsolver 2013). Based on life history theory, a faster life history will come at the cost of a reduced investment in other functions, including detoxification and repair (Sibly and Calow 1989; Congdon et al. 2001).

A powerful way to assess the potential of gradual thermal evolution (being direct or indirect) in shaping trait evolution is to study besides high-latitude populations at their current temperature and the predicted higher temperature under warming, also low-latitude populations currently living at the higher temperature predicted at the high latitude under global warming. Such space-for-time substitution approach (Fukami and Wardle 2005; De Frenne et al. 2013) has only been rarely applied in the context of predator–prey interactions (but see De Block et al. 2013) and ecotoxicology (but see Janssens et al. 2014). Instead, the few studies on warming effects on predator–prey interactions typically applied a 'step-increase' temperature experiment at one latitude (e.g., Rall et al. 2010; Miller et al. 2014; Hayden et al. 2015; Sentis et al. 2015). Such studies, however, do not allow assessing the role of long-term gradual evolution in mediating the impact of a temperature increase and the associated changes in sensitivity to contaminants.

To better understand how warming and pesticides will shape the outcome of predator–prey interactions, it is important to expose both predator and prey to these stressors. Yet, the few studies that manipulated both stressors only exposed the prey (e.g., Broomhall 2002, 2004) or the predators (Dinh Van et al. 2014a). More general, most studies on the effect of pesticides on predator–prey interactions only exposed the predators (e.g., Dinh Van et al. 2014b) or the prey (e.g., Brooks et al. 2009; Reynaldi et al. 2011). Yet, joint exposure of both predator and prey, the likely field scenario, may have strongly different outcomes (Junges et al. 2010; Englert et al. 2012; Rasmussen et al. 2013). Moreover, the relatively few studies that exposed both predator and prey to a pesticide, mostly scored the behavior of only one interactor, thereby precluding a full

understanding of how pesticides change the outcome of predator–prey interactions (Schulz and Dabrowski 2001; Rasmussen et al. 2013).

In the current study, we tested how warming and exposure to a pesticide shape predator–prey interactions in the larval stage between a vector mosquito and important invertebrate predators, damselfly larvae. We explicitly considered the potential of thermal evolution of the predator in high-latitude populations in modifying these effects by applying a space-for-time substitution approach where we studied triplicated low- and high-latitude populations of the damselfly predators. Moreover, to get a mechanistic understanding of how both stressors change the outcome of predator–prey interactions, we studied the behavior of both antagonists when they were exposed to the stressors in a factorial way. Damselfly larvae are important natural predators of mosquitoes (Klecka and Boukal 2012) and are used as biological control agent (e.g., Mandal et al. 2008). The predator was the damselfly *Ischnura elegans* (Vander Linden, 1820), whose latitudinal differentiation in life history is well characterized (e.g., Shama et al. 2011; Stoks et al. 2012). The prey species was *Culex pipiens* (Linaeus, 1758) form molestus, a member of the *C. pipiens* complex, which is an important vector of pathogens such as West Nile virus and St. Louis encephalitis virus (Becker et al. 2010; Farajollahi et al. 2011). We chose the pesticide endosulfan, an organochlorine insecticide, that has been widely used to control vector mosquitoes (Calamari and Naeve 1994). This pesticide has been reported to increase the vulnerability of aquatic invertebrates to predation (e.g., Janssens and Stoks 2012; Trekels et al. 2013).

Materials and methods

Experimental design
We investigated the combined impact of warming and pesticide exposure on predator–prey interactions between damselflies and mosquitoes using a full factorial design with two predator latitudes (low- versus high-latitude damselflies) × two temperature treatments (20°C vs 24°C) × two pesticide treatments (endosulfan absent versus present). To keep the experiment feasible, we did not study mosquito populations from different latitudes; all mosquitoes came from a temperature regime matching that of the high-latitude populations of the damselfly predators. This way we only tested for the effects of thermal evolution of the predators in high-latitude populations.

The chosen temperatures reflect the mean summer water temperatures in shallow ponds occupied by the damselfly *Ischnura elegans* in southern Sweden (20°C) and southern France (24°C) (De Block et al. 2013). Based on simulations using the model Flake (e.g., Kirillin et al. 2011; Dinh Van et al. 2014a), the mean summer water temperature of ponds where the mosquito culture originates is about 20°C (for details see Appendix S1). Note that high-latitude damselfly populations and the studied mosquito populations currently encounter daily summer water temperatures of 24°C, although this occurs infrequently. Indeed, based on the Flake model (Kirillin et al. 2011), the percentage of daily water temperatures during summer equal to or exceeding 24°C is ca. 3% in high-latitude damselfly populations and 11–19% in the mosquito populations. Importantly, the 4°C difference corresponds with the predicted temperature increase by 2100 according to IPCC (2013) scenario RPC 8.5. This allows a space-for-time substitution where the potential impact of gradual thermal evolution in the high-latitude predator populations can be evaluated. The comparison of the phenotypes of the high-latitude predators at 20°C and 24°C indicates the potential of the currently present thermal plasticity (without change in the genetic constitution, hence without thermal evolution) to deal with 4°C warming. The comparison of the high-latitude predators and the low-latitude predators at 24°C reflects the potential of gradual thermal evolution in response to 4°C warming to shift the phenotypes of the high-latitude populations. In addition, we also tested the low-latitude populations at 20°C, to obtain a full factorial design where populations from both latitudes are tested at their 'local' mean temperature and the mean temperature of the other latitude provides a powerful design to test for local thermal adaptation (Kawecki and Ebert 2004). Both predators and prey were reared at one of the two temperatures from the egg stage onwards and afterward tested only at their rearing temperature. This way we allowed developmental, long-term acclimatization to the experimental temperatures and avoided any abrupt thermal changes before exposing the animals to the pesticide and testing them in the predation trials. This mimics a more realistic scenario compared to testing animals directly after exposing them to a higher temperature (Seebacher et al. 2015).

The study consisted of two coupled experiments that both tested for single and combined effects of temperature increase and pesticide exposure. In the first experiment, the exposure experiment, we examined effects on survival and growth rate of predators and prey kept in isolation. In the second follow-up experiment, the predation experiment, we studied the survival of the mosquito larvae in the presence of a lethal damselfly predator and monitored the behaviors of both predators and prey. All predator and prey individuals were kept at the same temperature-by-pesticide treatment during both experiments.

Study animals and rearing

The laboratory culture of *Culex pipiens* was initiated from the stock culture housed at the Helmholtz Centre for

Environmental Research – UFZ, Germany. To start up the experiments, freshly hatched mosquito larvae were reared at 20°C or 24°C until they reached the final instar (L4) (for details see Appendix S2) after which they entered the exposure experiment. A rearing temperature of 24°C was provided by placing trays in temperature-controlled water baths in the same room.

We collected *Ischnura elegans* damselflies at two latitudes representing low-latitude (southern France) and high-latitude (southern Sweden and Denmark) regions of the species' distribution range in Europe (Gosden et al. 2011). At each latitude, three populations were randomly collected, namely Saint-Martin-de-Crau (43°38′16.57″N, 4°50′49.06″E), Camaret-sur-Aigues (44°9′1.47″N, 4°51′20.37″E) and Domaine de Valcros (43°10′9.02″N, 6°16′11.36″E) in southern France; Nöbbelövs mosse (55°44′5.98″N, 13°9′10.02″E) and Erikso (58°56′4.90″N, 17°39′21.50″E) in southern Sweden and Ahl Hage (56°10′59.64″N, 10°39′1.69″E) in Denmark. All collecting sites were shallow ponds with abundant aquatic vegetation. Except for the one French population Camaret-sur-Aigues, the collecting sites were not embedded by cropland and close to forest (Appendix S3) making it unlikely that they were affected by agriculture (Declerck et al. 2006). Further, damselfly larvae from Camaret-sur-Aigues did not differ in their response to the pesticide compared to the other two French populations (Appendix S3). Moreover, any local adaptation to pesticides would be unlikely in damselflies given the high levels of gene flow (Johansson et al. 2013).

In each damselfly population, eggs of eight mated females were collected and transferred to the laboratory in Belgium. Ten days after hatching, larvae were placed individually in 200-mL plastic cups filled with aerated tap water. Larvae were daily fed *ad libitum* with *Artemia* nauplii (mean ± SE: 305 ± 34 nauplii per food portion, $n = 10$ food portions) 6 days a week until they reached the final instar after which they entered the exposure experiment. During the exposure period, the larvae were daily fed the same amount of *Artemia* nauplii as during the pre-exposure period.

Pesticide concentration

We selected an endosulfan concentration of 28 µg/L based on a range finding experiment (for details see Appendix S4). In Europe, endosulfan concentrations up to 100 µg/L have been detected in surface waters (Brunelli et al. 2009). We daily prepared the endosulfan exposure solution based on a stock solution of 500 µg/mL dissolved in acetone (stored in the dark at 4°C). In the control treatment, we used aerated tap water instead of a solvent control, as the range finding experiment showed no significant difference in survival and growth between the water control and solvent control and this both in the mosquito larvae and in the damselfly larvae (for details see Appendix S4).

Exposure experiment

At the start of the exposure experiment, 25 freshly molted L4 mosquito larvae of the same rearing temperature were placed in 200-mL cups containing 125 mL control or pesticide medium. During the 5-day exposure period, mosquito larvae were reared under the same conditions as during the pre-exposure period. Damselfly larvae were exposed individually in the same type of cups as the mosquitoes and were daily fed the same amount of *Artemia* nauplii as during the pre-exposure period. The medium was renewed every other day for both species. For mosquito larvae, we used 25 replicates (sets of 25 larvae, total of 625 larvae) per temperature-by-pesticide treatment combination. For damselfly larvae, the number of replicates varied from 8 to 15 per latitude-by-temperature-by-pesticide treatment combination (total of 97 damselfly larvae); exact sample sizes are shown in the figures.

We daily checked mortality of the two study species and adjusted the food provided to each cup with mosquitoes based on the number of living larvae in the cup. We additionally quantified growth rate based on the increase in wet mass over the exposure period for the two study species. For mosquito larvae, we obtained an estimate of the initial mean wet mass per larva based on the fresh mass of 10 randomly selected larvae entering L4 at each temperature. These larvae were carefully blotted dry and weighed to the nearest 0.01 mg using an electronic balance (AB135-S, Mettler Toledo®, Zaventem, Belgium). At the end of the exposure period, three to five mosquito larvae per cup (depending on the survival) were randomly selected and weighed in the same way to obtain mean final wet mass per larva. For damselfly larvae, each larva was weighed at the start and end of the exposure period. Growth rates of both mosquito and damselfly larvae were calculated as ($\ln_{\text{final mass}} - \ln_{\text{initial mass}}$)/5 days (Dinh Van et al. 2013).

Predation experiment

After the exposure experiment, mosquito larvae and damselfly larvae were jointly tested in the predation experiment. Mosquito larvae were used directly after their exposure period. Damselfly larvae were first starved for 24 h at their temperature-by-pesticide condition before being used in the predation trial to equalize hunger levels. For each predation trial, ten mosquito larvae and one damselfly larva of the same temperature-by-pesticide treatment combination were placed together in a 2.5-L container (11 × 13 × 19 cm) filled with 1 L of their exposure medium and tested at their rearing temperature. Hence, both

predators and prey were tested at the condition they experienced during the preceding exposure experiment. The number of replicates varied from 8 to 15 per damselfly latitude-by-temperature-by-pesticide treatment combination (total of 93 trials). Each mosquito larva and each damselfly larva were used in only one predation trial.

At the start of each 1-h predation trial, the mosquito larvae were added 1 min before the introduction of the predator. Thereafter, we scored the position and activity of each mosquito larva every 10 min based on the protocol of Kesavaraju and Juliano (2010). Positions were classified into four categories: bottom, wall, water surface and water column. We also defined four activity categories (Kesavaraju and Juliano 2010): browsing (the mouthparts were in contact with the bottom or the wall of the container to graze for food), filtering (the larva was moving in the water column and made feeding movements with its mouthparts), thrashing (the larva was moving its body from side to side with vigorous flexion) and resting (the lava showed no movement). We calculated at each time point ($n = 6$) per container the percentage of mosquito larvae in each position and in each activity category and this throughout the predation trial (1 h).

During each predation trial, we also monitored the behavior of the damselfly larvae. Every 10 min we categorized the behavior as swimming, walking, head orientations toward the prey and inactivity (when the larva did not exhibit any of the other three categories) (see Janssens et al. 2014). At the end of the observation period, we calculated the frequency of each behavioral category per damselfly larva. Mass-corrected predation rates by the damselfly larvae were calculated as the number of mosquito larvae eaten by a damselfly larva during 1 h divided by its body mass (see De Block et al. 2013).

Statistical analyses

All statistical analyses were run in Statistica v.12 (StatSoft, Tulsa, OK, USA). To test for the effects of temperature, pesticide exposure and latitude of the damselflies on the response variables, we ran separate ANOVAs. Survival data of both mosquitoes and damselflies during the exposure experiment were analyzed using logistic regression models with a binomial error structure. When analyzing effects on the damselfly larvae, we initially included population nested in latitude as a random factor; however, it had no effect on any of the response variables and we removed it from the final models.

For analyzing the detailed behaviors scored during the predation experiment, we first extracted principal components. Prior to the PCA, the mosquito behavioral data, which were expressed as percentages, were arcsine-transformed while the damselfly behavioral data were log (x + 1)-transformed. The resulting PC axes were then analyzed using ANOVAs as mentioned above. When testing the effects of the temperature and pesticide treatments on mosquito behaviors, latitude of the damselfly predator was also included in the model; as it never had an effect, we removed it from the final models.

Results

Exposure experiment

Survival of mosquito larvae was ca. 100% at 20°C in the absence of the pesticide; survival was lower at the higher temperature and lower in the presence of the pesticide (Fig. 1A, Table 1). There was no interaction between the temperature and the pesticide treatments (Table 1). Growth rate was neither affected by the temperature nor by the pesticide treatment (Fig. 1B, Table 1).

Survival of the damselfly larvae was ca. 100% and not affected by the treatments (Fig. 1C,D, Table 1). Growth rate was higher in low-latitude than in high-latitude larvae (Fig. 1E,F, Table 1). The effects of both the temperature and the pesticide treatments differed between latitudes (Temperature × Latitude and Pesticide × Latitude, Fig. 1, Table 1). Follow-up ANOVAs indicated that growth rate was only higher at 24°C than at 20°C in low-latitude larvae ($F_{1,37} = 25.67$, $P < 0.001$), but not in high-latitude larvae ($F_{1,44} = 2.91$, $P = 0.095$). Growth rate only increased in larvae exposed to the pesticide compared to the control treatment in high-latitude larvae ($F_{1,44} = 13.22$, $P < 0.001$), but not in low-latitude larvae ($F_{1,37} = 0.52$, $P = 0.476$).

Predation experiment

The PCA on the eight behavioral variables of the mosquito larvae resulted in three PC axes accounting for 80.1% of the total variation (Appendix S5). Mosquitoes with more positive scores on PC1 spent more time browsing on the bottom and at the walls of the container, and spent less time at the surface. Larvae with higher scores on PC2 spent more time filtering and less time resting. Larvae with higher values on PC3 spent more time thrashing in the water column. Exposure to the pesticide significantly affected each behavioral PC (Fig. 2, Table 2). Mosquito larvae exposed to the pesticide spent more time at the surface and browsed less frequently on the bottom, and at the walls (PC1), they showed more filtering and less time resting (PC2), and they spent more time thrashing in the water column (PC3).

The PCA on the four behavioral variables of the damselfly larvae resulted in three PC axes that accounted for 97.8% of the total variation (Appendix S5). Larvae with more positive scores on PC1 spent more time walking and spent less time being inactive. Larvae with lower scores on

PC2 spent more time swimming. Larvae with lower scores on PC3 showed more head orientations toward prey. The ANOVAs showed that exposure to the pesticide affected behavioral PC1 and PC3 (Fig. 3, Table 2). Damselfly larvae exposed to the pesticide increased walking activity (PC1) but only at 24°C (Temperature × Pesticide). Pesticide-exposed larvae spent more time being inactive and showed less head orientations (PC3) but only in low-latitude damselfly larvae (Pesticide × Latitude).

Mass-corrected predation rates by the damselfly larvae on the mosquito larvae were lower at 24°C than at 20°C (Fig. 4, Table 3). Low-latitude damselfly larvae consumed more mosquito larvae than high-latitude damselfly larvae, but only in the absence of the pesticide (Pesticide × Latitude, Fig. 4, Table 3). This Pesticide × Latitude interaction also indicated that exposure to the pesticide reduced predation rates but only in trials with low-latitude damselfly larvae (Fig. 4).

Discussion

Our results indicate that both exposure to endosulfan and warming differentially affected life history and behavior of the mosquito prey and the damselfly predators, and shaped the outcome of their predator–prey interactions. Moreover, several of the treatment effects on the damselfly predators differed between high-latitude and low-latitude populations, likely driven by the evolution of faster growth rates (and associated higher vulnerability to the pesticide) and thermal adaptation in the low-latitude populations. Key results were that endosulfan and warming only imposed mortality in the mosquito larvae, while endosulfan induced a growth rate increase in the high-latitude damselfly larvae and temperature induced a growth rate increase in the low-latitude damselfly larvae. Most importantly, predation rates on the mosquito larvae were reduced under warming and, in interactions with low-latitude predators, also in the presence of the pesticide.

Pesticide effects

The used endosulfan concentration differentially affected life history and behavior of the mosquito prey and the damselfly predators. Endosulfan imposed mortality on the mosquito larvae while the damselfly larvae instead only showed sublethal effects on growth rate. Specifically, exposure to the pesticide generated latitude-specific effects consistent with the prediction that low-latitude damselfly populations evolved a higher vulnerability to pesticides (see also Dinh Van et al. 2014a for the pesticide chlorpyrifos). In the presence of the pesticide, only high-latitude larvae increased growth rate while only low-latitude larvae reduced foraging activity (number of orientations toward

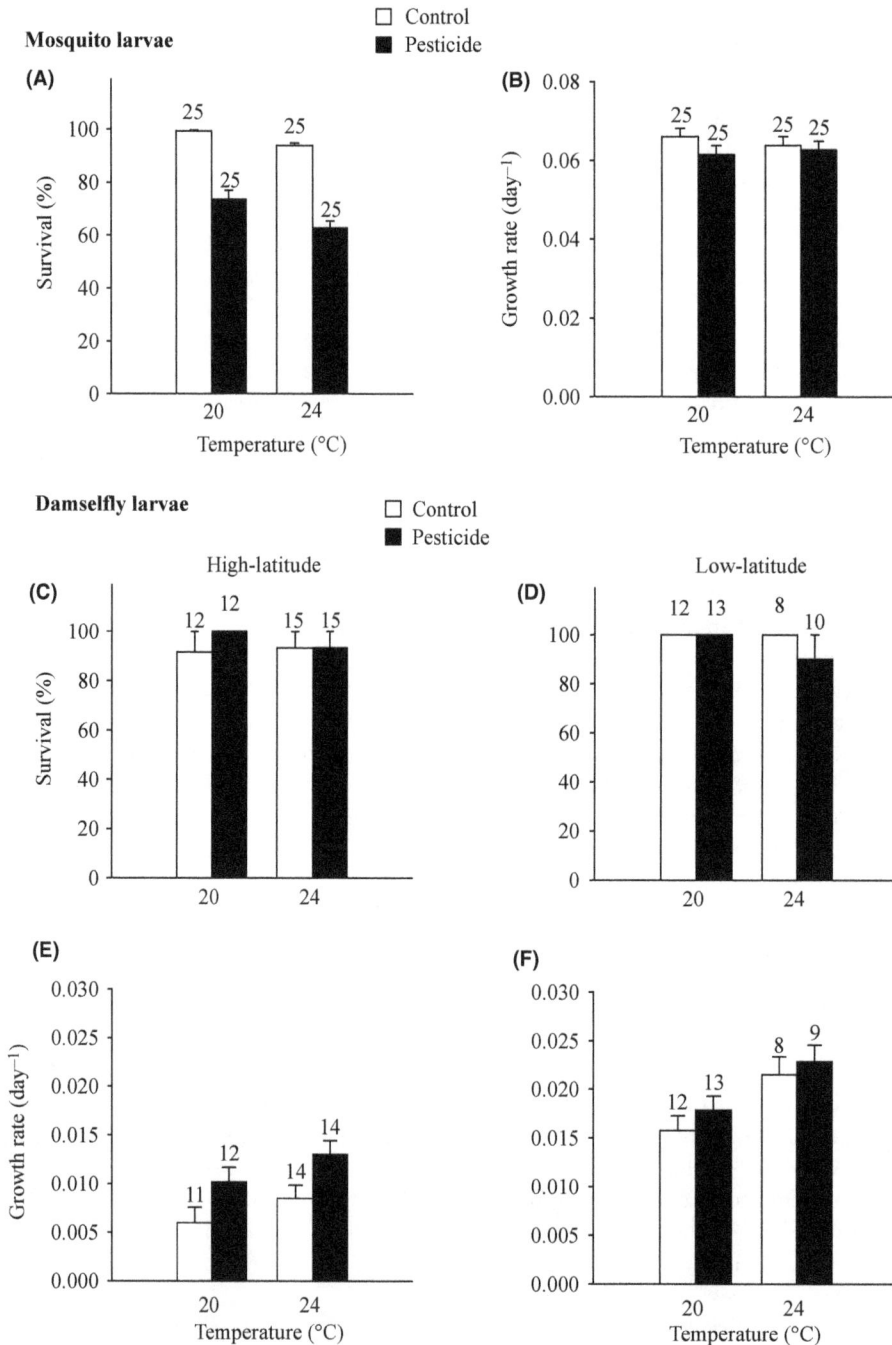

Figure 1 Survival (A, C, D) and growth rate (B, E, F) of *Culex pipiens* mosquito larvae (A, B) and *Ischnura elegans* damselfly larvae from low and high latitudes (C–F) as a function of the temperature and pesticide treatments. Given are least-squares means with 1 SE. Numbers above bars indicate sample sizes.

prey). Given that high-latitude larvae increased growth rate in the presence of the pesticide while their food intake did not change, the hormetic response was likely caused by a change in digestive efficiency. In line with this, endosulfan exposure caused an increase in growth rate in larvae of the damselfly *Coenagrion puella* which was not associated with

an increased food intake but an increased efficiency of assimilating food (Campero et al. 2007). Given that hormetic responses are costly (Forbes 2000; McClure et al. 2014), we interpret this as only the less vulnerable populations, here the high-latitude populations, being able to generate a hormetic growth response. This fits the general idea

Table 1. Results of ANOVAs testing for the effects of temperature, pesticide exposure and latitude of origin of the damselfly larvae on survival and growth rate of *Culex pipiens* mosquito larvae and *Ischnura elegans* damselfly larvae during the exposure experiment.

| Effect | Mosquito larvae | | | | | | Damselfly larvae | | | | | |
| | Survival | | | Growth rate | | | Survival | | | Growth rate | | |
	df	χ^2	P	df1, df2	F	P	df	χ^2	P	df1, df2	F	P
Temperature	1	16.09	**<0.001**	1, 96	0.05	0.819	1	0.86	0.352	1, 81	20.99	**<0.001**
Pesticide	1	170.43	**<0.001**	1, 96	1.66	0.200	1	0.01	0.913	1, 81	9.79	**0.0023**
Latitude							1	0.51	0.477	1, 81	124.42	**<0.001**
Temperature × Pesticide	1	3.70	0.055	1, 96	0.61	0.436	1	1.31	0.253	1, 81	0.01	0.922
Temperature × Latitude							1	0.10	0.746	1, 81	4.37	**0.040**
Pesticide × Latitude							1	1.31	0.235	1, 81	4.76	**0.032**
Temperature × Pesticide × Latitude							1	0.01	0.913	1, 81	0.15	0.701

Significant P values (P < 0.05) are indicated in bold.

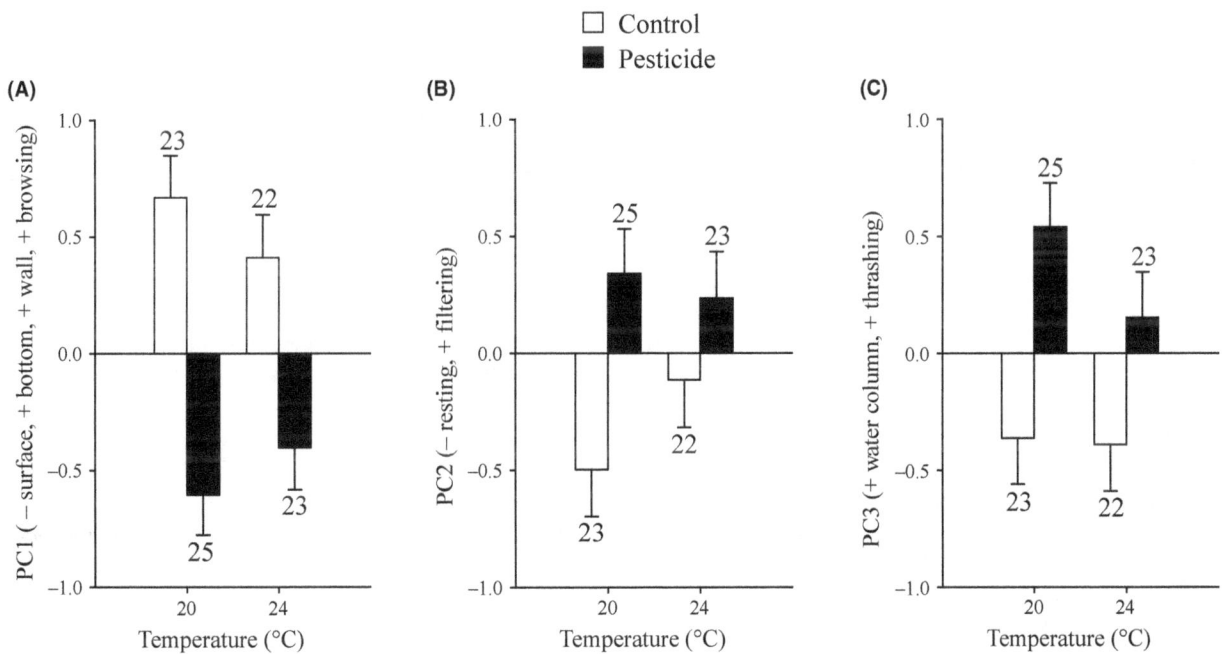

Figure 2 Behavioral PC scores (A: PC1; B: PC2; C: PC3) of *Culex pipiens* mosquito larvae during the predation experiment as a function of the temperature and pesticide treatments. Given are least-squares means with 1 SE. Numbers above bars indicate sample sizes.

that adaptation to a warmer climate (here at the low latitude) will come at the cost of a reduced tolerance to contaminants (Moe et al. 2013). The higher vulnerability to pesticides in low-latitude populations can be explained by their higher growth rates which through allocation trade-offs likely result in less energy being allocated to defense (Sibly and Calow 1989; Congdon et al. 2001). Low-latitude larvae of *I. elegans* evolved faster growth rates than high-latitude larvae as they have multiple generations per year, hence have less time available per generation to complete a generation (Shama et al. 2011). In line with their higher energy demand, low-latitude damselfly larvae consumed

more mosquito larvae compared to high-latitude larvae in the absence of the pesticide.

A key finding was that the evolution of different larval life histories and associated vulnerabilities to pesticides of the predators shaped predator–prey interactions in a latitude-specific way. Specifically, the pesticide reduced predation rates on the mosquitoes but only in the low-latitude damselfly larvae. Exposure to the pesticide had no main effect on the predation rates of damselfly larvae. Together with the observation that in the presence of the pesticide fewer mosquitoes were eaten, but only in interactions with low-latitude damselflies, this indicates that it were

Table 2. Results of ANOVAs testing for the effects of temperature, pesticide exposure and latitude of origin of the damselfly larvae on the behavioral factor scores of *Culex pipiens* mosquito larvae (a), and *Ischnura elegans* damselfly larvae (b) during the predation experiment.

Effect	PC1			PC2			PC3		
	df1, df2	F	P	df1, df2	F	P	df1, df2	F	P
(a) Mosquito larvae									
Temperature	1, 89	0.02	0.877	1, 89	0.49	0.487	1, 89	1.14	0.288
Pesticide	1, 89	34.62	**<0.001**	1, 89	8.85	**<0.004**	1, 89	13.87	**<0.001**
Temperature × Pesticide	1, 89	1.67	0.200	1, 89	1.50	0.224	1, 89	0.87	0.354
(b) Damselfly larvae									
Temperature	1, 85	1.21	0.275	1, 85	1.67	0.200	1, 85	1.33	0.252
Pesticide	1, 85	0.45	0.503	1, 85	1.70	0.196	1, 85	2.91	0.092
Latitude	1, 85	0.51	0.476	1, 85	0.00	0.952	1, 85	3.08	0.083
Temperature × Pesticide	1, 85	5.49	**0.021**	1, 85	1.74	0.191	1, 85	0.25	0.619
Temperature × Latitude	1, 85	0.55	0.460	1, 85	0.00	0.954	1, 85	0.19	0.663
Pesticide × Latitude	1, 85	0.16	0.692	1, 85	0.00	0.995	1, 85	4.88	**0.030**
Temperature × Pesticide × Latitude	1, 85	0.06	0.804	1, 85	0.00	0.965	1, 85	0.18	0.674

Significant *P* values (*P* < 0.05) are indicated in bold.

primarily the pesticide effects on the predators that were driving the outcome of predator–prey interactions. This was supported by the observation that the pesticide reduced foraging activity (number of head orientations) of the predators but only in the low-latitude populations. While many studies reported reduced predation rates in the presence of contaminants, very few tried to identify the underlying changes in the behaviors of predators and prey (reviewed in Fleeger et al. 2003; but see, e.g., Junges et al. 2012).

While the pesticide also affected all scored behaviors of the mosquito larvae, this apparently did not change their overall vulnerability to damselfly predators. Some of these behavioral changes (such as increased filtering and thrashing behaviors) likely made them easier to detect by the damselfly larvae. Yet, the pesticide-induced changes in the position of the mosquito larvae (more at the surface and in the water column) likely reduced the encounter probability with the damselfly larvae and therefore may have counteracted the higher detection probability. This increased occurrence at the water surface may be a response to the increased oxygen need associated with an increased metabolic rate in the presence of the pesticide (Srivastava and Misra 1981). Note, however, that the latter mechanism together with the pesticide-induced increase in thrashing behavior may make mosquito larvae more vulnerable to pelagic predators such as notonectids (Gimonneau et al. 2012).

Despite the mosquito prey suffering more from the pesticide than the damselfly predators in terms of survival, the pesticide, if anything, shaped the outcome of the predator–prey interactions in favor of the mosquito larvae. This seems counterintuitive and is in contrast with the prey stress model (Menge and Olson 1990) stating that when prey are more affected by the stressor than the

predator, prey are expected to suffer higher predation rates in the presence of the stressor (for empirical support, see, e.g., Schulz and Dabrowski 2001). Yet, deviations from the prey stress model may not be unexpected (Junges et al. 2010). Indeed, in our study the pesticide-induced mortality may have removed the mosquitoes with the slowest escape responses in the presence of the pesticide, so that the escape responses in the survivors that were used in the predation trials were no longer strongly affected by the pesticide.

Temperature effects

Warming affected the mosquito prey and the damselfly predators in opposite ways and thereby shaped the outcome of predator–prey interactions. Mosquitoes suffered at the higher temperature as indicated by their higher mortality. This matches a previous study showing a higher mortality of *C. pipiens* at 24°C compared to 20°C (Ciota et al. 2014). In our study, this may reflect local thermal adaptation given that 20°C corresponds with the mean summer water temperatures of the mosquito source populations (Appendix S1). Instead, the damselfly larvae were not negatively affected by warming. Moreover, low-latitude damselfly larvae were even growing faster at 24°C. This indicates a pattern of local thermal adaptation as previously observed for growth rate in this species (Shama et al. 2011; Dinh Van et al. 2014a).

Intriguingly, while only the prey suffered mortality at the high temperature, warming switched the outcome of predator–prey interactions in favor of the mosquitoes. This resembles the counterintuitive response pattern to the pesticide, yet here survival selection is a less likely explanation given that survival only slightly decreased under warming. The recorded behaviors of the mosquito prey and damselfly

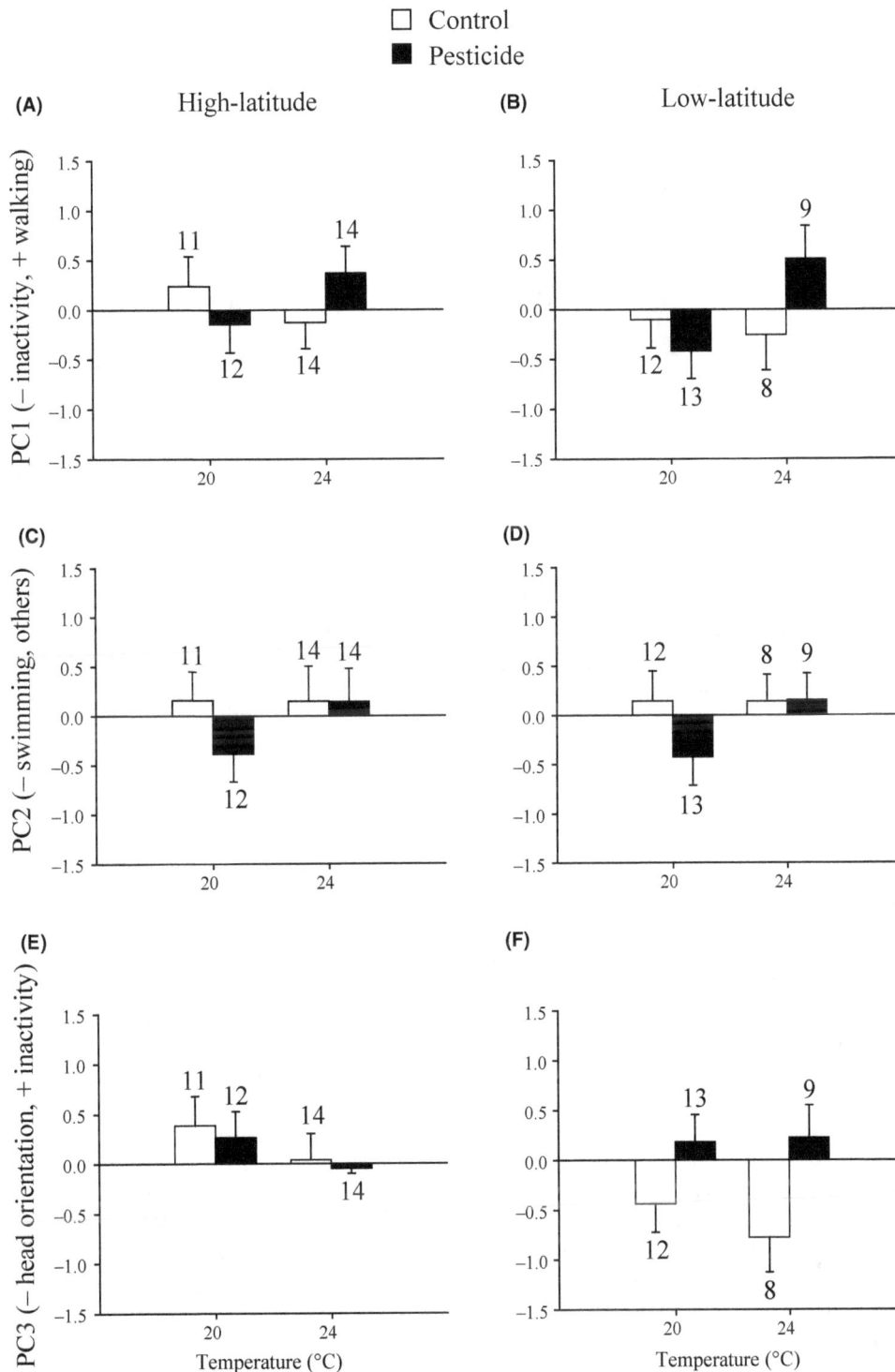

Figure 3 Behavioral PC scores of *Ischnura elegans* damselfly larvae from high (A, C, E) and low (B, D, F) latitudes during the predation experiment as a function of the temperature and pesticide treatments. Given are least-squares means with 1 SE. Numbers above bars indicate sample sizes.

predators can also not explain the reduced predation rates under warming: temperature did not affect the mosquito behaviors, and there was no overall main effect of warming on the damselfly behaviors. Potentially, the mosquitoes became more efficient at evading predator attacks at the higher temperature because their escape speed increased

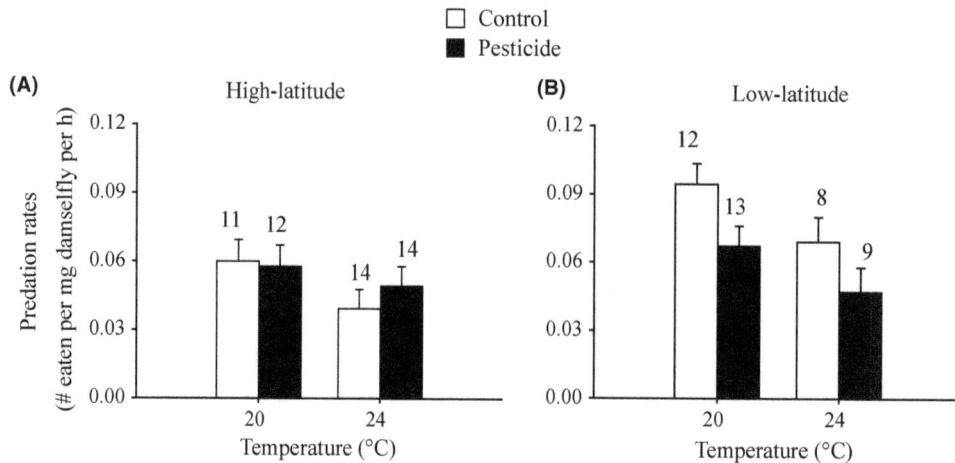

Figure 4 The number of *Culex pipiens* mosquito larvae eaten by *Ischnura elegans* damselfly larvae from high (A) and low (B) latitudes during the predation experiment as a function of the temperature and pesticide treatments. Given are least-squares means with 1 SE. Numbers above bars indicate sample sizes.

Table 3. Results of ANOVAs testing for the effects of temperature, pesticide exposure and latitude of origin of the damselfly larvae on the number of *Culex pipiens* mosquito larvae eaten in the predation experiment.

Effect	Predation rate		
	df1, df2	F	P
Temperature	1, 85	7.87	**0.006**
Pesticide	1, 85	2.36	0.128
Latitude	1, 85	7.31	**0.008**
Temperature × Pesticide	1, 85	0.43	0.513
Temperature × Latitude	1, 85	0.38	0.537
Pesticide × Latitude	1, 85	4.59	**0.035**
Temperature × Pesticide × Latitude	1, 85	0.06	0.799

Significant P values (P < 0.05) are indicated in bold.

more relative to the attack efficiency of the predators. Similarly, the stronger increase in escape speeds made mosquitofish less prone to predation by predatory bass under warming (Grigaltchik et al. 2012). In contrast to current findings, warming imposed higher predation rates of *I. elegans* on *Artemia* nauplii (Dinh Van et al. 2013, 2014a), and on *Daphnia magna* water fleas (De Block et al. 2013). Possibly, the latter two prey taxa do not increase escape speed to the same extent as mosquito larvae under warming hence cannot significantly lower the capture efficiency by the damselfly predators.

Evolutionary perspectives with regard to global warming and mosquito control

How pest species will cope with pesticides and with their predators will be a major factor in shaping their control under global warming. Our results tentatively suggest that

in the absence of evolution of the damselfly predators, a 4°C temperature increase as predicted by IPCC (2013) scenario RCP8.5 will change the outcome of predator–prey interactions at the high latitude in favor of the vector mosquitoes. In other words, all else remaining equal, biological control by damselfly predators would become less efficient. This is based on the general effect of decreased predation rates at 24°C. Note, however, that (assuming no thermal evolution of the mosquitoes) the higher temperature will also impose much higher direct mortality on the mosquitoes so that the changed predator–prey interactions likely will not translate into higher mosquito abundances. In case, high-latitude populations of the damselfly predators, however, evolve toward the phenotype of low-latitude populations currently living and adapted to 24°C, we may expect that the biological control of mosquitoes by damselfly larvae in the high latitudes will not change compared to the current situation. This is based on the observation that at 24°C the low-latitude larvae had the same predation rates as the high-latitude larvae currently living at 20°C. These predictions are, however, contingent on the limiting assumptions of the space-for-time substitution approach (Fukami and Wardle 2005; De Frenne et al. 2013; Elmendorf et al. 2015): (i) that besides temperature no other factors differ between latitudes that shape the studied traits (which is partly dealt with as we ran a common-garden warming experiment), (ii) that populations will respond to changes in temperature over time in the same way that they will over space (Fukami and Wardle 2005) and (iii) that no interfering factors slow down the trait responses. While comparisons with other approaches proved space-for-time substitutions to be a valid approach (e.g., Elmendorf et al. 2015), the listed assumptions may limit the extent to which the results of our experiment can be used to simulate actual warming scenarios.

Another prediction based on our results is that latitude-associated evolution may shape the outcome of predator–prey interactions under a scenario of invading low-latitude predators. Poleward movements are very common and pronounced in damselflies (Hickling et al. 2006). Our results indicate that predation rates on mosquitoes at the high latitude will increase when they encounter invading southern damselflies. Yet, this is only true in the absence of the pesticide. In the presence of the pesticide, the evolved higher vulnerability to pesticides in the low-latitude damselflies will result in equal predation rates compared to the high-latitude damselflies. These latitude-associated patterns are also directly relevant for current biological control of mosquitoes as they indicate that, all else being equal, predation rates by damselfly larvae will be higher at the low than at the high latitudes in the absence of pesticides.

Insights into how species interactions will change under global warming are outstanding applied evolutionary topics that are crucial to evaluate the potential of biological control in a warming world (Roderick et al. 2012). Specifically, we identified the potential role of evolution in shaping mosquito control by predators in a warming world, a largely overlooked aspect of how global warming may affect vector species and associated diseases (Kovats et al. 2001; Ramasamy and Surendran 2012; Parham et al. 2015). Our results indicate how the evolutionary differentiation of the damselfly predators between latitudes in life history and the associated differentiation in vulnerability to pesticides shape how a pesticide affects the current outcome of predator–prey interactions with a vector mosquito. Moreover, our results inform how *in situ* evolution and poleward movements of the predators may change these interactions at the high latitude under warming. Our results thereby illustrate the value of a space-for-time substitution approach (Fukami and Wardle 2005; De Frenne et al. 2013) to address applied evolutionary questions related to global warming.

Acknowledgements

We thank Jeremias Becker and Matthias Liess for providing mosquito eggs, Frank Johansson and Ulf Norling for providing the Swedish damselfly eggs, Kent Olsen and Nicolas Bell for providing the Danish damselfly eggs, Philippe Lambret, Sarah Oexle and Vincent Lemoine for providing the French damselfly eggs. Lieven Therry assisted with running the Flake model. TTT is an IRO PhD Fellow, LJ is a postdoctoral Fellow of FWO-Flanders, KVD is a postdoctoral Fellow of H.C. Ørsted, Technical University of Denmark and LODB is a PhD Fellow of IWT-Flanders. Financial support came from the Belspo project SPEEDY, KU Leuven Excellence Centre Financing PF/2010/07 and FWO research grant G.0943.15.

Literature cited

Becker, N., D. Petric, C. Boase, J. Lane, M. Zgomba, C. Dahl, and A. Kaiser 2010. Mosquitoes and Their Control. Springer, Berlin, Germany.

Brooks, A. C., P. N. Gaskell, and L. L. Maltby 2009. Sublethal effects and predator-prey interactions: implications for ecological risk assessment. Environmental Toxicology and Chemistry 28:2449–2457.

Broomhall, S. D. 2002. The effects of endosulfan and variable water temperature on survivorship and subsequent vulnerability to predation in *Litoria citropa* tadpoles. Aquatic Toxicology 61:243–250.

Broomhall, S. D. 2004. Egg temperature modifies predator avoidance and the effects of the insecticide endosulfan on tadpoles of an Australian frog. Journal of Applied Ecology 41:105–113.

Brunelli, E., L. Bernabò, C. Berg, K. Lundstedt-Enkel, A. Bonacci, and S. Tripepi 2009. Environmentally relevant concentrations of endosulfan impair development, metamorphosis and behaviour in *Bufo bufo* tadpoles. Aquatic Toxicology 91:135–142.

Calamari, D., and H. Naeve 1994. Review of Pollution in the African Aquatic Environment. Food and Agriculture Organisation, Rome, Italy.

Campero, M., S. Slos, F. Ollevier, and R. Stoks 2007. Sublethal pesticide concentrations and predation jointly shape life history: behavioral and physiological mechanisms. Ecological Applications 17:2111–2122.

Ciota, A. T., A. C. Matacchiero, A. M. Kilpatrick, and L. D. Kramer 2014. The effect of temperature on life history traits of *Culex mosquitoes*. Journal of Medical Entomology 51:55–62.

Congdon, J. D., A. E. Dunham, W. A. Hopkins, C. L. Rowe, and T. G. Hinton 2001. Resource allocation-based life histories: a conceptual basis for studies of ecological toxicology. Environmental Toxicology and Chemistry 20:1698–1703.

De Block, M., K. Pauwels, M. Van Den Broeck, L. De Meester, and R. Stoks 2013. Local genetic adaptation generates latitude-specific effects of warming on predator–prey interactions. Global Change Biology 19:689–696.

De Frenne, P., B. J. Graae, F. Rodríguez-Sánchez, A. Kolb, O. Chabrerie, G. Decocq, H. De Kort et al. 2013. Latitudinal gradients as natural laboratories to infer species' responses to temperature. Journal of Ecology 101:784–795.

Declerck, S., T. De Bie, D. Ercken, H. Hampel, S. Schrijvers, J. Van Wichelen, V. Gillard et al. 2006. Ecological characteristics of small farmland ponds: associations with land use practices at multiple spatial scales. Biological Conservation 131:523–532.

Dinh Van, K., L. Janssens, S. Debecker, M. De Jonge, P. Lambret, V. Nilsson-Örtman, L. Bervoets et al. 2013. Susceptibility to a metal under global warming is shaped by thermal adaptation along a latitudinal gradient. Global Change Biology 19:2625–2633.

Dinh Van, K., L. Janssens, S. Debecker, and R. Stoks 2014a. Temperature- and latitude-specific individual growth rates shape the vulnerability of damselfly larvae to a widespread pesticide. Journal of Applied Ecology 51:919–928.

Dinh Van, K., L. Janssens, S. Debecker, and R. Stoks 2014b. Warming increases chlorpyrifos effects on predator but not anti-predator behaviours. Aquatic Toxicology 152:215–221.

Elmendorf, S. C., G. H. R. Henry, R. D. Hollister, A. M. Fosaa, W. A.

Gould, L. Hermanutz, A. Hofgaard et al. 2015. Experiment, monitoring, and gradient methods used to infer climate change effects on plant communities yield consistent patterns. Proceedings of the National Academy of Sciences of the United States of America 112:448–452.

Englert, D., M. Bundschuh, and R. Schulz 2012. Thiacloprid affects trophic interaction between gammarids and mayflies. Environmental Pollution 167:41–46.

Farajollahi, A., D. M. Fonseca, L. D. Kramer, and K. A. Marm 2011. "Bird biting" mosquitoes and human disease: a review of the role of Culex pipiens complex mosquitoes in epidemiology. Infection, Genetics and Evolution 11:1577–1585.

Fleeger, J. W., K. R. Carman, and R. M. Nisbet 2003. Indirect effects of contaminants in aquatic ecosystems. Science of the Total Environment 317:207–233.

Forbes, V. E. 2000. Is hormesis an evolutionary expectation? Functional Ecology 14:12–24.

Fukami, T., and D. A. Wardle 2005. Long-term ecological dynamics: reciprocal insights from natural and anthropogenic gradients. Proceedings of the Royal Society B 272:2105–2115.

Gilman, S. E., M. C. Urban, J. Tewksbury, G. W. Gilchrist, and R. D. Holt 2010. A framework for community interactions under climate change. Trends in Ecology and Evolution 25:325–331.

Gimonneau, G., M. Pombi, R. K. Dabire, A. Diabate, S. Morand, and F. Simard 2012. Behavioural responses of Anopheles gambiae sensu stricto M and S molecular form larvae to an aquatic predator in Burkina Faso. Parasites Vectors 5:65.

Gosden, T. P., R. Stoks, and E. I. Svensson 2011. Range limits, large-scale biogeographic variation, and localized evolutionary dynamics in a polymorphic damselfly. Biological Journal of the Linnean Society 102:775–785.

Grigaltchik, V. S., A. J. Ward, and F. Seebacher 2012. Thermal acclimation of interactions: differential responses to temperature change alter predator-prey relationship. Proceedings of the Royal Society B 279:4058–4064.

Hayden, M. T., M. K. Reeves, M. Holyoak, M. Perdue, A. L. King, and S. C. Tobin 2015. Thrice as easy to catch! Copper and temperature modulate predator-prey interactions in larval dragonflies and anurans. Ecosphere 6:56.

Hickling, R., D. B. Roy, J. K. Hill, R. Fox, and C. D. Thomas 2006. The distributions of a wide range of taxonomic groups are expanding polewards. Global Change Biology 12:450–455.

IPCC 2013. The Physical Science Basis. In T. F. Stocker, D. Qin, G.-K. Plattner, M. Tignor, S. K. Allen, J. Boschung, A. Nauels, Y. Xia, V. Bex, and P. M. Midgley, eds. Contribution of Working Group I to the Fifth Assessment Report of the Intergovernmental Panel on Climate Change, pp. 1–27. Cambridge University Press, Cambridge, UK and New York, NY, USA.

Janssens, L., and R. Stoks 2012. How does a pesticide pulse increase vulnerability to predation? Combined effects on behavioral antipredator traits and escape swimming. Aquatic Toxicology 110:91–98.

Janssens, L., K. Van Dinh, S. Debecker, L. Bervoets, and R. Stoks 2014. Local adaptation and the potential effects of a contaminant on predator avoidance and antipredator responses under global warming: a space-for-time substitution approach. Evolutionary Applications 7:421–430.

Johansson, H., R. Stoks, V. Nilsson-Örtman, P. K. Ingvarsson, and F. Johansson 2013. Large-scale patterns in genetic variation, gene flow and differentiation in five species of European Coenagrionid damselfly provide mixed support for the central-marginal hypothesis. Ecography 36:744–755.

Junges, C. M., R. C. Lajmanovich, P. M. Peltzer, A. M. Attademo, and A. Bassó 2010. Predator–prey interactions between Synbranchus marmoratus (Teleostei: Synbranchidae) and Hypsiboas pulchellus tadpoles (Amphibia: Hylidae): importance of lateral line in nocturnal predation and effects of fenitrothion exposure. Chemosphere 81:1233–1238.

Junges, C. M., P. M. Peltzer, R. C. Lajmanovich, A. M. Attademo, M. C. Zenklusen, and A. Basso 2012. Toxicity of the fungicide trifloxystrobin on tadpoles and its effect on fish–tadpole interaction. Chemosphere 87:1348–1354.

Kamareddine, L. 2012. The biological control of the malaria vector. Toxins 4:748–767.

Kattwinkel, M., J. V. Kuhne, K. Foit, and M. Liess 2011. Climate change, agricultural insecticide exposure, and risk for freshwater communities. Ecological Applications 21:2068–2081.

Kawecki, T. J., and D. Ebert 2004. Conceptual issues in local adaptation. Ecology Letters 7:1225–1241.

Kesavaraju, B., and S. A. Juliano 2010. Nature of predation risk cues in container systems: mosquito responses to solid residues from predation. Annals of the Entomological Society of America 103:1038–1045.

Kirillin, G., J. Hochschild, D. Mironov, A. Terzhevik, S. Golosov, and G. Nützmann 2011. FLake-Global: Online lake model with worldwide coverage. Environmental Modelling and Software 26:683–684.

Klecka, J., and D. S. Boukal 2012. Who eats whom in a pool? A comparative study of prey selectivity by predatory aquatic insects. PLoS One 7: e37741.

Kovats, R. S., D. H. Campbell-Lendrum, A. J. McMichael, A. Woodward, and J. S. Cox 2001. Early effects of climate change: do they include changes in vector-borne disease? Philosophical Transactions of the Royal Society B 356:1057–1068.

Mandal, S. K., A. Ghosh, I. Bhattacharjee, and G. Chandra 2008. Biocontrol efficiency of odonate nymphs against larvae of the mosquito, Culex quinquefasciatus Say, 1823. Acta Tropica 106:109–114.

McClure, C. D., W. Zhong, V. L. Hunt, F. M. Chapman, F. V. Hill, and N. K. Priest 2014. Hormesis results in trade-offs with immunity. Evolution 68:2225–2233.

Menge, B. A., and A. M. Olson 1990. Role of scale and environmental factors in regulation of community structure. Trends in Ecology and Evolution 5:52–57.

Merilä, J., and A. P. Hendry 2014. Climate change, adaptation, and phenotypic plasticity: the problem and the evidence. Evolutionary Applications 7:1–14.

Millenium Ecosystem Assessment 2005. Ecosystems and Human Well-Being: Biodiversity Synthesis. World Resources Institute, Washington, DC.

Miller, L. P., C. M. Matassa, and G. C. Trussell 2014. Climate change enhances the negative effects of predation risk on an intermediate consumer. Global Change Biology 20:3834–3844.

Moe, S. J., K. De Schamphelaere, W. H. Clements, M. T. Sorensen, P. J. Van den Brink, and M. Liess 2013. Combined and interactive effects of global climate change and toxicants on populations and communities. Environmental Toxicology and Chemistry 32:49–61.

Parham, P. E., J. Waldock, G. K. Christophides, D. Hemming, F. Agusto, K. J. Evans, N. Fefferman et al. 2015. Climate, environmental and socio-economic change: weighing up the balance in vector-borne disease transmission. Philosophical Transactions of the Royal Society of

London B **370**:20130551.

Rall, B. C., O. Vucic-Pestic, R. B. Ehnes, M. Emmerson, and U. Brose 2010. Temperature, predator–prey interaction strength and population stability. Global Change Biology **16**:2145–2157.

Ramasamy, R., and S. N. Surendran 2012. Global climate change and its potential impact on disease transmission by salinity-tolerant mosquito vectors in coastal zones. Frontiers in Physiology **3**:198.

Rasmussen, J. J., U. Nørum, M. R. Jerris, P. Wiberg-Larsen, E. A. Kristensen, and N. Friberg 2013. Pesticide impacts on predator–prey interactions across two levels of organisation. Aquatic Toxicology **140**:340–345.

Reynaldi, S., M. Meiser, and M. Liess 2011. Effects of the pyrethroid fenvalerate on the alarm response and on the vulnerability of the mosquito larva *Culex pipiens molestus* to the predator *Notonecta glauca*. Aquatic Toxicology **104**:56–60.

Roderick, G. K., R. Hufbauer, and M. Navajas 2012. Evolution and biological control. Evolutionary Applications **5**:419–423.

Schmitz, O. J., and B. T. Barton 2014. Climate change effects on behavioral and physiological ecology of predator-prey interactions: implications for conservation biological control. Biological Control **75**:87–96.

Schulz, R., and J. M. Dabrowski 2001. Combined effects of predatory fish and sublethal pesticide contamination on the behavior and mortality of mayfly nymphs. Environmental Toxicology and Chemistry **20**:2537–2543.

Seebacher, F., C. R. White, and C. E. Franklin 2015. Physiological plasticity increases resilience of ectothermic animals to climate change. Nature Climate Change **5**:61–66.

Seiter, S., and J. Kingsolver 2013. Environmental determinants of population divergence in life-history traits for an invasive species: climate, seasonality and natural enemies. Journal of Evolutionary Biology **26**:1634–1645.

Sentis, A., J. Morisson, and D. S. Boukal 2015. Thermal acclimation modulates the impacts of temperature and enrichment on trophic interaction strengths and population dynamics. Global Chang Biology **21**:3290–3298.

Shama, L. N. S., M. Campero-Paz, K. M. Wegner, M. De Block, and R. Stoks 2011. Latitudinal and voltinism compensation shape thermal reaction norms for growth rate. Molecular Ecology **20**:2929–2941.

Sibly, R. M., and P. Calow 1989. A life-cycle theory of responses to stress. Biological Journal of the Linnean Society **37**:101–116.

Srivastava, V., and P. C. Misra 1981. Effect of endosulfan on plasma membrane function of the yeast *Rhodotorula gracilis*. Toxicology Letters **7**:475–480.

Stoks, R., I. Swillen, and M. De Block 2012. Behaviour and physiology shape the growth accelerations associated with predation risk, high temperatures and southern latitudes in *Ischnura* damselfly larvae. Journal of Animal Ecology **81**:1034–1040.

Stoks, R., A. N. Geerts, and L. De Meester 2014. Evolutionary and plastic responses of freshwater invertebrates to climate change: realized patterns and future potential. Evolutionary Applications **7**:42–55.

Trekels, H., F. Van de Meutter, and R. Stoks 2013. Predator cues magnify effects of the pesticide endosulfan in water bugs in a multi-species test in outdoor containers. Aquatic Toxicology **138**:116–122.

Zarnetske, P. L., D. K. Skelly, and M. C. Urban 2012. Biotic multipliers of climate change. Science **336**:1516–1518.

11

The evolutionary legacy of size-selective harvesting extends from genes to populations

Silva Uusi-Heikkilä,[1,2] Andrew R. Whiteley,[3] Anna Kuparinen,[4,†] Shuichi Matsumura,[5,†] Paul A. Venturelli,[6] Christian Wolter,[1] Jon Slate,[7] Craig R. Primmer,[2] Thomas Meinelt,[8] Shaun S. Killen,[9] David Bierbach,[1] Giovanni Polverino,[1] Arne Ludwig[10] and Robert Arlinghaus[1,11]

1 Department of Biology and Ecology of Fishes, Leibniz-Institute of Freshwater Ecology and Inland Fisheries, Berlin, Germany
2 Division of Genetics and Physiology, Department of Biology, University of Turku, Turku, Finland
3 Department of Environmental Conservation, University of Massachusetts, Amherst, MA, USA
4 Department of Environmental Sciences, University of Helsinki, Helsinki, Finland
5 Faculty of Applied Biological Sciences, Gifu University, Gifu, Japan
6 Department of Fisheries, Wildlife, and Conservation Biology, University of Minnesota, St Paul, MN, USA
7 Department of Animal and Plant Sciences, University of Sheffield, Western Bank, Sheffield, UK
8 Department of Ecophysiology and Aquaculture, Leibniz-Institute of Freshwater Ecology and Inland Fisheries, Berlin, Germany
9 Institute of Biodiversity, Animal Health and Comparative Medicine, College of Medical, Veterinary & Life Sciences, University of Glasgow, Glasgow, UK
10 Department of Evolutionary Genetics, Leibniz-Institute for Zoo and Wildlife Research, Berlin, Germany
11 Chair of Integrative Fisheries Management, Faculty of Life Sciences, Albrecht-Daniel-Thaer Institute of Agricultural and Horticultural Sciences, Humboldt-Universität zu Berlin, Berlin, Germany

Keywords

conservation, fisheries-induced evolution, life-history evolution, personality, population dynamics.

Correspondence

Silva Uusi-Heikkilä, Division of Genetics and Physiology, Department of Biology, 20014 University of Turku, Finland.

e-mail: silva.uusi-heikkila@utu.fi

†Authors contributed equally to this work.

Abstract

Size-selective harvesting is assumed to alter life histories of exploited fish populations, thereby negatively affecting population productivity, recovery, and yield. However, demonstrating that fisheries-induced phenotypic changes in the wild are at least partly genetically determined has proved notoriously difficult. Moreover, the population-level consequences of fisheries-induced evolution are still being controversially discussed. Using an experimental approach, we found that five generations of size-selective harvesting altered the life histories and behavior, but not the metabolic rate, of wild-origin zebrafish (*Danio rerio*). Fish adapted to high positively size selective fishing pressure invested more in reproduction, reached a smaller adult body size, and were less explorative and bold. Phenotypic changes seemed subtle but were accompanied by genetic changes in functional loci. Thus, our results provided unambiguous evidence for rapid, harvest-induced phenotypic and evolutionary change when harvesting is intensive and size selective. According to a life-history model, the observed life-history changes elevated population growth rate in harvested conditions, but slowed population recovery under a simulated moratorium. Hence, the evolutionary legacy of size-selective harvesting includes populations that are productive under exploited conditions, but selectively disadvantaged to cope with natural selection pressures that often favor large body size.

Introduction

Human harvest of wild populations is often intense and nonrandom with respect to phenotypes (e.g. Darimont et al. 2009). In most situations, individuals carrying certain fitness-related traits (e.g. large body size or explorative and bold behavior) are more vulnerable to harvest than others

(Allendorf et al. 2008; Alós et al. 2012; Sutter et al. 2012). A well-studied example of human harvest is fishing, which often targets the largest and oldest individuals and is thus positively size selective (Lewin et al. 2006; Jørgensen et al. 2007; Kuparinen and Merilä 2007; Law 2007). Life-history theory suggests that elevated adult mortality favors individuals that allocate energy to reproduction early in life through early maturation at small size and/or increased reproductive investment at the expense of postmaturation somatic growth (Stearns 1992). Such phenotypic changes could be magnified when harvesting is not only intensive but also positively size selective (Laugen et al. 2014). While early maturation increases the probability that an individual will reproduce before it is harvested, small body size at reproduction may confer fitness costs through a decrease in egg number (fecundity), reduced egg and offspring quality (Walsh et al. 2006; Arlinghaus et al. 2010; Uusi-Heikkilä et al. 2010), and increased natural mortality (Jørgensen and Fiksen 2010; Audzijonyte et al. 2013a; Heino et al. 2013; Jørgensen and Holt 2013). However, depending on a species' ecology and local harvesting patterns, evolution of late, rather than early, maturation (Poos et al. 2011) and fast, rather than slow, growth rate (Walters and Martell 2004; Matsumura et al. 2011; Enberg et al. 2012) can also occur in response to harvesting. Furthermore, if adult mortality is very high and there is thus little fitness to gain by allocating energy to future reproduction, fish might invest heavily in the first reproduction and produce high, rather than low (Walsh et al. 2006), quality eggs and offspring. In fact, despite one might intuitively expect a certain of change in response to size-selective harvesting (e.g. evolution of slower growth rate, Walters and Martell 2004), exact predictions of life-history changes in response to fisheries exploitation are challenging and require stock- and fishery-specific analyses (Arlinghaus et al. 2009; Laugen et al. 2014). Evolutionary changes of body size and related life-history traits can have important repercussions for species and community ecology (Peters 1983; de Roos and Persson 2002; Haugen et al. 2007), management reference points (Heino et al. 2013), and population productivity, recovery speed, and fisheries yield (Law and Grey 1989; Hutchings and Fraser 2008; Conover et al. 2009; Laugen et al. 2014) and thus may be of high relevance to contemporary fisheries management (Conover and Munch 2002; Jørgensen et al. 2007).

A common expectation and empirically reported effect of intensive and size-selective harvesting is the downsizing of body size (Conover and Munch 2002; Jørgensen et al. 2007; Swain et al. 2007; Alós et al. 2014). Beyond correlations between body size and a range of early life-history traits, any selection on body size may also affect underlying physiological and behavioral characteristics through correlated selection responses (Walsh et al. 2006; Uusi-Heikkilä et al. 2008; Diaz Pauli and Heino 2014). Several, not mutually exclusive, mechanisms may be at play. For example, adults may be large because they are efficient in converting energy into somatic growth (an energy conversion mechanism), because they mature late (an energy allocation mechanism) or because they are dominant, bold and aggressive in social interactions and hence superior in

securing and defending food resources (an energy acquisition mechanism; Enberg et al. 2012). Large individuals may be also more active and explore the environment more in search for food and they may be able to do so due to lower predation risk (Biro and Post 2008). Any changes in adult body size in response to size-selective fisheries can thus be a consequence of changes in juvenile growth rate (which is an unconfounded measure of growth rate capacity not affected by maturation changes, Enberg et al. 2012), altered maturation schedules (leading to altered energy allocation patterns), or represent an indirect response to selection due to direct selection on correlated behavioral or physiological traits (Walsh et al. 2006; Uusi-Heikkilä et al. 2008; Biro and Stamps 2010; Enberg et al. 2012). Therefore, size-selective fisheries might also induce changes in physiological traits (e.g. metabolism) or behavior (e.g. aggression, boldness) that contribute to energy acquisition and hence growth (Enberg et al. 2012; Sutter et al. 2012; Alós et al. 2015). Most empirical studies on fisheries-induced evolution (FIE) have so far focused on three key life-history traits, namely growth rate, age and size at maturation, and reproductive investment (e.g. Rijnsdorp 1993; Olsen et al. 2004; for reviews, see Policansky 1993; Heino and Godø 2002; Sharpe and Hendry 2009; Devine et al. 2012; Audzijonyte et al. 2013b). Currently, there is little doubt that FIE in the wild could be both plausible and potentially widespread (Jørgensen et al. 2007; Kuparinen and Merilä 2007). However, examination of the joint effects of size-selective fisheries on several traits, including physiology (e.g. metabolism), behavior (e.g. feeding activity) and life history (e.g. growth capacity), in terms of the resulting effects for population dynamics and fisheries has been largely confined to modeling studies (e.g. Thériault et al. 2008; Andersen and Brander 2009; Dunlop et al. 2009; Enberg et al. 2009; Matsumura et al. 2011), and the potential management consequences of FIE remain the least well-understood aspects of FIE (Jørgensen et al. 2007). Notwithstanding the ongoing controversy of whether harvesting causes genetic as opposed to mere phenotypic change, human-induced rapid phenotypic trait change may trigger equally fast ecological change and thereby shape populations, food webs, and ecosystems on a global scale (Darimont et al. 2009; Palkovacs et al. 2012).

Because life-history, morphological, and behavioral traits are at least moderately heritable (Mousseau and Roff 1987), intensive and size-selective fishing over multiple generations is expected to cause genetic (i.e. evolutionary) changes in a range of traits (e.g. Law 2007; Dunlop et al. 2009; Laugen et al. 2014; Marty et al. 2015). Genetic changes, as opposed to mere phenotypic change, may magnify the ecological challenges related to overfishing because they are usually slowly, if at all, reversible in the absence of similarly strong natural selection pressures working in opposite direction than harvest selection (Conover et al. 2009) and thus may have lasting effects on populations and consequently on fisheries. Indeed, meta-analyses, modeling studies, and experimental work have all shown that evolutionary effects of harvesting can impair biomass recovery of overharvested populations (Conover et al. 2009; Enberg et al. 2009; Kuparinen and Hutchings 2012; Neubauer et al. 2013), affect management reference points (Heino et al. 2013), and reduce catchability and hence catch rates and fisheries quality (Philipp et al. 2009; Alós et al. 2015). Fisheries-induced adaptive change might thus have multiple consequences for the population and the fishery, in particular when slowly reversible genetic, as opposed to plastic, changes are involved (Laugen et al. 2014). Despite increasing concern about the effects of size-selective harvesting on wild fish populations (Borrell 2013), the consequences of FIE for populations and fisheries continue to raise controversy (Browman et al. 2008; Jørgensen et al. 2008; Andersen and Brander 2009; see also Bunnefeld and Keane 2014). Although exploited fish populations can consist of individuals of reduced average adult body size for both demographic and evolutionary reasons (Jørgensen et al. 2007), they might remain biologically viable and highly productive precisely because of the evolution of 'fast' life histories, that is, early maturation and high reproductive investment (Hutchings 2009; Heino et al. 2013; Jørgensen and Zimmermann 2015). Modeling studies also suggest that if fishing pressure can be kept within optimal limits, FIE is not expected to cause major economic repercussions (Eikeset et al. 2013; Jørgensen and Zimmermann 2015).

The concern about fisheries-induced evolution was first raised at the beginning of the 20th century (Rutter 1902), but has only gained significant momentum since the 1990s when Law and coworkers published their groundbreaking research on FIE (e.g. Law and Grey 1989). Although the potential for rapid FIE is now theoretically, and also empirically, well founded (Jørgensen et al. 2007; Kuparinen and Merilä 2007; Laugen et al. 2014), conclusively detecting it in natural populations has remained a challenge due to the limited opportunities for disentangling plastic and genetic responses in a suite of phenotypic traits (Allendorf et al. 2008; Naish and Hard 2008; Therkildsen et al. 2010; Cuveliers et al. 2011; Pérez-Rodrígues et al. 2013). In theory, experiments on FIE could also be designed in the wild (McAllister and Peterman 1992), but the cause-and-effect mechanism of size-selective harvesting can best be studied experimentally in controlled laboratory environments (Conover and Baumann 2009; Diaz Pauli and Heino 2014). However, to date, only few experimental studies have conclusively reported harvest-induced genetic changes based on quantitative genetics (Conover and Munch 2002; Philipp et al. 2009) or molecular approaches (van Wijk et al. 2013). While there is little doubt that FIE might be occurring in practical fisheries, the magnitude of life-history and other phenotypic changes, the specific genes under selection, and the way how the phenotypic and genetic changes affect fisheries, population viability, productivity, and recovery remain largely unresolved (Jørgensen et al. 2007; Heino et al. 2013; Laugen et al. 2014).

From a conservation and management perspective, fisheries-induced phenotypic changes are of particular concern if they affect population dynamics, viability, and recovery (Hutchings and Fraser 2008; Dunlop et al. 2009; Heino et al. 2013; Laugen et al. 2014). There is ongoing debate whether FIE can significantly affect populations and ecosystem services on timescales that are relevant to fisheries managers (Andersen and Brander 2009; Laugen et al. 2014). Here, we present the results of a selection experiment that provides a comprehensive picture of the evolutionary legacy of size-selective harvesting by examining its phenotypic, genetic, and population-level consequences. A major part of this experiment was to quantify phenotypic and genetic changes in response to five generations of size-selective harvesting in wild zebrafish (*Danio rerio*) in the laboratory using functional genomic markers that can occur close by, within, or in the regulatory areas of genes under selection. Phenotypic changes were scaled up to the population level using a life-history model. Our results provide important insights into FIE, such as how quickly fishing might bring about evolutionary changes, and contribute to the ongoing debate over whether size-selective harvesting causes evolutionary changes in ecologically and economically important phenotypic traits (e.g. in adult body size), and whether these changes matter for biomass renewal, population stability, and conservation.

Materials and methods

Size-selective harvesting and breeding design

We used F_1-generation offspring from approximately 1500 wild-collected zebrafish (parental stock) from West Bengal in India (see Uusi-Heikkilä et al. 2010 for details) in our selection experiment to ensure maximum genetic variation. Our three experimental treatments, each with one replicate (i.e. two tanks per treatment) consisted of approximately 450 zebrafish per replicate tank. We reared individuals in each generation in identical environmental (Supporting Information S1) and density conditions in three replicated selection treatments similar to Conover and Munch's (2002) landmark study. In contrast to Conover and Munch's (2002) design where maturation was triggered by photoperiod and harvesting occurred at a fixed age, the timing of harvesting in our experiment was determined by the maturation schedule of the randomly harvested control treatment: Once 50% of the randomly harvested fish were mature (determined macroscopically from 20 lethally sampled females), we harvested 75% of individuals in all treatments size selectively or randomly according to size. Hence, as not all fish were mature at harvesting, the design represented a harvesting pattern that targeted both mature and immature fish. Prior to harvesting, we measured the standard length (SL) of all fish in all tanks to the nearest mm and wet mass (WM) to the nearest 0.1 g. During harvesting, we sorted the fish by SL and estimated the 75th and the 25th percentiles of the size distributions. To mimic a highly intensive lethal capture fishery (Lewin et al. 2006; Worm et al. 2009; Hilborn and Stokes 2010), we applied a 75% per-generation harvest rate. In the randomly selected line (random with respect to sizes that were harvested), we measured all fish and then assigned the fish randomly with respect to body size to either the harvested group or to the spawning stock. In a large-selected line, we assigned the 25% largest fish to the spawning stocks, and in the small-selected line we assigned the 25% smallest fish to the spawning stocks. The large-selected line hence represented a mortality schedule mimicking a maximum-length limit where the largest mature fish survived. The small-selected line instead represented a positive size-selective harvest scheme common to most fisheries, where young immature fishes are saved because they are not vulnerable to the gear (e.g. too small to become entangled in a fishing net) and/or have to be released due to the minimum-length limit regulations. There was no unharvested control for logistical reasons. A subsample of F_1-generation fish from each selection line ($N = 50$ per selection line) was measured at age 60 days to ensure that there were no initial differences in body size when the selection experiment was started (large-size selected: 15.2 ± 2.79 mm; randomly selected: 15.0 ± 2.26 mm; small-size selected: 15.8 ± 2.58 mm; mean $\pm$ SD; $\chi^2 = 1.8105$, $P = 0.4044$).

After harvesting, we kept the spawners from each of the six experimental populations in separate aquaria for 14 days to ensure that most fish reached maturity before initiating the spawning trial. To increase the odds of all spawners contributing to the next generation, we mimicked natural spawning conditions (Hutter et al. 2010) by transferring small groups of individuals to spawning boxes. The boxes contained a mesh structure that prevented egg cannibalism (Uusi-Heikkilä et al. 2010, 2012a). We used two sizes of spawning boxes. Five-liter spawning boxes were each stocked with two females and four males, and three-liter spawning boxes were each stocked with one female and two males (altogether 40 females and 80 males per selection line from F_1- to F_6-generation). We measured the SL and WM of each spawner before placing it into a spawning box. Females were swapped among boxes once during the spawning trial to ensure a high number of parental combinations and to sustain genetic variation. Spawning trials lasted for 5 days. Each day we cleaned the spawning boxes, placed fertilized eggs on petri dishes, and transferred the petri dishes to an incubator (Tintometer GmbH; $T = 26°C$). We reared hatchlings in the spawning boxes (adults were removed) for 30 days and then transferred them to the rearing tank that their parents were from. The selective harvesting was repeated again when 50% of the randomly selected fish were mature.

We continued the size-selective harvesting for five generations (F_1–F_6) and then halted size selection up to four generations (F_7–F_{10}). To estimate the strength of selection during the first five generations, per-generation standard deviation-standardized selection differentials (also known as selection intensity), which describes the difference in the average body size between the spawners and the entire experimental population standardized by the phenotypic standard deviation, were estimated for each generation separately following Matsumura et al. (2012). Note that given our harvesting design, where age at harvest varied from generation to generation in line with potential changes in age at 50% maturation of the random line, did not allow the estimation of selection responses with respect to size at age. From the F_6-generation onward, 100 individuals per selection line were randomly selected for spawning, and, as in previous generations, their offspring were reared in identical environmental and density conditions. Comparisons of life-history traits among selection lines were conducted two to three generations postselection, and physiological and behavioral traits four generations postselection, in trials where the rearing and growth conditions were strictly standardized among lines. This was performed to remove all confounding maternal and paternal effects and any potential epigenetic effects, thereby increasing the odds that life-history evolution was measured in a comparative way. For example, producing the next generation of fish was time-consuming, and in some cases, populations of our selection treatments were started a couple of weeks apart and consisted of slightly different-aged offspring. Uncontrolled environmental effects during the subsequent holding phase might affect life histories, so the comparison of different traits among the selection lines was performed only after keeping the selected fish for at least two generations without any further selection under controlled and time-matched conditions. Life-history, physiological, and behavioral traits were assessed using offspring from parents that had experienced at least two generations of no selection. For the among-line comparisons, experimental fish of all lines were produced and reared at the same time and in identical conditions. This approach was similar to earlier work in male guppies (*Poecilia reticulata*; van Wijk et al. 2013). Admittedly, delaying the assessment of life-history, physiological, and behavioral traits might not be always desirable in terms of experimental design because the fish might have started converging back to their original life histories as shown by Conover et al. (2009). At the same time, our experimental design can give us some insight into the persistence of potential harvest-induced changes and certainly allowed comparisons among lines that were unconfounded by uncontrolled tank or rearing effects.

Assessment of life-history, physiological, and behavioral traits

We assessed a range of life-history, physiological, and behavioral traits expected to change in response to size selection after five generations of size-selective harvesting. We focused on life-history traits commonly studied in the context of FIE, namely juvenile and adult growth rates, reproductive investment and maturation schedule (Heino et al. 2013) as well as early life-history traits (Walsh et al. 2006). Moreover, any changes in energy allocation might be related to metabolic changes (e.g. routine metabolism) or changes in energy acquisition patterns related to behavior (exploration and boldness; Enberg et al. 2012). Hence, we also measured standard metabolic rate and measures of risk-taking behavior and exploration in juvenile zebrafish.

Growth
To study the growth differences among the selection lines after five generations of size-selective harvesting, we used replicate boxes (each stocked with 10 fish) for each of the experimental populations and their replicate lines (altogether 48 rearing boxes). We measured the SL and wet mass (WM) of the F_9-generation fish (i.e. third generation after the selection was halted) every 15 days from age 30 days to age 210 days. To derive growth and other growth-related life-history traits, we fitted a biphasic growth model (Lester et al. 2004) to length-at-age data (sexes combined) from each experimental population. The biphasic growth model produced estimates for several key life-history traits, such as juvenile growth rate (h), reproductive investment (g), age at maturity (T), length at maturity (L), and asymptotic length (L_∞), and expected instantaneous mortality rate (M). For more information about the growth experiment and the growth model, see Supporting Information S2.

Maturation
To estimate the plasticity in age and size at maturation that stemmed from growth variation among the selection lines, we conducted a maturation experiment in which F_8-generation fish (i.e. second generation after the selection was halted) were reared under three different feeding conditions (1%, 2%, and 4% of body weight in dry food daily) following the protocol in Uusi-Heikkilä et al. (2011). We used the demographic estimation method (Barot et al. 2004) for estimation of the probabilistic maturation reaction norms (PMRNs; Dieckmann and Heino 2007) for each selection treatment (see Supporting Information S3).

Reproductive performance and early life-history traits
We estimated the reproductive performance and potential differences in early life-history traits among the selection lines using F_9-generation fish. Variables of interest included spawning frequency, clutch size (number of fertilized eggs produced by female, that is, absolute fecundity) and the relative fecundity (number of fertilized eggs per gram of female WM; these three were measures of reproductive performance), egg size, egg survival, larval hatching probability, larval age at hatch, larval length at hatch, larval yolk sac volume, swim bladder inflation probability, and larval survival (these were measures of early life-history traits; Supporting Information S4). In all statistical analyses, selection line was treated as a predictive variable, and selection line replicate, spawning day and couple (i.e. the spawning female and male) were treated as random variables. If there was virtually no variance associated with the random variables, they were excluded from the model. We modeled count data with a Poisson and probability data (e.g. hatching probability) with a binomial error structure. To generate predictions of a potential selection line-specific trait divergence and to control for the effect of female body size on early life-history traits (i.e. size-dependent maternal effects, Hixon et al. 2014), female size was added as predictive variable in a second set of models (except in the larval survival probability where larvae from different spawning couples were pooled). Our approach of running models with and without female size as covariate allowed us to examine whether differences in reproductive traits were associated with maternal body size or whether the traits had evolved independently of maternal size.

Metabolic rate
We used juvenile zebrafish of the F_{10}-generation (i.e. fourth generation after the selection was halted) to study differences in mass-specific standard metabolic rate (SMR) among the selection lines. Juveniles were used to achieve a measure of base metabolism unaffected by maturation. SMR was measured as rates of oxygen uptake calculated according to a previously published protocol (Dupont-Prinet et al. 2010). We ensured that assumptions of homogeneity and normality of residuals were met, and examined differences in SMR among treatments using a linear mixed model, with selection line as a predictive variable, and selection line replicate as a random variable (Supporting Information S5).

Behavior
To study differences in boldness and exploratory behavior as measures of energy acquisition-related behaviors (Enberg et al. 2012) among the selection lines, we used the same

fish that were tested for SMR. Each individual was tested twice for its exploration behavior in an open-field test in a novel environment similar to Ariyomo and Watt (2012). In each trial, a single focal fish was introduced into a transparent plastic cylinder in the center of the arena (Supporting Information S5). After a brief acclimatization period, the cylinder was carefully removed and the fish movement was videotaped for 5 min (1st trial). Measurements were repeated for each individual after a break of 30 min (2nd trial). Test fish were measured for total length (TL) and WM after the tests were completed to not stress the fish. As a first proxy for exploration, we calculated the individual mean velocity during each trial (distance moved in 250 s). We further scored the time that a fish spent freezing (defined as not moving faster than 20 mm^{-s}) as another proxy for its exploration and degree of boldness (assuming that fish that freeze less are more explorative and bold).

To analyze the data, we first searched for correlations between both response variables (i.e. velocity and time spent freezing) using a principle component analysis (PCA) for both test trials (Supporting Information S5). We used the first principal component (PC1) as a response variable in a linear mixed model in which selection treatment, experimental trial, TL, and WM were predictive variables and individual and selection line replicate were random variables. Model fitting was performed by first evaluating the random effects through likelihood ratio tests. We then excluded all covariates with $P > 0.1$ and refitted the model. In the final model, all other explanatory variables could be excluded, except the selection treatment (predictive variable) and the individual (random variable). More details are given in the Supporting Information S5.

Evolutionary rate

We estimated evolutionary rates in body size at age 90 days for each selection line using haldanes (Haldane 1949). Haldanes were calculated as:

$$((\chi_2/s_p) - (\chi_1/s_p))/t,$$

where χ is the mean body length at age 90 days after one generation of selection (F_2-generation, χ_1) and after five generations of selection and three generation of no selection (F_8-generation, χ_2), s_p is the pooled standard deviation of trait values across time, and t is the number of generations. Mean body lengths at age 90 days were measured from a subsample of fish collected at F_2- and F_8-generations.

Genetic analyses

To determine whether size-selective harvesting induced genetic changes in the experimentally exploited zebrafish populations, we used 371 genomewide, evenly distributed single nucleotide polymorphisms (SNPs) that were chosen from a previously analyzed wild zebrafish dataset (Whiteley et al. 2011) in 502 individuals (Supporting Information S6). Outlier analysis was conducted for F_6-generation individuals (i.e. first generation after the selection was halted) using the Fdist method (Beaumont and Nichols 1996) implemented in software LOSITAN – Selection Workbench (Antao et al. 2008). We studied the differences in allele frequencies of outlier loci among selection lines with PCA. Because outliers detected by the outlier test can be caused by allele frequency divergence in any number of experimental replicates, we directly examined allele frequency variation for the outlier loci to further characterize the nature of the parallel adaptive divergence. This was performed by creating 95% confidence intervals for the allele frequencies of the selected lines by bootstrapping the allele frequency data. We then determined whether the confidence intervals of the selection lines and selection line replicates overlapped. By doing so for each outlier locus, we identified the selection treatment responsible for the allele frequency differences (and the detection of the outlier loci). We identified genes that were nearby outlier loci on the same linkage group using a SNP database (www.ncbi.nlm.nih.gov) and the Zebrafish Model Organism Database (www.zfin.org).

Population growth model

We estimated the finite rate of population growth (λ) of each selection line under different harvest scenarios (with and without size-dependent harvesting) via a density-dependent Leslie matrix model that incorporated evolved life-history traits (i.e. h, g, T, and M, see section 'Growth' above), fecundity estimates that were based on empirical zebrafish egg weight data, and age-dependent survival probabilities that were determined, with certain adjustments, using empirical estimates of early life-history traits, that is, fertilization rate, egg survival, hatching probability, and larval survival (Supporting Information S7). For simplicity, we modeled only females. We studied the performance of individuals from the large- and the small-selected treatments by comparing the population growth rate of a variant individual in an equilibrium population represented by the randomly selected life history. We introduced three prototypical size-dependent harvest mortality scenarios in the population dynamical model: (i) small-size harvested, which mimicked a harvest slot as harvesting only started on maturing fish and large mature fish were saved, (ii) randomly harvested, which represented unselective harvesting with respect to body size, and (iii) large-size harvested, which represented a standard positive size-selective fishery and/or a fishery managed with a minimum landing size or minimum-size limit (i.e. small-selected experimental fish) (Supporting Information S7; Fig. S2). Actual values of maximum daily instantaneous mortalities and length limits were determined to mimic our experimental 75% per-generation harvest rate. Accordingly, the maximum daily instantaneous mortality (F_{max}) and the lower (L_1) and upper (L_2) limits of the harvesting scheme for each mortality scenario were as follows: small-size harvested: $F_{max} = 0.01825$ per day, $L_1 = 20.8$ mm, and $L_2 = 24.1$ mm; randomly harvested: $F_{max} = 0.009125$ per day, $L_1 = 20.8$ mm, and $L_2 = \infty$; and large-size harvested: $F_{max} = 0.01825$, $L_1 = 23.7$ mm, and $L_2 = \infty$ (see Supporting Information S7; equation 8, Fig. S2, and Table S10 for details). To reveal potential costs of evolution, we compared the population recovery between the three selection lines after exposure to fishing. To that end, during the first 4000 days (about 30 generations), we introduced size-selective fishing mortality of a similar selectivity as in the experiment (represented by the three harvesting schemes mentioned above) and then stopped fishing and allowed the populations to recover up to 8000 days (about 60 generations). For detailed methodological information of the population growth model, including equations, see Supporting Information S7.

Results

Selection intensities

As expected, we found the harvesting mortality that mimicked positive size-selective fisheries mortality (i.e. small-selected fish) exerted negative standard deviation-standardized selection differentials (i.e. selection intensity) on body length, while the random fish experienced selection intensity close to zero (Fig. 1). By contrast, selection for large body size (large-selected fish) exerted a consistently positive standardized selection differential on body length (Fig. 1). In the last generation of the selection experiment (F_5-generation), the selection intensity on body size in the small-selected fish was also close to zero. The reason was that in the last selected generation, the average body size of the small-selected fish in the experimental population was too small (18.7 ± 3.21 mm) for a timely reproduction after selection. To not risk the experiment, it was necessary to apply a slightly higher size threshold (20 mm) when selecting the spawners of the small-selected line. Thus, the small-selected fish only experienced four rather than five generations of intensive size-selective harvest.

Life-history changes

After five generations of size selection followed by three generations of no selection, the small-selected fish had evolved a significantly lower asymptotic length (L_∞) (27.4 ± 0.40 mm) compared to the other selection lines

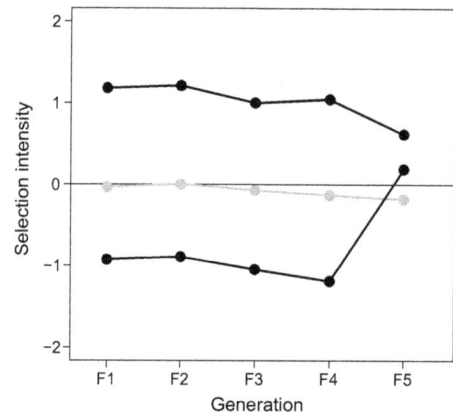

Figure 1 Standard deviation-standardized selection differential (S; also known as the selection intensity) estimated for each generation separately. Red, gray, and blue symbols and lines represent populations of small, random, and large fish, respectively.

(random 29.2 ± 0.29 mm; large-selected 29.5 ± 0.59 mm; Fig. 2A, Table 1). The small-selected fish were also significantly smaller (9.7 ± 1.96 mm) when the growth experiment started (at age 30 days) compared to random (10.4 ± 2.09 mm) and large-selected fish (10.5 ± 1.87 mm; $\chi^2 = 15.20$; df = 4,6; $P = 0.0005$; Fig. 2A). Despite the lower maximum length, small-selected (and random) fish exhibited a somewhat, yet statistically not significantly, higher juvenile growth rate (h; Fig. 2B; Table 1). Small-selected and random fish also invested more energy in reproduction (g) (Fig. 2C, Table 1) and matured earlier (T) than large-selected fish in the growth experiment (Fig. 2A, Table 1). Despite the similar age at maturation, small-selected fish matured at a smaller size (L) than random and large-selected fish (Fig. 2A, Table 1). The instantaneous natural mortality (M) estimated from the parameters of the biphasic growth model (Supporting Information S2) was higher among small-selected and random (0.018 and 0.017/day, respectively) than among large-selected fish (0.015/day).

The probabilistic maturation reaction norm (PMRN) describes the 50% probability of maturation as a function of age and size (and potentially other traits affecting maturation) while controlling for the effect of growth on maturation. The maturity ogive used in the estimation of the age-, length-, and condition-based PMRNs included the main effects of age, length, and condition. Condition was a significant factor determining maturity among small- and randomly selected but not among large-selected fish (Supporting Information S3; Table S1). None of the interaction terms in the ogive models were significant. The three-dimensional PMRNs estimated for large- and small-selected fish largely overlapped on the right-hand side of the curve (low growth rate) but not on the left (high

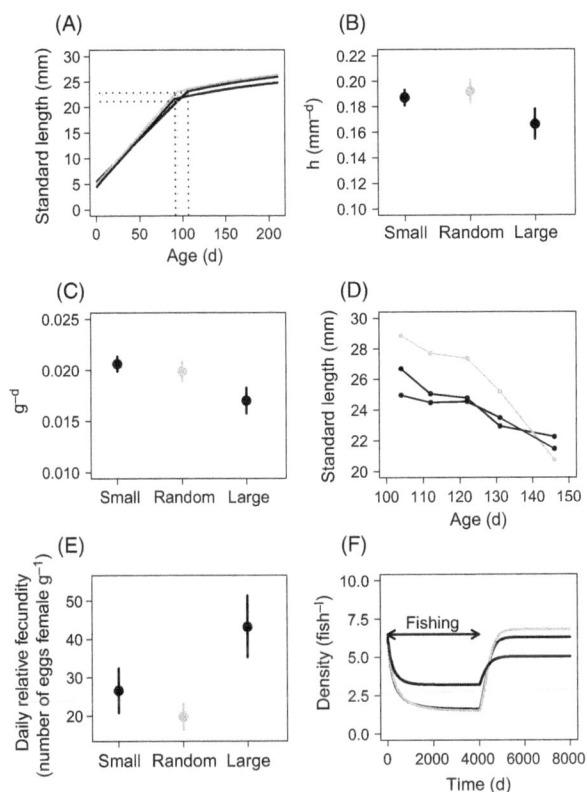

Figure 2 Differences among selection lines in life-history traits and reproductive output. Data are mean ± SEM for panels (B), (C), and (E). Red, gray, and blue symbols and lines represent populations of small-, randomly, and large-selected fish, respectively. (A) Biphasic growth curves (dotted lines refer to treatment-specific age *T* and length at maturity *L*). (B) Average juvenile growth rate (*h*) (small-selected 0.190 ± 0.045; random 0.194 ± 0.040; large-selected 0.167 ± 0.025 mm^{-d}). (C) Average daily reproductive investment (*g*) (small-selected 0.021 ± 0.003; random 0.020 ± 0.004; large-selected 0.017 ± 0.005). (D) 50% quantiles for the probabilistic maturation reaction norms (PMRNs; note that most plastic responses in maturation shift the phenotype along the reaction norm, whereas an evolutionary response in maturation shifts the reaction norm itself). (E) Average relative fecundity (small-selected 26.6 ± 5.83; random 19.8 ± 3.44; large-selected 43.3 ± 8.03 eggs/g female). (F) Simulated population-level consequences of life-history changes induced by size-selective harvesting. Population recovery was monitored after large-size harvested fishing was operated for the first 4000 days.

growth rate) where small-selected fish exhibited somewhat reduced age and size at maturation (Fig. 2D). The PMRN intercept estimated for the random fish was higher than the ones estimated for the small- and large-selected fish, particularly on the left-hand side of the curve, indicating that under fast growth random fish matured at older age and larger compared to the small- and large-selected fish. Variation (measured in standard deviation) in body size across all ages in restricted growth conditions was higher among random fish (SD 3.4 mm; range of SL 9–25 mm) com-

pared to the large-selected (SD 2.7 mm; range of SL 9.0–23 mm) and small-selected fish (SD 2.6 mm; range of SL 8.0–23 mm).

Changes in reproductive performance and early life-history traits

Without controlling for female body size, small-selected fish of the F_9-generation had a significantly lower spawning probability (0.28 ± 0.05; mean ± S.E.) than large- (0.51 ± 0.05) and randomly selected fish (0.44 ± 0.05; Table 2). Furthermore, small-selected fish produced significantly fewer eggs than random and large-selected fish (lower absolute fecundity), also relative to body size (i.e. relative fecundity, Fig. 2E), and overall smaller eggs than large-selected fish (Table 2). In terms of relative fecundity, small-selected fish exhibited higher values compared to random fish, but these differences were not significant. Large-selected fish produced slightly larger larvae (3.45 ± 0.013 mm) than small- (3.38 ± 0.021 mm) and randomly selected fish (3.41 ± 0.023 mm), but neither these differences were statistically significant (Table 2). Unexpectedly, the offspring produced by random fish had lower hatching probability (0.58 ± 0.03) than offspring of either large- (0.85 ± 0.02) or small-selected fish (0.80 ± 0.03; Table 2). The offspring of random fish also took longer to hatch (5.2 ± 0.08 days) than the offspring of large- (4.9 ± 0.05 days) and small-selected fish (4.7 ± 0.08 days; Table 2). Other early life-history traits, in particular the larval traits, (Table 2) did not differ significantly among the selection lines.

When female body size was added as a predictive variable in the analyses, it explained a significant amount of variation in spawning probability, in absolute and relative fecundity (particularly in small-selected fish), in egg size, and in hatching probability (Table 3). Across all selection lines, larger females were more likely to produce eggs more frequently and at higher numbers, but the eggs they produced were smaller and suffered from lower hatching probability. In all of these analyses, except spawning probability, selection treatment remained a significant explanatory variable even after controlling for female size, but again there were few significant effects of either selection treatment or female body size on larval traits (Table 3).

Changes in metabolic rate

Standard metabolic rate (SMR) did not differ among the selection lines (linear mixed model, $F_{2,119} = 0.157$, $P = 0.855$). The average SMR of a standardized fish weighing 0.1 g was 0.0738 ± 0.0053 mg h^{-1} for small-selected, 0.0781 ± 0.0057 mg h^{-1} for random, and 0.0753 ± 0.0053 mg h^{-1} for large-selected fish.

Table 1. Differences in life-history parameters among the selection lines as estimated by the biphasic growth model.

Trait	Selection line	Parameter estimates (SE)	Chi-square value* (df)	P-value
Juvenile growth rate (h; mm)	Large	0.167 (0.013)	4.8999 (3,5)	0.0863
	Random	0.194 (0.012)		
	Small	0.189 (0.013)		
Maximum length (L_∞; mm)	Large	29.53 (0.592)	29.946 (3,4)	<0.0001
	Random	29.24 (0.371)		
	Small	27.37 (0.398)		
Age at maturity (T; d)	**Large**	106.2 (3.001)	12.490 (3,4)	<0.0001
	Random	94.14 (4.027)		
	Small	91.27 (4.315)		
Length at maturity (L; mm)	Large	22.89 (0.449)	0.9742 (3,4)	<0.0001
	Random	22.61 (0.300)		
	Small	21.15 (0.322)		
Daily reproductive investment (g)	**Large**	0.017 (0.001)	6.4809 (3,4)	0.0109
	Random	0.020 (0.001)		
	Small	0.021 (0.002)		

Parameter estimates for each treatment for each trait are shown. *P*-values were derived from chi-square statistics. The selection line differing significantly from the other treatments is indicated in bold.
*Chi-square value from the deletion of the variable from the full model.

Changes in behavior

In the behavioral analysis, the PC1 scores for the behavioral traits differed significantly among the selection lines (Supporting Information S5; Table S3). The PC1 captured behaviors (swimming velocity and time spent freezing) that were suggestive of risk taking and boldness. Large-selected zebrafish were significantly more explorative and bolder (i.e. swam with higher mean velocities and spent less time freezing) than small- ($P = 0.047$) and randomly selected fish ($P = 0.01$). Based on the high repeatability value (0.47) for the PC1 score, individuals were highly consistent in their boldness behavior between the 1st and the 2nd trial ($\chi^2 = 16.54$, df = 1, $P < 0.001$). The consistency may be an indicator of personality.

Evolutionary rate

The evolutionary rate in body size at age 90 days, estimated as haldanes, was 0.165 for randomly selected fish, 0.053 for the large-selected fish, and -0.116 for small-selected fish.

Genetic changes

Among the 371 SNPs, we identified 22 outlier loci that responded to divergent selection as indicated by high genetic differentiation ($P < 0.025$) (Supporting Information; Table S4). There was also evidence of balancing selection at 12 loci (Supporting Information; Table S5). However, loci under divergent selection are of greatest relevance to studies such as ours and were thus explored in more detail. A PCA on the outlier SNPs found substantial

evidence for similar amount and direction of genetic change within each size-selected replicate and relative to the random replicates after five generations of size-selective harvesting (Fig. 3).

In eight of the 22 outliers, differences in allele frequencies were significant and consistent between the selection line replicates (Supporting Information; Table S6) as there was virtually no overlap in 95% confidence intervals for both size-selected replicates of one selection line (e.g. large-selected) relative to other treatment replicates (e.g. small-selected and random; Supporting Information; Table S6, Fig. 4). Parallel allele frequency divergence at these eight loci (hereafter parallel outlier loci) made drift an unlikely explanation and revealed that the mechanistic response underlying adaptive divergence was similar for a subset of loci. For an additional six outlier loci, the significant difference in allele frequency occurred in one of the treatment replicates (Supporting Information S1). Selection may be responsible for these single replicate-specific results, but it is more difficult to rule out genetic drift in this case. Six of the eight parallel outlier loci were in significant linkage disequilibrium (LD; $P < 0.05$) with a nearby SNP on the same linkage group (Supporting Information; Table S7).

Five of the eight parallel outlier SNPs occurred in or close to a gene or in a regulatory region of a gene that has a known function, such as serotonin synthesis, ion transport, regulation of transcription, and collagen formation (Table 4). Furthermore, six of the parallel outlier SNPs were in significant LD with another SNP that occurred in a gene or in a regulatory region of a gene with a known function, such as embryonic yolk processing, immune response system, and stress response (Table 4).

Table 2. The effect of the selection treatment on zebrafish reproductive performance and early life-history traits.

Trait	Selection line	Parameter estimates* (SE)	Chi-square value† (df)	P-value‡
Spawning probability	Large	0.4911 (0.2764)	12.138 (3,4)	0.0005
	Random	0.4437 (0.3160)		
	Small	0.2614 (0.3086)		
Fecundity (eggs/female/day)	**Large**	9.1831 (0.4543)	779.16 (3,5)	<0.0001
	Random	5.3896 (0.0396)		
	Small	3.6352 (0.0354)		
Relative fecundity (eggs/g of female WM/day)	**Large**	14.050 (0.4698)	498.52 (3,4)	<0.0001
	Random	8.5289 (0.0315)		
	Small	8.6313 (0.0246)		
Egg yolk size (mm)	**Large**	0.7286 (0.0116)	82.959 (3,4)	<0.0001
	Random	0.6611 (0.0077)		
	Small	0.6618 (0.0076)		
Egg survival probability	Large	0.9490 (0.5148)	0.958 (3,5)	0.6194
	Random	0.9609 (0.5993)		
	Small	0.9310 (0.5810)		
Hatching probability	Large	0.9037 (0.4083)	8.943 (3,4)	0.0028
	Random	0.6177 (0.5684)		
	Small	0.8365 (0.6003)		
Larval age at hatch (days)	Large	4.4977 (0.3788)	8.7742 (5,7)	0.0124
	Random	4.9623 (0.5411)		
	Small	4.5052 (0.5820)		
Larval size (mm)	Large	3.6893 (0.1457)	3.2586 (4,6)	0.1877
	Random	3.5411 (0.1218)		
	Small	3.4732 (0.1247)		
Yolk sac volume (mm³)	Large	0.0234 (0.0052)	2.5359 (5,7)	0.2814
	Random	0.0171 (0.0045)		
	Small	0.0237 (0.0039)		
Swim bladder inflation probability	Large	0.7561 (0.3926)	1.0741 (4,6)	0.5845
	Random	0.7282 (0.3188)		
	Small	0.7999 (0.3573)		
Larval survival probability	Large	0.9094 (0.2916)	3.4567 (3,5)	0.1776
	Random	0.8284 (0.3893)		
	Small	0.8935 (0.4169)		

WM, wet mass.

Parameter estimates values are given for each treatment and for each trait. Selection treatment differing significantly from the other lines indicated in bold.

*Logit-transformed estimated parameters for binomial distributed data and log-transformed estimated parameters for Poisson distributed data.

†Chi-square value from the deletion of the variable from the full model.

‡P-values derived from the chi-square statistics.

Population-level consequences of harvesting-induced life-history evolution

The population model revealed that, in the absence of fishing, small-selected zebrafish had lower population growth rates per day than random or large-selected fish (0.18% and 0.20% lower, respectively; Table 5). By contrast, when fish in the model were exploited in a similar positively size-selective manner as in our experiment (as would for example be typical in a minimum-length limit scenario, i.e. small-size selection), the population growth rate per day of small-selected fish exceeded that of the random and large-selected fish (by 0.12% and 0.14%, respectively; Table 5). When standardizing the population growth rate by genera-

tion time of the random fish, these differences revealed that the population growth rate of small-selected fish was 25.5% lower than the one of random fish in the absence of fishing, while it was 40.5% higher than the one of random fish when exploited with a minimum-length limit. Accordingly, during the period when positively size-selective fishing was operating, small-selected fish showed the slowest population decline among the treatments but when fishing was stopped, the speed of recovery by small-selected fish was slower than that of the other two selection lines (Fig. 2F). Moreover, small-selected fish did not recover to the pre-exploitation densities when we assumed no potential for life-history evolution during population recovery (Fig. 2F). Finally, small-selected fish did not perform well when ran-

Table 3. The effect of female body size (standard length; SL mm) and selection treatment on zebrafish reproductive performance and early life-history traits.

Trait	Variable	Parameter estimates* (SE)	Chi-square value† (df)	P-value‡
Spawning probability	Treatment	0.0047 (0.0454)	0.0249 (3,5)	0.9876
	Female SL		18.693 (2,3)	<0.0001
Fecundity (eggs/female per day)	**Treatment**		197.92 (4,6)	<0.0001
	Large-selected	0.0076 (0.5096)		
	Random	0.0049 (0.0404)		
	Small-selected	0.0110 (0.0065)		
	Female SL	0.2448 (0.0107)	571.57 (5,6)	<0.0001
Relative fecundity (eggs per g of female WM per day)	**Treatment**		398.15 (4,6)	<0.0001
	Large-selected	0.0748 (0.4892)		
	Random	0.0488 (0.0318)		
	Small-selected	0.1204 (0.0461)		
	Female SL	0.1805 (0.0075)	607.74 (5,6)	<0.0001
Egg yolk size (mm)	**Treatment**		21.691 (5,6)	<0.0001
	Large-selected	0.7362 (0.0430)		
	Random	0.7110 (0.0073)		
	Small-selected	0.6793 (0.0087)		
	Female SL	−0.0048 (0.0014)	11.817 (6,7)	0.0006
Egg survival probability	Treatment		0.9581 (3,5)	0.6194
	Female SL		0.0011 (5,6)	0.9740
Hatching probability	**Treatment**		12.984 (5,7)	0.0015
	Large-selected	0.9999 (3.4546)		
	Random	0.9998 (0.5575)		
	Small-selected	0.9997 (0.8234)		
	Female SL	−0.2796 (0.1146)	5.9779 (6,7)	0.0145
Larval age at hatch (d)	**Treatment**		8.7742 (5,7)	0.0124
	Large-selected	4.4977 (0.3788)		
	Random	4.9623 (0.5411)		
	Small-selected	4.5052 (0.5820)		
	Female SL		1.7207 (7,8)	0.1896
Larval size (mm)	Treatment		3.3456 (4,6)	0.9585
	Female SL		0.0847 (6,8)	0.1877
Yolk sac volume (mm³)	Treatment		2.5359 (5,7)	0.2814
	Female SL		1.3175 (7,8)	0.2510
Swim bladder inflation probability	Treatment		1.0741 (4,6)	0.5845
	Female SL		0.1670 (6,7)	0.6828

Parameter estimates are given for the significant covariates, which are indicated in bold.

*Logit-transformed estimated parameters for binomial distributed data and log-transformed estimated parameters for Poisson distributed data.

†Chi-square value from the deletion of the variable from the full model.

‡P-values derived from the chi-square statistics.

dom harvesting (with respect to body size) or dome-shaped size-selective harvesting (representing large-selection with a harvest slot-length limit) was operating (Table 5, Supporting Information S7; Fig. S3). The modeling results of the relative performance of each of the three life histories were robust to parameter uncertainties (Table 5). Although the performance of small-selected fish relative to the other selection lines varied with fishing mortality, small-selected fish outperformed the other selection lines as long as positively size-selective fishing was operating with moderate or high intensity (Fig. S4). Another noteworthy finding was that the population dynamics of the large-selected and the random fish were often quite similar when fishing was

operating, suggesting that these two life histories were performing functionally equivalent.

Discussion

Our experimental approach in zebrafish demonstrated changes in genotypes, phenotypes, and population dynamics in response to just five generations of size-selective harvesting. Thus, our results present a comprehensive picture of the evolutionary legacy of size-selective exploitation. Controlled laboratory environment and a specific harvesting design allowed controlling for size-dependent and other parental and epigenetic effects. This helped us to establish

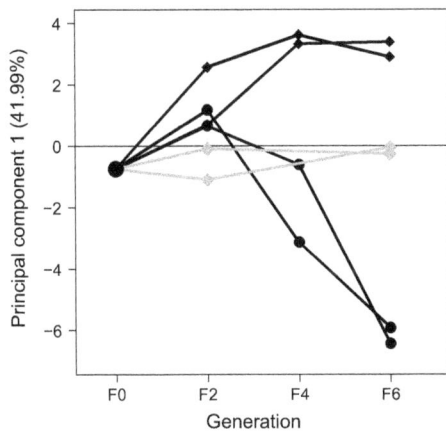

Figure 3 A principal component analysis on allele frequencies of the 22 outlier loci illustrates a similar amount and direction of genetic change occurred among the selection lines induced by size selection after five generations of harvesting. Red lines and symbols indicate small-, gray randomly, and blue large-selected replicates.

unambiguous cause (size-selective harvesting)-and-effect (phenotypic and genetic changes) relationships (Diaz Pauli and Heino 2014) reinforcing the possibility that intensive harvesting of wild populations can indeed lead to fisheries-induced evolution (FIE). Despite introducing obvious simplifications by maintaining discrete generations and allowing only single reproductive events, our selection experiment has value because it allowed the assessment of various phenotypic traits ranging from life-history traits to physiology and behavior and applying a genomic approach to discover specific genes under selection. Five generations of size-selective harvesting of wild zebrafish affected life history and behavior by elevating reproductive investment, decreasing mean maximum body size and reproductive output, and reducing boldness. The evolved phenotypic changes in the small-selected fish were overall relatively subtle and often statistically nonsignificant in relation to the random fish, but they were accompanied by genetic changes and large population dynamical effects. These results collectively showed that contemporary harvest-induced evolution is conceivable in response to intensive size-selective exploitation. The population model further revealed that the phenotypic and genetic changes induced by positive size selection allowed fish to adapt to harvesting, but hindered population recovery in the absence of exploitation. Our results overall highlight the potential for large, harvest-induced population-level consequences to emerge from rather subtle phenotypic changes in response to positively size-selective exploitation that might easily go unnoticed when monitoring natural populations.

We examined the outcome of five generations of size-selective harvesting by comparing the phenotypes of individuals among selection treatments several generations

after selection was halted. Hence, our results are conservative because delaying the trait assessment up to four generations without harvesting probably had resulted in some recovery of phenotypic traits due to fecundity selection similar to the case in the famous Atlantic silverside (*Menidia menidia*) experiment (Conover et al. 2009; Salinas et al. 2012). Moreover, for logistical reasons, we only exerted negative selection differentials on body size for four rather than five generations in our fishing treatment (the small-selected line, Fig. 1), further reducing the potential for phenotypic (and genetic) change. We nevertheless documented effects of positively size-selective harvesting on a range of traits and also presented molecular evidence of directional selection altering the genotypes, reinforcing the previously expressed notion that rapid evolutionary change is possible over very short time periods of intensive size-selective harvesting (Conover and Munch 2002; van Wijk et al. 2013).

Evolved differences in growth and behavior

According to the biphasic growth model, small-selected fish (under selection similar to most capture fisheries managed with minimum-length limits) reached significantly lower mean maximum body size (L_∞) than large-selected and random fish, but there was no significant difference in early growth rate (h) among the selection lines (Fig. 1A,B). These findings were in line with recent field evidence in heavily exploited coastal marine fish species (Alós et al. 2014) and have been also reported elsewhere (Nússle et al. 2009). Early maturation at small size and high reproductive investment can together explain the lower L_∞ of the small-selected fish given the fundamental energetic trade-off between growth and reproduction (Enberg et al. 2012). In addition to such differences in energy allocation, juveniles of the small-selected fish evolved differences in energy acquisition because they were significantly less bold in an open-field experiment compared to large-selected fish. These results suggest that the small-selected zebrafish evolved a more cautious behavioral type and personality, likely in relation to feeding behavior, which might have contributed to their lower body size at harvesting. Similarly, small-selected fish in the Atlantic silverside study evolved lower food consumption and were less willing to forage under threat of predation (Walsh et al. 2006). Variation in exploration and boldness can have fitness consequences because these behaviors can facilitate foraging success (Stamps 2007; Klefoth et al. 2012), dispersal (Cote et al. 2010), cognitive performance (Vital and Martins 2011), reproduction (Ariyomo and Watt 2012), and survival (Smith and Blumstein 2007; Biro and Stamps 2008). Moreover, boldness and exploration relate directly to vulnerability to fishing gear (Alós et al. 2012, 2015; Härkönen

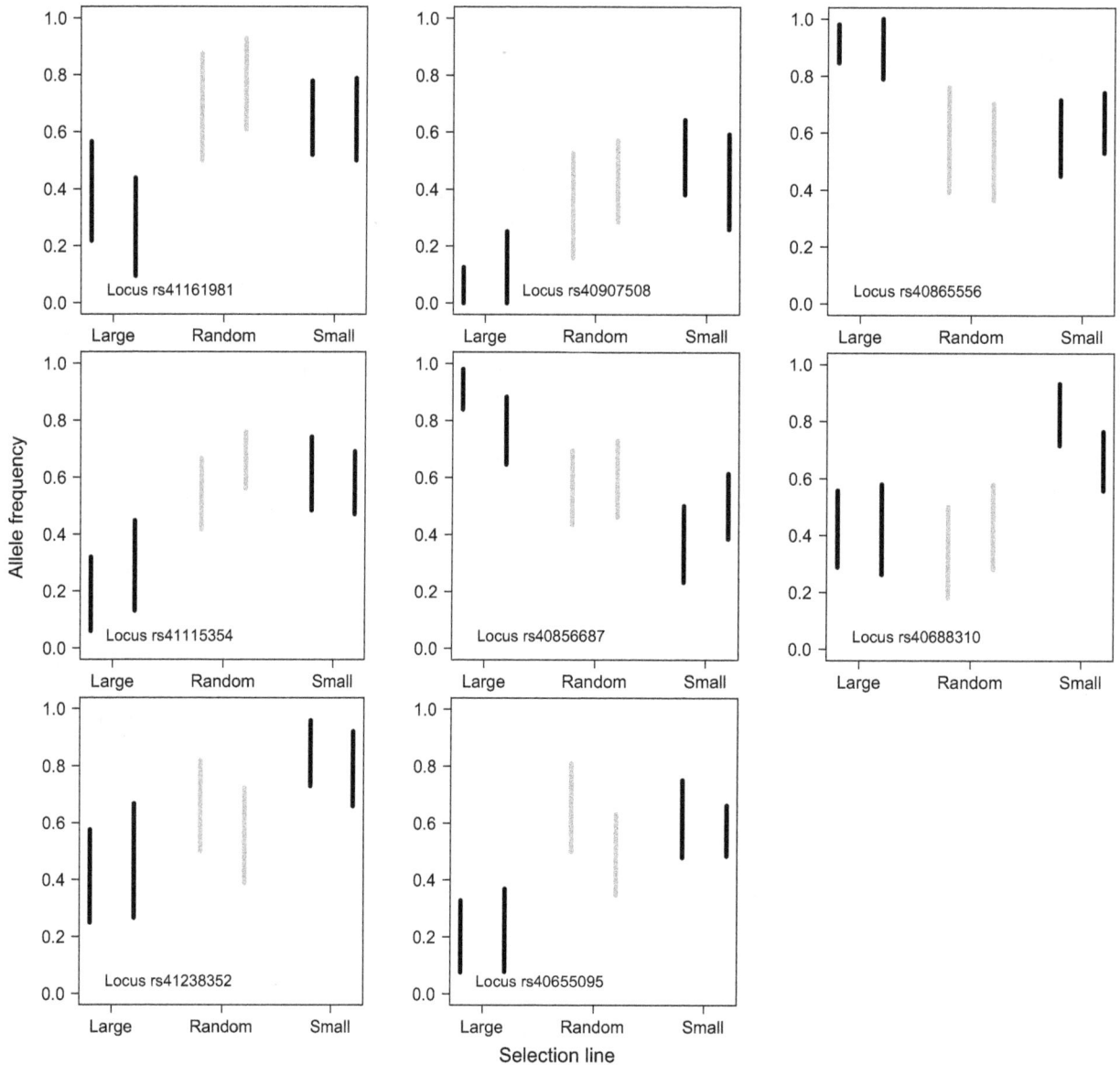

Figure 4 Replicable genetic changes in differently selected zebrafish lines induced by size-selective harvesting indicated by nonoverlapping 95% confidence intervals estimated for allele frequencies for each outlier locus. Large-selected fish indicated with blue lines, random fish with gray lines, and small-selected fish with red lines.

et al. 2014; Diaz Pauli et al. 2015). Therefore, increased timidity as an evolutionary response to size-selective fishing will negatively affect catch rates (Philipp et al. 2009), reduce angler satisfaction (Arlinghaus et al. 2014), and potentially also affect the economic value of a fishery. Moreover, when catchability declines so does the value of fishery-dependent information to index stock size (Alós et al. 2015).

A common assumption in the fisheries literature is that size-selective mortality should reduce growth rate (Conover and Munch 2002; but see Walters and Martell 2004 for a

critical view of this apparently intuitive prediction). By contrast, we found that small-selected zebrafish did not differ significantly in their juvenile growth rate compared to random and large-selected fish, and there were also no physiological differences in standard metabolic rate (and, therefore, in physiological growth capacity) among the selection lines. In fact, there was a tendency for small-selected and random fish to grow faster than large-selected fish (Fig. 1B). This agrees with modeling work (Matsumura et al. 2011) and empirical data in positively size selectively exploited fisheries (van Walraven et al. 2010). The lack of

Table 4. Outlier loci with the most pronounced, replicable allele frequency divergence among selection treatments and which occur in or close to a gene with known function.

SNP name	Treatment	Type	Gene name	Gene function
rs40907508	Large ≠ small	UTR	LysM	The chemical reactions and pathways resulting in the breakdown of macromolecules that form part of a cell wall
rs40688310	Large and random ≠ small	UTR	Tryptophan hydroxylase 2	The chemical reactions and pathways involving aromatic amino acid family. Controls brain serotonin synthesis in human and mice
rs40655095	Large ≠ small	S	eph receptor B2b	The process of introducing a phosphate group on to a protein. Regulates transcription
rs40856687	Large ≠ small	In	atp1b3b	Ion transport
rs41238352	Large ≠ small	In	col5a1	Collagen formation. Collagen strengthens and supports many tissues, such as bones and muscles
rs41141381		*UTR*	*Cathepsin L 1 a*	*Involved in embryonic yolk processing*
rs40878095		*NS*	*Interleukin-1 receptor-associated kinase 4*	*Involved in innate immune response system in zebrafish (Stein et al. 2007)*
rs40784742		*UTR*	*lims1*	*Involved in cardiac contraction and heart beating*
rs40769175		*UTR*	*Guanylate cyclase activator 1C*	*Involved in calcium signaling in the eye and in light regulation, circadian rhythms and stress response (Scholten and Koch 2011, Wegener et al. 2011)*
rs40739415		*D*	*Ddt*	*Inner ear development*

Also shown are the type of the variant (S, synonymous, UTR, untranslated region, D, downstream gene variant, In, intron variant), gene name, and gene function.
SNPs in LD with an outlier loci occurring within or close to a gene with a known function are in *italics*.

significant differences in juvenile growth (which, unlike the adult growth rate, is a clean measure of growth capacity; Enberg et al. 2012) could have been caused by our harvesting design where each fish could only spawn once. Hence, in the random line, larger females, which also carried more eggs due to the positive relationship of female size and fecundity (hereafter referred to as fecundity selection), after harvesting were likely selectively favored, in turn likely creating a selection pressure on fast juvenile growth rate despite a nonselective harvest pattern relative to size. Similarly, in the small-selected line, the fastest growing fish probably contributed more eggs to the next generation and this maintained positive selection pressure on fast juvenile growth. Thus, harvesting alone, even when nonselective, could have had a similarly strong effect as size-selective harvesting on the evolution of fast juvenile growth through fecundity selection (Engen et al. 2014). In other words, the lack of substantial differences in juvenile growth rate among random and the small lines could be indicative of a lack of additional evolution of size-selective harvest compared to unselective harvesting. Not having a nonharvest control precludes our ability to fully understand the relative effects of selection pressures on juvenile growth rate caused by unselective versus size-selective harvesting. Nevertheless, our results support theoretical arguments and empirical data that one should not generally assume that size-selective harvesting will cause evolution toward low growth rates (Walters and Martell 2004; van Walraven et al. 2010; Enberg et al. 2012). In fact, the opposite can and will occur in many situations because when adult mortality rate is elevated, it is advantageous to be as large as possible on the first spawning attempt to outpace the high mortality rate with higher reproductive output (Boukal et al. 2008; Dunlop et al. 2009; Matsumura et al. 2011), as in our experiment.

The weak response of juvenile growth rate of small-selected fish after five generations of size-selective harvesting is in contrast to a groundbreaking experimental study on harvest-induced selection in Atlantic silversides by Conover and Munch (2002), who reported a steep decline in (juvenile) growth rate after four generations of size-selective harvesting. Although silverside and zebrafish share many life-history and behavioral characteristics (e.g. high fecundity, small egg size, external fertilization, and schooling behavior), silverside are semelparous, while zebrafish are iteroparous batch spawners. These differences in life-history strategies could strongly affect energy allocation patterns and thus juvenile growth. Silverside is a capital-breeding species that uses stored energy to make large investments into reproduction, and females typically repro-

Table 5. Comparison of the population growth rate among selection lines under different fishing scenarios.

Without fishing		With fishing					
		Small-harvested		Random		Large-harvested	
Large	Small	Large	Small	Large	Small	Large	Small
1.00020	0.99823	0.99991	0.99710	0.99985	0.99915	0.99985	1.00125
(1.00026 ± 0.00081)	(0.99860 ± 0.00120)	(1.00040 ± 0.00295)	(0.99691 ± 0.00337)	(1.00021 ± 0.00063)	(0.99917 ± 0.00081)	(1.00039 ± 0.00145)	(1.00076 ± 0.00192)

Values of large-selected and small-selected fish relative to the random fish are shown. Estimated uncertainty bounds are shown in parentheses (mean ± SD).

duce in years when they have accumulated a threshold level of stored energy reserves (Bull and Shine 1979). By contrast, income breeders, such as zebrafish, spend energy on reproduction as it is gained (Jönsson 1997). Furthermore, Conover and Munch (2002) exerted a greater harvesting pressure (90% per generation) compared to the present experiment in zebrafish (75% per generation), and this might have affected the results by increasing the selection response in juvenile growth rate. Most importantly, however, in the silverside study, juvenile traits were exclusively under selection because maturation was induced by photoperiod after the experimental harvesting (Diaz Pauli and Heino 2014). This experimental procedure channelized selection differentials on juvenile growth rate, in contrast to the present case in zebrafish where reproductive traits, in particular reproductive investment, were allowed to be under selection in addition to juvenile growth rate. A recent selection experiment in male guppies (*Poecilia reticulata*) similarly demonstrated evolution in maturation and only a minor change in juvenile growth rate after only three generations of selection (van Wijk et al. 2013). However, that study differed from ours because it focused on determinately (rather than indeterminately) growing males (rather than males and females combined). In general, however, responses of juvenile growth rate to harvesting vary among species and fisheries, and one should not necessarily expect juvenile growth rate to decline in response to positively size-selective harvesting (Walters and Martell 2004; Enberg et al. 2012).

Maturation

While reproductive investment increased in response to selection for small body size in our experiment, we saw little differentiation in the maturation schedule (represented by the probabilistic maturation reaction norm, PMRN; Heino et al. 2002; Dieckmann and Heino 2007) between small- and large-selected zebrafish after five generation of size-selective harvesting. It is noteworthy, however, that although the PMRNs of small- and large-selected fish largely overlapped, investigation of the left part of the PMRNs, that is, the area where food was abundant and growth rate was high, indicated that small-selected fish matured somewhat earlier and at smaller size than large-selected zebrafish (Fig. 2D). This pattern was consistent with the predicted reduced age at maturation (T) that we estimated from growth under *ad libitum* food conditions (Table 1). The high size- and age-specific maturation probability of random fish could have been affected by unselective harvesting. Despite we lacked a nonharvest control, the evolved differences in the random fish still represent adaptation to unselective fishing. However, we based the timing of harvesting on the 50% maturation status of the random fish; thus, one should have expected a lower PMRN intercept compared to the large-selected fish as indicative of earlier maturation. It is more likely that the higher size- and age-specific maturation probability in the random line was caused by the large-size variation compared to large- and small-selected fish. In zebrafish, social dominance is size dependent (Paull et al. 2010) and in random fish the variation (measured as standard deviation) in body size during the maturation experiment across all ages was substantially higher (3.0 mm), particularly in growth-restricted conditions, compared to small- (2.5 mm) and large-selected fish (2.5 mm). Accordingly, the higher size- and age-specific maturation probability of the random fish could have been caused by size-dependent social hierarchies or other social factors, which have been shown to inhibit and delay reproduction for example in guppies (Diaz Pauli and Heino 2013).

While the evolution of elevated reproductive investment g (Fig. 2C) and lower L_∞ (Table 1) in response to size-selective harvesting among small-selected fish was consistent with empirical and theoretical studies (e.g. Rijnsdorp 1993; Jørgensen et al. 2007; Sharpe and Hendry 2009; van Walraven et al. 2010), the lack of substantial difference in the maturation schedule caused by selective harvesting (as inferred from the PMRN) may seem counterintuitive. However, our experiment was based on nonoverlapping generations; hence, each selected spawner could contribute to the next generation just once during one spawning event at a fixed age. Therefore, our experiment prevented early-maturing fish from benefiting from the increase in spawning frequency, which is a key fitness benefit associated with early maturation when adults face a high risk of mortality (Poos et al. 2011). The conditions in our experiment were thus artificial and in contrast with the conditions in the wild. In the wild, fisheries maturation schedules have been found to readily respond to harvest selection as indicated by large changes in the PMRN's intercepts and slopes in many stocks, and these changes were often much more pronounced than changes in other life-history traits, such as reproductive investment (Hutchings and Fraser 2008; Sharpe and Hendry 2009; Devine et al. 2012; Audzijonyte et al. 2013b). Therefore, it is important not to misinterpret our results as evidence that elevated or size-selective mortality does not alter the maturation schedule of exploited fish species. In fact, evolutionary response in maturation is expected under most exploited conditions when generations overlap (Devine et al. 2012). Further experiments with overlapping generations are needed to fully understand how maturation will evolve in response to size-selective harvesting (Diaz Pauli and Heino 2014).

Evolved differences in reproductive success and early life-history traits

Evolutionary downsizing of adult body size, especially of females, can have large consequences for offspring production and larval viability (Johnson et al. 2011), for example through direct fecundity decline associated with the reductions in adult body size or indirectly through size-dependent maternal effects on egg and offspring quality (Walsh et al. 2006; Arlinghaus et al. 2010; Hixon et al. 2014). In our study, we demonstrated a positive association between maternal body size and reproductive output (spawning probability and fecundity; Table 3). Larger females had a higher spawning probability, which was exclusively determined by female body size and not affected by the selection treatment. Female body size was also positively associated with fecundity. Large females of many fish species have been found to have higher amount of energetic resources to allocate to reproduction compared to small females and thus are able to spawn more frequently and produce a higher number of eggs (Hixon et al. 2014), in line with our findings in zebrafish reported here and elsewhere (Uusi-Heikkilä et al. 2010). Maybe somewhat surprisingly egg size was negatively associated with maternal body size. This could represent a fundamental egg number – egg size trade-off and relate to smaller females compensating their lower fecundity by producing larger eggs (Hendry et al. 2001; Uusi-Heikkilä et al. 2010). Similarly, the negative relationship between female size and hatching probability could be related to the above-mentioned trade-off. However, the biological relevance of egg size as a trait of egg quality has been questioned before in zebrafish (Uusi-Heikkilä et al. 2010). Irrespectively, when early life-history or reproductive traits correlate with the focal trait under selection (i.e. adult body size), these traits can respond via correlated selection response due to genetic covariance (Munch et al. 2005). We found significant differences in fecundity (but not in spawning probability) and in several early life-history traits (egg size, hatching probability, and larval age at hatch) among the three selection lines even after statistically controlling for maternal body size, suggesting an evolutionary response unrelated to size-dependent maternal effects.

After just five generations of selection, large-selected fish produced more and larger eggs compared to small-selected and random zebrafish, and there was a modest nonsignificant increase in relative fecundity in small-selected fish compared to random fish (Fig. 2E). As reproductive investment increased in the small-selected line compared to the random line, this likely compensated for the evolution of smaller adult body size and maintained fecundity high and largely unaltered with respect to the random line. It is surprising that there were such large differences in fecundity between random and large-selected zebrafish despite their similar maximum length (Fig. 2A). These differences could have been caused, at least partly, by size-dependent dominance hierarchies. In zebrafish, dominance and aggression are positively associated with increasing body size (Paull et al. 2010). Larger differences in spawner body sizes in random females (SD 1.99 mm; range 24–32 mm) compared to large-selected females (SD 1.37 mm; range 27–31 mm) could have maintained higher level of aggression between females and males in the spawning boxes occupied with random fish and resulted in lower egg production. Also, males being larger than females, which by chance should have been occurred more often in the random line given the larger size range of the spawners, could have affected egg production due to female stress caused by male dominance (D. Bierbach, S. Uusi-Heikkilä, P. Tscheligi, C. Wolter and R. Arlinghaus, unpublished data). Furthermore, zebrafish females allocate more reproductive resources to more preferred, large males (Uusi-Heikkilä et al. 2012b), and because in large-selected line males were generally larger, large-selected females could have released more eggs toward them compared to random line, where fewer females were coupled with a large male.

Similar to the fecundity assessments, it is also noteworthy that there were no large differences in egg traits between the small-selected and random fish. Earlier studies in zebrafish showed that egg size might not be a major determinant of larval quality (Uusi-Heikkilä et al. 2010). Instead, offspring quality may be better indicated by larval traits such as age at hatch, size at hatch, and the amount of nutrient reserves (yolk sac volume). Large-selected and random fish produced slightly larger larvae than small-selected fish, but the differences were not statistically significant and the differences neither translated into shorter hatching time as an indicator of better larval condition and faster development rate (Kimmel et al. 1995). Hence, size-selective fisheries selection did not substantially alter egg and larval traits when comparing the small-selected fish with the random line. However, random fish had a significantly greater spawning probability than small-selected fish, suggesting that the random fish still might have greater reproductive fitness compared to the small-selected fish.

Despite the lack of differences in early life-history traits and juvenile growth rate, our results suggest that random and large-selected fish exhibited faster larval growth than small-selected fish. This was indicated by the large differences in body at age 30 days (Fig. 2A) when the growth experiment started. Despite their similar sizes at hatch (Table 2), random and large-selected fish were significantly larger at age 30 days compared to small-selected fish. Fast larval growth has fitness benefits because it allows larvae to pass the most vulnerable life-history stages quickly, and although there might be some costs related to fast growth

(e.g. increased larval mortality; Pepin 1991), it has also been suggested that even slightly enhanced rates of early survival and growth can translate into increased probability of subsequent recruitment (Houde 1987; Hare and Cowen 1997).

Overall, the differences in early life-history traits among zebrafish selection lines were substantially smaller (and often nonsignificant) compared to those previously reported in the silverside study (Walsh et al. 2006). The inconsistency between the silverside and our study is probably related to the steeper decline in the body size of mature silversides after four generations of harvesting and to the larger difference in body size between small- and large-selected fish during the spawning trials compared to our zebrafish study. In fact, large-selected silversides were more than six times heavier than small-selected ones, whereas in our study the differences were less than twofold. Moreover, the differences between these two studies could again be related to the different life-history strategies of the two species. Semelparous silversides may invest a much larger proportion of surplus energy into a single reproductive season compared to zebrafish, which spread their reproductive effort over multiple batches. Different results among experimental evolutionary studies reinforce the species and environment specificity of FIE, which complicates the formulation of general predictions about the type and magnitude of phenotypic changes as a result of size-selective harvesting (Heino et al. 2013; Laugen et al. 2014).

Evolutionary rate

The degree of decline in adult body size of small-selected fish in our experiment (7.8% over five generations of selection) was similar to recent experimental work in male guppies exposed to three generations of size-selective mortality (7% over three generations; van Wijk et al. 2013) but differed substantially from the silverside study (25% over four generations; Conover and Munch 2002). Although the change in adult body size in our study was significant, such a subtle phenotypic change in body length might easily go unnoticed in phenotypic time series from the wild because fish growth has a large plastic component (Lorenzen and Enberg 2002). In addition, the rate of evolution of adult body size that we observed was lower (−0.116 to 0.165 haldanes) than the rate estimated in the male guppy experiment (0.3 haldanes; van Wijk et al. 2013) and much lower than those estimated for size at maturation from actual fisheries data (−2.2 to 0.9 haldanes; Devine et al. 2012). Thus, the phenotypic changes that we observed in the laboratory were conservative relative to the data from the wild, possibly because the latter include both genetic and plastic changes and because selection in overlapping generations may lead to stronger responses in maturation traits than revealed in our experiment. Our finding underscores an important dilemma for FIE studies: While FIE can be widespread in exploited systems (Jørgensen et al. 2007; Devine et al. 2012; Laugen et al. 2014), it is very difficult to detect when one is confined to just phenotypic analysis, given that fish life-history traits are extremely plastic and vary in response to the environment (Kuparinen and Merilä 2007; Law 2007). This issue has potentially contributed to the lack of broad acceptance of FIE among fishers, fisheries managers, and some fisheries scientists (Jørgensen et al. 2007; Kuparinen and Merilä 2007; Law 2007; Hilborn and Minte-Vera 2008).

Harvest-induced genetic changes

In the genetic analyses, we identified 22 outlier loci responding to divergent selection (Table S6) and 12 outliers indicative of balancing selection (Table S7). The outliers indicative of balancing selection were linked to genes involved in processes such as movement of metal ions within a cell or between cells (Varshney et al. 2013), proteolysis, neuronal development (Ahrens et al. 2012), and glycopeptide hormone activity (Alderman and Bernier 2007). These loci with signatures of balancing selection in F_7-generation might have responded to laboratory rearing environment (i.e. captivity). Another possibility is that they are associated with fecundity selection likely experienced by all experimental lines. Among the 12 loci that exhibited significant signatures of balancing selection, three can be considered as candidates for a fecundity selection hypothesis (Table S7). None of these loci was directly associated with genes related to reproduction or fecundity (Table S7), although they could be in linkage disequilibrium (LD) with genes that are. Despite the fact that the loci with signatures of balancing selection might have been indicative of adaptation to the laboratory environment, at least some of them might have been false positives. Inaccurate detection of balancing selection is an inherent weakness of outlier approaches (Narum and Hess 2011), and the method employed here, in particular, has been shown to have relatively high type I error rate for balancing selection (Beaumont and Nichols 1996).

Our results provide conclusive evidence that size-selective harvesting can lead to genetic change in timescale relevant to fisheries. We conservatively focused on the eight outliers that showed parallel divergence in both size-selected replicates of one selection line. These outliers and adjacent loci emerged as the strongest candidates of adaptive divergence. Five of the eight parallel SNPs occurred within or close to a gene with a known function (Table 4) and six of them were in significant LD with a nearby SNP on the same linkage group (Table 4). Two of the eight outliers were in LD with a SNP that occurs within a regulatory

area of a gene associated with zebrafish embryological metabolism (Tingaud-Sequeira and Cerdà 2007) and two other outliers with SNPs occurring within regulatory areas of the genes or within genes associated with zebrafish circadian rhythms, stress response, and immune system (Stein et al. 2007; Scholten and Koch 2011; Weger et al. 2011). These traits might be important in determining adaptive responses related to fitness in juvenile and adult fish. Another parallel outlier was located within a regulatory area of a gene, which has been found to control brain serotonin synthesis in humans and mice. Serotonin is a key element in the synthesis of melatonin, a hormone that affects feeding behavior and aggression in fish (Falcón et al. 2010). Differences in melatonin production could relate to differences in fish exploration tendency, which was found to evolve in our experiment. Denser SNP panels and mapping approaches would be needed to test the functional role of these genes along with other genes with which they are in LD.

Population-level consequences

The consequences of even subtle phenotypic changes for populations could potentially be severe. For example, early maturation and high investment in reproduction cumulatively reduce life span (Jørgensen and Fiksen 2010). Indeed, the estimated instantaneous natural mortality (M) of zebrafish was higher among small-selected and random fish than among large-selected fish. According to our population model, under positively size-selective fishing (minimum-length limit scenario), the small-selected fish population would exhibit a substantially greater population growth rate than the random and large-selected fish populations. This finding supports the idea that life-history changes are compensatory in that they allow individuals (and therefore populations) to remain productive in the face of positively size-selective harvest mortality (Hutchings 2009; Matsumura et al. 2011; Kuparinen and Hutchings 2012; Heino et al. 2013). Hence, from a fisheries perspective, FIE is not necessarily negative (Eikeset et al. 2013; Jørgensen and Zimmermann 2015). However, in our model during a simulated fishing moratorium, the small-selected fish population exhibited a substantially lower population growth rate than the random and large-selected fish populations (Fig. 2F and Fig. S2). Our modeling results therefore suggest that FIE impedes population recovery during a moratorium, which is in line with previous empirical and theoretical research (Conover et al. 2009; Enberg et al. 2009; Eikeset et al. 2013; Hutchings and Kuparinen 2014; Kuparinen et al. 2014; Laugen et al. 2014; Marty et al. 2015). Thus, seemingly subtle changes in life-history traits could have a strong effect on the recovery rate and rebound potential of exploited fish populations. Moreover, evolutionary downsizing in body size of only 0.1% per year over 50 years has been predicted to reduce biomasses up to 35% in some species (Audzijonyte et al. 2013a). Hence, FIE matters for the management and conservation of exploited fish populations, even if phenotypic responses are modest and seemingly unimportant at an individual level.

Our population model was simplified by design and therefore subject to caveats. Although the model included knowledge of density dependence of vital rates in zebrafish, it used the dominant eigenvalue of the Leslie matrix as a fitness metric although the dominant Lyapunov exponent has been suggested as an appropriate measure in density-dependent population models (Roff 2010). In addition, the model did not incorporate the potential effects of dominance hierarchies and female differential allocation on zebrafish reproductive output (Uusi-Heikkilä et al. 2012b). Moreover, model results were based on the assumption that the parameters for growth, maturation, and reproductive investment that we measured in the laboratory would translate to field conditions. Our approach to fitting the biphasic growth model (Lester et al. 2004) allowed us to predict numerous life-history traits that could be incorporated into the model. However, our approach assumed that these traits were optimally adapted to treatment conditions. Our experimental populations might not have reached an evolutionary stable state after five generations of selection. Nevertheless, results of the among-population comparison of population growth rate should be robust to this omission because the population model itself does not require the assumption of evolutionary equilibrium and the estimated population growth rates were rather insensitive to uncertainty of the parameter values of the Lester growth model (i.e. growth h, reproductive investment g, and maturation T). Finally, we were not able to perform an evolutionary impact assessment (Jørgensen et al. 2007; Laugen et al. 2014) in a strict sense because we lacked preharvest life-history data and were unable to compare population-level effects over time with and without evolution. We thus could not evaluate the full implications of FIE. Nevertheless, there is value in comparing fitness of evolved life-histories (both small- and large-selected fish) relative to the random fish, and we can interpret our population dynamical results as showing the effect of size-selective harvesting relative to unselective harvesting.

Conclusions and implications

Much of the current debate around the prevalence of FIE has centered on whether the observed phenotypic changes are genetic (Jørgensen et al. 2007; Kuparinen and Merilä 2007; Law 2007) and if so, whether these changes matter for population dynamics and hence management (Hutchings and Fraser 2008; Andersen and Brander 2009; Kupari-

nen and Hutchings 2012; Laugen et al. 2014; Marty et al. 2015). The strength of our experimental study is that it establishes an unambiguous cause-and-effect relationship by showing that (i) size-selective harvesting can lead to genetic and a range of phenotypic changes in contemporary timescales, (ii) a relatively low evolutionary rate, and (iii) seemingly subtle phenotypic changes in individual life-history traits can cumulatively have a strong effect on population growth rate and recovery potential. FIE can help to maintain a productive population while harvesting is intensive, but our results suggest that the same population adapted to exploitation is expected to recover slowly and may not reach pre-exploitation levels when fishing is relaxed. Our work on the evolutionary legacy of size-selective harvesting thus reinforces the notion that the potential for FIE and its population-level consequences are of relevance to fisheries management and conservation. Negative consequences of FIE will be particularly large for stocks that have been poorly managed in ecological and economic terms (Eikeset et al. 2013; Jørgensen and Zimmermann 2015) for a long period of time (Neubauer et al. 2013), and in such cases, it is critical that the ecological and evolutionary consequences of fishing are being carefully evaluated and mitigated.

A straightforward measure that can help curtail the largely inevitable FIE (Matsumura et al. 2011) is to carefully control fishing mortality to keep it within ecologically sustainable and economically optimal bounds as shown in two recent modeling studies in a FIE context (Eikeset et al. 2013; Jørgensen and Zimmermann 2015). A second complementary measure could be to manage the fishing-induced selectivity, which may produce positive outcomes from a human perspective (e.g. evolution of large adult body size as opposed to downsizing of adults, Boukal et al. 2008; Jørgensen et al. 2009; Matsumura et al. 2011). For example, we found that the population dynamics of large-selected fish, which evolved large asymptotic adult body size, did not differ from the random fish in any of the modeled fishing scenarios (Fig. 2F and Fig. S2). Large-selected fish were exposed to a maximum-size harvest; thus, our results could be interpreted that saving large fish selects for life histories that are more similar to unselectively exploited fish compared to a strictly positively size-selective exploitation common with minimum-length limit regulations and in most other real fisheries. Previous modeling studies have also emphasized a superior performance of harvest slots (i.e. dome-shaped selectivity where large fish and small fish are saved from harvesting) over standard minimum-length limits in terms of reducing selection responses in maturation and other traits while facilitating evolution of large adult size under certain conditions (Hutchings 2009; Jørgensen et al. 2009; Matsumura et al. 2011). Therefore, when feasible and desired by stakeholders, the implementation of maximum-size limits or harvest slots at the expense of using minimum-size limits could be recommended as an additional measure of altered selectivity patterns to complement management measures directed at controlling fishing mortality.

Acknowledgements

We thank Karena Kuntze, Asja Vogt, Marcus Ebert, Yvonne Klaar, Sylvia Werner, Theresa Arlt, Sarah Becker, and Julie Ménard for fish husbandry, care taking, and data collection; Rachel Tucker for helping in genotyping the samples; and Henrik Zwadlo for technical assistance. Funding for this study was received through the Adaptfish grant to RA and CW, through the BTypes grant to RA, both funded by the Gottfried-Wilhelm-Leibniz-Community in the Leibniz-Competition, through the Besatzfisch grant to RA funded by the German Federal Ministry for Education and Research (Grant # 01UU0907), through grant-in-aid for Scientific Research (C) from the Japanese Ministry of Education, Culture, Sports, Science and Technology (Grant # 25440190) to SM and through the Kone Foundation to SUH. We express our gratitude to two anonymous reviewers and thank them for their unusually rich and constructive comments that helped to improve the manuscript.

Literature cited

Ahrens, M. B., J. M. Li, M. B. Orger, D. N. Robson, A. F. Schier, F. Engert, and R. Portugues 2012. Brain-wide neuronal dynamics during motor adaptation in zebrafish. Nature 485:471–477.

Alderman, S. L., and N. J. Bernier 2007. Localization of corticotropin-releasing factor, urotensin I, and CRF-binding protein gene expression in the brain of the zebrafish, Danio rerio. The Journal of Comparative Neurology 502:783–793.

Allendorf, F. W., P. R. England, G. Luikart, P. A. Ritchie, and N. Ryman 2008. Genetic effects of harvest on wild animal populations. Trends in Ecology & Evolution 23:327–337.

Alós, J., M. Palmer, and R. Arlinghaus 2012. Consistent selection towards low activity phenotypes when catchability depends on encounters among human predators and fish. PLoS ONE 7:e48030.

Alós, J., M. Palmer, I. A. Catalan, A. Alonso-Fernandez, G. Basterretxea, A. Jordi, L. Buttay et al. 2014. Selective exploitation of spatially structured coastal fish populations by recreational anglers may lead to evolutionary downsizing of adults. Marine Ecology Progress Series 503:219–233.

Alós, J., M. Palmer, P. Trías, C. Díaz-Gil, R. Arlinghaus, and M.-J. Rochet 2015. Recreational angling intensity correlates with alteration of vulnerability to fishing in a carnivorous coastal fish species. Canadian Journal of Fisheries and Aquatic Sciences 72:217–225.

Andersen, K. H., and K. Brander 2009. Expected rate of fisheries-induced evolution is slow. Proceedings of the National Academy of Sciences of the United States of America 106:11657–11660.

Antao, T., A. Lopes, R. J. Lopes, A. Beja-Pereira, and G. Luikart 2008. LOSITAN: a workbench to detect molecular adaptation based on a Fst-outlier method. BMC Bioinformatics 9:323.

Ariyomo, T. O., and P. J. Watt 2012. The effect of variation in boldness and aggressiveness on the reproductive success of zebrafish. Animal Behaviour 83:41–46.

Arlinghaus, R., S. Matsumura, and U. Dieckmann 2009. Quantifying selection differentials caused by recreational fishing: development of modeling framework and application to reproductive investment in pike (Esox lucius). Evolutionary Applications 2:335–355.

Arlinghaus, R., S. Matsumura, and U. Dieckmann 2010. The conservation and fishery benefits of protecting large pike (Esox lucius L.) by harvest regulations in recreational fishing. Biological Conservation 143:1444–1459.

Arlinghaus, R., B. Beardmore, C. Riepe, J. Meyerhoff, and T. Pagel 2014. Species-specific preferences of German recreational anglers for freshwater fishing experiences, with emphasis on the intrinsic utilities of fish stocking and wild fishes. Journal of Fish Biology 85:1843–1867.

Audzijonyte, A., A. Kuparinen, R. Gorton, and E. A. Fulton 2013a. Ecological consequences of body size decline in harvested fish species: positive feedback loops in trophic interactions amplify human impact. Biology Letters 9:20121103.

Audzijonyte, A., A. Kuparinen, and E. A. Fulton 2013b. How fast is fisheries-induced evolution? Quantitative analysis of modelling and empirical studies. Evolutionary Applications 6:585–595.

Barot, S., M. Heino, L. O'Brien, and U. Dieckmann 2004. Estimating reaction norms for age and size at maturation when age at first reproduction is unknown. Evolutionary Ecology Research 6:659–678.

Beaumont, M. A., and R. A. Nichols 1996. Evaluating loci for use in the genetic analysis of population structure. Proceedings of the Royal Society B: Biological Sciences 263:1619–1626.

Biro, P. A., and J. R. Post 2008. Rapid depletion of genotypes with fast growth and bold personality traits from harvested fish populations. Proceedings of the National Academy of Sciences of the United States of America 105:2919–2922.

Biro, P. A., and J. A. Stamps 2008. Are animal personality traits linked to life-history productivity? Trends in Ecology & Evolution 23:361–368.

Biro, P. A., and J. A. Stamps 2010. Do consistent individual differences in metabolic rate promote consistent individual differences in behavior? Trends in Ecology & Evolution 25:653–659.

Borrell, B. 2013. Ocean conservation: a big fight over little fish. Nature 493:597–598.

Boukal, D. S., E. S. Dunlop, M. Heino, and U. Dieckmann 2008. Fisheries-induced evolution of body size and other life history traits: the impact of gear selectivity. ICES CM 2008/F:07.

Browman, R., R. Law, and C. Marshall 2008. The role of fisheries-induced evolution. Science 320:47.

Bull, J. J., and R. Shine 1979. Iteroparous animals that skip opportunities for reproduction. American Naturalist 114:296–303.

Bunnefeld, N., and A. Keane 2014. Managing wildlife for ecological, socioeconomic, and evolutionary sustainability. Proceedings of the National Academy of Sciences of the United States of America 111:12964–12965.

Conover, D. O., and S. B. Munch 2002. Sustaining fisheries yields over evolutionary time scales. Science 297:94–96.

Conover, D. O., S. B. Munch, and S. A. Arnott 2009. Reversal of evolutionary downsizing caused by selective harvest of large fish. Proceedings of the Royal Society B: Biological Sciences 267:2015–2020.

Conover, D. O., and H. Baumann 2009. The role of experiments in understanding fishery-induced evolution. Evolutionary Applications 2:276–290.

Cote, J., S. Fogarty, K. Weinersmith, T. Brodin, and A. Sih 2010. Personality traits and dispersal tendency in the invasive mosquitofish (*Gambusia affinis*). Proceedings of the Royal Society B: Biological Sciences 277:1571–1579.

Cuveliers, E. L., F. A. M. Volckaert, A. D. Rijnsdorp, M. H. D. Larmuseau, and G. E. Maes 2011. Temporal genetic stability and high effective population size despite fisheries-induced life-history trait evolution in the North Sea sole. Molecular Ecology 20:3555–3568.

Darimont, C. T., S. M. Carlson, M. T. Kinnison, P. C. Paquet, T. E. Reimchen, and C. C. Wilmers 2009. Human predators outpace other agents of trait change in the wild. Proceedings of the National Academy of Sciences of the United States of America 106:952–954.

de Roos, A. M., and L. Persson 2002. Size-dependent life-history traits promote catastrophic collapses of top predators. Proceedings of the National Academy of Sciences of the United States of America 99:12907–12912.

Devine, J. A., P. J. Wright, H. E. Pardoe, M. Heino, and D. J. Fraser 2012. Comparing rates of contemporary evolution in life-history traits for exploited fish stocks. Canadian Journal of Fisheries and Aquatic Sciences 69:1105–1120.

Diaz Pauli, B., and M. Heino 2013. The importance of social dimension and maturation stage for the probabilistic maturation reaction norm in *Poecilia reticulata*. Journal of Evolutionary Biology 26:2184–2196.

Diaz Pauli, B., and M. Heino 2014. What can selection experiments teach us about fisheries-induced evolution? Biological Journal of the Linnean Society 111:485–503.

Diaz Pauli, B., M. Wiech, M. Heino, and A. C. Utne-Palm 2015. Opposite selection on behavioral types by active and passive fishing gears in a simulated guppy *Poecilia reticulata* fishery. Journal of Fish Biology 86:1030–1045.

Dieckmann, U., and M. Heino 2007. Probabilistic maturation reaction norms: their history, strengths, and limitations. Marine Ecology Progress Series 335:253–269.

Dunlop, E. S., M. Heino, and U. Dieckmann 2009. Eco-genetic modeling of contemporary life-history evolution. Ecological Applications 19:1815–1834.

Dupont-Prinet, A., B. Chatain, L. Grima, M. Vandeputte, G. Claireaux, and D. J. McKenzie 2010. Physiological mechanisms underlying a trade-off between growth rate and tolerance of feed deprivation in the European sea bass (*Dicentrarchus labrax*). The Journal of Experimental Biology 213:1143–1152.

Eikeset, A. M., A. Richter, E. S. Dunlop, U. Dieckmann, and N. Chr. Stenseth 2013. Economic repercussions of fisheries-induced evolution. Proceedings of the National Academy of Sciences of the United States of America 110:12259–12264.

Enberg, K., C. Jørgensen, E. S. Dunlop, M. Heino, and U. Dieckmann 2009. Implications of fisheries-induced evolution for stock rebuilding and recovery. Evolutionary Applications 2:394–414.

Enberg, K., C. Jørgensen, E. S. Dunlop, Ø. Varpe, D. S. Boukal, L. Baulier, S. Eliassen et al. 2012. Fishing-induced evolution of growth: concepts, mechanisms and the empirical evidence. Marine Ecology 33:1–25.

Engen, S., R. Lande, and B.-E. Saether 2014. Evolutionary consequences of nonselective harvesting in density-dependent populations. American Naturalist 184:714–726.

Falcón, J., H. Migaud, J. A. Muñoz-Cueto, and M. Carrillo 2010. Current knowledge on the melatonin system in Teleost fish. General and Comparative Endocrinology 165:469–482.

Haldane, J. B. S. 1949. Suggestions as to quantitative measurement of rates of evolution. Evolution 3:51–56.

Hare, J. A., and R. K. Cowen 1997. Size, growth, development, and survival of the planktonic larvae of *Pomatomus saltatrix* (Pisces: Pomatomidae). Ecology 78:2415–2431.

Haugen, T. O., I. J. Winfield, L. A. Vøllestad, J. M. Fletcher, J. B. James, and N. Chr. Stenseth 2007. Density dependence and density independence in the demography and dispersal of pike over four decades. Ecological Monographs 77:483–502.

Heino, M., and O. R. Godø 2002. Fisheries-induced selection pressures in the context of sustainable fisheries. Bulletin of Marine Science 70:639–656.

Heino, M., U. Dieckmann, and O. R. Godø 2002. Measuring probabilistic reaction norms for age and size at matuartion. Evolution 56:669–678.

Heino, M., L. Baulier, D. S. Boukal, B. Ernande, F. D. Johnston, F. M. Mollet, H. Pardoe et al. 2013. Can fisheries-induced evolution shift reference points for fisheries management? ICES Journal of Marine Science 70:707–721.

Hendry, A. P., T. Day, and A. B. Cooper 2001. Optimal size and number of propagules: allowance for discrete stages and effects of maternal size on reproductive output and offspring fitness. American Naturalist 157:387–407.

Hilborn, R., and C. V. Minte-Vera 2008. Fisheries-induced changes in growth rates in marine fisheries: are they significant? Bulletin of Marine Science 83:95–105.

Hilborn, R., and K. Stokes 2010. Defining overfished stocks: have we lost the plot? Fisheries 35:113–120.

Hixon, M. A., D. W. Johnson, and S. M. Sogard 2014. BOFFFFs: on the importance of conserving old-growth age structure in fishery populations. ICES Journal of Marine Science 71:2171–2185.

Houde, E. D. 1987. Fish early life dynamics and recruitment variability. American Fisheries Society Symposium 2:17–29.

Hutchings, J. A., and D. J. Fraser 2008. The nature of fisheries- and farming-induced evolution. Molecular Ecology 17:294–313.

Hutchings, J. A. 2009. Avoidance of fisheries-induced evolution: management implications for catch selectivity and limit reference points. Evolutionary Applications 2:324–334.

Hutchings, J. A., and A. Kuparinen 2014. Ghosts of fisheries-induced depletions: do they haunt us still? ICES Journal of Marine Sciences 71:1467–1473.

Hutter, S., D. J. Penn, S. Magee, and S. M. Zala 2010. Reproductive behaviour of wild zebrafish (*Danio rerio*) in large tanks. Behaviour 147:641–660.

Härkönen, L., P. Hyvärinen, J. Paappanen, and A. Vainikka 2014. Explorative behavior increases vulnerability to angling in hatchery-reared brown trout (*Salmo trutta*). Canadian Journal of Fisheries and Aquatic Sciences 71:1900–1909.

Johnson, D. W., M. R. Christie, J. Moye, and M. A. Hixon 2011. Genetic correlations between adults and larvae in a marine fish: potential effects of fishery selection on population replenishment. Evolutionary Applications 4:621–633.

Jönsson, K. I. 1997. Capital and income breeding as alternative tactics of resource use in reproduction. Oikos 78:57–66.

Jørgensen, C., K. Enberg, E. S. Dunlop, R. Arlinghaus, D. S. Boukal, K. Brander, B. Ernande, et al. 2007. Managing evolving fish stocks. Science 318:1247–1248.

Jørgensen, C., K. Enberg, E. S. Dunlop, R. Arlinghaus, D. S. Boukal, K. Brander, B. Ernande, et al. 2008. The role of fisheries-induced evolution – response. Science 320:48–50.

Jørgensen, C., B. Ernande, and Ø. Fiksen 2009. Size-selective fishing gear and life history evolution in the Northeast Arctic cod. Evolutionary Applications 2:356–370.

Jørgensen, C., and Ø. Fiksen 2010. Modelling fishing-induced adaptations and consequences for natural mortality. Canadian Journal of Fisheries and Aquatic Sciences 67:1086–1097.

Jørgensen, C., and R. E. Holt 2013. Natural mortality: its ecology, how it shapes fish life histories, and why it may be increased by fishing. Journal of Sea Research 75:8–18.

Jørgensen, C., and F. Zimmermann 2015. Bioeconomic consequences of fishing-induced evolution: a model predicts limited impact on net present value. Canadian Journal of Fisheries and Aquatic Sciences 72:612–624.

Kimmel, C. B., W. W. Ballard, S. R. Kimmel, B. Ullman, and T. F. Schilling 1995. Stages of embryonic development of the zebrafish. Developmental Dynamics 203:253–310.

Klefoth, T., C. Skov, J. Krause, and R. Arlinghaus 2012. The role of ecological context and predation risk-stimuli in revealing the true picture about the genetic basis of boldness evolution in fish. Behavioral Ecology and Sociobiology 66:547–559.

Kuparinen, A., and J. Merilä 2007. Detecting and managing fisheries-induced evolution. Trends in Ecology & Evolution 22:652–659.

Kuparinen, A., and J. A. Hutchings 2012. Consequences of fisheries-induced evolution for population productivity and recovery potential. Proceedings of the Royal Society B: Biological Sciences 279:2571–2579.

Kuparinen, A., N. Chr. Stenseth, and J. A. Hutchings 2014. Fundamental population-productivity relationships can be modified through density-dependent feedbacks of life-history evolution. Evolutionary Applications 7:1218–1225.

Laugen, A. T., G. H. Engelhard, R. Whitlock, R. Arlinghaus, D. J. Dankel, E. S. Dunlop, A. M. Eikeset et al. 2014. Evolutionary impact assessment: accounting for evolutionary consequences of fishing in an ecosystem approach to fisheries management. Fish and Fisheries 15:65–96.

Law, R. 2007. Fisheries-induced evolution: present status and future directions. Marine Ecology Progress Series 335:271–277.

Law, R., and D. R. Grey 1989. Evolution of yields from populations with age-specific cropping. Evolutionary Ecology 3:343–359.

Lester, N. P., B. J. Shuter, and P. A. Abrams 2004. Interpreting the von Bertalanffy model of somatic growth in fishes: the cost of reproduction. Proceedings of the Royal Society B: Biological Sciences 271:1625–1631.

Lewin, W. C., R. Arlinghaus, and T. Mehner 2006. Documented and potential biological impacts of recreational fishing: insights for management and conservation. Reviews in Fisheries Science 14:305–367.

Lorenzen, K., and K. Enberg 2002. Density-dependent growth as a key mechanism in the regulation of fish populations: evidence from among-population comparisons. Proceedings of the Royal Society B: Biological Sciences 269:49–54.

Marty, L., U. Dieckmann, and B. Ernande 2015. Fisheries-induced neutral and adaptive evolution in exploited fish populations and consequences for their adaptive potential. Evolutionary Applications 8:47–63.

Matsumura, S., R. Arlinghaus, and U. Dieckmann 2011. Assessing evolutionary consequences of size-selective recreational fishing on multiple life-history traits, with an application to Northern pike (Esox lucius). Evolutionary Ecology 25:711–735.

Matsumura, S., R. Arlinghaus, and U. Dieckmann 2012. Standardizing selection strengths to study selection in the wild: a critical comparison and suggestions for the future. BioScience 62:1039–1054.

McAllister, M. K., and R. M. Peterman 1992. Decision analysis of a large-scale fishing experiment designed to test for a genetic effect of size-selective fishing on British Columbia pink salmon (Oncorhynchus gorbuscha). Canadian Journal of Fisheries and Aquatic Sciences 49:1305–1314.

Mousseau, T. A., and D. A. Roff 1987. Natural selection and the heritability of fitness components. Heredity 59:181–197.

Munch, S. B., M. R. Walsh, and D. O. Conover 2005. Harvest selection, genetic correlations and evolutionary changes in recruitment: one less thing to worry about? Canadian Journal of Fisheries and Aquatic Sciences 62:802–810.

Naish, K. A., and J. J. Hard 2008. Bridging the gap between the genotype and the phenotype: linking genetic variation, selection and adaptation in fishes. Fish and Fisheries 9:396–422.

Narum, S. R., and J. E. Hess 2011. Comparison of F_{ST} outlier tests for SNP loci under selection. Molecular Ecology Resources 11(Suppl. 1):184–194.

Neubauer, P., O. P. Jensen, J. A. Hutchings, and J. K. Baum 2013. Resilience and recovery of overexploited marine populations. Science 340:347–349.

Nússle, S., C. N. Bornand, and C. Wedekind 2009. Fishery-induced selection on an Alpine whitefish: quantifying genetic and environmental effects on individual growth rate. Evolutionary Applications 2:200–208.

Olsen, E., M. Heino, G. R. Lilly, M. J. Morgan, J. Brattey, B. Ernande, and U. Dieckmann 2004. Maturation trends indicative of rapid evolution preceded the collapse of northern cod. Nature 428:932–935.

Palkovacs, E. P., M. T. Kinnison, C. Correa, C. M. Dalton, and A. P. Hendry 2012. Fates beyond traits: ecological consequences of human-induced trait change. Evolutionary Applications 5:183–191.

Paull, G. C., A. L. Filby, H. G. Giddins, T. S. Coe, P. B. Hamilton, and C. R. Tyler 2010. Dominance hierarchies in zebrafish (Danio rerio) and their relationship with reproductive success. Zebrafish 7:109–117.

Pepin, P. 1991. Effect of temperature and size on development, mortality, and survival rates of the pelagic early life history stages of marine fish. Canadian Journal of Fisheries and Aquatic Sciences 48:503–518.

Pérez-Rodrígues, A., M. J. Morgan, M. Koen-Alonso, and F. Saborido-Rey 2013. Disentangling genetic change from phenotypic response in reproductive parameters of Flemish Cap cod Gadus morhua. Fisheries Research 138:62–70.

Peters, R. H. 1983. The Ecological Implications of Body Size. American Journal of Physical Anthropology, Vol. 66. Cambridge University Press, Cambridge.

Philipp, D. P., S. J. Cooke, J. E. Claussen, J. B. Koppelman, C. D. Suski, and D. P. Burkett 2009. Selection for vulnerability to angling in largemouth bass. Transactions of the American Fisheries Society 138:189 199.

Policansky, D. 1993. Fishing as a cause of evolution in fishes. In T. K. Stokes, J. M. McGlade, and R. Law, eds. The Exploitation of Evolving Resources, pp. 2–18. Springer-Verlag, Heidelberg, Germany.

Poos, J. J., Å. Brännström, and U. Dieckmann 2011. Harvest-induced maturation evolution under different life-history trade-offs and harvesting regimes. Journal of Theoretical Biology 279:102–112.

Rijnsdorp, A. D. 1993. Fisheries as a large-scale experiment on life-history evolution, disentangling phenotypic and genetic effects in changes in maturation and reproduction of North Sea place, Pleuronectes platessa L. Oecologia 96:391–401.

Roff, D. A. 2010. Modelling Evolution – An Introduction to Numerical Methods. Oxford University Press, Oxford.

Rutter, C. 1902. Natural history of the *Quinnat salmon*. A Report on Investigations in the Sacramento River, 1886–1901. Bulletin of the U.S. Fisheries Commission **22**:65–141.

Salinas, S., K. O. Perez, T. A. Duffy, S. J. Sabatino, L. A. Hice, S. B. Munch, and D. O. Conover. 2012. The response of correlated traits following cessation of fishery-induced selection. Evolutionary Applications **5**:657–663.

Scholten, A., and K. W. Koch 2011. Differential calcium signaling by cone specific guanylate cyclase-activating proteins from the zebrafish retina. PLoS ONE **6**:e23117.

Sharpe, D. M. T., and A. P. Hendry 2009. Life history change in commercially exploited fish stocks: an analysis of trends across studies. Evolutionary Applications **2**:260–275.

Smith, B. R., and D. T. Blumstein 2007. Fitness consequences of personality: a meta-analysis. Behavioral Ecology **19**:448–455.

Stamps, J. A. 2007. Growth-mortality tradeoffs and 'personality traits' in animals. Ecology Letters **10**:355–363.

Stearns, S. C. 1992. The Evolution of Life Histories. Oxford University Press, Oxford.

Stein, C., M. Caccamo, G. Laird, and M. Leptin 2007. Conservation and divergence of gene families encoding components of innate immune response systems in zebrafish. Genome Biology **8**: R251.

Sutter, D. A. H., C. D. Suski, D. P. Philipp, T. Klefoth, D. H. Wahl, P. Kersten, S. J. Cooke et al. 2012. Recreational fishing selectively captures individuals with the highest fitness potential. Proceedings of the National Academy of Sciences of the United States of America **109**:20960–20965.

Swain, D. P., A. F. Sinclair, and J. M. Hanson 2007. Evolutionary response to size-selective mortality in an exploited fish population. Proceedings of the Royal Society B: Biological Sciences **274**:1015–1022.

Thériault, V., E. S. Dunlop, U. Dieckmann, L. Bernatchez, and J. J. Dodson 2008. The impact of fishing-induced mortality on the evolution of alternative life-history tactics in brook charr. Evolutionary Applications **1**:409–423.

Therkildsen, N., E. E. Nilsen, D. P. Swain, and J. S. Pedersen 2010. Large effective population size and temporal genetic stability in Atlantic cod (*Gadus morhua*) in the southern Gulf of St. Lawrence. Canadian Journal of Fisheries and Aquatic Sciences **67**:1585–1595.

Tingaud-Sequeira, A., and J. Cerdà 2007. Phylogenetic relationships and gene expression pattern of three different cathepsin L (Ctsl) isoforms in zebrafish: Ctsla is the putative yolk processing enzyme. Gene **386**:98–106.

Uusi-Heikkilä, S., C. Wolter, T. Klefoth, and R. Arlinghaus 2008. A behavioral perspective on fishing-induced evolution. Trends in Ecology & Evolution **23**:419–421.

Uusi-Heikkilä, S., C. Wolter, T. Meinelt, and R. Arlinghaus 2010. Size-dependent reproductive success of wild zebrafish *Danio rerio* in the laboratory. Journal of Fish Biology **77**:552–569.

Uusi-Heikkilä, S., A. Kuparinen, C. Wolter, T. Meinelt, A. C. O'Toole, and R. Arlinghaus 2011. Experimental assessment of the probabilistic maturation reaction norm: condition matters. Proceedings of the Royal Society B: Biological Sciences **278**:709–717.

Uusi-Heikkilä, S., A. Kuparinen, C. Wolter, T. Meinelt, and R. Arlinghaus 2012a. Paternal body size affects reproductive success in laboratory-held zebrafish (*Danio rerio*). Environmental Biology of Fishes **93**:461–474.

Uusi-Heikkilä, S., L. Böckenhoff, C. Wolter, and R. Arlinghaus 2012b. Differential allocation by female zebrafish (*Danio rerio*) to different-

sized males - an example in a fish species lacking parental care. PLoS ONE **7**:e48317.

Walsh, M. R., S. B. Munch, S. Chiba, and D. O. Conover 2006. Maladaptive changes in multiple traits caused by fishing: impediments to population recovery. Ecology Letters **9**:142–148.

Walters, C. J., and S. J. D. Martell 2004. Fisheries Ecology and Management. Princeton University Press, Princeton.

van Walraven, L., F. M. Mollet, C. J. G. van Damme, and A. D. Rijnsdorp 2010. Fisheries-induced evolution in growth, maturation and reproductive investment of the sexually dimorphic North Sea plaice (*Pleuronectes platessa* L.). Journal of Sea Reasearch **64**:84–93.

van Wijk, S. J., M. I. Taylor, S. Creer, C. Dreyer, F. M. Rodrigues, I. W. Ramnarine, C. van Oosterhout et al. 2013. Experimental harvesting of fish populations drives genetically based shifts in body size and maturation. Frontiers in Ecology and the Environment **11**:181–187.

Varshney, G. K., J. Lu, D. E. Gildea, H. Huang, W. Pei, Z. Yang, S. C. Huang et al. 2013. A large-scale zebrafish gene knockout resource for the genome-wide study of gene function. Genome Research **23**:727–735.

Weger, B. D., M. Sahinbas, G. W. Otto, P. Mracek, O. Armant, D. Dolle, K. Lahiri et al. 2011. The light responsive transcriptome of the zebrafish: function and regulation. PLoS ONE **6**:e17080.

Whiteley, A. R., A. Bhat, E. P. Martins, R. L. Mayden, M. Arunachalam, S. Uusi-Heikkilä, A. T. Ahmed et al. 2011. Population genomics of wild and laboratory zebrafish (*Danio rerio*). Molecular Ecology **20**:4259–4276.

Vital, C., and E. P. Martins 2011. Strain differences in zebrafish (*Danio rerio*) social roles and their impact on group task performance. Journal of Comparative Psychology **125**:278–285.

Worm, B., R. Hilborn, J. K. Baum, T. A. Branch, J. S. Collie, C. Costello, M. J. Fogarty et al. 2009. Rebuilding global fisheries. Science **325**:578–585.

12

Women in evolution – highlighting the changing face of evolutionary biology

Maren Wellenreuther[1,2] and Sarah Otto[3]

1 Department of Biology, University of Lund, Lund, Sweden
2 Institute for Plant and Food Research, Lund, New Zealand
3 Department of Zoology & Biodiversity Research Centre, University of British Columbia, Vancouver, BC, Canada

Keywords
equality, evolutionary biology, gender, women in science.

Correspondence
Maren Wellenreuther, Department of Biology, University of Lund, Sölvegatan 37, 223 62, Lund, Sweden.

e-mail: Maren.wellenreuther@biol.lu.se

Abstract

The face of science has changed. Women now feature alongside men at the forefront of many fields, and this is particularly true in evolutionary biology. This special issue celebrates the outstanding achievements and contributions of women in evolutionary biology, by highlighting a sample of their research and accomplishments. In addition to original research contributions, this collection of articles contains personal reflections to provide perspective and advice on succeeding as a woman in science. By showcasing the diversity and research excellence of women and drawing on their experiences, we wish to enhance the visibility of female scientists and provide inspiration as well as role models. These are exciting times for evolutionary biology, and the field is richer and stronger for the diversity of voices contributing to the field.

Introduction

In the century and a half since Charles Darwin published the Origin of Species, the face of the field has changed dramatically. Emerging from an era with virtually no women practicing as evolutionary biologists, the contributions of women today are many and varied. This special issue highlights the creativity and diversity of women's voices active in evolutionary biology today. No longer are the eminent scientists all male. The leaders of the field are shifting, from a history dominated by male figures – Darwin, Fisher, Haldane, Wright, Dobzhansky, Waddington, Falconer, Stebbins, Mayr, etc. – to a field more balanced by perspectives from both sexes.

While the face of science has been changing, subtle barriers that dissuade women from a career in science remain widespread. Not only is there a paucity of women in leadership positions, but many successful women in science are not as visible as their contributions deserve. An abundance of research demonstrates that having few women represented in science creates a lack of role models to attract and retain young women in scientific professions (e.g. Latu et al. 2013). Women are at the forefront of many scientific advances but their role in driving science forward is not represented equally in courses and textbooks. This was the motivation for Evolutionary Applications to publish this special issue co-edited by Maren Wellenreuther (Guest editor) and Louis Bernatchez (Editor-in-Chief), entitled: Women's contribution to basic and applied evolutionary biology. In particular, we hope that by highlighting the accomplishments of women scientists, this publication may inspire young women scientists, who may be questioning themselves regarding the feasibility and realism of a scientific career, to conclude: 'Yes we can!'

In this introductory essay, we provide a historical perspective of women in science, followed by an overview of the research contributions in this special issue and closing with a synthesis of the personal reflections contained in each article. In addition, Drs. Rosemary Grant, Mary Jane West-Eberhard, Josephine Pemberton and Michelle Tseng contributed personal reflections, which are included in this introductory essay. The women invited to contribute to this special issue include both junior and senior researchers in the field, and thus, their reflections also provide a temporal perspective on how barriers have changed. The diversity of their voices provides a great deal of insight into the achievements of these women and their advice for lowering the remaining barriers to women in evolutionary biology.

A short history of women in science

Evolutionary biology, as a field, developed at a time when women were prohibited from contributing to scientific research. Before the early 19th century, most women were prevented from accessing formal scientific training, so that only women with independent means could pursue scientific study (Orr 2014). This discrimination was founded on centuries of prejudice about women's intellectual abilities. Ironically, these assumptions were often reinforced by pseudoscientific data and paradigms emanating from evolutionary biology. In *The Descent of Man*, with its significant subtitle *Selection in Relation to Sex*, Darwin stated that gender equality was impossible to achieve because the lesser female brain power presented an inescapable consequence of nature (Darwin 1871). Men are simply more intelligent, Darwin argued, because over the millennia their brains became superior due to the need to be effective hunter-gatherers. Reflecting the beliefs of his time, Darwin wrote 'The chief distinction in the intellectual powers of the two sexes is shown by man's attaining to a higher eminence in whatever he takes up, than can woman - whether requiring deep thought, reason, or imagination, or merely the use of the senses or the hands'. Scientific arguments were also used to justify why women were expected to stay home. Francis Galton (Darwin's half cousin) and other advocates of positive eugenics had, for example, warned that if women of distinguished pedigree turn their back on being nurturing mothers, then the quality of the next generation would decline (Fara 2015). Consequently, it was expected that women divert all their energy into childrearing and not indulge themselves in education or employment. While suffragists rallied for women's rights and counterargued that evolutionary principles emphasize the fundamental similarities among members of a species, their opponents emphasized Darwin's ideas on the role of sexual selection, which supported the cultural prejudices of female intellectual inferiority (Richards 1997).

The laws that prohibited women from seeking education in Europe and North America were gradually overturned during the mid-19th century, with the opening of the first women's colleges (Etzkowitz et al. 2000). While these colleges granted women access to both education and employment, they unfortunately also applied strict employment rules, which came at a considerable personal price: all faculty had to remain unmarried, a practice that in some parts of the Western World continued even through World War I and II (Barnett and Sabattini 2009). Employment chances of women were low and reduced by the requirement by many colleges that faculty hold a PhD degree, despite the fact that most European and North American universities refused to allow women into their graduate programmes (Barnett and Sabattini 2009). In the 1890s, a small number of institutions allowed women to matriculate into advanced degree programmes, but most institutions were slow to follow this lead.

It was only in the second half of the 20th century that women gained access to graduate education in large numbers and had the potential to become professors. Some traditional all-male colleges circumvented the need to grant official access to women by opening all-female sister institutions that coexisted alongside the prestigious all-male colleges. Harvard, for example the oldest institution of higher education in the United States, was founded in 1836 to educate an all-male clergy. Nevertheless, women showed a persistent interest to attend courses. When lectures at Harvard were opened to the public in 1863, women immediately flocked to attend (Ulrich 2004). Indeed, Harvard proved so popular with women that by 1870 they presented the majority of all students (Ulrich 2004). Public access to the open lectures was suspended when Harvard established its own Graduate Department in 1872 (Ulrich 2004). This move was met by considerable resistance, which Harvard solved by simply opening a sister institution called 'The Society for the Collegiate Instruction of Women' in 1879 (later to become Radcliffe College). This allowed women to receive instruction from Harvard professors who were willing to earn extra income by teaching their courses twice, once for men and once for women. It was only in 1963 that Radcliffe Graduate School merged with Harvard Graduate School, after having awarded 750 PhD degrees (Horowitz 1986). Likewise, Princeton only became co-educational in 1969 (Selden 2000). An early attempt to establish a parallel institute for women (the Evelyn College for Women) closed in 1897, ten years after its founding, due to financial problems and a lack of support from Princeton (Selden 2000).

In addition to a lack of access, the scientific achievements of women have traditionally been underrecognized. Women have historically been limited to secondary roles in science production and authorship, as translators, illustrators and popularizers of science, and these by virtue of marriage or kinship relations with eminent male scientists (Orr 2014). As a result, women were all too often portrayed as 'volunteer' faculty members, with credit for their significant discoveries assigned to male colleagues. Esther Lederberg (1922–2006), a microbiologist, conducted ground-breaking research in the field of genetics, most notably on bacteriophages. She discovered lambda phage, a virus that infects *E. coli* bacteria and published the first report of this in Microbial Genetics Bulletin (Lederberg 1950). Her work helped her husband, Joshua Lederberg, win a Nobel prize in 1958, which he shared with Edward Tatum and George Beadle (Harvey 2012). Likewise, Rosalind Franklin (1920–1958), a pioneering X-ray crystallographer, developed images of DNA molecules that were critical to deciphering its structure – one of the biggest and most important scientific breakthroughs of the 20th century. Recognition for her contribution to the discovery of the DNA structure remained limited during her lifetime (Jones and Hawkins 2014; Orr 2014).

Even less attention has been given to the discovery of Dr. Marthe Gautier, who first identified aneuploidy as the cause of Down syndrome (Pain 2014). As a young French researcher, she received a prestigious scholarship in 1955 to undergo medical training at the paediatric cardiology unit at Harvard to examine cell cultures. Upon her return to work in Paris in 1956, she became involved in genetic research on human diseases. In the same year, the human chromosome number was corrected to 46 (Harper 2006), allowing researchers for the first time to detect chromosomal abnormalities (Gartler 2006). The head of the paediatric unit, Dr. Raymond Turpin, was interested in the possibility that Down syndrome was caused by a chromosomal abnormality and provided Gautier with a tissue sample from a patient and a disused laboratory equipped with a poor-quality microscope (and no funding). Despite these inauspicious conditions, Gautier discovered that the karyotype from the Down syndrome patient contained an extra chromosome, but she was unable to establish which chromosome was involved using her microscope (Gautier and Harper 2009). She entrusted the slides to Dr. Jérôme Lejeune, who established the karyotype using a higher quality photomicroscope and announced at a conference that he had discovered the cause of Down syndrome. He went on to publish the story with Gautier's name as middle author (Lejeune et al. 1959) – a paper she did not get to see and knew nothing about until the day before publication (Gautier and Harper 2009).

'La découverte de la trisomie n'ayant pu être faite sans les contributions essentielles de Raymond Turpin et Marthe Gautier[;] il est regrettable que leurs noms n'aient pas été systématiquement associés à cette découverte tant dans la communication que dans l'attribution de divers honneurs'.

[The discovery of trisomy would not have been possible without the essential contributions of Raymond Turpin and Marthe Gautier[;] it is regrettable that their names have not been systematically associated with the discovery in both the communication and attribution of various honors.]

– Ethics Committee, French Institute of Health and Medical Research (2014)[1]

Recognition of women's achievements through membership in academies was also restricted. Indeed, three of the most prestigious scientific societies, the *Royal Society of London* (the oldest continuous society of science, founded 1660), the Parisian *Académie royale des Sciences* (founded 1666) and the *Akademie der Wissenschaften* in Berlin (founded 1700), did not permit women to become members for almost 300 years following their establishment (Schiebinger 1993). Marie Curie, probably the best known female scientist, who earned two Nobel Prizes, was turned down for membership of the prestigious *Académie royale des Sciences* in 1911, the very year she went on to win her second Nobel Prize. In fact, it was only in 1979 that the first woman, the physicist and mathematician Yvonne Choquet-Bruhat, was elected as a fellow (Schiebinger 1993). The mathematician Hertha Ayrton (1854–1923) became the first woman nominated as fellow of the *Royal Society of London* in 1902, but she was refused this distinction on grounds that she was married (Mason 1991; Fara 2015). More than four decades were to elapse before the next women were nominated. Finally, in 1945, Kathleen Lonsdale (1903–1971), an early pioneer of X-ray crystallography, along with microbiologist Marjory Stephenson (1885–1948) were the first women to be admitted as fellows (Glazer 2015). It took the German *Akademie der Wissenschaften* even longer to open its doors to women. Only in 1964 was Elisabeth Welskopf-Henrich (1901–1979), a participant in the resistance during World War II and professor of history, elected as its first full fellow (Wobbe 2002).

While inequalities in these three academies were strong and prevented the admission of women for over 300 years, gender disparities exist even nowadays in many of the scientific societies and academies. For example, the National Academy of Sciences (NAS) in the United States was founded in 1863, yet despite this comparably much younger age, the membership remains heavily skewed towards men. The total number of active and emeritus members is 2113, of which a mere 219 are women (similar numbers hold if we include foreign associates: 2508 total versus 251 women[2]). The Association for Women in Science (AWIS) has examined the NAS membership distribution over the last 20 years and found that for many years women were underrepresented relative to the available pool of Ph.D. holders worthy of consideration.[3] An even stronger division can be seen if we look at the gender distribution of Nobel Prize laureates. The Nobel Prize is by many regarded as the most prestigious award given for intellectual achievement in the world, yet between 1901 and 2014 only 3% of all Nobel awardees in Medicine, Physics or Chemistry have been women.[4]

When looking more specifically at the field of evolutionary biology, these biases persist. The Society for the Study of Evolution has one annual prize to recognize the accomplishments of outstanding young evolutionary biologists called the Theodosius Dobzhansky Prize. The prize was established in 1981 and has only been awarded to a woman four times since its establishment.[5] A similarly distorted sex ratio is evident when looking at presenters at evolutionary biology meetings. A recent study analysed data on the sex ratio of presenters from the European Society for Evolutionary Biology (ESEB) biannual congress 2011 and found that women were underrepresented among invited speakers at symposia (15% women) compared to all presenters (46%), regular oral presenters (41%) and plenary speakers (25%) (Schroeder et al. 2013). This underrepresentation of women was found to be partly attributable to a larger proportion of women, than men, declining invitations, yet the reasons for this are not known. A possible cause for the higher decline of invitations by women may be related to family responsibilities, such as child rearing or looking after elderly people, which are both activities that are predominately carried out by women (Namrata et al. 2005). Similarly, managing domestic responsibilities poses a challenge for research projects that involve extensive travel for fieldwork.

The representation of women in science improved considerably in the second half of the 20th century (England and Li 2006) and continues to do so. Particularly, notable pioneering women in evolutionary biology emerged during this time, including Barbara McClintock, Tomoko Ohta, Mary Leakey, Lynn Margulis and Margaret Kidwell. Over this period, the numbers of women and men pursuing science degrees have converged and are now nearly equal. Despite a significant increase in women graduating in science degrees, the proportion of women in science leadership or as research professionals decreases dramatically with seniority. In the United States, for example, even though women earned approximately half of all science bachelor (55.6%), master (54.0%) and doctoral (46.7%) degrees in 2010, only 34.5% of all professors are women, and this representation declines further to 21% among full professors (NSF 2013). While the number of women advancing through the ranks has risen as more and more women earn graduate degrees, the percentage of women seen at the more senior levels is not mirroring the increase seen at the graduate level. The 'leaky pipeline' is particularly acute at two stages: as women enter academic positions and as women reach the highest echelons in academia (the equivalent of Full Professorship) (Council of Canadian

Academies 2012). As concluded by a recent report of the Council of Canadian Academies (2012), 'these data indicate that time alone will probably not be enough to balance the proportion of women and men at the highest levels of academia'.

Among the challenges women face advancing through careers in science, one factor is systematic bias in evaluating the sexes (Reuben et al. 2014; Leslie et al. 2015). Research shows that both men and women tend to overrate men and underrate women in competence, particularly when women are in nontraditional fields such as science, and that this bias occurs unconsciously (Zuk and Rosenqvist 2005). For example, a 2012 study showed that both men and women tend to undervalue women candidates for a research technician position, even when the applications are identical except for the name of the candidate (Moss-Racusin et al. 2012). Empirical studies of gender equality document a broad bias against representation of women in conference presentations (Schroeder et al. 2013; Jones et al. 2014), scholarly authorship (West et al. 2013) and invited journal articles (Conley and Stadmark 2012). Additionally, research has demonstrated that more stringent criteria are used to measure women's qualifications in grant evaluation panels (Wennerås and Wold 1997; Reuben et al. 2014; Leslie et al. 2015) and that women receive smaller grants and fewer nominations for awards (Cho et al. 2014).

As an example, Tomoko Ohta – best known for her development of the nearly neutral theory – reflected on her initial struggles against implicit bias when first working in the laboratory of Motoo Kimura. '*Kimura was a typical Japanese man of his time, who regarded women's scientific activities as insignificant. After two years or so, I had convinced him that I should continue to do research*' (Ohta 2012).

On top of the systematic bias and more stringent criteria lies outright sexual harassment. Studies in the United States and Europe demonstrate that a considerable number of women in STEM had to face sexual harassment at work. For example, Sonnert and James Holton (1995) surveyed 191 female fellowship recipients in the United States and found that 12% of them had been sexually harassed during their graduate school or early professional experience. Likewise, a study on Finnish University academic staff found that about 7% of employees had suffered sexual harassment, 78% of whom were women (Mankkinen 1999). Based on an internet survey of scientists' experiences at field sites, Clancy et al. (2014) found that gender was a significant predictor of having personally experienced sexual harassment, with women respondents being 3.5 times more likely to have experienced sexual harassment than men. These studies also highlight that sexual harassment may take multiple and diverse forms, from serious harassment to the overemphasis of sexual roles, and can provoke a deep feeling of isolation and professional discouragement.

Another contributing factor is that the 'paucity of women in leadership positions makes it difficult for other women to envision themselves as leaders' (Council of Canadian Academies 2012). To understand that science careers are realistic options, women need to see the evidence that those they identify with, people like them, can and do succeed (Latu et al. 2013). Seeing women with children in science leadership positions, and how they balance their careers and personal lives, is particularly important for the younger generation of women who struggle to imagine how to combine family life with a career in science (Shen 2013). They also need to know that the people around them see a career in science as a valid choice for them.

Revising perceptions of women in evolutionary biology

A powerful way to counteract this skewed and biased representation of women in science is to showcase their contributions openly (Jones and Hawkins 2014). By requesting articles from women in evolutionary biology, the goal of this special issue is to highlight the diversity of research performed by women in the field today. Fortunately, nowadays there are so many women succeeding in evolutionary biology that it would be impossible to feature all of their research, but we asked for contributions from a representative sample of women, at various stages in their careers, working in a variety of subfields, and across a diversity of countries.

The authors were asked to contribute both research articles and personal reflections (see next section). The research they describe reflects the strength and diversity of evolutionary biology today, regardless of gender. The articles in this special issue span the breadth of areas in evolutionary biology and touch upon many of the hottest topics, including sex chromosome evolution (Charlesworth 2016), sexual selection (Servedio 2016; Wellenreuther and Sánchez-Guillén 2016), behavioural evolution (Aubin-Horth 2016; Charmantier et al. 2016), evolution of cooperation (Aktipis 2016), diversification and speciation (Charmantier et al. 2016; Johannesson 2016; Qvarnström et al. 2016; Servedio 2016; Wellenreuther and Sánchez-Guillén 2016), host–parasite interactions (Leftwich et al. 2016; Myers and Cory 2016), evolutionary change accompanying invasion (Aktipis 2016; Gillespie 2016; Haig et al. 2016; Lee 2016; Olivieri et al. 2016; Sork 2016), genomics and genome evolution (Aitken and Bemmels 2016; Charlesworth 2016; Charmantier et al. 2016; Johannesson 2016; Sork 2016), and climate change (Aitken and Bemmels 2016; Qvarnström et al. 2016). Although there are undoubtedly

some areas of evolutionary biology that have further than others to go to achieve gender equality, women are strong contributors to all subfields, from theory to genomics, from behaviour to systematics.

Among the research contributions, some articles summarize classic work by the authors that has shaped our understanding of the field. Charlesworth (2016) describes her theoretical research to understand when and how nonrecombining gene complexes ('supergenes') are formed, as well as empirical efforts to determine the relevance of supergenes (e.g. to sex chromosomes, mimicry and distyly). Myers and Cory (2016) describe long-term studies to understand how insect population dynamics are shaped by pathogens, particularly the cyclic outbreaks seen in species such as western tent caterpillars. Gillespie (2016) describes her now-classic work in the Hawaiian archipelago on spider diversification, which spans a range of low levels of 'nonadaptive radiation' in some groups (without ecological diversification) to high levels of adaptive radiation (new species formation with ecological diversification). Aubin-Horth (2016) describes the exciting emergence of ecological genomics as a subfield, which has opened up the black box linking molecules and environmental cues to phenotypic and behavioural variation, focusing on her work on salmonid alternative reproductive tactics.

The papers also explore new topics and altered perspectives. Qvarnström et al. (2016), for example, investigate the possibility that adaptation to different climates may drive speciation, even if this divergence is difficult to detect morphologically. Servedio (2016), in her recent theoretical work, has explored the interplay between sexual selection and speciation, yielding many counterintuitive results that challenge our understanding of these processes (e.g. finding that stronger sexual selection can hinder rather than promote divergence between populations and that mating preferences can reduce local adaptation of traits).

Several research articles delve into the evolutionary forces acting in particular species, describing the evolutionary insights that come from an in-depth understanding of a group of organisms. Wellenreuther and Sánchez-Guillén (2016) describe the laboratory's research on damselflies, which leads them to conclude that divergence in reproductive traits, not ecological differences, has driven speciation in this group. Johannesson (2016) describes her work to rationalize the messy taxonomic treatment of *Littorina saxatilis* snails, while previously split into many species due to shell polymorphisms that arise between crab-rich and wave-swept microenvironments, her work has shown that these ecotypes arise repeatedly but are likely prevented from speciating due to gene flow and the high fitness of hybrids in intermediate environments. Penczykowski et al. describe how their work in various host–parasite systems is revealing the spatial and temporal scales at which co-evolutionary (mal)adaptation occurs and the insights this yields about the underlying dynamics (e.g. fluctuating selection versus arms race). Charmantier et al. (2016) describe over 40 years of monitoring blue tits (*Cyanistes caeruleus)* that has revealed the strong connection between habitat heterogeneity and diversity in life-history, behavioural and reproductive strategies.

Women are also at the forefront of evolutionary applications. Aitken and Bemmels (2016) describe work to improve reforestation practices, switching from current 'local is best' reseeding strategies towards more diversified stocks ('assisted gene flow'), incorporating seeds that are adapted to warmer environments to better protect future forests. Chapman and colleagues review efforts to improve biocontrol of insects, including work in her laboratory to use antagonistic seminal fluid proteins as a means of population control (Leftwich et al. 2016). From their decades of work on endemic plants, Olivieri et al. (2016) conclude that conserving the landscape that generates genetic diversity is critical for the maintenance of evolutionary potential, a factor that must be incorporated into management strategies to better protect extant biodiversity and biodiversification. Haig and colleagues review their efforts to conserve species at risk via the application of genetics to taxonomic delimitation, to determine landscape use and migratory connections, and to guide breeding programmes for recovery from severe bottlenecks (Haig et al. 2016). Applied research by Lee's group is investigating evolutionary changes that accompany the invasive spread of a species by comparing parallel invasions into freshwater of the copepod *Eurytemora affinis* (Lee 2016).

Learning from personal reflections

In addition to requesting research articles, the editors of this issue asked women to contribute a section or box

Figure 1 A word cloud of the personal reflections in this volume (tagul.com).

describing personal reflections (see Fig. 1). The content was left open: 'your personal opinions or experiences regarding any professional aspects of the life of female evolutionary biologists that you would feel relevant to mention'. While not all women approached could contribute full-length research articles, many were keen to support this special issue in other ways. Three contributed their personal reflections to this chapter (Box 1: Rosemary Grant; Box 2: Mary Jane West-Eberhard; Box 3: Josephine Pemberton). We also requested the personal reflections of Michelle Tseng, who conceptualized, launched and managed *Evolutionary Applications* from 2008 to 2013 (Box 4).

The more senior women writing in this series entered into evolutionary biology when women were still strongly dissuaded from entering science. These women speak of loved ones and teachers warning them against careers in science. For Rosemary Grant's school head mistress, mumps was evidence of 'God's will you should not go to university'. Deborah Charlesworth writes of 'astonishing episodes of explicit prejudice' early on in her career. Judy Myers recalls being told by a senior scientist that 'women don't use their Ph.D's but just get married and have babies'. Victoria Sork writes that 'several professors advised that academia was not a place for a woman and a potential faculty mentor unabashedly informed me that he did not accept women as graduate students'. But these women's reflections also speak of persistence, resilience and a deep belief in their right to pursue a scientific career:

> 'these moderate disadvantages, and the odd faculty member who tried to ignore female students, merely seemed ridiculous, and I expected them to disappear shortly (which they did)' – Deborah Charlesworth

These senior evolutionary biologists also speak of gratitude to previous generations of women who forged a path in academia. We, in turn, are grateful to them for continuing to widen and smooth this path and for providing younger women with inspiration and role models.

Several contributors decried the fact that the very attributes that facilitate success in science are often seen as negatives traits in women. Positive descriptors including '*determined, motivated, persistent, stubborn, rebellious, irrepressible or independent-minded*' become twisted when exhibited by women into negative traits such as '*pushy, strident, aggressive, selfish, obnoxious, mannish, unbecoming and less mentionable words*' (West-Eberhard; Box 2). Women face accusations of being 'too independent' or working 'too hard' (Wellenreuther and Sánchez-Guillén 2016), precisely when these qualities are also seen as critical for success.

Similarly, collaboration can be an essential element in breakthrough science, allowing a team to do what an individual cannot, yet several women voiced aggravation at

Box 1: Personal reflections of Rosemary Grant, Princeton University, USA.

I was born in 1936 in a small, coastal village, in the Lake District of North West England. It is an area of carboniferous limestone fossil-rich cliffs, backed by wooded valleys and high grass fells where rare species of butterflies and plants can be found.

Even before I knew the word biology I was fascinated by the diversity of organisms around me. My parents told me that some of the fossils I had found were of plants and animals that were now extinct, and this added to my excitement and curiosity. A delightfully perceptive old gardener, (Jerry) Jeremiah Swindlehurst, who had never been to school but taught himself to read and write, was a constant inspiration and source of local biological wisdom. When I was twelve my Father, who was a physician, suggested I should read Darwin's 'Origin of Species'.

In my teens I thought that genetics would provide some basic understanding of this variation, and I desperately wanted to study genetics at Edinburgh University. The head mistress of the all-girl's boarding school I attended dissuaded me from taking all the necessary university entrance examinations, saying: 'A girl with two brothers should not go to university'. This was the norm in 1950s Britain. When it was time to take my Scottish higher leaving certificate in 1954, I had mumps with pancreatic complication…leaning over my bed the mistress's words were: 'It is God's will you should not go to university'.

I left school, determined rather than deterred, took a job, followed a correspondence course and applied on my own to university. Although it did not seem so at the time, this course must have immeasurably enhanced my chance of getting good grades in the examinations, because the school I had attended was academically weak.

My good fortune in entering Edinburgh University was enhanced when in my third year I was accepted into the genetics diploma course, consisting of a very small group of national and international students studying under Professor C.H. Waddington. My mother and Jerry gave me the love of nature, my father showed me how to study it, and Conrad Waddington, Douglas Falconer and Charlotte Auerbach inspired me to be a scientist.

My first project was an undergraduate thesis on soil amoebae. Among many identical looking amoebae living in the soil, one was thought to be pathogenic. My genetics training had introduced me to the possible importance of differences in cell surface proteins. I made antibodies (from rabbits) to the pathogenic one that had been collected from mammalian tissue. I then dropped serum containing these antibodies onto slides with soil amoebae. It worked! The dangerous amoebae clumped together; the benign ones were unaffected. Spurred on by this small success, I wanted to tackle a much larger question: How and why do populations of organisms diverge to the point of becoming different lineages? At the time, I thought that char, landlocked in fjords of known ages in Iceland, would be perfect to study this. Before I started my PhD, I had the opportunity to teach embryology for initially one year, 1960–1961, at UBC in Canada. Here I met Peter. It was thrilling to find we had similar interests and similar goals. At that time, Peter approached the same questions from an ecologist's viewpoint, while I approached them more from a geneticist's point of view and the synergism was electrifying. We were married in January 1962.

With two small children and in those days a lack of good day care facilities in Montreal, I stayed at home until they went to school. One day a week, on Mondays, I had a babysitter, and instead of spending that day catching up on household chores, I spent it in the McGill library catching up on research articles. This put me in a good position when I finally was able to return to full-time research, slowly via an intermediate step of high school teaching in Montreal.

Our two children always accompanied us while doing fieldwork on Darwin's finches in the Galápagos. We homeschooled them during the lunch hours. One on one is more effective and creative than a classroom, and they were always ahead of their class on return to school. Looking back neither remembers receiving 'lessons' on the islands. Now they are mother's themselves and both say that taking them into the field was the best thing we ever did.

We moved to the University of Michigan in Ann Arbor in 1978, and I returned to full-time research and worked on my PhD project on the Large Cactus Finch on Genovesa Island in the Galápagos. Staffan Ulfstrand at the University of Uppsala in Sweden was my supervisor. He was invaluable, allowing me flexibility in pursuing my own research interests and at the same time immensely stimulating with his provocative questions.

Thus, I entered a fully professional career through false starts (at high school) that did not put me off, with a hiatus (childrearing), which I enjoyed, and a husband (Peter) who gave me immense support when the time came to get a PhD and return to full-time research and teaching.

Research:

How do new species form? How do populations of organisms diverge and become different enough that they no longer interbreed, or do so rarely? On the Galápagos Islands, we tackled the problem of species formation from many different angles, examining the ecology, behaviour and genetics.

Climatic swings caused by El Nino Southern Oscillation phenomenon brought years of torrential rain interspersed with drought years, and it was these drought years when 80–90% of individuals died that allowed us to measure the strength of selection and evolutionary responses to this selection in the next generation.

An unexpected finding was rare genetic exchange between closely related species of similar body size. In Darwin's finches, the premating barrier is based on species-specific song learned from the father in association with parental morphology during a brief receptive period early in life. Being based on learning, this barrier is vulnerable to disruption. On rare occasions, a young bird learns the song of another species, as a result of nest take over, or death of the father, and this can lead to hybridization. Whether or not this hybrid survives depends on ecological conditions, and on whether there is appropriate food for birds of intermediate beak size. When suitable food is available, backcrossing to one or other parental species according to song type leads to genetic exchange between species. We were fortunate to witness an unprecedentedly severe El Nino in 1983 that changed the ecological conditions of the island to one that allowed the survival of hybrids. It led to genetic exchange between the medium ground finch (*Geospiza fortis*) and cactus finch (*G. scandens*) populations over the next 30 years. Although hybridization was never more than 1–2% of any breeding attempt, over time the genetic and phenotypic variation of both populations was increased to a measurable and noticeable extent. This is a situation that allows a rapid evolutionary response to natural selection events in a new environment.

The most unanticipated finding in our many years of research was the formation of a new lineage as a consequence of genetic exchange, which we followed from its inception over the next six generations. His descendants formed a population that was unique in song and morphology and became reproductively isolated from all other finch species on the island.

Having found that genetic exchange played a key role in forming a new lineage in contemporary time we were gratified to learn of genetic exchange throughout the whole of the Darwin's Finch radiation. This discovery, published this year, came from analyses of genome sequences by Leif Andersson and his group from Uppsala, using blood samples that we had collected. It suggests that the fuelling of genetic variation through hybridization has contributed not only to a new lineage in contemporary time but also to the formation of new phenotypes and the development of new species in the past.

Advice to women scientists today:

In the 1950s, two pieces of advice were given to me. First, never use your full name, only initials on examinations. Second, a male professor with daughters and no sons is often the most supportive! Today, I would say to all young scientists, follow your passion and the direction that most interests you. Value your exceptions, be open to alternative explanations and do not hunt by expectation. The road will not be smooth, but there will be magic in it.

Box 2: Personal reflections of Mary Jane West-Eberhard, Smithsonian Tropical Research Institute, Costa Rica.

Every scientist seems to have definite ideas about women in science based on personal impressions. Here, I express some of my own, knowing that real data would be better but unable to resist the opportunity to say what I think: that a woman scientist who wishes to have children and is conscientious about their care still faces special challenges if she also wishes to be competitive in her career and enjoy the excitement of a life in science – of creative discovery, of international travel and collaborations, and of teaching and writing that may be of lasting benefit to humanity.

In my own reflections on women in science, one not-very-surprising generalization leaps forth: the career trajectories of women are often different from those of men. The reason of course is that the child-rearing role of women and their shorter reproductive lifespan more often pull them away from an ascendant career. Sexual asymmetries in reproductive lifespan and parental investment will continue to affect humans, at least in the foreseeable future, along with sexual asymmetries in parental role: certain evolved morphological, physiological and psychological characteristics enhance the ability of most mothers to sense and respond to the needs of offspring, to the benefit of all concerned (offspring and both parents). Despite great variation in the parental commitments, inclinations and abilities of men and women alike, the evolved sex asymmetries are biological facts that demand special accommodation if a society aspires to take advantage of both the maternal and the scientific abilities of women. Fortunately, reproductive biology need not limit intellectual destiny if social arrangements resolutely accommodate both.

The *historical* result of the sex differences just mentioned is that employment policies reflect male life-history trajectories. So a man's career is automatically more easily harmonized with family life, whereas each woman scientist has to find some idiosyncratic pathway to having both a career and a family. A particular woman's pathway depends on individual variables, such as her career progress relative to her age, attitudes of family and teachers towards an unusual female role, whether or not she falls in love and when, or decides to have children and when, or comes from a culture that condones work by women outside of the home. This background of asymmetrical conditions and adjustments has not changed much through the four generations familiar to me personally – my grandmother's, my mother's, my own and my daughter's.

As a result of these fundamental biological and social asymmetries, women who have been successful in demanding fields such as science have always been some combination of unusually *determined, motivated, persistent, stubborn, rebellious, irrepressible or independent-minded* – qualities also heightened in professionally successful men but not always seen as virtues in women. One sign of progress is that women scientists now do not have to endure those adjectives so often used as euphemisms for *pushy, strident, aggressive, selfish, obnoxious, mannish, unbecoming and less mentionable words*. Early in my own career – in the 1960s – I felt that avoiding such epithets required a sufficiently feminine manner of behaviour and dress to mitigate the qualities of assertiveness and independence needed to compete successfully as a scientist, while at the same time avoiding stereotypical female behaviours that could lead to being dismissed as 'just a woman'. I had some rules. For example, do not keep your notes for a plenary lecture in a dainty ladies purse and pull them out at the podium (as I once saw a woman do). To get an idea of the image of femininity that permeated the atmosphere of my generation where I lived in the Midwestern USA, and more broadly including in the world of science, look at some romantic movies from the 1950s or TV comedies like 'George Burns and Gracie Allen' and 'I love Lucy' – in which female eccentricity and independent mindedness are portrayed as comical frivolity that can be harmonized with domesticity only by a forbearing husband.

It is interesting to ask particular women why they went into science despite the challenges of doing it. I once heard someone say that 'men branch out, but one woman leads to another'. If you ask a successful woman scientist what her mother did, you often get a revealing reply, and the replies have changed over time. Women of my grandmother's generation would often describe a mother with strong character and an independent mind, not often a career. For my mother's generation, the answer often referred to a mother's hobby, like a love of botany or natural history. For my generation – growing up in the 1950s (I was born in 1941) – the answer more commonly described a mother's career outside of the home, such as primary school teaching, or work as a nurse or a secretary, but still careers oriented towards children or support rather than leadership, and only very rarely a career in science or the professions that would require long training and long hours away from home, except as a collaborating wife. Still, the generational shift in the replies I have heard suggest progress for women in science. My own genealogy has a good dose of the right stuff for motivating generational change in the right direction: my mother was a schoolteacher with a master's degree in primary education and was the first of her family to attend college. Before her was my favourite grandmother, a tough and stubborn farm woman known for working and cussing like a man, whose own mother (my great-grandmother) was so independent-minded that, even though mild mannered and kind-hearted, she divorced an abusive husband, an act so rebellious and unusual for her time that it was never mentioned even within the family when I was small, except in hushed tones.

Despite obvious progress, especially in the increased collaboration of partners in child rearing and ample parental leave in a few countries, the central problem for women in science – still incompletely solved – is how to harmonize relationships and family life with a full-blown scientific career. This largely boils down to childcare. Even today woman scientists are obliged to adopt extreme solutions to this everyday human task. Some of them decide not to have children. Others postpone motherhood until they feel established in their careers or postpone their careers until they feel their children are sufficiently launched. Others have a family fortune or a wealthy husband that allows them to hire expert help; others spend most of what they earn to obtain it. Some decide that children are so important and enjoyable that they drop out of science and then find that they cannot readily get back in. Others, with unusual stamina and determination, become 'superwomen' who seem able to do everything at once. My own extreme solution was to leave the United States to live in Latin America. Before leaving, I cleverly married Bill Eberhard, a prolific evolutionary biologist with wide interests that overlap broadly with mine – in effect, a portable home biology department and my closest scientific colleague, who also loves kids (we have three). And then, I serendipitously got a wonderful job with the Smithsonian Tropical Research Institute which allowed doing research literally in our back yard, in countries (Colombia and Costa Rica) where we could afford a full-time helper for household chores and babysitting.

It should not be necessary for woman scientists to adopt any of these extreme solutions, or for their partners to be unusually committed to childcare. I hope that the situation for women will continue to change until there is no longer a need for exceptional arrangements for living a balanced life. No issue is more important now than working towards a true accommodation of the child-bearing and child-rearing biology of women and men, through, for instance, ample parental leave, temporarily part-time early-career appointments that still get full respect, policies that permit ample participation of supportive partners, reassignment of academic duties to allow more work at home and a smoothing of the on-ramp to a full life of science after the temporary off-ramp required by maternity and family life. Scandinavian-style free public day care is only a partial solution because even at its best it does not achieve the goal of allowing, and *encouraging*, parents to spend time with their children, including consideration of the long hours of overtime work currently required to sustain a career in science.

I realize that women face barriers not *seen* as child-rearing problems, such as differences in classroom treatment, hiring, salaries and promotions. But I believe that family demands are the root causes of the irregular career paths of women and of the generalized biases that can accumulate during a lifetime to reduce the likelihood of a woman being in the top echelons of science. So I believe that measures that help to harmonize pregnancy and child rearing with scientific careers are fundamental for the progress of women in science, whether they have children or not.

Box 3: Personal Reflections of Josephine Pemberton, University of Edinburgh, UK.

Raising children has not happened for me. Instead, I have gone through a scientific career at a pace where I feel very well rewarded and have spent the last 5 years as head of an institute containing about 32 principal investigators (i.e. faculty and independent fellows), as well as being heavily involved in running two long-term studies of wild mammals on Scottish islands (red deer on the Isle of Rum and Soay sheep on St. Kilda).

I was born in 1957 and brought up in Wimbledon, London. Rearing tadpoles through to frogs in a tank in my bedroom is one of my earliest memories, and there followed many pets including fish of all varieties, terrapins, tortoises, budgies, cockatiels and guinea pigs as well as the much-loved family dog. I was an early member of the Young Ornithologist's Club, the youth wing of the UK's largest bird conservation charity, the Royal Society for the Protection of Birds. I was animal-mad, although perhaps unusually, my equestrian phase was only brief. Wimbledon High School had a nature study prize, named after a previous pupil called Jennifer Crafter. My dismay on discovering that the main exercise set involved observing a plantain plant through the reproductive season was intense, but

I won it anyway. With five like-minded school friends, I formed the Wimbledon Society for the Protection of Nature (WSPN) which was an excuse for a weekly postschool tea party.

This early life was disrupted by my being sent to an all-girl boarding school for 6.5 years. This forced me along academically so well that I won a scholarship to read Zoology at Oxford for which I am profoundly grateful, but I suspect it also stunted my social development. At Oxford, I lived in college (a reflection of my lack of social development?). I worked hard and got stressed: who would not with a biochemistry tutor who reminded one that all previous holders of this particular scholarship got first class degrees? I did not, which I hope has made it easier for my successors! Summer vacation volunteering in Kenya got me hooked on large mammals, and a PhD opportunity in Reading pushed me in the direction of genetics. The principle finding of my thesis was that the fallow deer (an abundant introduced species in the UK) lacks genetic variation as studied by allozyme electrophoresis.

I had always been interested in the opportunities offered by individual-based studies, because of the potential to understand sources of variation in survival and breeding success. Since hearing about it as an undergraduate, I had been interested in joining the red deer study on Rum. Starting in 1972, the remarkable Fiona Guinness had lived very remotely on Rum, recognizing each individual deer in the study area from its facial and other features (we use tags and collars these days) and following their lives in extraordinary detail. Tim Clutton-Brock realized the potential of such data to answer many questions in ecology and behaviour, wrote many key papers and raised funds for continued data collection and analysis, the latter being mostly performed by Steve Albon. I was lucky to be funded to investigate genetic variation in the deer, at first using allozyme electrophoresis. Alec Jeffreys' discovery of the class of loci called minisatellites, which enabled DNA fingerprinting to identify parentage in natural populations, provided me with a key break, and ever since I have been engaged in reconstructing pedigrees and investigating sources of variation in traits, including fitness components, in both the red deer, and since the inception of the sister project in 1985, the Soay sheep. After a very happy nine years of postdoc work in London and Cambridge, I won a lectureship in Edinburgh and have been here ever since.

My colleagues and I in Edinburgh now run both the deer and Soay sheep projects, ably assisted by a wide range of collaborators at other universities. Like the study of Rosemary and Peter Grant in the Galapagos, these are long-term studies straddling the disciplines of ecology and evolution, in which the data sets become more and more valuable as complete individual life histories accumulate and can be interrogated to answer different questions. Indeed, we share our data with many people who suggest new data analyses that we do not plan to do. The major challenge with such projects is, of course, to keep them running in a funding landscape which is hugely competitive and where grants last for just three or at best 5 years. It is a constant exercise in scanning the horizon for ideas.

With regard to women in science, I have two reflections. First, as well as addressing the difficulties women in science face, I think we should be on a mission to identify what advantages there are to the progress of science to have more women in the system, especially at senior levels. In business, studies have found a correlation between having women on company boards and company profits (Luckerath-Rovers 2013). Whether this is a causal relationship and why it occurs are both unclear. Nevertheless, the internet is full of evidence of initiatives to increase representation of women on boards, and progress seems to be fast in some cases (e.g. Women on Boards: Davies Review Annual Report 2015). What about women in science? Strictly anecdotally, my impression is we are more predisposed to try and help with an identified problem; we treat students more sympathetically; we are more likely to be cautious about publishing an unexpected result; and we are perhaps more interested in a long game than in short-term glamour. I have not read much research on any of this, although there is some evidence that women are underrepresented in cases of scientific misconduct (Fang et al. 2013). Are these characteristics good for science? Clearly, a mix of talent is required, but they certainly make running a department much easier and my bet is that they are.

My other reflection is about promotion. The criteria for promotion in UK universities are strongly biased towards research prowess, for example rewarding high profile papers and grant acquisition more than teaching success (although emphasis is changing towards the latter). As discussed in Mary Jane West-Eberhard's piece, women traverse this process at widely different rates, depending on their investment in family life. This sets up a sex difference in the underlying cause of status variation. Men by and large hold positions reflecting their research prowess. Women hold positions reflecting their maternal status (I exaggerate somewhat, of course) because women engaged in childcare cannot simultaneously fulfil their full potential as researchers. It is surely hugely frustrating for women to see themselves being overtaken, and as a head of institute, I have not enjoyed seeing this happen. I do not necessarily think it would be fair to either men or childless women to change the promotion system (beyond incorporating far more reward for good teaching). Rather, we should continuously remind ourselves that having children is a career in itself. I am always incredibly impressed by anyone who holds down a scientific career, however, part time, at the same time as rearing children. With the passage of time, they will very likely win through on both careers, so they will have both a scientific and a genetic legacy, whereas my direct fitness will be zero!

automatically being given less credit than a man would have been.

'a male scientist with a well-developed network of collaborators is often judged to have excellent management skills, while a woman in the same position is more likely to get her independence questioned'. – Qvarnström et al. 2016.

Work done collaboratively 'was treated as if it 'didn't count' towards my record, even for projects that I had initiated' – Aktipis 2016

Box 4: Personal Reflections of Michelle Tseng, University of British Columbia, Canada. Founding Editor, Evolutionary Applications.

The idea for a journal on applied evolutionary biology started from casual conversations I had as a graduate student with friends and colleagues who were largely unfamiliar with the many practical uses of evolutionary biology. After a comprehensive survey of the literature, I found that applied evolutionary biology papers were often published in high profile journals and were well cited, but that there was no 'home' journal for these types of papers. After consultation with established academic editors Loren Rieseberg and the late Harry Smith, I pitched the idea to Blackwell Publishing (now Wiley), and the journal was born. Wiley has been very supportive of *Evolutionary Applications* from the start, and I am especially grateful to Liz Ferguson at Wiley for taking a chance on this idea and for entrusting in me (at that time a recently graduated Ph.D. student) the responsibility of building the editorial team and managing the day-to-day operations. We implemented double-blind reviewing from the start to help minimize biases of peer reviewers, and we consistently aimed to have a balanced gender ratio on the editorial board.

As an early-career female in academics and academic publishing, I have been fortunate to be have been surrounded by colleagues (women and men) who have been very supportive of women in academics. Perhaps, my department is an exception, but it seems to be the norm to have a family and to excel at your career. Undoubtedly, the climate has not always been like this, and I am indebted to the many people who have fought, and continue to fight, to bring equality to the workplace. I advocate that there is room for improvement in both gender and ethnic diversity in academics. We all need to be acutely aware that our subconscious or conscious stereotypes of certain genders or ethnicities bias our ability to objectively evaluate academic achievement.

In this minefield where criteria for success in science mismatch gender stereotypes, several contributors provide guidance. Many women describe the usefulness of a rebellious spirit that allows you to define yourself and to forge your own path, dismissing notions of what you should be as a woman (e.g. Charlesworth 2016; West-Eberhard in Box 2). Sometimes the road to success requires freeing yourself from those who limit you, so that you can follow your own passions. Carol Eunmi Lee, for example, describes the long and painful process of liberating herself from the limitations imposed upon her by others so that she could find and pursue her own passions and success (Lee 2016). Several authors emphasize the importance of building faith in yourself, particularly by surrounding yourself with supporters – friends and advisors – who validate your opinions, boost your confidence and expand your network (e.g. Aitken and Bemmels 2016; Johannesson

2016; Penczykowski et al. 2016; Wellenreuther and Sánchez-Guillén 2016).

The connection between work and family was discussed in the majority of reflections. Mary Jane West-Eberhard (Box 2) reminds us that there are some real differences that face women who seek to balance science with having children. Men do not bear, birth or breastfeed, and these and potentially other biological differences can create strong differences in family roles and commitments, both emotional and temporal. This generates, in her view, 'the central problem for women in science' of how to 'harmonize relationships and family life with a full-blown scientific career'. She writes of the many creative, if extreme, ways in which our colleagues have done so.

Several contributors with children expressed positive opinions about being both a parent and a scientist that suggest that we need to reframe the concept of 'work–life balance'. Often this phrase is used to emphasize an inherent (negative) trade-off between time spent doing science and time spent on other pursuits, including child rearing. But many authors did not see it this way. Women spoke of a different, more positive interpretation of this balance: the advantage of a life that is not overly focused and the benefits of stepping back occasionally from work to see the broader picture. Raising children can enforce this balance.

'I truly believe that my two children make me a better scientist. Motherhood has helped me recognize my priorities and manage my time. Also, there is no better way of decompressing after work than being with kids, as they demand 100% of your attention'. – Britt Koskella (2016)

'I believe that working fewer hours can actually benefit research, so long as the actual working hours allow time for thinking – the ideas then sometimes mature in the "non-working" hours'. – Deborah Charlesworth (2016)

The reverse point was also made that children benefit from exposure to their parents' work, especially when fieldwork accommodates children (Charmantier et al. 2016; Grant in Box 1; Gillespie 2016). The work–life balance should thus be framed as finding your personal optimum: the combination of work, sports, music, art, family and/or adventure that personally makes you content and that allows you to work to the best of your ability. It is also important to remember not to let life pass too fast. If you want to make sure that you reach old age having a particular experience, then do not repeatedly put off that experience for the sake of work. Decide when to do it and make it happen.

Many women also emphasized the many positives of being a scientist, particularly the freedom, stimulation and deep satisfaction that comes from discovery. The freedom is truly wonderful – we can choose what to study, where and when. Academia also allows more freedom than most jobs to alternate time at home and at work, as needed. Indeed, the ability to strike one's own balance is a major perk, as Victoria Sork notes: 'I cannot imagine a career that is more family-friendly than being a university professor'.

The life-long intellectual stimulation from students, collaborators and research discovery is another tremendous privilege. Johannesson (2016) urges us to remember 'that the most important driver in research is to have fun at work', and many contributors expressed how much they love what they do. The rewards of collaboration and working as a part of a team were also frequently raised.

> 'This is our experience of research, as female scientists at various stages of their career: we feel very lucky having worked in an environment where cooperation was always valued above competition, and where friendship was intricately mixed with high intellectual stimulation'. – Olivieri et al. (2016)

Ways forward

The contributions in this special issue also provide recommendations to improve the lives of scientists in ways that will encourage women to enter science, stay in science and help repair the leaky pipeline. A recurrent theme is the need for more supportive policies for academic parents, which can include parental leave, at-work day care, help with care for sick children, travel support for young children, day care at conferences, part-time early-career appointments and policies that smooth the return to science following parental leave. As emphasized by Susan Haig (2016), 'it is clear that as a profession, we have work to do to address lifestyle demands or we risk losing the best and brightest researchers and laboratory workers'.

Women benefit from learning from each other how to navigate family life and work–life. One way forward is to develop improved university-wide policies that pool together the best of the practices in various departments – learning also from forward-thinking policies at other Universities; this especially benefits women isolated in male-dominated departments. As an example of best practices, it has been suggested that reviewers of job and grant applications should use the 'academic', rather than chronological, age when it comes to the evaluation of applicants to avoid penalizing women who have taken parental leave (Lane 1999).

Many women emphasized the benefits of networking, to open doors to new opportunities as well as to build the circle of collaborators and advisors who will support you in your career. Several caution against the time demands of administration and service tasks. This issue is particularly acute for women, as Universities, granting agencies, and journals strive for gender balanced committees even though the source pool (especially among senior academics) remains imbalanced. This pull towards service can limit research time and research success, and so you should evaluate carefully requests for service (e.g. setting a fair time budget for such service and sticking to it).

Other advice describes how to lessen implicit biases. For example, Leftwich et al. (2016) suggest a greater reliance on computer-based citation searches to develop references for a paper, rather than relying solely on memory, which is more susceptible to bias. Strategies also exist to reduce bias in publishing itself, which have been described by Budden et al. (2008): Keeping authors of scientific papers anonymous has been shown to improve women's odds of acceptance. The practice can also block bias against minorities and bias in favour of authors and institutions with big reputations. Double-blind peer review, in which the identities of both authors and reviewers are hidden from one another, is already common in the social sciences and humanities but still rarely practiced in ecology or evolution journals. As highlighted by Tseng (Box 4), Evolutionary Applications has been a pioneer in the field by adopting a double-blind strategy since its foundation in 2008, with the aim to avoid our implicit biases and to ensure the fair assessment of research quality. Cognitive biases are so numerous and universal that, at the very least, we should make ourselves aware of how deep they can run. For faster change, we all need to act as exemplars – correcting our own mistakes, and monitoring biases around us. The stakes are higher than most of us realize. Biases in academia stifle the insight we could be gaining from a more diverse set of collaborators.

Indeed, Pemberton (Box 3) points out that the number of women on corporate boards has been shown to be positively correlated with business success (e.g. profits). She suggests that we investigate and document the ways in which progress in science is also advanced by having more diversity in sciences. Such evidence would provide further incentive to stop the leaky pipeline and promote gender equality.

In closing, this issue is an exciting opportunity to paint a new portrait of the changing face of evolutionary biology, as illustrated by our front cover. Women are altering the way we think about evolutionary biology, charting new research directions, smoothing the path for other women, and loving their jobs. Obstacles and hiccups remain, but there also are incredible opportunities for women entering the field. To young women considering a career in academia, find your passion in research, have faith in yourself (or construct it if needed), and strive to accomplish all that you can accomplish. As obscure as the path can seem, the journey is deeply satisfying.

'love what you do with a passion - and do what you love with equal passion' – Rosemary Gillespie (2016)

Acknowledgements

The idea for this special issue is to celebrate the achievements of women in evolutionary biology. We like to thank all authors that have contributed and shared their insights. MW thanks Louis Bernatchez for suggesting the idea of this special issue to her and for supporting it all the way. We also thank Wei Mun Chan and Alice Ellingham for assistance during the editorial process. MW would like to thank the network 'Women in Great Sciences' (WINGS) for providing a support platform for female scientists at Lund University in Sweden. SPO would like to thank Judy Myers for paving the way for women at UBC and thank her many supportive colleagues who have made it both fun and exhilarating to do science. Mary Jane West-Eberhard and Louis Bernatchez provided helpful comments on earlier drafts of this manuscript.

Literature cited

Aitken, S. N., and J. B. Bemmels 2016. Time to get moving: assisted gene flow of forest trees. Evolutionary Applications **9**:271–290.

Aktipis, A. 2016. Principles of cooperation across systems: from human sharing to multicellularity and cancer. Evolutionary Applications **9**:17–36.

Aubin-Horth, N. 2016. Using an integrative approach to investigate the evolution of behaviour. Evolutionary Applications **9**:166–180.

Barnett, R. C., and L. Sabattini 2009. A short history of women in science: from stone walls to invisible walls. In: T. Science, ed. The Science on Women and Science. Enterprise Institute, Washington, DC.

Budden, A. E., T. Tregenza, L. W. Aarssen, J. Koricheva, R. Leimu, and C. J. Lortie 2008. Double-blind review favours increased representation of female authors. Trends in Ecology & Evolution **23**:4–6.

Charlesworth, D. 2016. The status of supergenes in the 21st century: recombination suppression in Batesian mimicry and sex chromosomes and other complex adaptations. Evolutionary Applications **9**:74–90.

Charmantier, A., C. Doutrelant, G. Dubuc Messier, A. Fargevieille, and M. Szulkin 2016. Mediterranean blue tits as a case study of local adaptation. Evolutionary Applications **9**:135–152.

Cho, A. H., S. A. Johnson, C. E. Schuman, J. M. Adler, O. Gonzalez, S. J. Graves, J. R. Huebner et al. 2014. Women are underrepresented on the editorial boards of journals in environmental biology and natural resource management. PeerJ **2**:e542.

Clancy, K. B. H., R. G. Nelson, J. N. Rutherford, and K. Hinde 2014. Survey of Academic Field Experiences (SAFE): trainees report harassment and assault. PLoS One **9**:e102172.

Conley, D., and J. Stadmark 2012. Gender matters: a call to commission more women writers. Nature **488**:590.

Council of Canadian Academies. 2012. Strengthening Canada's research capacity: the gender dimension.

Darwin, C. R. 1871. The Descent of Man, and Selection in Relation to Sex, 1st edn. Murray, London.

England, P., and S. Li 2006. Desegregation stalled the changing gender composition of college majors, 1971–2002. Gender & Society **20**:657–677.

Etzkowitz, H., C. Kemelgor, and B. Uzzi 2000. Athena Unbound: The Advancement of Women in Science and Technology, 1 edn. Cambridge University Press, Cambridge.

Fang, F. C., J. W. Bennett, and A. Casadevall 2013. Males are overrepresented among life science researchers committing scientific misconduct. MBio **4**:e00640–12.

Fara, P. 2015. Women, science and suffrage in World War I. Notes and Records of the Royal Society **69**:11–24.

Gartler, S. M. 2006. The chromosome number in humans: a brief history. Nature Reviews Genetics **7**:655–660.

Gautier, M., and P. S. Harper 2009. Fiftieth anniversary of trisomy 21: returning to a discovery. Human Genetics **126**:317–324.

Gillespie, R. G. 2016. Island time and the interplay between ecology & evolution in species diversification. Evolutionary Applications **9**:53–73.

Glazer, A. M. 2015. There ain't nothing like a Dame: a commentary on Lonsdale (1947)'Divergent beam X-ray photography of crystals'. Philosophical Transactions of the Royal Society of London A: Mathematical, Physical and Engineering Sciences **373**:20140232.

Haig, S. M., R. M. Bellinger, H. M. Draheim, D. M. Mercer, M. P. Miller, and T. D. Mullins 2016. An integrative approach to single species conservation genetics over the past 35 years. Evolutionary Applications **9**:181–195.

Harper, P. S. 2006. The discovery of the human chromosome number in Lund, 1955–1956. Human Genetics **119**:226–232.

Harvey, J. 2012. The mystery of the nobel laureate and his vanishing wife. In For Better or For Worse? Collaborative Couples in the Sciences Volume 44 of the series Science Networks. Historical Studies, chapter 4, pp. 57–77 Springer, Basel.

Horowitz, H. L. 1986. The 1960s and the transformation of campus cultures. History of Education Quarterly **26**:1–38.

Johannesson, K. 2016. What can be learnt from a snail? Evolutionary Applications **9**:153–165.

Jones, C. G., and S. Hawkins. 2014. Women and science. Notes and Records of the Royal Society:rsnr20140056.

Jones, T. M., K. V. Fanson, R. Lanfear, M. R. E. Symonds, and M. Higgie 2014. Gender differences in conference presentations: a consequence of self-selection? PeerJ **2**:e627.

Lane, N. J. 1999. Why are there so few women in science. Nature debates **19**:1999.

Latu, I. M., M. S. Mast, J. Lammers, and D. Bombari 2013. Successful female leaders empower women's behavior in leadership tasks. Journal of Experimental Social Psychology **49**:444–448.

Lederberg, E. M. 1950. Lysogenicity in *Escherichia* coli strain K-12. Microbial Genetics Bulletin **1**:5–9.

Lee. 2016. Evolutionary mechanisms of habitat invasions, using the copepod *Eurytemora affinis* as a model system. Evolutionary Applications **9**:248–270.

Leftwich, P. T., M. Bolton, and T. Chapman. 2016. Evolutionary biology and genetic techniques for insect control. Evolutionary Applications **9**:212–230.

Lejeune, J., M. Gauthier, and R. Turpin. 1959. Les chromosomes humains en culture de tissus: Gauthier-Villars/Editions Elsevier 23 RUE LINOIS, 75015 PARIS, FRANCE.

Leslie, S.-J., A. Cimpian, M. Meyer, and E. Freeland 2015. Expectations of brilliance underlie gender distributions across academic disciplines. Science 347:262–265.

Luckerath-Rovers, M. 2013. Women on boards and firm performance. Journal of Management and Governance 17:491–509.

Mankkinen, T. 1999. Walking the academic tightrope: sexual harassment at the University of Helsinki. In P. Fogelberg, ed. Hard Work in the Academy: Research and Interventions on Gender Inequalities in Higher Education, pp. 219–221. Helsinki University Press, Helsinki.

Mason, J. 1991. Hertha Ayrton (1854–1923) and the admission of women to the Royal Society of London. Notes and Records of the Royal Society of London 45:201–220.

Moss-Racusin, C. A., J. F. Dovidio, V. L. Brescoll, M. J. Graham, and J. Handelsman 2012. Science faculty's subtle gender biases favor male students. Proceedings of the National Academy of Sciences 109:16474–16479.

Myers, J. H., and J. S. Cory 2016. Ecology and evolution of pathogens in natural populations of Lepidoptera. Evolutionary Applications 9:231–247.

Namrata, G., C. Kemelgor, S. Fuchs, and H. Etzkowitz 2005. Triple burden on women in science: a cross-cultural analysis. Current Science 89.

NSF. 2013. Women, minorities, and persons with disabilities in science and engineering. In Special Report NSF 13-304: NSF Arlington, VA.

Ohta, T. 2012. Tomoko Ohta. Current Biology 22:R618–R619.

Olivieri, I., J. Tonnabel, O. Ronce, and A. Mignot 2016. Why evolution matters for species conservation: perspectives from three case studies of plant metapopulations. Evolutionary Applications 9:196–211.

Orr, M. 2014. Women peers in the scientific realm: Sarah Bowdich (Lee)'s expert collaborations with Georges Cuvier, 1825–33. Notes and Records of the Royal Society of London.

Pain, E. 2014. After more than 50 years, a dispute over Down Syndrome discovery. Science 343:720–721.

Penczykowski, R. M., A.-L. Laine, and B. Koskella 2016. Understanding the ecology and evolution of host–parasite interactions across scales. Evolutionary Applications 9:37–52.

Qvarnström, A., M. Ålund, S. Eryn McFarlane, and P. M. Sirkiä 2016. Climate adaptation and speciation: particular focus on reproductive barriers in Ficedula flycatchers. Evolutionary Applications 9:119–134.

Reuben, E., P. Sapienza, and L. Zingales 2014. How stereotypes impair women's careers in science. Proceedings of the National Academy of Sciences 111:4403–4408.

Richards, E. 1997. Redrawing the boundaries: darwinian science and Victorian women intellectuals. Victorian science in context 6:119–142.

Schiebinger, L. 1993. Women in science: historical perspectives. Paper read at Women at work: a meeting on the status of women in astronomy held at the Space Telescope Science Institute, September 8–9, 1992.

Schroeder, J., H. L. Dugdale, R. Radersma, M. Hinsch, D. M. Buehler, J. Saul, L. Porter et al. 2013. Fewer invited talks by women in evolutionary biology symposia. Journal of Evolutionary Biology 26:2063–2069.

Selden, W. K. 2000. Women of Princeton: 1746–1969. Princeton University, Princeton, NJ.

Servedio, M. R. 2016. Geography, assortative mating, and the effects of sexual selection on speciation with gene flow. Evolutionary Applications 9:91–102.

Shen, H. 2013. Mind the gender gap. Nature 495:22–24.

Sonnert, G., and G. James Holton. 1995. Who Succeeds in Science?: The Gender Dimension: Rutgers University Press, New Brunswick, NJ.

Sork, V. L. 2016. Gene flow and natural selection shape spatial patterns of genes in tree populations: a career perspective. Evolutionary Applications 9:291–310.

Ulrich, L. T. 2004. Yards and Gates: Gender in Harvard and Radcliffe History. New York: Palgrave Macmillan.

Wellenreuther, M., and R. Sánchez-Guillén 2016. Non-adaptive radiation in damselflies. Evolutionary Applications 9:103–118.

Wennerås, C., and A. Wold 1997. Nepotism and sexism in peer review. Nature 387:341–343.

West, J. D., J. Jacquet, M. M. King, S. J. Correll, and C. T. Bergstrom 2013. The role of gender in scholarly authorship. PLoS One 8:e66212.

Wobbe, T. 2002. Die longue durée von Frauen in der Wissenschaft: orte, Organisationen, Anerkennung. In Frauen in Akademie und Wissenschaft: Arbeitsorte und Forschungspraktiken 1700–2000, vol 10. pp. 1–28. Forschungsberichte der interdisziplinären Arbeitsgruppen der Berlin-Brandenburgischen Akademie der Wissenschaften, Berlin.

Women on Boards: Davies Review Annual Report. 2015. https://www.gov.uk/government/publications/women-on-boards-2015-fourth-annual-review (accessed on 17 October 2015).

Zuk, M., and G. Rosenqvist 2005. Evaluation bias hits women who aren't twice as good. Nature 438:559.

13

Evolution of age and length at maturation of Alaskan salmon under size-selective harvest

Neala W. Kendall,[1,†] Ulf Dieckmann,[2] Mikko Heino,[2,3,4] André E. Punt[1] and Thomas P. Quinn[1]

1 School of Aquatic and Fishery Sciences, University of Washington, Seattle, WA, USA
2 International Institute of Applied Systems Analysis, Laxenburg, Austria
3 Department of Biology, University of Bergen, Bergen, Norway
4 Institute of Marine Research, Bergen, Norway
† Present address: Washington Department of Fish and Wildlife, Olympia, WA, USA

Keywords

fishery selection, harvest-induced evolution, Oncorhynchus nerka, phenotypic plasticity, probabilistic maturation reaction norms

Correspondence

Neala W. Kendall, School of Aquatic and Fishery Sciences, University of Washington, Box 355020, Seattle, WA 98195 USA.

e-mail: neala.kendall@dfw.wa.gov

Abstract

Spatial and temporal trends and variation in life-history traits, including age and length at maturation, can be influenced by environmental and anthropogenic processes, including size-selective exploitation. Spawning adults in many wild Alaskan sockeye salmon populations have become shorter at a given age over the past half-century, but their age composition has not changed. These fish have been exploited by a gillnet fishery since the late 1800s that has tended to remove the larger fish. Using a rare, long-term dataset, we estimated probabilistic maturation reaction norms (PMRNs) for males and females in nine populations in two basins and correlated these changes with fishery size selection and intensity to determine whether such selection contributed to microevolutionary changes in maturation length. PMRN midpoints decreased in six of nine populations for both sexes, consistent with the harvest. These results support the hypothesis that environmental changes in the ocean (likely from competition) combined with adaptive microevolution (decreased PMRNs) have produced the observed life-history patterns. PMRNs did not decrease in all populations, and we documented differences in magnitude and consistency of size selection and exploitation rates among populations. Incorporating evolutionary considerations and tracking further changes in life-history traits can support continued sustainable exploitation and productivity in these and other exploited natural resources.

Introduction

Age and size at maturation help to determine an individual's reproductive success and thus its fitness and are also important in the dynamics of populations (Stearns 1992). Locally adapted populations reproducing in different habitats exhibit different patterns of age and length at maturation (e.g., fishes: Beacham 1983; Quinn et al. 2001). Life-history traits, including age and length, at maturation can change rapidly in exploited populations because they are exposed to novel regimes of mortality (Darimont et al. 2009), and exploitation often selectively removes larger individuals (Coltman et al. 2003; Carlson et al. 2007; Kendall and Quinn 2012). The nature of hunting and fishing suggests that this force may lead to microevolutionary trait changes (Policansky 1993; Law 2000; Allendorf et al. 2008;

Allendorf and Hard 2009). Such changes take longer to reverse than those associated with phenotypic plasticity alone (Law and Grey 1989).

Few studies have quantified patterns of harvest selection and compare these with associated trait changes over time, especially for multiple stocks of the same species that are differentially harvested. Thus, scientists and managers often lack the ability to correlate harvest with trait changes, understand the mechanisms for these trait changes, and evaluate if and how to modify harvest practices associated with the trait changes. These objectives were accomplished in our study, wherein we estimated average length at age at maturity, age at maturation, and probabilistic maturation reaction norms (PMRNs) of nine heavily exploited sockeye salmon (*Oncorhynchus nerka* Walbaum) populations from two different Alaskan lake systems over five decades. We then compared size-selective fishing mortality with the PMRNs. These results can help scientists and managers address gaps in our understanding and management of harvest-induced selection and evolution and how life-history diversity can be maintained across populations.

PMRNs can help to understand population dynamics and sustainable management by estimating changes in length at age at maturation (Heino et al. 2002a,b; Heino and Dieckmann 2008). PMRNs help to disentangle, to some degree (Heino and Dieckmann 2008; Morita et al. 2009; Uusi-Heikkilä et al. 2011), phenotypic plasticity of life-history traits caused by environmental changes affecting growth and mortality from microevolutionary trait changes associated with size-selective fishing (Olsen et al. 2004; Mollet et al. 2007).

Scientists and managers have recognized that age and length at maturation in Pacific salmon (*Oncorhynchus* sp.) have changed over the past half-century (Ricker 1981; Bigler et al. 1996). These traits are heritable (Carlson and Seamons 2008), and studies demonstrated selection by some fisheries against large size (e.g., Kendall and Quinn 2012). Genetic changes due to size-selective fishing could cause changes in exploited populations (Ricker 1981; Fukuwaka and Morita 2008), but size is influenced by an intricate combination of genetic and environmental factors (Pyper and Peterman 1999).

Sockeye salmon of Bristol Bay, Alaska (see supporting information Fig. S1) are ideal for studying long-term changes in age and length at maturation and shifts in PMRNs as possible microevolutionary changes associated with size-selective fishing. There are large and phenotypically diverse sockeye salmon runs, no stocking from hatcheries, breeding and feeding environments largely unaltered by humans (Hilborn et al. 2003), and size-selective commercial gillnet fisheries that have operated for over 100 years (Kendall et al. 2009; Kendall and Quinn 2012). Sockeye salmon spawn in diverse habitats, and age and length at maturation vary consistently among populations, so they differ in vulnerability to size-selective fishing (Kendall and Quinn 2009).

In this study, we hypothesized that changes in the sockeye salmon life-history traits and PMRN midpoints would be correlated with fishery selectivity patterns, specifically that fish would become shorter at a given age and PMRN midpoints decrease (greater probability of maturing at a shorter length at age) under higher fishing pressure that removes larger than average fish. This would suggest that fisheries-induced evolution is consistent with changes in PMRNs. While a number of other studies have also estimated PMRNs for harvested fish stocks (e.g., Olsen et al. 2004; Mollet et al. 2007; Fukuwaka and Morita 2008; Pardoe et al. 2009) and many have found trends toward maturation at younger ages and/or smaller sizes, our study is rare in two respects.

First, we estimate changes in PMRNs for multiple stocks of the same species. This is important because length and age at maturation and PMRNs may vary among populations (or stocks), due to local adaptation (Taylor 1991) via population-specific selection pressures on the spawning grounds. Because of these differences and variation in fishery selection, PMRNs may evolve in different ways for different populations. Examining these differences can shed light on how selection can act across populations with differing traits, which can inform managers about the potential for life-history evolution in natural populations. Such analyses can also help to understand whether and how life-history diversity can be maintained, supporting the portfolio effect, whereby a diverse 'portfolio' of populations and traits among populations increases long-term stability (Schindler et al. 2010), among harvested populations. However, few studies have been able to perform such analyses due to the lack of data or difficulty in differentiating stocks or populations.

Second, our study compares trends in PMRNs with fisheries selection and intensity patterns for the multiple sockeye salmon populations. Again, this is rare given the inability of many studies to accurately estimate size-selective fishing patterns. Such comparisons allow scientists and managers to more confidently associate changes in PMRNs to selective fishing. Previous work by Sharpe and Hendry (2009) related PMRN changes to fishery exploitation rates for multiple fish species. Changes were strongly correlated with fishing intensity, supporting the finding that fishing can play an important role in life-history changes and that such changes may have a genetic basis.

Methods

Study site

We studied sockeye salmon populations in two lake systems of Bristol Bay, Alaska. Returning Iliamna Lake sockeye salmon are fished in the Naknek-Kvichak district, whereas Wood River lakes sockeye salmon are fished in the Nushagak district (Fig. S1). Both fisheries have used gillnets since the late 1800s. Fishery size selection has varied over time, but in most years (93% of years for males and 91% for females for the Naknek-Kvichak fishery; 62% of years for males and 89% for females in the Nushagak fishery) since 1963 fish longer than average have been caught, leaving shorter fish to breed (Kendall et al. 2009; Kendall and Quinn 2012).

Data have been collected on two spatial scales since the early 1960s. On a larger scale (all populations together within a fishery), the total catch and escapement (i.e., fish that escape the fishery and can spawn) are estimated, and age, sex, and length (ASL) data have been collected on individual fish for both fisheries by the Alaska Department of Fish and Game as detailed by Kendall and Quinn (2012). On a finer scale (population specific), ASL data have been collected on the Iliamna Lake spawning grounds in most years since the early 1960s and in the Wood River lakes spawning grounds from 1960–1965 and 1990–2009 by the University of Washington Alaska Salmon Program. In general, 110 males and females from each population were sampled, measured for length, and otoliths were collected to age the fish. We analyzed data on five populations from Iliamna Lake and four from the Wood River lakes with the most complete records, spanning the range of spawning sites and fish body sizes and ages (Fig. S1 and Table S1; Quinn et al. 2001). These spawning sites ranged from small streams (<4 m wide) to larger rivers (>75 m wide) with a range of depths and also included beaches. Fish body size and age are correlated with spawning site type, width, and depth (Quinn et al. 2001), with shorter and younger fish spawning in smaller and shallower streams, longer and older fish spawning in larger and deeper rivers, and beach spawners spanning a wider range of sizes and ages.

Analyses

We first estimated the average length at ocean ages 2 and 3 years of males and females and the proportion of fish of ocean ages 2 and 3 (age composition) in each population over time. We examined temporal differences using linear models. Second, we calculated population-specific PMRNs for ocean age 2 Iliamna Lake sockeye salmon from 41 cohorts since 1960 of sexes and ocean age classes. From the Wood River lakes populations, we estimated ocean age 2 PMRNs over 14 cohorts from 1958–1962 and 1994–2004. This age was chosen because the necessary data were most abundant.

For PMRN estimation, the number and length distribution of immature fish must be compared with those of mature fish at a given age and in a given cohort (Heino et al. 2002a,b). However, length at age distributions of immature salmon are unknown because the fish are only measured at maturity. Therefore, we reconstructed the immature fish length distributions based on those of mature fish following methods used previously (Heino et al. 2002a,b).

Length reconstruction was completed separately for each population and cohort. We back-projected the lengths of ocean age 3 years mature fish measured on the spawning grounds 1 year before they matured, thus estimating immature lengths after fish had spent 2 years in the ocean. Salmon marine growth is not linear; Burgner (1991) reported that the increase in body length was convex over time and that length increased most during

the first year at sea. Thus, we back-calculated immature lengths 1 year before maturation (l') for fish in a given cohort (c) that matured at a given ocean age (a; 3 years in this case).

$$l'_{c,a} = l_{c,a} - \frac{(l_{c,a} - h_c)}{a} * f_{a,y-1} \qquad (1)$$

Here, h_c is the average smolt length in a particular cohort for fish leaving Iliamna Lake or the Wood River lakes of a given ocean age (Crawford et al. 1992; Crawford and Fair 2003); $l_{c,a}$ is the mature fish length by cohort and ocean age; and $f_{a,y-1}$ is a growth factor specific to ocean age and represents the proportion of growth associated with the year prior to maturation (from age 2 to 3 years). Growth factors were estimated empirically by Ruggerone et al. (2005) and represent the percent of growth during each year of marine residence (Table S2). Each of the alternate growth factors (Table S2; Lander and Tanonaka 1964; Lander et al. 1966; French et al. 1976) was used in all years in the sensitivity analyses and we also modeled PMRN midpoints using different marine growth factors at different points in the time series. Specifically, we mimicked either a long-term increase in first-year growth conditions (starting with growth factor 1 in 1958–1970, and ending with growth factor 4 in 1991–2004), or a long-term decline in first-year growth conditions (the reverse). It is very difficult to know how temperature trends, changes in fish density in the ocean, and other environmental factors would affect whether sockeye salmon grow more in their first year in the ocean versus in their second and/or third years, and further research is needed on this topic. In these analyses, $l'_{c,a}$ values were re-estimated using each cohort- and age-specific growth factor and used to calculate PMRNs midpoints as described in detail below.

We projected the number of immature fish 1 year before they matured by adjusting the number of mature fish to account for natural mortality, high seas fishing (Myers et al. 1993), and terminal area fishing (Kendall et al. 2009; Kendall and Quinn 2012). Annual cohort-specific offshore mortality rates ($M_c yr^{-1}$) were estimated as a combination of mortality due to high seas fishing and natural mortality associated with the last year that salmon were in the ocean (between ocean age 2 and 3 years). Furnell and Brett (1986) modeled marine growth and mortality of sockeye salmon and estimated that 90% of the natural mortality at sea occurs in the first 4 months in the ocean. On this basis, we estimated that 10% of the smolts from a given cohort that die in the ocean do so between the end of their first year at sea and their return to spawn following their second or third year in the ocean (Fig. S2).

For Iliamna Lake, but not the Wood River lakes, data on the total number of smolts outmigrating per cohort (S_c) were available in many (but not all) years between 1961 and 1998 (e.g., Crawford et al. 1992; Crawford and Fair 2003). Using these data and the total number of adults returning to spawn for each cohort by ocean age ($A_{c,a}$, a = 2 or 3 years), we first estimated that:

$$0.9 = \frac{S_c(1 - x_c)}{S_c - A_{c,2} - A_{c,3}}, \qquad (2)$$

Where X_c is the cohort-specific survival in their first year in the ocean. Equation 2 can be re-written as:

$$X_c = \frac{0.1S_c - 0.9(A_{c,2} + A_{c,3})}{S_c}. \qquad (3)$$

Therefore, the number of age 1 year immature sockeye salmon ($N_{c,1}$) alive in the ocean was:

$$N_{c,1} = S_c X_c. \qquad (4)$$

We then estimated both the cohort-specific number of number of sockeye salmon alive after their second year in the ocean ($N_{c,2}$; before a fraction of them matured, so immature and mature combined) by multiplying $N_{c,1}$ by a cohort-specific annual survival rate (Y_c) from year 1 to 2 and 2 to 3 in the ocean. We assumed a constant Y_c after age 1 year based on Ricker's (1976) finding of similar values for sockeye salmon in their penultimate and ultimate years in the ocean.

$$N_{c,2} = N_{c,1} Y_c. \qquad (5)$$

Because 10% of the mortality occurs after the first year in the ocean:

$$0.1 = \frac{N_{c,1}(1 - Y_c) + (N_{c,2} - A_{c,2})(1 - Y_c)}{S_c - A_{c,2} - A_{c,3}}. \qquad (6)$$

Substituting equation 4 for $N_{c,1}$ and equation 5 for $N_{c,2}$ gives:

$$0.1 = \frac{S_c X_c(1 - Y_c) + (S_c X_{c,1} Y_c - A_{c,2})(1 - Y_c)}{S_c - A_{c,2} - A_{c,3}}. \qquad (7)$$

For each cohort, we solved this equation for Y_c and then estimated annual cohort-specific instantaneous offshore mortality rates for Iliamna Lake sockeye salmon as:

$$M_c yr^{-1} = -ln(Y_c). \qquad (8)$$

The values for M_c ranged from 0.01 to 1.565 year^{-1} with an average value of 0.588 year^{-1}. The large range is likely

due to uncertainty in the smolt count estimates; smolt counts are known to be difficult to quantify. In years, for which total smolt counts were not available, we used the average instantaneous M_c year^{-1} values estimated in adjacent years where data were available. Annual smolt counts were not available for any years for the Wood River lakes, and thus, we used the average M_c estimated for Iliamna Lake sockeye salmon (0.588 year^{-1}). A range of offshore instantaneous mortality rates for sockeye salmon that were deemed unbiased and specific to the ultimate year of life in the ocean (0.1 and 0.3 year^{-1}; Ricker 1976) along with larger values (0.5 and 0.8 year^{-1}) were used in the sensitivity analyses. For this analysis, M_c yr^{-1} values were re-estimated using each offshore instantaneous mortality rate and then used to calculate PMRNs midpoints as described in detail below.

Inshore fishing mortality rates by the Naknek-Kvichak and Nushagak fishing districts were calculated as the proportion of fish caught (u) per year (y), by 10 mm length bins (l), as in Kendall et al. (2009). These proportions, ranging from 0.005 to 0.98, were calculated for all fish of both sexes.

We then estimated the number of immature individuals (i) by length (l'), ocean age (a; 2 in this case), sex (s), cohort (c), and spawning population (p) one year (y) prior to their maturation (for sockeye salmon maturing at ocean age 3) using the number of mature fish (n) that were measured on the spawning grounds using equation:

$$i_{l',a-1,c,y-1,s,p} = \frac{n_{l,a,c,y,s,p}}{(1 - u_{l,y})e^{-(M_c yr^{-1})}}. \qquad (9)$$

PMRNs were calculated for sex-cohort-population groupings of ocean age 2 years that had ten or more length at age data points available for both mature and immature fish (so that small samples sizes would not skew the results). The probability of a fish maturing (o) was calculated from the individual mature and immature fish data using logistic regression with a binomial error distribution, as maturation is a binary response variable (Heino et al. 2002a,b). We used the generalized linear model (GLM) framework in the program R (R Development Core Team 2011). Different GLMs were used to fit o (Table 1) based on length, population, cohort, and sex (e.g., Equation 10).

$$logit(o) \sim \beta_0 + \beta_1 l + \beta_p + \beta_c + \beta_s \qquad (10)$$

We also fitted models that included two-way interactions between the predictor variables. We selected the best models using calculated AIC$_c$ values. Utilizing the best model, we calculated the length at which the probability of maturing was 50% (L_{P50}) to illustrate the midpoint of the PMRN:

Table 1. Models used to predict maturation of Iliamna Lake and Wood River lakes sockeye salmon, and thus estimate PMRNs, along with their ΔAICc values (the difference between each model's AICc value and that of the model with the lowest value).

Variables in model	Iliamna Lake		Wood River Lakes	
	# parameters	ΔAICc	# parameters	ΔAICc
Length + cohort	43	2554	15	943
Length * cohort	84	2397	28	901
Length + sex	3	3912	3	5552
Length * sex	4	3589	4	5339
Length + population	6	4097	5	3565
Length * population	10	4031	8	3564
Length + cohort + sex	44	1436	16	477
Length + cohort * sex	85	1350	29	478
Length * cohort + sex	85	1317	29	442
Length + population + sex	7	2741	6	3270
Length + population * sex	11	2660	9	3214
Length * population + sex	11	2681	9	3270
Length * sex + population	8	2482	7	3153
Length * sex + cohort	45	1184	17	437
Length + population + cohort	47	1690	18	517
Length + population * cohort	211	102	57	405
Length * population + cohort	51	1647	21	456
Length * cohort + population	88	1544	31	455
Length * cohort + sex	85	1317	29	442
Length + population + cohort + sex	48	640	19	129
Length + population * cohort + sex	212	236	58	53
Length + population + cohort * sex	89	558	32	125
Length * population + cohort + sex	52	581	22	114
Length * population + cohort * sex	93	479	35	106
Length + cohort + population * sex	52	596	22	113
Length * cohort + population + sex	89	539	32	85
Length * cohort + population * sex	93	486	35	79
Length * sex + cohort + population	49	412	20	78
Length * sex + cohort * population	213	0	59	0

$$L_{P50_{a,c,s,p}} = -\frac{\beta_0 + \beta_c + \beta_s + \beta_p}{\beta_l} \quad (11)$$

Temporal variation in L_{P50} values was evaluated by assessing the significance of the coefficients for the cohort terms in the GLMs and by regressing the predicted L_{P50} values against cohort. We also examined the significance of the population and sex terms in the GLMs to evaluate differences in L_{P50} values among populations and between males and females.

We estimated the uncertainty associated with the L_{P50} values by bootstrapping the original data. For each cohort and population, by sex, and ocean age, we sampled the length data 1000 times with replacement, used these data to recalculate the immature lengths and counts, fitted the 'best' model to the generated proportions-by-length and predicted L_{P50} values.

We calculated two metrics describing the fishing mortality and size selectivity experienced by fish in each population over time using methods detailed in Kendall and Quinn (2009): 1) each population's annual exploitation rate ($V_{p,y}$) and 2) population-specific length-based standardized selection differentials by year ($SSD_{p,y}$). Population-specific exploitation rates and SSDs in most years for the Wood River system populations were previous presented in Kendall and Quinn (2009), but those for the Iliamna Lake populations have not been previously estimated. We examined the relationships between each population's L_{P50} trends and population-specific fishing exploitation rates and $SSDs$.

Results

The average lengths at ocean ages 2 and 3 years of male and female sockeye salmon spawning in tributaries of Iliamna Lake and the Wood River lakes have decreased over time (Fig. S3). The slopes of average length of fish for all ocean age-sex-population groups over time were negative, and significantly so for 10 of the 20 Iliamna Lake groups (linear models, $P < 0.01$ required by Šidák correction for multiple comparisons, $F = 9.8$–80.3 for significant groups) and all 16 of the Wood River lakes groups (linear models, $P < 0.01$, $F = 25.4$–228.4). Linear regression slopes of age compositions over time for the various populations included a range of negative and positive values, and no statistically significant trends in age composition were detected for fish of either sex in any Iliamna (linear models, $P > 0.01$, $F = 0.0$–3.9) or Wood River lakes population (linear models, $P > 0.01$, $F = 0.0$–6.4).

The GLMs indicated that length, population, cohort, and sex all affected maturation (Table 1). The GLM P-values associated with many cohorts were <0.05 for the best-fit model and the AIC_c value of a model not including the cohort term was much larger than the model including it (Table 1), emphasizing variation in PMRNs over time. Linear regression models showed that L_{P50} values for ocean age 2 fish decreased over time for males and females in all populations. These decreases were statistically significant for both males and females in two of five Iliamna Lake populations (linear models, slope $P < 0.01$, $F = 9.0-11.6$ for significant populations) and for males and females in all four Wood River lakes populations (linear models, slope $P < 0.01$, $F = 31.5-188.7$). Most populations in the best-fit GLM also had corresponding P-values <0.05, suggesting significant differences in L_{P50} values among them, and that of sex was <0.001 for both Iliamna and Wood River lakes, signifying that this term was very important to understand L_{P50} differences and thus that males and females had different L_{P50} values. AIC_c values of GLM models not including the population or sex term were much larger than models with them (Table 1), emphasizing variation among populations and that female PMRNs were significantly different than those of males. Iliamna Lake L_{P50} values (determined by linear regression; details above) decreased by 0.1–0.4 mm per cohort for females (Fig. 1A) and 0.2–0.7 mm per cohort for males (Fig. 1B) between the 1960 and 2004 cohorts. Wood River lakes L_{P50} values declined even more, by 0.8–1.3 mm per cohort for females (Fig. 1C) and 1.1–1.7 mm per cohort for males (Fig. 1D) between the 1958 and 2004 cohorts.

L_{P50} values changed little across the range of ocean mortality rates used in the sensitivity analysis (Table S3). Even

when the ocean mortality rate of 0.8 year^{-1} was used, there were no differences in the overall conclusions. L_{P50} values did depend on marine growth factors, decreasing with time for all populations and growth factor combinations by 4.7–45.4 mm for Iliamna Lake populations and 5.4–45.6 mm for Wood River lakes populations. Using growth factor combination 4 (Table S2), we found differences in the statistical significance of L_{P50} declines for Iliamna Lake populations but not for Wood River lakes populations. Specifically, for Iliamna Lake fish L_{P50} values still declined for all populations but were significant for only one sex-populations group, whereas originally this was seen for four groups. When growth factor 1 was used in the early years and growth factor 4 in the later years, the decline in the L_{P50} values was not statistically significant for any population-sex group, but when growth factor 4 was used in the early years and growth factor 1 in the later years, a statistically significant (sharp) decline in L_{P50} values was seen for all population-sex groups. Thus, the overall patterns of decline were, for most but not all ocean growth models, robust to growth rate variation.

Estimated average standardized selection differentials (SSDs) were negative for all Iliamna (one-sided t-test, $P < 0.0001$ for each, $t = -9.5$ to -6.7; Fig. S4 and Table S4) and Wood River lakes populations (one-sided t-test, $P < 0.0001$, $t = -5.1$ to -4.5). Additionally, linear regressions showed that exploitation rates increased over time for all populations (though only the increase for Gibraltar Creek females was statistically significant at the 0.01 level

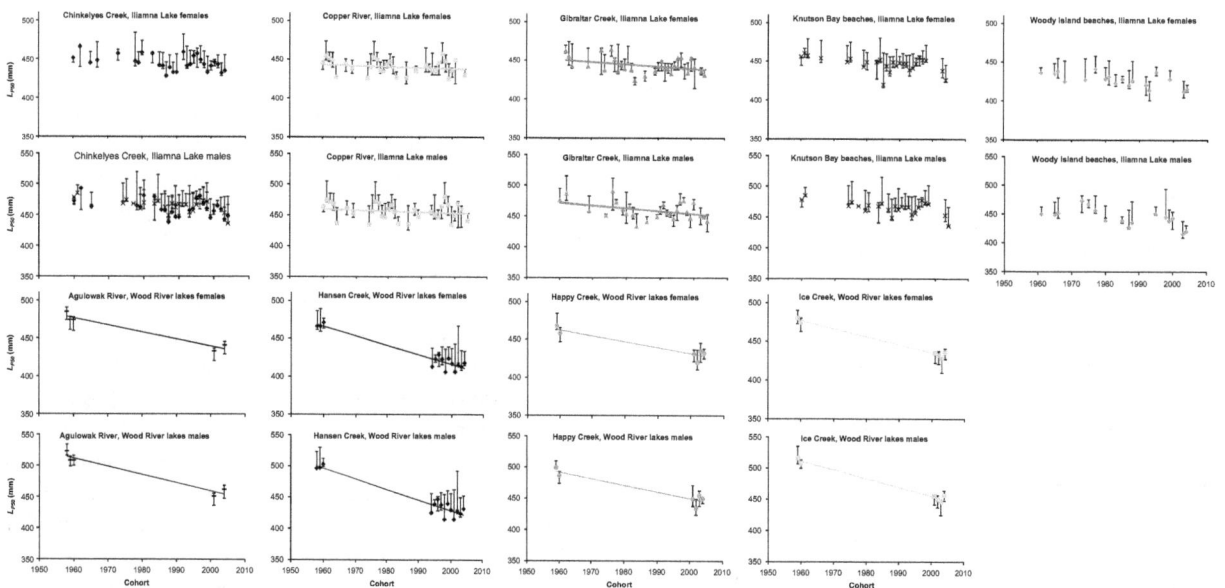

Figure 1 Ocean age 2 sockeye salmon L_{P50} values for females and males of Iliamna Lake and Wood River lakes populations. Error bars are 95% CIs estimated from bootstrap analysis. Best-fit lines are for the populations where L_{P50} values decreased significantly over time (males and females from Copper River and Gibraltar Creek in Iliamna Lake and males and females from all populations in the Wood River lakes).

for Iliamna population-sex groups [$P = 0.005$, $F = 9.2$], while the exploitation rate increases for half of the population-sex groups were significant for the Wood River lakes [$P < 0.01$, $F = 18–59.8$]). Both of these findings were consistent with the overall decreases in the PMRNs across all populations. Exploitation rates were higher for Wood River lakes fish (average = 0.61 for all populations and cohorts vs. Iliamna Lake average = 0.45; t-test, $P < 0.00001$, $t = 7.5$; Fig. S4 and Table S4) and $SSDs$ were more negative for Wood River lakes fish (average = −0.26 vs. Iliamna Lake average = −0.14; t-test, $P = 0.00003$, $t = −4.4$). Consistent with these differences, L_{P50} values decreased more for Wood River populations than for Iliamna populations.

The Iliamna Lake populations did not vary significantly in estimated exploitation rates or $SSDs$ and thus, not surprisingly, these features were not linked to differences in PMRNs. For the Wood River lakes populations, though, size-selective fishing may have influenced changes in PMRNs over time more than overall exploitation. Specifically, we found larger declines in L_{P50} values for sockeye salmon from Hansen Creek (average of 1.3 mm decline per cohort for females and 1.7 mm for males) than Agulowak River (average of 0.9 mm decline per cohort for females and 1.3 mm for males; Fig. 1). Accordingly, since the early 1960s, $SSDs$ were more frequently negative and greater in magnitude for Hansen Creek and other shorter-bodied populations than populations with longer fish (e.g., Agulowak River and Ice Creek; Kendall and Quinn 2009).

Discussion

All 36 sockeye salmon age-sex-population groups from Iliamna Lake and the Wood River lakes have become shorter at maturity, 26 (72%) statistically significantly so, since the early 1960s. The significant decreases in length at age ranged from 22 to 37 mm for Iliamna Lake populations and 62–106 mm for Wood River lakes populations. However, the proportions maturing after three versus 2 years at sea have not changed significantly in these populations. Morita et al. (2005) concluded that decreases in Pacific salmon length at maturation, concurrent with increases in age, may be adaptive, plastic responses to reduced growth rate. However, our results suggested that factors besides growth, most likely size-selective fishing, have contributed to trait shifts in Bristol Bay sockeye salmon.

We applied the PMRN methodology for the first time to multiple spawning populations of an exploited salmonid stock, estimating PMRNs for sockeye salmon populations using five decades of data from mature fish to back-calculate the number and length of immature fish. L_{P50} values declined over time for all populations and decreased significantly in 4 of 10 Iliamna Lake sex-population groups and in all 8 Wood River lakes groups. These reductions in L_{P50}

values indicated that the declines in length at age were not only related to changes in growth or mortality over the decades.

Bristol Bay sockeye salmon have experienced heavy but variable size-selective gillnet fisheries since the late 1800s (Kendall et al. 2009; Kendall and Quinn 2012) with significantly negative $SSDs$ for all populations (i.e., fish longer than average have been removed, leaving shorter individuals to spawn). The decreases in L_{P50} values over time are consistent with this size-selective fishing. Our results support the work of Bromaghin et al. (2011), whose individual based model indicated that age and length of harvested western Alaska Chinook salmon were likely to decline with continued harvest.

Decreases in L_{P50} values were not significant for all Bristol Bay sockeye salmon spawning populations, and this may be due to the large variation in fishery selection over time (Kendall and Quinn 2012). Fewer Iliamna Lake populations, with lower exploitation rates and less size selectivity, showed significant changes in PMRNs than Wood River lakes fish. Overall, our findings support the hypothesis that observed declines in length at maturation of fish of a given age were microevolutionary responses to size-selective exploitation and thus represent fisheries-induced evolution. However, significant changes in PMRNs in some populations were not detected, perhaps due to variation in the size-selective exploitation, the exploitation being less size selective, or lower exploitation rates.

In contrast to the length at age and L_{P50} patterns, we did not find significant changes in age composition in Iliamna and Wood River lakes sockeye salmon. Decreases in the PMRNs suggest that if growing conditions (related to food availability from production or competition, temperature, or other factors) had remained the same over time, we would have seen the fish maturing at younger ages in recent years. Because such shifts in the age structure have not been realized, concurrent changes in growth and in the PMRN could account for the observed patterns. With overall slower growth in the ocean (Seo et al. 2011; Zavolokin et al. 2012), fewer sockeye salmon would have reached the (lower) PMRNs at younger ages, and thus, age composition did not change. Additional factors may have affected age and length at age at maturation such as environmental conditions including freshwater and sea-surface temperatures (Pyper and Peterman 1999), density of salmon at sea including hatchery fish (Bigler et al. 1996; Pyper and Peterman 1999), and changes in species distributions (Hinch et al. 1995).

The sensitivity analysis showed that our PMRN findings were generally insensitive to the marine mortality rate, but PMRNs varied with the different marine growth factors used to estimate immature length distributions. However, even with the most extreme growth factors applied for each cohort and applying growth factors that varied over time, the slopes for PMRNs still decreased, indicating that the overall conclusions are robust to this factor. Further research to understand Bristol Bay sockeye salmon marine growth, and variation in growth over time, could clarify the L_{P50} trends. Past studies of Pacific salmon PMRNs directly estimated fish length at certain time periods, and thus growth, by measuring annual growth rings on salmon scales (Morita et al. 2005; Fukuwaka and Morita 2008). This was simply not possible in our study due to the number of fish included and because historical scales were not available for measurement. Thus, we used different ocean growth factors to simulate a variety of growth patterns during a fish's marine residence and also varied these growth factors over time to understand how temporal trends in growth conditions, affecting the growth factors, could impact our results. Uncertainty in our estimates of immature fish growth and our inability to model how growth may have changed over time and reflect such changes in our growth factors (specifically how growth rate variation is reflected in the proportion of growth that a fish experienced during its first, second, and third year in the ocean) are limitations in this study.

The degree to which shifts in PMRNs can indicate microevolution remains somewhat uncertain; the methodology has been criticized for not disentangling genetic and environmental effects on maturation other than through length at age (Kraak 2007; Uusi-Heikkilä et al. 2011), and environmental factors can affect PMRNs directly, not just through growth (Morita et al. 2009). For example, temperature can also directly influence maturation, with increasing temperatures being linked to decreasing age and size at maturation (Tobin and Wright 2011). However, offshore waters of the North Pacific Ocean, in which Bristol Bay sockeye salmon reside during their maturation decision period, is one of the few places where temperatures have decreased slightly since the 1950s (Cane et al. 1997; Mantua 2009), inconsistent with the decreases in size at maturation observed for these fish. PMRNs are not a perfect tool but can help track changes in life-history traits and understand the contribution of harvest to microevolutionary changes, and in this case, the conclusion is broadly supported by the data. The trends we observed are unlikely to have resulted only from a progressive shift in environmental conditions because of the spatial heterogeneity and complex temporal variation in ocean conditions affecting salmon growth and survival over the past decades.

Additionally, recent research has found that differences in growth in Chinook salmon in New Zealand under selection in novel environmental conditions can drive evolutionary changes in life-history traits such as age at maturation rather than evolution of the maturation thresholds defining PMRNs (Kinnison et al. 2011). Such evolu-

tionary forcing is not considered by the PMRN approach because PMRNs tease apart life-history trait changes in maturation correlated with changes in growth (often assumed to be phenotypic plasticity) from those unrelated with growth changes, potentially caused by size-selective fishing (Olsen et al. 2004; Mollet et al. 2007). Thus, we must consider that differences in growth rate over time or among populations could also influence evolutionary age and size at maturation trends in Iliamna and Wood River lakes sockeye salmon populations.

Thus, for these fish, changes in growth have likely interacted with size-selective fishing pressures and resulted in the maturation schedules and age and length compositions seen on the spawning grounds. Both phenotypic plasticity, resulting from changing environmental conditions, and adaptive evolution, due to size-selective fishing and environmental and other forces, can contribute to life-history trait changes (Fukuwaka and Morita 2008), and our study is consistent with the interaction of these effects in shaping age and length at maturation.

Our work supports the findings of Sharpe and Hendry (2009) and points to the importance of considering fisheries-induced evolution as an important mechanism affecting life-history traits in exploited species. Fishery managers should be aware of genetic changes associated with size-selective harvest (Allendorf et al. 2008; Allendorf and Hard 2009) and might use data on changes in age and length at maturation to adjust fishing strategies. For example, managers could reduce exploitation rates or change gear regulations to reduce selectivity (Kendall et al. 2009; Garcia et al. 2012; Kendall and Quinn 2012). Overall, Bristol Bay sockeye salmon stocks are quite healthy (Hilborn et al. 2003; Schindler et al. 2010), but managers should be aware that microevolutionary changes in life-history traits may make these populations less able to respond to future environmental or management changes. Reversing trends toward shorter lengths at age may be difficult, while removing the selective pressure on larger fish may slow or stop the changes in maturation length, selection toward the original genotype in the absence of fishing may be weaker than selection caused by intensive fishing (Fukuwaka and Morita 2008; Enberg et al. 2009).

Acknowledgements

We gratefully acknowledge the Alaska Sustainable Salmon Fund, the School of Aquatic and Fishery Sciences at the University of Washington, the Gordon and Betty Moore Foundation, the National Science Foundation (Biocomplexity in the Environment, and Dynamics of Coupled Natural and Human Systems programs), the International Institute for Applied Systems Analysis, and the National Academy of Sciences for funding this research. Alaska

Department of Fish and Game (especially Fred West) provided access to long-term data. Spawning ground data were
collected and organized by many University of Washington
Fisheries Research Institute and Alaska Salmon Program
members, supervised by Donald Rogers, Ole Mathisen, Ray
Hilborn, Daniel Schindler, Thomas Quinn, Chris Boatright,
Jackie Carter, and Harry Rich, Jr. Helpful comments and
discussion were provided by Jeff Hard, Ray Hilborn, David
Policansky, Davnah Urbach, Andrew Hendry, and two
anonymous reviewers.

Literature cited

Allendorf, F. W., and J. J. Hard 2009. Human-induced evolution caused
    by unnatural selection through harvest of wild animals. Proceedings
    of the National Academy of Sciences of the United States of America
    106:9987–9994.

Allendorf, F. W., P. R. England, G. Luikart, P. A. Ritchie, and N. Ryman
    2008. Genetic effects of harvest on wild animal populations. Trends in
    Ecology and Evolution **23**:327–336.

Beacham, T. D. 1983. Variability in median size and age at sexual matu-
    rity of Atlantic cod, *Gadus morhua*, on the Scotian Shelf in the north-
    west Atlantic Ocean. Fishery Bulletin **81**:303–321.

Bigler, B. S., D. W. Welch, and J. H. Helle 1996. A review of size trends
    among North Pacific salmon (*Oncorhynchus* spp.). Canadian Journal
    of Fisheries and Aquatic Sciences **53**:455–465.

Bromaghin, J. F., R. M. Nielson, and J. J. Hard 2011. A model of Chi-
    nook salmon population dynamic incorporating size-selective exploi-
    tation and inheritance of polygenic correlated traits. Natural Resource
    Modeling **24**:1–47.

Burgner, R. L. 1991. Life history of sockeye salmon (*Oncorhynchus ner-
    ka*). In: C. Groot, and L. Margolis, eds. Pacific Salmon Life Histories.
    UBC Press, Vancouver.

Cane, M. A., A. C. Clement, A. Kaplan, Y. Kushnir, D. Pozdnyakov, R.
    Seager, S. E. Zebiak et al. 1997. Twentieth-century sea surface temper-
    ature trends. Science **275**:957–960.

Carlson, S. M., and T. R. Seamons 2008. A review of quantitative genetic
    components of fitness in salmonids: implications for adaptation to
    future change. Evolutionary Applications **1**:222–238.

Carlson, S. M., E. Edeline, L. A. Vøllestad, T. O. Haugen, I. J. Winfield, J.
    M. Fletcher, J. B. James et al. 2007. Four decades of opposing natural
    and human-induced artificial selection acting on Windermere pike
    (*Esox lucius*). Ecology Letters **10**:512–521.

Coltman, D. W., P. O'Donoghue, J. T. Jorgenson, J. T. Hogg, C. Stro-
    beck, and M. Festa-Blanchet 2003. Undesirable evolutionary conse-
    quences of trophy hunting. Nature **426**:655–658.

Crawford, D. L., and L. F. Fair 2003. Bristol Bay Sockeye Salmon
    Smolt Studies Using Upward-Looking Sonar, 2002. Alaska Depart-
    ment of Fish and Game, Division of Commercial Fisheries,
    Anchorage, AK.

Crawford, D. L., J. D. Woolington, and B. A. Cross 1992. Bristol Bay
    Sockeye Salmon Smolt Studies for 1990. Alaska Department of Fish
    and Game, Division of Commercial Fisheries, Juneau, AK.

Darimont, C. T., S. M. Carlson, M. T. Kinnison, P. C. Paquet, T. E. Re-
    imchen, and C. C. Wilmers 2009. Human predators outpace other
    agents of trait change in the wild. Proceedings of the National Acad-
    emy of Sciences of the United States of America **106**:952–954.

Enberg, K., C. Jørgensen, E. S. Dunlop, M. Heino, and U. Dieckmann

2009. Implications of fisheries-induced evolution for stock rebuilding
    and recovery. Evolutionary Applications **2**:394–414.

French, R., H. T. Bilton, M. Osako, and A. Hartt. 1976. Distribution and
    origin of sockeye salmon (Oncorhynchus nerka) in offshore waters of
    the North Pacific Ocean. International North Pacific Fisheries Com-
    mission Bulletin **34**:1–113.

Fukuwaka, M., and K. Morita 2008. Increase in maturation size after the
    closure of a high seas gillnet fishery on hatchery-reared chum salmon
    Oncorhynchus keta. Evolutionary Applications **1**:376–387.

Furnell, D. J., and J. R. Brett 1986. Model of monthly marine growth and
    natural mortality for Babine Lake sockeye salmon (*Oncorhynchus ner-
    ka*). Canadian Journal of Fisheries and Aquatic Sciences **43**:999–1004.

Garcia, S. M., J. Kolding, J. Rice, M.-J. Rochet, S. Zhou, T. Arimoto, J. E.
    Beyer, L. Borges, A. Bundy, D. Dunn, E. A. Fulton, M. Hall, M. Heino,
    R. Law, A. D. Rijnsdorp, F. Simard, and A. D. M. Smith 2012. Recon-
    sidering the consequences of selective fisheries. Science **335**:1045–1047.

Heino, M., and U. Dieckmann 2008. Detecting fisheries-induced life-his-
    tory evolution: an overview of the reaction norm approach. Bulletin
    of Marine Science **83**:69–93.

Heino, M., U. Dieckmann, and O. R. Godø 2002a. Estimating reaction
    norms for age and size at maturation with reconstructed immature
    size distributions: a new technique illustrated by application to North-
    east Arctic cod. ICES Journal of Marine Science **59**:562–575.

Heino, M., U. Dieckmann, and O. R. Godø 2002b. Measuring probabi-
    listic reaction norms for age and size at maturation. Evolution
    56:669–678.

Hilborn, R., T. P. Quinn, D. E. Schindler, and D. E. Rogers 2003.
    Biocomplexity and fisheries sustainability. Proceedings of the National
    Academy of Sciences of the United States of America **100**:6564–6568.

Hinch, S. G., M. C. Healey, R. E. Diewert, K. A. Thomson, R. Hourston,
    M. A. Henderson, and F. Juanes 1995. Potential effects of climate
    change on marine growth and survival of Fraser River sockeye salmon.
    Canadian Journal of Fisheries and Aquatic Sciences **52**:2651–2659.

Kendall, N. W., and T. P. Quinn 2009. Effects of population-specific var-
    iation in age and length on fishery selection and exploitation rates of
    sockeye salmon. Canadian Journal of Fisheries and Aquatic Sciences
    66:896–908.

Kendall, N. W., and T. P. Quinn 2012. Comparative size-selectivity
    among Alaskan sockeye salmon fisheries. Ecological Applications
    22:804–816.

Kendall, N. W., J. J. Hard, and T. P. Quinn 2009. Quantifying six
    decades of fishery selection for size and age at maturity in sockeye
    salmon. Evolutionary Applications **2**:523–536.

Kinnison, M. T., T. P. Quinn, and M. J. Unwin 2011. Correlated
    contemporary evolution of life history traits in New Zealand Chinook
    salmon, *Oncorhynchus tshawytscha*. Heredity **106**:448–459.

Kraak, S. B. M. 2007. Does the probabilistic maturation reaction norm
    approach disentangle phenotypic plasticity from genetic change? Mar-
    ine Ecology Progress Series **335**:295–300.

Lander, A. H., and G. K. Tanonaka. 1964. Marine growth of western
    Alaskan sockeye slamon (*Oncorhynchus nerka*). International North
    Pacific Fisheries Commission Bulletin **14**:1–31.

Lander, A. H., G. K. Tanonaka, K. N. Thorson, and T. A. Dark. 1966.
    Ocean mortality and growth. International North Pacific Fisheries
    Commission Annual Report **1964**:105–111.

Law, R. 2000. Fishing, selection, and phenotypic evolution. ICES Journal
    of Marine Science **57**:659–668.

Law, R., and D. R. Grey 1989. Evolution of yields from populations with
    age-specific cropping. Evolutionary Ecology **3**:343–359.

Mantua, N. J. 2009. Patterns of change in climate and Pacific salmon production. American Fisheries Society Symposium **70**:1143–1157.

Mollet, F. M., S. B. M. Kraak, and A. D. Rijnsdorp 2007. Fisheries-induced evolutionary changes in maturation reaction norms in North Sea sole *Solea solea*. Marine Ecology Progress Series **351**:189–199.

Morita, K., S. H. Morita, M. Fukuwaka, and H. Matsuda 2005. Rule of age and size at maturity of chum salmon (*Oncorhynchus keta*): implications of recent trends among *Oncorhynchus* spp. Canadian Journal of Fisheries and Aquatic Sciences **62**:2752–2759.

Morita, K., J. Tsuboi, and T. Nagasawa 2009. Plasticity in probabilistic reaction norms for maturation in a salmonid fish. Biology Letters **5**:628–631.

Myers, K. W., C. K. Harris, Y. Ishida, L. Margolis, and M. Ogura 1993. Review of the Japanese landbased driftnet salmon fishery in the western North Pacific Ocean and the continent of origin of salmonids in this area. International North Pacific Fisheries Commission Bulletin **52**:1–78.

Olsen, E. M., M. Heino, G. R. Lilly, M. J. Morgan, J. Brattey, B. Ernande, and U. Dieckmann 2004. Maturation trends indicative of rapid evolution preceded the collapse of northern cod. Nature **428**:932–935.

Pardoe, H., A. Vainikka, G. Thordarson, G. Marteinsdottir, and M. Heino 2009. Temporal trends in probabilistic maturation reaction norms and growth of Atlantic cod (*Gadus morhua*) on the Icelandic shelf. Canadian Journal of Fisheries and Aquatic Sciences **66**:1719–1733.

Policansky, D. 1993. Fishing as a cause of evolution in fishes. In: T. K. Stokes, J. M. McGlade, and R. Law, *eds.* The Exploitation of Evolving Resources. Springer-Verlag, Berlin.

Pyper, B. J., and R. M. Peterman 1999. Relationship among adult body length, abundance, and ocean temperature for British Columbia and Alaska sockeye salmon (*Oncorhynchus nerka*), 1967–1997. Canadian Journal of Fisheries and Aquatic Sciences **56**:1716–1720.

Quinn, T. P., L. Wetzel, S. Bishop, K. Overberg, and D. E. Rogers 2001. Influence of breeding habitat on bear predation and age at maturity and sexual dimorphism of sockeye salmon populations. Canadian Journal of Zoology **79**:1782–1793.

R Development Core Team. 2011. R: A Language and Environment for Statistical Computing. R Foundation for Statistical Computing, Vienna, Austria. ISBN 3-900051-07-0. Available from http://www.R-project.org.

Ricker, W. E. 1976. Review of the rate of growth and mortality of Pacific salmon in salt water, and noncatch mortality caused by fishing. Journal of the Fisheries Research Board of Canada **33**:1483–1524.

Ricker, W. E. 1981. Changes in the average size and average age of Pacific salmon. Canadian Journal of Fisheries and Aquatic Sciences **38**:1636–1656.

Ruggerone, G. T., E. Farley, J. L. Nielsen, and P. Hagen 2005. Seasonal marine growth of Bristol Bay sockeye salmon (*Oncorhynchus nerka*) in relation to competition with Asian pink salmon (*O. gorbuscha*) and the 1977 ocean regime shift. Fishery Bulletin **103**:355–370.

Schindler, D. E., R. Hilborn, B. Chasco, C. P. Boatright, T. P. Quinn, L. A. Rogers, and M. S. Webster 2010. Population diversity and the portfolio effect in an exploited species. Nature **465**:609–612.

Seo, H., H. Kudo, and M. Kaeriyama 2011. Long-term climate-related changes in somatic growth and population dynamics of Hokkaido chum salmon. Environmental Biology of Fishes **90**:131–142.

Sharpe, D. M. T., and A. P. Hendry 2009. Life history change in commercially exploited fish stocks: an analysis of trends across studies. Evolutionary Applications **2**:260–275.

Stearns, S. C. 1992. The Evolution of Life Histories. Oxford University Press, Oxford.

Taylor, E. B. 1991. A review of local adaptation in Salmonidae, with par-ticular reference to Pacific and Atlantic salmon. Aquaculture **98**:185–207.

Tobin, D., and P. J. Wright 2011. Temperature effects on female maturation in a temperate marine fish. Journal of Experimental Marine Biology and Ecology **403**:9–13.

Uusi-Heikkilä, S., A. Kuparinen, C. Wolter, T. Meinelt, A. C. O'Toole, and R. Arlinghaus 2011. Experimental assessment of the probabilistic maturation reaction norm: condition matters. Proceedings of the Royal Society Biological Sciences Series B **278**:709–717.

Zavolokin, A. V., V. V. Kulik, I. I. Glebov, E. N. Dubovets, and Y. N. Khokhlov 2012. Dynamics of body size, age, and annual growth rate of Anadyr chum salmon *Oncorhynchus keta* in 1962–2010. Journal of Ichthyology **52**:207–225.

14

Rapid evolution of increased vulnerability to an insecticide at the expansion front in a poleward-moving damselfly

Khuong Van Dinh,[1,2] Lizanne Janssens,[2] Lieven Therry,[2] Hajnalka A. Gyulavári,[2] Lieven Bervoets[3] and Robby Stoks[1]

1 Institute of Aquaculture, Nha Trang University, Nha Trang, Vietnam
2 Laboratory of Aquatic Ecology, Evolution and Conservation, University of Leuven, Leuven, Belgium
3 Systemic, Physiological and Ecotoxicological Research Group, University of Antwerp, Antwerp, Belgium

Keywords
agriculture, carryover effects, energy storage, evolutionary ecotoxicology, flight muscles, latitude, pyrethroids, range expansion.

Correspondence
Khuong Van Dinh, Institute of Aquaculture, Nha Trang University, Nha Trang, Vietnam.

e-mail: khuongaquatic@gmail.com

Abstract

Many species are too slow to track their poleward-moving climate niche under global warming. Pesticide exposure may contribute to this by reducing population growth and impairing flight ability. Moreover, edge populations at the moving range front may be more vulnerable to pesticides because of the rapid evolution of traits to enhance their rate of spread that shunt energy away from detoxification and repair. We exposed replicated edge and core populations of the poleward-moving damselfly *Coenagrion scitulum* to the pesticide esfenvalerate at low and high densities. Exposure to esfenvalerate had strong negative effects on survival, growth rate, and development time in the larval stage and negatively affected flight-related adult traits (mass at emergence, flight muscle mass, and fat content) across metamorphosis. Pesticide effects did not differ between edge and core populations, except that at the high concentration the pesticide-induced mortality was 17% stronger in edge populations. Pesticide exposure may therefore slow down the range expansion by lowering population growth rates, especially because edge populations suffered a higher mortality, and by negatively affecting dispersal ability by impairing flight-related traits. These results emphasize the need for direct conservation efforts toward leading-edge populations for facilitating future range shifts under global warming.

Introduction

Global warming is causing widespread poleward range expansions where species try to keep pace with their moving climate niche (Hickling et al. 2006; Chen et al. 2011). Under ongoing and more intense global warming, range-expanding species are expected to continue to move more poleward to track their optimal thermal niche (Hickling et al. 2006; Chen et al. 2011). There is large variation in the rates at which different species' geographic ranges expand in response to climate warming (Moritz and Agudo 2013; Mair et al. 2014), yet only part of the variation in these rates can be explained by species differences in intrinsic dispersal abilities (Angert et al. 2011; Fordham et al. 2013). Understanding factors shaping the speed of range expansion is timely as there is increasing concern that many species are too slow to track their moving climate niche (Razgour et al. 2013).

Pesticide exposure may be one notable factor that may affect range expansion as individuals have to cross-agricultural landscapes with extensive use of pesticides. Moreover, the frequency of pesticide application is likely to increase under global warming, particularly at higher latitudes (Kattwinkel et al. 2011) where many edge populations are migrating to (Hickling et al. 2006). How species will deal with pesticides under global warming is becoming a major topic in ecotoxicology (Noyes et al. 2009; Moe et al. 2013), yet the expected interplay of range expansions and contaminants on organisms has been ignored. The vulnerability of edge populations at the moving range front to pesticides may slow down the range expansion in two ways. Firstly, pesticide exposure may impair the locomotory performance of animals by negatively affecting energy storage (e.g., Janssens et al. 2014) and muscles (e.g., Mehlhorn et al. 1999). Secondly, pesticides may reduce population growth rates by reducing larval growth rate and imposing mortality, thereby slowing down further range expansion.

During range expansions, edge populations may show rapid evolution as they experience novel evolutionary pressures because edge populations are assorted by dispersal ability and have a lower density of conspecifics than do core populations (Phillips et al. 2010). This rapid evolution entails a broad range of traits, including morphology, physiology, and behavior that are selected toward values that increase the rate of spread (Phillips 2009; Burton et al. 2010; Phillips et al. 2010; Shine et al. 2011; Brown et al. 2015). For example, edge populations at moving range fronts typically evolve a faster life history, a higher investment in reproduction (Phillips 2009; Phillips et al. 2010), higher activity levels (Therry et al. 2014b), an increased investment in locomotory ability (Hill et al. 2011) and a higher investment in immune function to avoid a reduction in dispersal rates (Therry et al. 2014c). Note that these evolutionary changes are not driven by adaptation to the range edge or any new biotic conditions met, but are driven by the dynamic process of range expansion itself. Therefore, these effects are only to be expected in edge populations at moving range fronts and not in edge populations at stable range fronts. These evolutionary changes require a higher allocation of energy toward growth and development and costly structures (such as muscles) and functions (such as immune function). Given that these investments in costly traits to accelerate range expansion will imply trade-offs with other costly processes such as investment in detoxification and repair (Sibly and Calow 1989; Congdon et al. 2001), it is to be expected that edge populations at moving range fronts will be more vulnerable to stressors such as pesticides. This novel hypothesis needs explicit testing and would provide an extra dimension to the insight that ecotoxicology needs a macroecological (Beketov and Liess 2012; Clements et al. 2012) and evolutionary (Coutellec and Barata 2011; Hammond et al. 2012) perspective.

We tested for the potential role of pesticide exposure in slowing down range expansion and whether evolutionary processes during range expansion increase the vulnerability to pesticides. We studied this in a currently poleward-moving damselfly by comparing replicated core and edge populations at low and high densities in an outdoor container experiment. Damselflies are among the taxa showing the strongest poleward range expansions (Hickling et al. 2006). They have a complex life cycle with an aquatic larval stage where growth occurs and a terrestrial flying adult stage where reproduction and dispersal occur (Stoks and Cordoba-Aguilar 2012). We included a density treatment as densities are initially lower at the expansion front while pesticide effects may be apparent or stronger at high densities (e.g., Jones et al. 2011; Knillmann et al. 2012). We tested for effects on larval survival, growth, and develop-

ment and for potential carryover effects bridging metamorphosis on a set of flight-related traits (body mass, relative flight muscle mass, and fat content), that may be especially relevant for dispersal ability. As study species we chose the poleward range-expanding damselfly *Coenagrion scitulum* (Swaegers et al. 2013). We have previously shown that this species evolved a faster life history (Therry et al. 2014b) and a higher investment in flight muscles and immune response (Therry et al. 2014c) at the range front. As pesticide we used esfenvalerate, a widely applied pyrethroid insecticide (Spurlock and Lee 2008; Stehle and Schulz 2015) that is highly toxic to aquatic invertebrates (Rasmussen et al. 2013), including damselfly larvae (Beketov 2004).

Materials and methods

Study populations and rearing experiment

Coenagrion scitulum is a Mediterranean damselfly preferring small ponds (Dijkstra 2006). Up to the 1990s the northern range limit was situated in northern France, after which a north-eastward range expansion has occurred (Swaegers et al. 2013). In 2010, the northern-most limit of the expanding range margin was situated in the southern parts of the Netherlands, and the northeastern limit in Western Germany. We studied two core populations and two edge populations. The two core populations, both in France, were situated in Nord-Pas-de-Calais (+50°26′34.37″N, +1°35′08.81″E) and Indre (+46°43′14.03″N, +1°10′20.22″E). Both core populations are within the historical distribution of the species (Therry et al. 2014c). Note that the Nord-Pas-de-Calais population is situated at the edge of the historical range as it is bordering the Atlantic Ocean, making our setup conservative as we only hypothesize a higher vulnerability to pesticides in edge populations at moving range fronts. The two edge populations were situated in Saarland (Germany, +49°14′52.96″N, +7°16′20.08″E) and in Zeeland (the Netherlands, +51°21′25.99″N, +3°40′01.37″E), both at the moving range front (further on we just call them 'edge populations'). The distances between populations are ca. 420 km between Nord-Pas-de-Calais and Indre, ca. 420 km between Nord-Pas-de-Calais and Saarland, ca. 180 between Nord-Pas-de-Calais and Zeeland, ca. 350 km between Saarland and Zeeland, and ca. 550 km between Indre and Saarland and between Indre and Zeeland. Despite the relatively small spatial scale, the edge populations are clearly differentiated from each other and from the core populations as indicated by neutral genetic markers (Swaegers et al. 2016). Moreover, common-garden rearing experiments from the egg stage showed the evolution of a faster life history (Therry et al. 2014b) and a higher investment in flight muscles and immune response (Therry et al. 2014c) in the edge populations at the range front compared to the nearby core populations.

The study populations at Nord-Pas-de-Calais, Indre, and Zeeland are in natural areas without agriculture and therefore were unlikely to be affected by pesticides (Coors et al. 2009; Cothran et al. 2013). The edge population in Saarland is within an agricultural area. This could have affected our results in two opposing ways: (i) animals in the Saarland population may have developed tolerance to the pesticide (e.g., Hua et al. 2015), or (ii) animals in the Saarland population may have suffered stress in the parental generation due to contamination of the habitat making them more vulnerable to pesticide exposure in the laboratory. The local adaptation option is unlikely given that polluted, and unpolluted ponds are intermixed in the landscape and the high levels of gene flow at a regional scale in *Coenagrion* damselflies (Johansson et al. 2013) and given the Saarland population was founded recently when sampled (<5 years, Therry et al. 2014c). Also, any local adaptation to pesticides in this edge population would make our results of increased vulnerability to pesticides in edge populations conservative. Furthermore, we did not detect differential effects of the pesticide on any response variable between the Saarland edge population and the other edge population in Zeeland, which was situated in a natural area (Appendix S1). This also suggests that any effects on the experimental larvae in the Saarland population working through stress due to pesticide contamination of the habitat is unlikely as it would have generated a higher vulnerability in the Saarland compared to the Zeeland edge population.

Mated females (Nord-Pas-de-Calais: 8, Indre: 12, Saarland: 12, and Zeeland: 11) were collected in June–July 2012 and allowed to oviposit *in situ*. Eggs were transported to the laboratory in Belgium. After hatching, larvae of each female were kept together in a plastic tank (15 × 10 × 12.5 cm) filled with ca. 500 mL dechlorinated tap water for 3 weeks to enhance survival. During this period, larvae were kept at 20°C and a photoperiod of 16:8 h light:dark. Larvae were fed *Artemia* nauplii *ad libitum* 5 days per week. After this 3-week period larvae were introduced in the container experiment.

Outdoor container experiment

To test whether evolutionary processes during the range expansion affect the vulnerability of *Coenagrion scitulum* damselflies to a pesticide and how density may play a role in modifying these effects, we set up a full factorial outdoor container experiment with 2 population types (edge and core, each represented by two populations) × 2 densities (low and high) × 3 esfenvalerate concentrations (0, 0.1 and 0.2 μg/L). Each treatment combination had 8 repli-

cated containers (10 L polypropylene cylindrical tanks, height of 22 cm, diameter of 24 cm) giving a total of 96 containers. The container experiment consisted of three periods: (i) a pre-exposure period that started when larvae were ca. 3 weeks old and that spanned fall and winter, (ii) a pesticide exposure period of 4 weeks in spring during which the larvae experienced four pulses of esfenvalerate, and (iii) a postexposure period that ended with adult metamorphosis. The initially installed larval densities were 15 and 45 larvae per container, corresponding to low (332 larvae per m^2) and high (995 larvae per m^2) densities of coenagrionid damselfly larvae in suitable habitats (Corbet 1999), respectively.

Due to higher mortality in the pre-exposure stage of the experiment in edge (62.89%) than in core populations (58.1%) (Loglinear model, $\chi_1^2 = 6.88$, $P < 0.0088$) and in high-density (64.10%) than in low-density containers (49.13%) (Loglinear model, $\chi_1^2 = 48.15$, $P < 0.001$) and the resulting density variation among containers of the same density treatment, we re-installed the density treatments after winter. This was carried out by redistributing larvae among containers (cf. Liess et al. 2013), thereby keeping larvae at their combination of population and density. Note this was carried out just before the pesticide exposure period started. The new densities were 8 (low density) and 20 (high density) larvae per container. The resulting number of containers per density treatment varied from 5 to 8 per population (exact numbers are shown in the figures). See Appendix S2 for more details of the experimental setup.

Application of esfenvalerate

The esfenvalerate concentrations were chosen based on a 48 h acute toxicity test in which *C. scitulum* damselfly larvae were individually exposed to concentrations of 0, 0.25, 0.5, 1, and 2 μg/L at 18°C (close to the temperature in the containers at the start of the exposure period, see Fig. S1G–H). After 48 h the survival was 100% in the control, 82% at 0.25 μg/L, 33% at 0.5 μg/L and 0% at 1 and 2 μg/L. In another acute toxicity test in which *Daphnia pulex*, the food source of the damselfly larvae in the containers, was exposed in groups of 10 individuals to the same esfenvalerate concentrations at 18°C, none of the *Daphnia* died after 48 h, even at the highest tested esfenvalerate concentration of 2.0 μg/L. Because we were also interested in sublethal effects, we selected concentrations of 0.1 μg/L (the lowest observed effect concentration for invertebrates, European Commission 2000) and 0.2 μg/L (below the lowest lethal concentration in our acute toxicity test) for our experiment. Both concentrations fall within the range of concentrations found in water bodies nearby agricultural areas, which go up to 0.76 μg/L (Stampfli et al. 2013). A 1 mg/

mL stock solution was prepared by dissolving esfenvalerate powder (purity >99%, Sigma-Aldrich) in absolute ethanol. This stock solution was further diluted with filtered water from the containers to obtain concentrations of 12 and 24 µg/L esfenvalerate (the spraying solutions), respectively. Fifty millilitre of each spraying solution (12 and 24 µg/L) was gently poured over the surface of the containers (Stampfli et al. 2013) to obtain the nominal esfenvalerate concentrations of 0.1 and 0.2 µg/L. In the control treatment, we added 50 mL ethanol (24 µL/L), using the ethanol concentration of the high esfenvalerate treatment, in the same manner as in the pesticide treatments.

Esfenvalerate was applied four times in spring, with 1 week between pulses, starting on 2 May 2013 with the last pulse on 23 May 2013. This mimics the realistic scenario of exposure to several pesticide pulses in spring through runoff (Van Drooge et al. 2001). The measured concentrations in the containers, based on a pooled sample from all containers of each of the exposure concentrations, were 0.072 and 0.084 µg/L 2 h after spraying (the expected peak concentration, Knillmann et al. 2013), for the nominal concentrations of 0.1 and 0.2 µg/L, respectively. After 1 week, just before applying a new pulse, the concentrations were below the detection limit of 0.005 µg/L. Esfenvalerate concentrations were analyzed by the research laboratory Lovap NV, Geel, Belgium using gas chromatography in combination with mass spectrometry.

Abiotic and biotic parameters

Temperature, dissolved oxygen, pH, and conductivity were measured in a subsample of 24 containers, 2 containers per combination of population type × density × esfenvalerate concentration. These parameters were quantified biweekly throughout the exposure and postexposure periods. Chlorophyll a concentrations were measured in all containers on a biweekly basis during the exposure and postexposure periods. The abundance of $D.$ $pulex$ was quantified in each container at the start of the pesticide exposure period to obtain the initial density, and after 7 days to obtain the lowest density. Thereafter, $Daphnia$ abundance was quantified every 2 weeks just before (lowest density) and after (highest density) the weekly addition of $Daphnia$. See Appendix S2 for detailed overviews of the temporal patterns of abiotic and biotic parameters in the experimental containers under the different treatment combinations.

Response variables

To estimate larval growth rate during the 4-week exposure period, all larvae from each container were collected and weighted on 26–29 April 2013 (just before the start of the exposure period) and on 29–30 May 2013 (end of the expo-

sure period). Mean per capita mass per container was used to calculate growth rate as $(\ln_{final\ mass} - \ln_{initial\ mass})$/duration exposure period. Based on the number of larvae counted at the start and at the end of the exposure period, we calculated mortality (%) during the exposure period as (initial number − number of survived larvae)/initial number of larvae × 100. After the exposure period, we daily checked for emergence of adult damselflies. The larval development time was calculated as the time from egg hatching to adult emergence. To quantify mass at emergence, all freshly-emerged adults were kept in the dark for ca. 16 h to harden their exoskeleton where after their wet mass was weighted to the nearest 0.01 mg using an electronic balance (AB135-S, Mettler Toledo®, Zaventem, Belgium). Afterward, all adults were stored at −80°C until the analyses of flight muscle mass and fat content. For each adult that emerged, flight muscle mass and fat content, two important flight-related traits (Therry et al. 2014c), were quantified based on protocols described in Swillen et al. (2009) (see Appendix S3 for more detail).

Statistical analyses

To test for effects of the population type, larval density, and esfenvalerate concentration on the response variables mortality, growth rate, and development time during the exposure period, and adult mass at emergence, flight muscle mass, and total fat content, we ran separate AN(C)OVAs using the mixed procedure of SAS v9.3 (SAS Institute Inc., Cary, NC, USA). In all models, population nested in population type was included as a random factor. When testing effects on flight muscle mass and fat content, we included the exoskeleton mass as covariate to correct for size differences (see Therry et al. 2014c). All models use containers as the unit of replication. We will here report results for total development time, the patterns for the duration of the postexposure period (relevant for potential recovery) are similar and shown in Appendix S4.

In damselflies, sexes may differ in their response to pesticide exposure (e.g., Campero et al. 2008). We therefore sexed all adults at emergence and analyzed the traits scored at emergence separately by sex (development time, mass at emergence, flight muscle mass, and fat content). Note that given the large number of larvae (>1000 larvae) involved, it was logistically not possible to sex all larvae at the start and the end of the exposure period, so we could not separately analyze larval traits by sex.

Results

Larval life history traits
Surprisingly, in total 71 adults emerged before winter during the months of October and November 2012, hence

before the pesticide exposure period started. These were all edge animals (ANOVA on numbers emerged per container, Population type: $F_{1,92} = 26.35$, $P < 0.0001$), and numbers did not differ between containers at low density ($n = 28$ adults) and containers at high density ($n = 43$ adults) (Density: $F_{1,92} = 1.18$, $P = 0.28$).

Overall mortality during the spring exposure period did not differ between edge and core populations ($F_{1,2} = 0.4$, $P = 0.59$) and between low and high density ($F_{1,68} = 1.13$, $P = 0.29$). Exposure to esfenvalerate increased mortality ($F_{2,68} = 34.34$, $P < 0.001$). Notably, the effect of esfenvalerate differed between edge and core populations (Population type × Pesticide, $F_{2,68} = 4.1$, $P = 0.021$). Contrasts analyses showed that at the high esfenvalerate concentration the pesticide-induced mortality was stronger in edge populations than in core populations ($F_{1,70} = 8.03$, $P = 0.006$, Fig. 1), while mortality did not differ between edge and core populations in the control ($F_{1,70} < 0.01$, $P = 0.99$) and at the low concentration ($F_{1,70} = 0.37$, $P = 0.54$). This pattern of increased mortality of edge compared to core populations at the high concentration was similar at both densities (Population type × Pesticide × Density: $F_{2,68} = 0.75$, $P = 0.48$, Fig. S4A,B in Appendix S5). The pesticide effect did not depend upon density (Density × Pesticide, $F_{2,68} = 1.29$, $P = 0.28$).

Growth rates differed neither between edge and core populations ($F_{1,2} = 0.02$, $P = 0.91$ Fig. S4C,D) nor between low and high densities ($F_{1,68} = 0.2$, $P = 0.65$). Growth rate during the exposure period strongly decreased with increasing esfenvalerate concentrations: the growth reductions were ca. 27% at the low and ca. 36% at the high esfenvalerate concentration ($F_{2,68} = 7.03$, $P = 0.0017$, Fig. S4C, D in Appendix S5). The pesticide effect did not depend

upon population type or density (all interactions: $P > 0.25$).

Exposure to esfenvalerate tended to result in a slightly later emergence of ca. 3 days (Males: $F_{2,67} = 3.1$, $P = 0.051$; Females: $F_{2,66} = 2.73$, $P = 0.072$, Fig. 2A,D). Development times were longer at high density (Males: $F_{1,67} = 69.1$, $P < 0.001$; Females: $F_{1,66} = 47.7$, $P < 0.001$, Fig. 2B,D). Development times tended to be slightly shorter in edge females than in core females in the control and at the high esfenvalerate concentration (Females: Population type × Pesticide, $F_{1,66} = 3.12$, $P = 0.051$, Fig. 2C, D).

Adult flight-related traits

Mass at emergence decreased with increasing esfenvalerate concentrations (Males: $F_{2,67} = 7.64$, $P = 0.001$; Females: $F_{2,65} = 8.4$, $P < 0.001$, Fig. 2E–H) and was lower at high density (Males: $F_{1,67} = 55.65$, $P < 0.001$; Females: $F_{1,65} = 47.17$, $P < 0.001$). Mass at emergence did not differ between edge and core animals (Males: $F_{1,2} = 0.04$, $P = 0.86$; Females: $F_{1,2} = 0.08$, $P = 0.80$, Fig. 2E–H).

Exposure to esfenvalerate negatively affected the relative flight muscle mass (Males: $F_{2,66} = 2.79$, $P = 0.068$; Females: $F_{2,64} = 9.42$, $P < 0.001$, Fig. 3A–D). Edge animals tended to have a higher flight muscle mass than core animals at high density in the absence of the pesticide, while the opposite was observed at low density (Males: Population type × Density, $F_{1,66} = 4.75$, $P = 0.033$; Females: Population type × Density × Pesticide, $F_{2,64} = 6.18$, $P = 0.0035$). High density resulted in a lower flight muscle mass (Males: $F_{1,66} = 29.35$, $P < 0.001$; Females: $F_{1,64} = 21.29$, $P < 0.001$).

Exposure to esfenvalerate strongly decreased the fat content (Males: $F_{2,66} = 5.58$, $P = 0.0058$; Females: $F_{2,64} = 7.61$, $P = 0.0011$). In males, this pesticide effect was density-dependent (Density × Pesticide, $F_{2,66} = 4.89$, $P = 0.011$, Fig. 3E–H): at low-density fat content was only reduced at the high esfenvalerate concentration while at high-density fat content was already reduced at the low esfenvalerate concentration. In both sexes, fat content was lower at high density than at low density (Males: $F_{1,66} = 18.89$, $P < 0.001$; Females: $F_{1,64} = 10.94$, $P = 0.0015$); this pattern was stronger in core animals than in edge animals (Population type × Density, Males: $F_{1,66} = 8.06$, $P = 0.006$; Females: $F_{1,64} = 7.07$, $P = 0.0099$, Fig. 3E–H).

Discussion

Main effects of the pesticide

We found strong negative effects of larval exposure to the ecologically realistic esfenvalerate concentrations on all studied traits not only in the larval but also in the adult

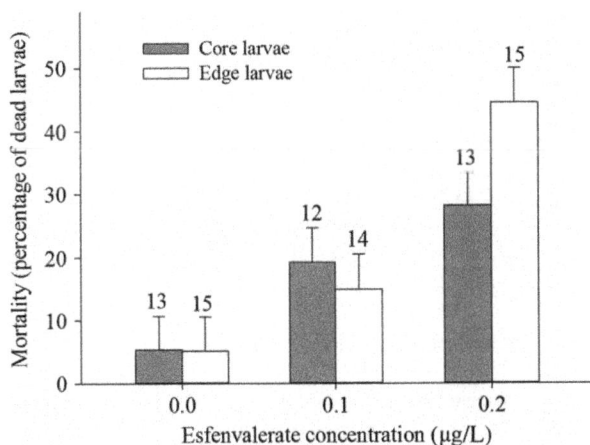

Figure 1 Mortality of *Coenagrion scitulum* damselfly larvae during the exposure period as a function of esfenvalerate concentration and population type. Numbers above the bars represent the number of container replicates. Least-square means are given with 1 SE.

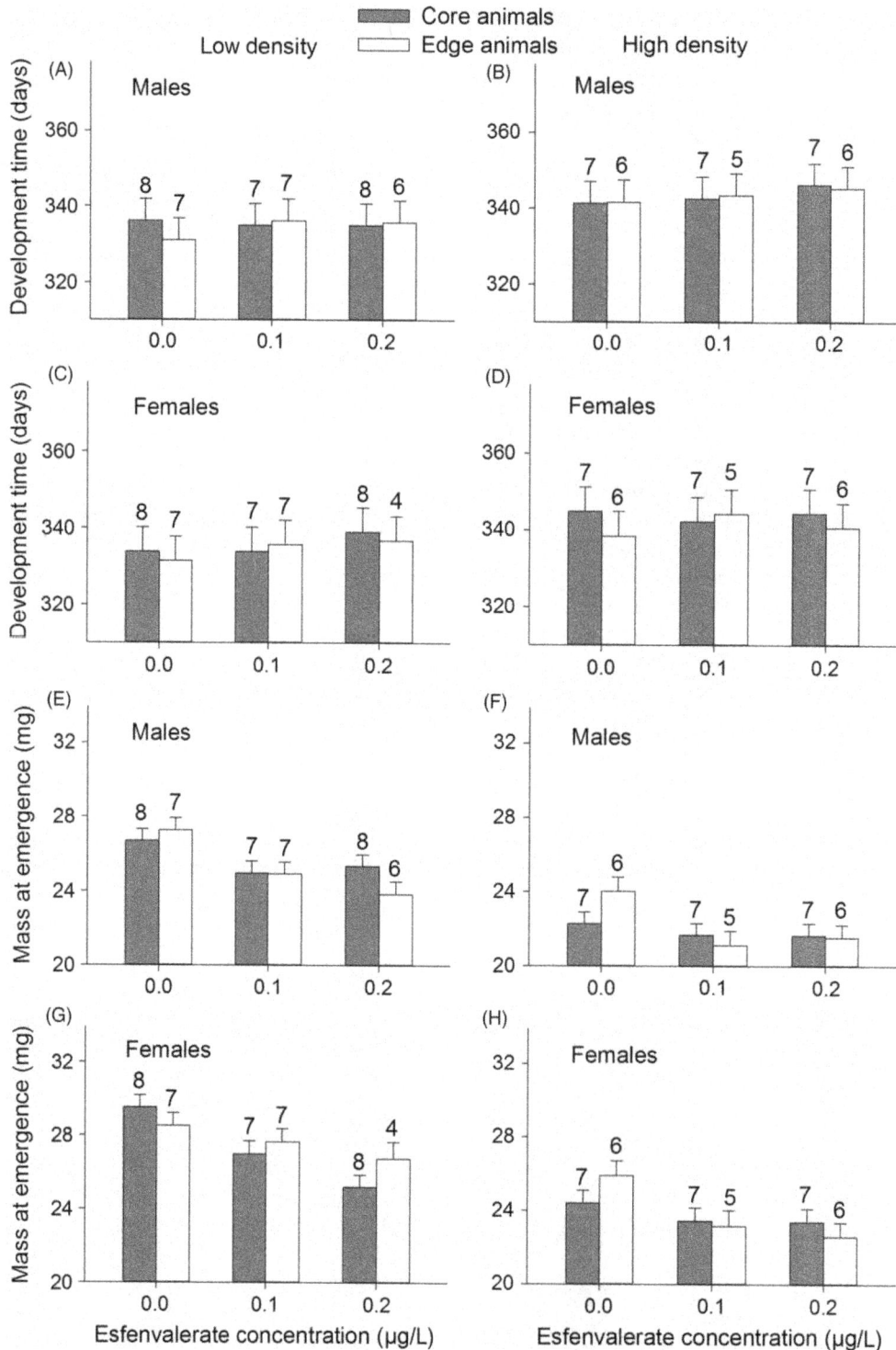

Figure 2 Development time of males (A, B) and females (C, D) and mass at emergence of males (E, F) and females (G, H) of the damselfly *Coenagrion scitulum* as a function of esfenvalerate concentration, density, and population type. Numbers above the bars represent the number of container replicates. Least-square means are given with 1 SE.

stage, and this despite the long period (ca. 25–30 days, see Appendix S4) that larvae were able to recover from pesticide exposure. Esfenvalerate-imposed mortality fits the pattern of lethal effects imposed by pyrethroids in aquatic insects (Liess 2002; Beketov and Liess 2005; Rasmussen et al. 2013), which result from damage to the nervous

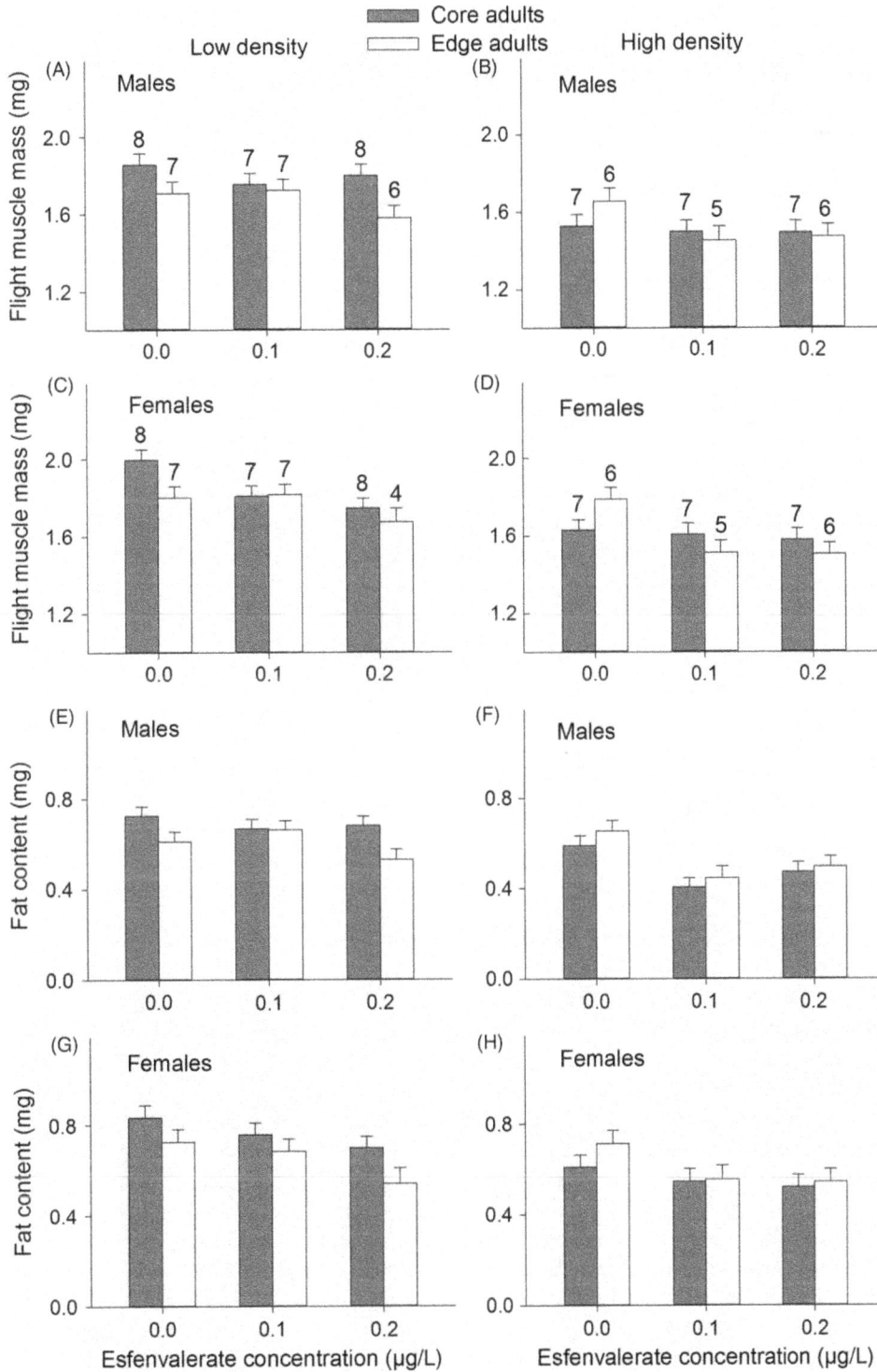

Figure 3 Flight muscle mass of males (A, B) and females (C, D), fat content of males (E, E) and females (G, H) of the damselfly *Coenagrion scitulum* as a function of esfenvalerate concentration, density, and population type. Numbers above the bars represent the number of container replicates. Least-square means corrected for size are given with 1 SE.

system (Cold and Forbes 2004). The negative effects of exposure to esfenvalerate on growth and flight-related traits likely were mediated by energy shortage as pesticide-

exposed animals need more energy for detoxification and repair, resulting in less energy allocation toward other functions (Campero et al. 2007). Note that these esfen-

valerate effects are likely direct effects on the damselfly larvae and no indirect effects working through the *Daphnia*
food because *D. pulex* survival was not affected by a 10×
higher esfenvalerate concentration and because the pesticide did not affect the abundance of *D. pulex* in the containers (Appendix S2).

A key finding was that esfenvalerate negatively affected
body mass, fat content, and relative flight muscle mass,
three traits known to shape flight performance in *Coenagrion* damselflies (Gyulavári et al. 2014; Therry et al.
2014c). Delayed effects of esfenvalerate across metamorphosis have also been documented in the caddisfly *Brachy-
centrus americanus* where adults that had been exposed to
esfenvalerate in the pupal stage invested less in egg mass
(Palmquist et al. 2008). The pesticide-induced reductions
in the flight-related traits, especially flight muscle mass is a
highly relevant trait for key functions such as flight ability
(e.g., Therry et al. 2014a), and therefore important for
shaping foraging, predator evasion, mating success, and
dispersal ability in damselflies (Stoks and Cordoba-Aguilar
2012).

Edge-core differentiation mediating the effect of the pesticide

We found some evidence for the expected faster life history
and increased investment in flight morphology in edge
populations. Animals at an expanding range are expected
to show a faster life history because of spatial sorting and r-
selection associated with the initial lower population densities at the expansion front (Phillips 2009; Burton et al.
2010; Phillips et al. 2010). For the study species this also
includes selection for a fast development to avoid having
less generations per year at the higher latitudes at the
expansion front (Nilsson-Örtman et al. 2012) which would
slow down the range expansion (Therry et al. 2014b).
Therry et al. (2014b) indeed reported higher growth and
development rates in edge larvae of the study species. In
line with this, we observed that the subset of animals that
were able to emerge before winter were all edge animals.
Yet, within the subset of larvae that overwintered (hence
those that were used in the spring exposure experiment),
no faster life history in edge animals was observed. The latter may be a result of the fastest animals already emerging
before winter. Moreover, the higher mortality during winter in edge populations (see methods) may have mainly
removed the faster growing larvae. Indeed, rapid growth
has been associated with reduced energy storage (Stoks
et al. 2006) and reduced cold resistance (Stoks and De
Block 2011) in damselflies, which may have reduced the
ability to survive winter. More general, a faster life history
has been associated with a higher mortality in damselfly
larvae (De Block et al. 2008; Sniegula et al. 2014). Edge

animals are also expected to have a higher relative flight
muscle mass as only the best dispersers may reach the
expansion front (Shine et al. 2011). Indeed, edge populations of poleward-moving insects, including the study species (Therry et al. 2014c), show a higher investment in
flight muscle mass (reviewed in Hill et al. 2011). Yet, in
current study this was only observed at high density (in the
control without the pesticide) suggesting that the higher
investment in flight muscles in edge populations may be
density-dependent.

Our data suggested that edge animals had a higher vulnerability to the pesticide in term of a higher mortality at
the high esfenvalerate concentration. Note that this higher
vulnerability in edge populations did not occur at the low
pesticide concentration as at the low concentration our
contrast analysis suggested no significant difference of mortality between edge and core populations. Notably, we
observed the higher vulnerability to the pesticide in edge
populations despite no indication of a faster life history in
the overwintered larvae that were exposed to the pesticide
and without a consistent higher investment in flight morphology. Edge populations, however, are expected to show
rapid evolutionary changes in a wide range of traits, including life history, morphology, behavior, and physiology
(Phillips 2009; Burton et al. 2010; Phillips et al. 2010; Shine
et al. 2011; Brown et al. 2015) that all require a higher allocation of energy and therefore are expected to be traded off
against investment in detoxification and repair (Sibly and
Calow 1989; Congdon et al. 2001). For example, the edge
animals may have invested more in immune function to
avoid parasite-driven reductions in dispersal ability, as has
been observed in the study species (Therry et al. 2014b; K.
V. Dinh, L. Janssens, L. Therry, L. Bervoets and R. Stoks,
unpublished data). Additionally, in another outdoor container experiment (L. Therry, and R. Stoks in prep.) edge
larvae of the study species showed a faster growth during
the winter period and as a result had a lower fat content
after winter, which may have made them more vulnerable
to the pesticide. Whatever the mechanism, our results suggest that evolutionary changes associated with range expansion, made edge populations more vulnerable to
esfenvalerate during spring application. Admittedly, the
increase in pesticide-induced mortality in edge compared
to core populations was relatively small (ca. 17%), yet will
translate into extra reductions in population growth rates if
edge populations are exposed to pesticides.

Larval density mediating the effects of the pesticide

While the high-density treatment did not influence larval
survival and growth during the pesticide exposure period,
larvae reared at high density showed longer larval development times, and reductions in mass at emergence, flight

muscle mass, and fat content. These negative density effects are in line with a higher exploitation competition for food at the high-density treatment. Additionally, at high densities there may have been more physical encounters among larvae, thereby imposing stress; this is especially likely in damselfly larvae as they are cannibalistic and impose predator stress on each other (De Block and Stoks 2004). Another important finding was that high density only reduced flight muscle mass in core adults but not in edge adults which is in line with the hypothesis of a stronger selection for flight performance in edge populations (Hill et al. 2011; Therry et al. 2014c). Negative effects of larval competition on adult flight muscle mass have not previously been documented and provide a rare empirical example of how the conditions encountered during the larval stage may shape the adult dispersal ability (Benard and McCauley 2008). In males, the pesticide-induced reduction in fat content was stronger at high density than at low density; this is in line with the stronger negative effect of pesticides at higher density in other aquatic animals (e.g., Jones et al. 2011; Knillmann et al. 2012).

Implications for ecological risk assessment and range expansions

Current ecological risk assessment (ERA) of pesticides is not effectively protecting biodiversity as strong losses in biodiversity are being detected at concentrations that current legislation considers as environmentally protective (Beketov et al. 2013; Malaj et al. 2014). Our study adds to this by identifying two reasons why current ERA may underestimate the impact of pesticides, and thereby points to concrete actions to improve legislation to make toxicity testing more effective toward management and protection of freshwater biodiversity under global warming. Firstly, we build further on previous insights that standard toxicity testing limited to one life stage may not capture the full impact of a pesticide (see, e.g., Campero et al. 2008; Distel and Boone 2010; Janssens et al. 2014). We thereby made an important extension by providing evidence that larval exposure to ecologically relevant concentrations of pesticides may negatively affect locomotory performance in the adult stage. This ignored delayed effect of pesticides may have major fitness consequences as locomotion is crucial for key functions such as foraging, escaping predation, securing matings, and dispersal (Stoks and Cordoba-Aguilar 2012). Secondly, we provide the first test and some supporting evidence that edge populations at an expanding range front are more vulnerable to high pesticide concentrations than core populations in term of a higher pesticide-induced mortality, thereby adding an evolutionary component to the emerging insight that we need spatially explicit ERA (Van den Brink 2008; Clements et al. 2012; Dinh Van et al. 2014).

Both the effect of larval pesticide exposure on mortality and its delayed effects on adult flight-related traits also are highly relevant to understand the impact of global warming on organisms as they highlight two overlooked pathways of how pesticides may slow down range expansions. Firstly, exposure to esfenvalerate at ecologically realistic concentrations caused mortality and thereby decreases in population growth rates, hence it is expected to reduce the rate of further range expansion. In addition, our data indicated rapid evolution of a slightly increased pesticide-induced mortality at the range front, which has the potential to magnify this effect, and thereby to slow down the range expansion even more. As species may show considerable population declines in core regions under global warming, researchers highlighted the need for direct conservation efforts toward leading-edge populations for spearheading future range shifts (Razgour et al. 2013). Our results thereby underscore the importance of considering pesticide exposure in such conservation programs. Secondly, esfenvalerate negatively affected three flight-related traits (body mass, relative flight muscle mass, and fat content) known to shape flight performance in *Coenagrion* damselflies (Gyulavári et al. 2014; Therry et al. 2014c), thereby reducing the dispersal ability. Any reductions in dispersal rates may have major implications as there is increasing concern that poleward range expansions do not allow timely tracking of the moving climate niche (La Sorte and Jetz 2012). These two overlooked mechanisms how pesticides may slow down range expansion, together with the expected increase in pesticide application at higher latitudes under global warming (Kattwinkel et al. 2011), raise concern about the potential for edge populations to act as potent sources for further range expansion in a polluted world.

Despite recent progress in identifying factors underlying species differences in range expansion rates (Angert et al. 2011; Mair et al. 2014), it is largely unknown why there is so much variation in the rates at which different species' geographic ranges expand in response to climate warming (Moritz and Agudo 2013). Yet, this information is crucial to identify species that may potentially be too slow to track their moving climate niche, thereby being more at risk under global warming and to understand the likely success of different conservation strategies for facilitating range shifts (Moritz and Agudo 2013; Mair et al. 2014). Some of the current models predicting future ranges already include estimates of dispersal ability to predict which species may be better at tracking their climate envelope (e.g., Thomas et al. 2001; Hughes et al. 2007). Species may differ considerably in their sensitivity to pesticides (e.g., Beketov 2004; Rasmussen et al. 2013; Weston et al. 2013). Our results therefore generate the hypothesis that besides dispersal ability also the degree to which survival and dispersal ability are affected by widely used contaminants and how the vulnerability to pesticides evolves at expanding range fronts may be key factors in shaping species differences in range expansion in a polluted world.

Acknowledgements

We thank Sara Debecker, Sarah Oexle, Janne Swaegers and Ria Van Houdt for providing assistance during the experiment. KVD was a PhD fellow of VIED and benefited an IRO Supplement. Financial support for this research came from FWO grants G.0419.08 and G.0610.11 and the KU Leuven Research Fund grants GOA/2008/06 and Excellence Center Financing PF/2010/07 to RS.

Literature cited

Angert, A. L., L. G. Crozier, L. J. Rissler, S. E. Gilman, J. J. Tewksbury, and A. J. Chunco 2011. Do species' traits predict recent shifts at expanding range edges? Ecology Letters 14:677–689.

Beketov, M. A. 2004. Comparative sensitivity to the insecticides deltamethrin and esfenvalerate of some aquatic insect larvae (Ephemeroptera and Odonata) and *Daphnia magna*. Russian Journal of Ecology 35:200–204.

Beketov, M. A., and M. Liess 2005. Acute contamination with esfenvalerate and food limitation: chronic effects on the mayfly, *Cloeon dipterum*. Environmental Toxicology and Chemistry 24:1281–1286.

Beketov, M. A., and M. Liess 2012. Ecotoxicology and macroecology – Time for integration. Environmental Pollution 162:247–254.

Beketov, M. A., B. J. Kefford, R. B. Schafer, and M. Liess 2013. Pesticides reduce regional biodiversity of stream invertebrates. Proceedings of the National Academy of Sciences of the USA 110:11039–11043.

Benard, M. F., and S. J. McCauley 2008. Integrating across life-history stages: consequences of natal habitat effects on dispersal. American Naturalist 171:553–567.

Brown, G. P., B. L. Phillips, S. Dubey, and R. Shine 2015. Invader immunology: invasion history alters immune system function in cane toads (*Rhinella marina*) in tropical Australia. Ecology Letters 18:57–65.

Burton, O. J., B. L. Phillips, and J. M. J. Travis 2010. Trade-offs and the evolution of life-histories during range expansion. Ecology Letters 13:1210–1220.

Campero, M., S. Slos, F. Ollevier, and R. Stoks 2007. Sublethal pesticide concentrations and predation jointly shape life history: behavioral and physiological mechanisms. Ecological Applications 17:2111–2122.

Campero, M., M. De Block, F. Ollevier, and R. Stoks 2008. Correcting the short-term effect of food deprivation in a damselfly: mechanisms and costs. Journal of Animal Ecology 77:66–73.

Chen, I. C., J. K. Hill, R. Ohlemuller, D. B. Roy, and C. D. Thomas 2011. Rapid range shifts of species associated with high levels of climate warming. Science 333:1024–1026.

Clements, W. H., C. W. Hickey, and K. A. Kidd 2012. How do aquatic communities respond to contaminants? It depends on the ecological context. Environmental Toxicology and Chemistry 31:1932–1940.

Cold, A., and V. E. Forbes 2004. Consequences of a short pulse of pesti-

cide exposure for survival and reproduction of *Gammarus pulex*. Aquatic Toxicology 67:287–299.

Congdon, J. D., A. E. Dunham, W. A. Hopkins, C. L. Rowe, and T. G. Hinton 2001. Resource allocation-based life histories: a conceptual basis for studies of ecological toxicology. Environmental Toxicology and Chemistry 20:1698–1703.

Coors, A., J. Vanoverbeke, T. De Bie, and L. De Meester 2009. Land use, genetic diversity and toxicant tolerance in natural populations of *Daphnia magna*. Aquatic Toxicology 95:71–79.

Corbet, P. 1999. Dragonflies: Behavior and Ecology of Odonata. Cornell University Press, London.

Cothran, R. D., J. M. Brown, and R. A. Relyea 2013. Proximity to agriculture is correlated with pesticide tolerance: evidence for the evolution of amphibian resistance to modern pesticides. Evolutionary Applications 6:832–841.

Coutellec, M. A., and C. Barata 2011. An introduction to evolutionary processes in ecotoxicology. Ecotoxicology 20:493–496.

De Block, M., and R. Stoks 2004. Cannibalism-mediated life history plasticity to combined time and food stress. Oikos 106:587–597.

De Block, M., M. A. McPeek, and R. Stoks 2008. Life history plasticity to combined time and biotic constraints in *Lestes* damselflies from vernal and temporary ponds. Oikos 117:908–916.

Dijkstra, K.-D. B. 2006. Field Guide to the Dragonflies of Britain and Europe. British Wildlife Publishing, Gillingham, Dorset, UK.

Dinh Van, K., L. Janssens, S. Debecker, and R. Stoks 2014. Temperature and latitude-specific individual growth rates shape the vulnerability of damselfly larvae to a widespread pesticide. Journal of Applied Ecology 51:919–928.

Distel, C. A., and M. D. Boone 2010. Effects of aquatic exposure to the insecticide carbaryl are species-specific across life stages and mediated by heterospecific competitors in anurans. Functional Ecology 24:1342–1352.

European Commission. 2000. Review report for the active substance esfenvalerate: 6846/VI/97-final.

Fordham, D. A., C. Mellin, B. D. Russell, R. H. Akcakaya, C. J. A. Bradshaw, M. E. Aiello-Lammens, J. M. Caley et al. 2013. Population dynamics can be more important than physiological limits for determining range shifts under climate change. Global Change Biology 19:3224–3237.

Gyulavári, H. A., L. Therry, G. Devai, and R. Stoks 2014. Sexual selection on flight endurance, flight-related morphology and physiology in a scrambling damselfly. Evolutionary Ecology 28:639–654.

Hammond, J. I., D. K. Jones, P. R. Stephens, and R. A. Relyea 2012. Phylogeny meets ecotoxicology: evolutionary patterns of sensitivity to a common insecticide. Evolutionary Applications 5:593–606.

Hickling, R., D. B. Roy, J. K. Hill, R. Fox, and C. D. Thomas 2006. The distributions of a wide range of taxonomic groups are expanding polewards. Global Change Biology 12:450–455.

Hill, J. K., H. M. Griffiths, and C. D. Thomas 2011. Climate change and evolutionary adaptations at species' range margins. Annual Review of Entomology 56:143–159.

Hua, J., D. K. Jones, B. M. Mattes, R. D. Cothran, R. A. Relyea, and J. T. Hoverman 2015. The contribution of phenotypic plasticity to the evolution of insecticide tolerance in amphibian populations. Evolutionary Applications 8:586–596.

Hughes, C. L., C. Dytham, and J. K. Hill 2007. Modelling and analysing evolution of dispersal in populations at expanding range boundaries. Ecological Entomology 32:437–445.

Janssens, L., K. Dinh Van, and R. Stoks 2014. Extreme temperatures in the adult stage shape delayed effects of larval pesticide stress: a comparison between latitudes. Aquatic Toxicology 148:74–82.

Johansson, H., R. Stoks, V. Nilsson-Örtman, P. K. Ingvarsson, and F. Johansson 2013. Large-scale patterns in genetic variation, gene flow and differentiation in five species of European Coenagrionid damselfly provide mixed support for the central-marginal hypothesis. Ecography 36:744–755.

Jones, D. K., J. I. Hammond, and R. A. Relyea 2011. Competitive stress can make the herbicide roundup® more deadly to larval amphibians. Environmental Toxicology and Chemistry 30:446–454.

Kattwinkel, M., J.-V. Kuehne, K. Foit, and M. Liess 2011. Climate change, agricultural insecticide exposure, and risk for freshwater communities. Ecological Applications 21:2068–2081.

Knillmann, S., N. C. Stampfli, M. A. Beketov, and M. Liess 2012. Intraspecific competition increases toxicant effects in outdoor pond microcosms. Ecotoxicology 21:1857–1866.

Knillmann, S., N. C. Stampfli, Y. A. Noskov, M. A. Beketov, and M. Liess 2013. Elevated temperature prolongs long-term effects of a pesticide on Daphnia spp. due to altered competition in zooplankton communities. Global Change Biology 19:1598–1609.

La Sorte, F. A., and W. Jetz 2012. Tracking of climatic niche boundaries under recent climate change. Journal of Animal Ecology 81:914–925.

Liess, M. 2002. Population response to toxicants is altered by intraspecific interaction. Environmental Toxicology and Chemistry 21:138–142.

Liess, M., K. Foit, A. Becker, E. Hassold, I. Dolciotti, M. Kattwinkel, and S. Duquesne 2013. Culmination of low-dose pesticide effects. Environmental Science & Technology 47:8862–8868.

Mair, L., J. K. Hill, R. Fox, M. Botham, T. Brereton, and C. D. Thomas 2014. Abundance changes and habitat availability drive species' responses to climate change. Nature Climate Change 4:127–131.

Malaj, E., P. C. von der Ohe, M. Grote, R. Kuehne, C. P. Mondy, P. Usseglio-Polatera, W. Brack et al. 2014. Organic chemicals jeopardize the health of freshwater ecosystems on the continental scale. Proceedings of the National Academy of Sciences of the USA 111:9549–9554.

Mehlhorn, H., N. Mencke, and O. Hansen 1999. Effects of imidacloprid on adult and larval stages of the flea Ctenocephalides felis after in vivo and in vitro application: a light- and electron-microscopy study. Parasitology Research 85:625–637.

Moe, S. J., K. De Schamphelaere, W. H. Clements, M. T. Sorensen, P. J. Van den Brink, and M. Liess 2013. Combined and interactive effects of global climate change and toxicants on populations and communities. Environmental Toxicology and Chemistry 32:49–61.

Moritz, C., and R. Agudo 2013. The future of species under climate change: resilience or decline? Science 341:504–508.

Nilsson-Örtman, V., R. Stoks, M. De Block, and F. Johansson 2012. Generalists and specialists along a latitudinal transect: patterns of thermal adaptation in six species of damselflies. Ecology 93:1340–1352.

Noyes, P. D., M. K. McElwee, H. D. Miller, B. W. Clark, L. A. Van Tiem, K. C. Walcott, K. N. Erwin et al. 2009. The toxicology of climate change: environmental contaminants in a warming world. Environment International 35:971–986.

Palmquist, K. R., P. C. Jepson, and J. J. Jenkins 2008. Impact of aquatic insect life stage and emergence strategy on sensitivity to esfenvalerate exposure. Environmental Toxicology and Chemistry 27:1728–1734.

Phillips, B. L. 2009. The evolution of growth rates on an expanding range edge. Biology Letters 5:802–804.

Phillips, B. L., G. P. Brown, and R. Shine 2010. Life-history evolution in range-shifting populations. Ecology 91:1617–1627.

Rasmussen, J. J., P. Wiberg-Larsen, E. A. Kristensen, N. Cedergreen, and N. Friberg 2013. Pyrethroid effects on freshwater invertebrates: a meta-analysis of pulse exposures. Environmental Pollution 182:479–485.

Razgour, O., J. Juste, C. Ibanez, A. Kiefer, H. Rebelo, S. J. Puechmaille, R. Arlettaz et al. 2013. The shaping of genetic variation in edge-of-range populations under past and future climate change. Ecology Letters 16:1258–1266.

Shine, R., G. P. Brown, and B. L. Phillips 2011. An evolutionary process that assembles phenotypes through space rather than through time. Proceedings of the National Academy of Sciences of the USA 108:5708–5711.

Sibly, R. M., and P. Calow 1989. A life-cycle theory of responses to stress. Biological Journal of the Linnean Society 37:101–116.

Sniegula, S., S. M. Drobniak, M. J. Golab, and F. Johansson 2014. Photoperiod and variation in life history traits in core and peripheral populations in the damselfly Lestes sponsa. Ecological Entomology 39:137–148.

Spurlock, F., and M. Lee 2008. Synthetic pyrethroid use patterns, properties, and environmental effects. In: J. Gan, F. Spurlock, P. Hendley, and D. P. Weston, eds. Synthetic Pyrethroids. American Chemical Society, Washington, DC.

Stampfli, N. C., S. Knillmann, M. Liess, Y. A. Noskov, R. B. Schäfer, and M. A. Beketov 2013. Two stressors and a community – Effects of hydrological disturbance and a toxicant on freshwater zooplankton. Aquatic Toxicology 127:9–20.

Stehle, S., and R. Schulz 2015. Agricultural insecticides threaten surface waters at the global scale. Proceedings of the National Academy of Sciences of the USA 112:5750–5755.

Stoks, R., and A. Cordoba-Aguilar 2012. Evolutionary ecology of Odonata: a complex life cycle perspective. Annual Review of Entomology 57:249–265.

Stoks, R., and M. De Block 2011. Rapid growth reduces cold resistance: evidence from latitudinal variation in growth rate, cold resistance and stress proteins. PLoS ONE 6:e16935.

Stoks, R., M. De Block, and M. A. McPeek 2006. Physiological costs of compensatory growth in a damselfly. Ecology 87:1566–1574.

Swaegers, J., J. Mergeay, L. Therry, M. H. D. Larmuseau, D. Bonte, and R. Stoks 2013. Rapid range expansion increases genetic differentiation while causing limited reduction in genetic diversity in a damselfly. Heredity 111:422–429.

Swaegers, J., J. Mergeay, A. Van Geystelen, L. Therry, M. H. D. Larmuseau, and R. Stoks. 2016. Neutral and adaptive genomic signatures of rapid poleward range expansion. Molecular Ecology 24:6163–6176.

Swillen, I., M. De Block, and R. Stoks 2009. Morphological and physiological sexual selection targets in a territorial damselfly. Ecological Entomology 34:677–683.

Therry, L., H. A. Gyulavári, S. Schillewaert, D. Bonte, and R. Stoks 2014a. Integrating large-scale geographic patterns in flight morphology, flight characteristics and sexual selection in a range-expanding damselfly. Ecography 37:1012–1021.

Therry, L., E. Lefevre, D. Bonte, and R. Stoks 2014b. Increased activity and growth rate in the non-dispersive aquatic larval stage of a damselfly at an expanding range edge. Freshwater Biology 59:1266–1277.

Therry, L., V. Nilsson-Ortman, D. Bonte, and R. Stoks 2014c. Rapid evolution of larval life history, adult immune function and flight muscles in a poleward-moving damselfly. Journal of Evolutionary Biology **27**:141–152.

Thomas, C. D., E. J. Bodsworth, R. J. Wilson, A. D. Simmons, Z. G. Davies, M. Musche, and L. Conradt 2001. Ecological and evolutionary processes at expanding range margins. Nature **411**:577–581.

Van den Brink, P. J. 2008. Ecological risk assessment: from book-keeping to chemical stress ecology. Environmental Science & Technology **42**:8999–9004.

Van Drooge, H. L., C. N. Groeneveld, and H. J. Schipper 2001. Data on application frequency of pesticide for risk assessment purposes. Annals of Occupational Hygiene **45**:S95–S101.

Weston, D. P., H. C. Poynton, G. A. Wellborn, M. J. Lydy, B. J. Blalock, M. S. Sepulveda, and J. K. Colbourne 2013. Multiple origins of pyrethroid insecticide resistance across the species complex of a nontarget aquatic crustacean, *Hyalella azteca*. Proceedings of the National Academy of Sciences of the USA **110**:16532–16537.

15

Can sexual selection theory inform genetic management of captive populations?

Rémi Chargé,[1] Céline Teplitsky,[2] Gabriele Sorci[3] and Matthew Low[4]

1 Department of Biological and Environmental Science, Centre of Excellence in Biological Interactions, University of Jyväskylä, Jyväskylä, Finland
2 Centre d'Ecologie et de Sciences de la Conservation UMR 7204 CNRS/MNHN/UPMC, Muséum National d'Histoire Naturelle, Paris, France
3 Biogéosciences, UMR CNRS 6282, Université de Bourgogne, Dijon, France
4 Department of Ecology, Swedish University of Agricultural Sciences, Uppsala, Sweden

Keywords
conservation biology, evolutionary theory, sexual selection.

Correspondence
Rémi Chargé Department of Biological and Environmental Science, Centre of Excellence in Biological Interactions, University of Jyväskylä, P.O Box 35, Jyväskylä, 40014, Finland.

e-mail: remi.r.charge@jyu.fi

Abstract

Captive breeding for conservation purposes presents a serious practical challenge because several conflicting genetic processes (i.e., inbreeding depression, random genetic drift and genetic adaptation to captivity) need to be managed in concert to maximize captive population persistence and reintroduction success probability. Because current genetic management is often only partly successful in achieving these goals, it has been suggested that management insights may be found in sexual selection theory (in particular, female mate choice). We review the theoretical and empirical literature and consider how female mate choice might influence captive breeding in the context of current genetic guidelines for different sexual selection theories (i.e., direct benefits, good genes, compatible genes, sexy sons). We show that while mate choice shows promise as a tool in captive breeding under certain conditions, for most species, there is currently too little theoretical and empirical evidence to provide any clear guidelines that would guarantee positive fitness outcomes and avoid conflicts with other genetic goals. The application of female mate choice to captive breeding is in its infancy and requires a goal-oriented framework based on the needs of captive species management, so researchers can make honest assessments of the costs and benefits of such an approach, using simulations, model species and captive animal data.

Introduction

Because of increasingly imperiled wildlife habitats (Pimm et al. 1995; Barnosky et al. 2011), wildlife conservation managers often incorporate *ex-situ* conservation policies to mitigate species loss (e.g., captive breeding programs). In these programs, species may be 'preserved' in captivity awaiting release at an unspecified future date or captive breeding used in a supportive role to supplement dwindling wild populations (Fa et al. 2011). Reintroductions (or supplementations) from captive populations have increased exponentially in recent years and are a valuable tool in many species conservation programs (Allendorf and Luikart 2007; Ewen 2012) and commercial systems (Laikre et al. 2010; Neff et al. 2011). However, there is compelling evidence that captivity-induced genetic changes of these populations contribute to reduce rates of reintroduction/

supplementation success (Ford 2002; Woodworth et al. 2002; Milot et al. 2013).

Because the main goal of supportive breeding is to release individuals that not only reinforce the population in terms of its size but also its evolutionary potential, captive breeding and release strategies must consider the dual issue of quantity and quality of the individuals released (Fa et al. 2011; Neff et al. 2011). Enough individuals need to be released to overcome small-population limiting factors (e.g., environmental and demographic stochasticity, Allee effects), in addition to being well adapted to their environment and able to respond to future selection pressures. Thus, for these reintroduced individuals to have a good chance at positively impacting on the population (or succeeding in establishing), the potential negative genetic consequences of captive breeding should be minimized: that is, inbreeding depression, the loss of genetic diversity, and genetic adaptation to captivity (Lacy 1994; Ballou and Lacy 1995; Frankham 2008).

Inbreeding and random genetic drift are consequences of small populations, like those in captive breeding programs or endangered wild populations (Allendorf and Luikart 2007). Inbreeding arises because mating among relatives is more likely in small populations, and this allows the expression of recessive deleterious alleles (Charlesworth and Willis 2009). Genetic drift is the main process by which captive populations lose genetic variation (Lacy 1987), and occurs because allele frequencies randomly fluctuate between generations, with the increasing potential for some alleles being lost completely in small populations through this random process. Thus, the fitness consequences can be dramatic if it means the loss of beneficial alleles or the fixation of deleterious mutations. Captive populations face an additional genetic risk because selection on traits vital for survival in the wild is relaxed: there are no predators, diseases are treated, food is provided *ad libitum* and mate choice is often circumvented. Rare alleles that are deleterious in the wild may thus become more frequent in captive populations (Laikre 1999; Ralls et al. 2000), and the captive environment itself will select for adaptations beneficial to captivity (Frankham 2008). In general, such adaptations do not favor survival and fecundity when organisms are released in to the wild (reviewed in Williams and Hoffman 2009).

Traditional management of the genetics of captive populations largely focuses on minimizing inbreeding and the loss of genetic variation, with occasional attention being given to ways of mitigating adaptation to captivity (see below). A cornerstone of this management is the equalization of founder representation in the population: this decreases selection (no variance in fitness) and slows the loss of genetic diversity. In practice, this is achieved using pedigree studbook information and 'match-making' sexual pairings that minimize the mean kinship between pairs. Despite the relatively beneficial population genetic outcome of such pairings, there has been little attention paid to potential genetic consequences of removing mate choice and sexual selection in captive breeding (but see Chargé et al. 2014; Quader 2005; Wedekind 2002). Sexual selection occurs through the competition for mates by one sex (usually males) and/or discriminating mate choice by the other (usually females). By allowing sexual selection in captive breeding, females would be able to choose among several males based on their secondary sexual characters. It has been suggested that sexual selection could improve purging of deleterious mutations and increase fitness in captivity because of mating with compatible individuals or individuals with 'good genes' (Whitlock and Agrawal 2009). In addition, the removal of mate choice in captivity will relax selection on female mate choice; potentially adding to the

issues associated with genetic adaptation to captivity when individuals are released (e.g., females may become less adept at choosing the best males resulting in a general reduction in fitness). Behavioral biologists have promoted sexual selection as a potential tool for captive breeding management for over 15 years (e.g., Asa et al. 2011; Grahn et al. 1998; Quader 2005; Wedekind 2002). In 1998, Grahn et al. suggested that mate choice be given more consideration in conservation breeding programs, and in 2011, it was emphasized that the zoo community carefully considers mate choice implications for captive breeding (Asa et al. 2011). The zoo community is becoming increasingly interested in this discussion, especially when faced with reproductive failure of breeding pairs due to mate incompatibility or aggression which can lead to injury or death (Wielebnowski et al. 2002). More recently, the integration of sexual selection into captive breeding programs has been promoted through symposia that bring together researchers in the field of mate choice and zoo population managers (e.g., St. Louis Zoo, USA, 2010). Despite this, practical implementation of mate choice methods by the zoo community is very limited because they are '*interested in including mate choice but simply do not know how to go about it and/or unsure of the implications for genetic management*' (Asa et al. 2011). Thus, there is an urgent research need to assess the costs and benefits of allowing mate choice in breeding programs. However, the relative benefit of including management strategies that account for sexual selection in captive population evolution are uncertain and have received little attention.

In this paper, we briefly review current genetic management guidelines in captive breeding and the potential for conflict between these as a baseline for exploring how management techniques could be informed by sexual selection and mate choice theory, and what benefits these insights could bring to captive breeding and reintroduction biology.

Current genetic management guidelines

Breeding histories and conservation goals vary for each species in captivity, and although this suggests genetic management should be tailored to each population relative to its specific short- and long-term program goals (Earnhardt 1999; Fa et al. 2011), most captive breeding programs for conservation utilize similar guidelines aimed at minimizing the rate of loss of genetic variability and inbreeding depression (Frankham et al. 2000; Fraser 2008; Wang and Ryman 2001; Williams and Hoffman 2009; see Fig. 1a).

Maximizing N_e/N ratio

The effective population size (N_e) is generally smaller than the absolute population size (N), with N_e being the size of

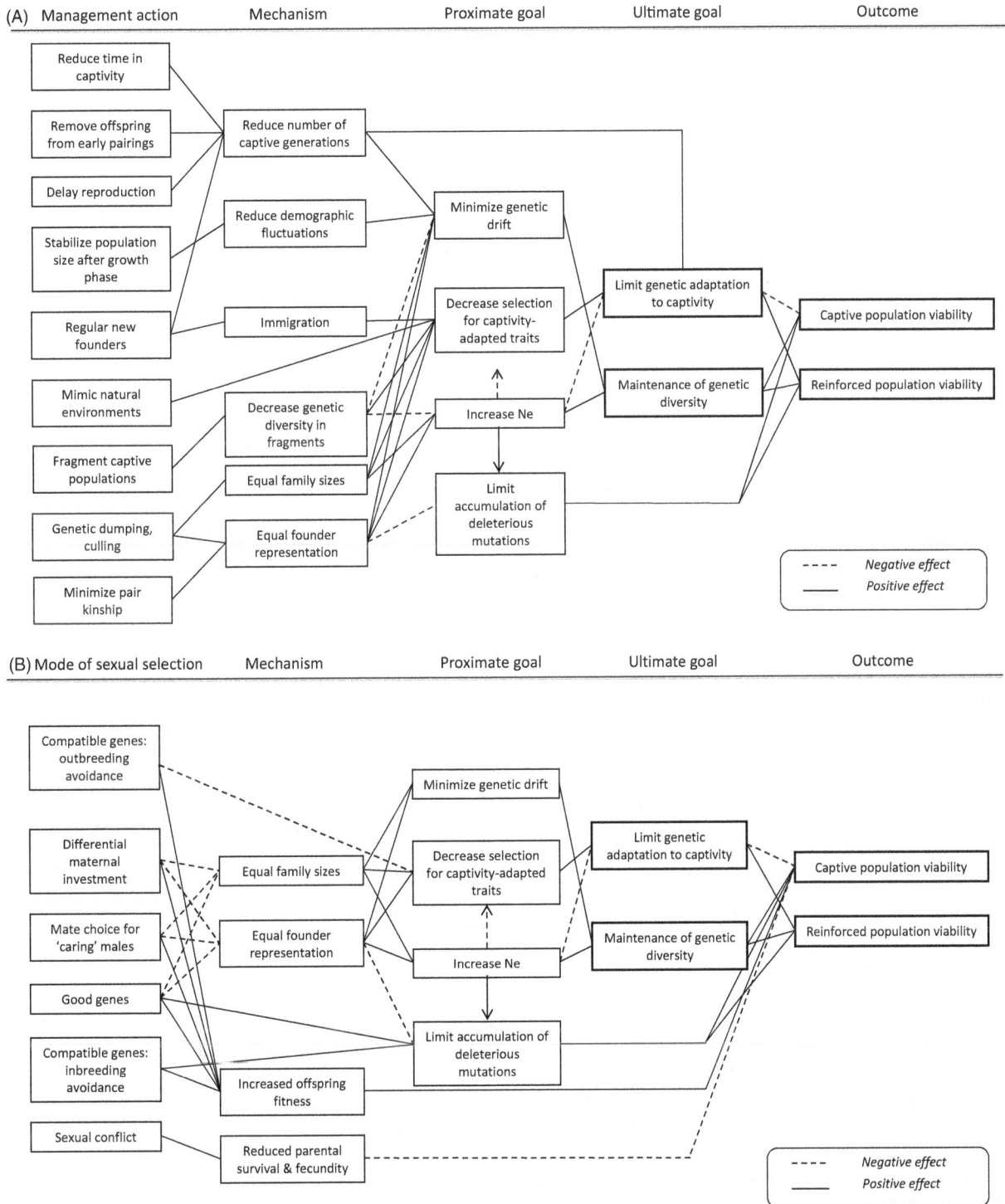

Figure 1 Interactions between management actions, goals, and outcome for the viability of captive and reinforced populations (A) and potential additional effects of sexual selection theories if female mate choice would be integrated to captive breeding programs (B). The direction of the linkages is from left to right unless otherwise specified by an arrowhead. Positive effects are indicated by a black line, negative effects by a red dashed line.

an idealized population with the same measure or rate of loss of some genetic quantity as that in the population under study (Allendorf and Luikart 2007). Because inbreeding depression and loss of heterozygosity are negatively related to N_e (Soulé 1980), one of the most important captive management aims for limiting loss of genetic diversity is to maximize N_e by equalizing family size, the sex ratio of breeders (Fa et al. 2011), and stabilizing population size after the initial population growth phase (Frankham 1995).

Equalizing founder representation and minimizing inbreeding

Another important strategy to limit genetic change is to equalize the representation of each founder in the captive population by minimizing kinship of mated pairs (Ballou and Lacy 1995; Frankham et al. 2000; Lacy 2000) or by removing offspring from breeders with the highest mean kinship (e.g., culling or 'genetic dumping' in Earnhardt 1999). Mean kinship is high when individuals are over-represented in the population, and low when they represent rare founder genetic lines (Ballou and Lacy 1995; Grahn et al. 1998; Saura et al. 2008; Asa et al. 2011). When founder contributions are equal, this increases N_e; thus reducing inbreeding and loss of genetic variation (Woodworth et al. 2002).

Minimizing the rate of random genetic drift

Demographic fluctuations increase genetic drift in captive populations (Frankham 1995); thus population sizes are usually stabilized after the initial population growth phase (Fa et al. 2011). Another method to slow the rate of genetic drift over time is to increase the generation length by delaying reproduction of breeders or the removal of offspring from early pairings (Williams and Hoffman 2009).

Limiting genetic adaptation to captivity

Organisms destined for subsequent reintroduction from captivity require genotypes suited to the reintroduction environment; however, genetic management for population viability in captivity does not take this into account. Indeed, there is increasing evidence that genotypes selected for under captive conditions are generally disadvantaged in natural environments (see (Frankham 2008; Williams and Hoffman 2009 for recent reviews). This has resulted in recent recommendations on how to manage genetic adaptation to captivity based on Frankham's (2008) equation, which positively relates the cumulative genetic change in reproductive fitness in captivity to selection, heritability, effective population size and, number of generations in captivity (see Box 1; Fig. 1a). Based on this, four options for minimizing genetic adaption to captivity have been recommended; however, not all of these are practical and some are in conflict with recommendations designed to limit losses of genetic variability (for more discussion of conflicts see below). First is minimizing the number of gen-

erations in captivity, either by reducing the length of the captive period, using cryopreservation or increasing generation length (Frankham 2008). This is seen as the most efficient method available because of the exponential relationship between number of generations and adaptation (Box 1), but it is not often feasible. Second is minimizing selection by creating captive environments that mimic natural habitats and/or through breeding strategies that reduce selection: such as equalizing founder representation through managing kinship of mated pairs and equalizing family sizes (Allendorf 1993; Frankham 2008). Minimizing variability in reproductive success removes the between-family component of selection, potentially halving the rate of genetic adaptation to captivity (Frankham and Loebel 1992; Saura et al. 2008). Third is minimizing the effective population size. Because this is in direct conflict with recommendations to preserve genetic variability (see above), it has been suggested that both goals can be achieved through fragmenting the captive population in to smaller management units (Frankham 2008; Margan et al. 1998; see section below). Finally is managing the captive population as a 'scmi-closed' system, and allowing the occasional recruitment of immigrants from wild populations to slow genetic adaptation (Frankham and Loebel 1992).

Box 1: Factors determining genetic adaptation to captivity

Frankham (2008) postulated that the cumulative genetic change in reproductive fitness in captivity over t generations (GA_t) can be derived from the breeder's equation (Lynch & Walsh 1998) and is a function of the selection differential (S), heritability (h^2), the effective population size (N_e), and number of generations in captivity (t):

$$GA_t \sim sh^2 \sum \left(1 - \frac{1}{2N_e}\right)^{t-1}$$

Thus, genetic adaptation to captivity will be positively related to the intensity of selection, genetic diversity, the effective population size, and number of generations in captivity (Frankham 2008; Williams and Hoffman 2009).

Conflicts, trade-offs, and fitness losses

Several approaches for managing genetic adaptation to captivity are incidental to already established practices for managing genetic variability in captive populations (e.g., minimizing the number of generations, equalizing family sizes and founder representation, and allowing the occasional recruitment of wild genotypes). However, there is a direct conflict between the recommendation concerning ideal captive population size for minimizing genetic drift and inbreeding (large N_e) and that for genetic adaptation to captivity (small N_e; Fig. 1a). This conflict between genetic goals is not a trivial concern as there is increasing evidence that adaptation to captivity in large populations can occur within very few generations (De Mestral and Herbinger 2013; Milot et al. 2013), resulting in serious fitness losses when organisms are released into the wild (reviewed in Williams and Hoffman 2009). Although much of the empirical evidence is still restricted to laboratory (e.g., Frankham and Loebel 1992; Lacy 2013) and commercial species (e.g., Laikre et al. 2010; Neff et al. 2011), it is well known that the reintroduction of organisms from captive breeding programs have lower fitness and lower probability of reintroduction success than those from the wild (Griffith et al. 1989; Wolf et al. 1996; Fischer and Lindenmayer 2000).

We see the potential for conflict in genetic management recommendations leading to three key decision making steps in captive breeding programs to limit fitness losses. First is assessing how long the captive population is expected to persist, how often and from where it may be reinforced, and if, when and how it may act as a source for reinforcing other populations (Lacy 2013). These program goals will largely determine how best to trade-off genetic variability against adaptation to captivity (step 2; see below), and how the needs of the captive population may be traded-off against the needs of reintroductions/wild supplementations (step 3). For example, if the captive population is being kept for reasons other than conservation reintroductions (e.g., public education), then adaptation to captivity effects can possibly be ignored or even promoted. There are fitness benefits to being well adapted to the local environment, so if populations will be permanently housed in captivity, behavioral and physiological adaptations suited to captivity may improve the fitness of captive animals (Woodworth et al. 2002) and thus the probability of long-term captive population persistence. The trade-off here being that organisms will change in some way from their wild counterparts, which may not be ideal if the purpose is to display or study 'natural' behaviors and morphologies, but they will change anyway; current genetic management in captivity is not a means of stopping genetic change, but simply slowing it (Lacy 2013).

Second, for those captive populations likely to be used for reintroductions or supplementations, how should effective population size in captivity be managed? Woodworth et al. (2002) show that fitness in captivity is expected to increase with increasing N_e because all genetic correlates with fitness operate in this direction. However, fitness in the wild after release shows a curvilinear pattern because of stronger adaptation to captivity with large N_e, while inbreeding and mutational accumulation reduce fitness for small N_e; thus, fitness is maximized in the wild after release from captive populations of a moderate size (Woodworth et al. 2002). Because of this relationship, it is now recommended that populations be managed in captivity through fragmentation (Margan et al. 1998; reviewed in Frankham 2008 and Williams and Hoffman 2009). This approach attempts to account for the opposing effects of N_e on fitness after release, whereby adaptation to captivity is reduced by fragmenting populations across institutions and allowing the small N_e to reduce genetic diversity at a local level (managing genetic adaptation to captivity), while retaining it at the metapopulation level (managing the loss of genetic diversity). Although the idea has theoretical and some empirical support, evidence from captive populations is extremely limited (Williams and Hoffman 2009).

Third, reintroducing captive animals to the wild is likely to involve a genetic trade off that is often not discussed, but one that may play a large role in reintroduction success (and future captive population viability)—that is, which animals should be released and which should remain in captivity? Earnhardt (1999) shows that the decision depends upon the relative value placed on the captive versus the reintroduced subpopulation. For example, one strategy (i.e., genetic dumping) promotes genetic diversity in the captive population at the expense of the reintroduced cohort; while minimizing kinship among released animals provides the greatest evolutionary potential for the release cohort, at the expense of the genetic health of the captive population. Thus, every reintroduction is a trade-off between the long-term persistence of the release and captive subpopulations and needs to be assessed on a case-by-case basis.

Incorporating mate choice into captive management

Current captive breeding programs primarily focus on limiting the loss of genetic diversity through the careful management of sexual pairings (see above). The nonrandom access to breeding partners usually increases the among-individual variance in reproductive success with few individuals securing most of the fertilizations and therefore reducing effective population size and increasing inbreeding. For these reasons, captive breeding programs are mostly based on enforced monogamy. However, because of concerns that such management may increasingly limit population evolvability and fitness (e.g., for animals released back into the wild; Frankham 2008; Neff et al. 2011), it has been suggested that integrating sexual selection into the genetic management of captive populations, by allowing reproductive partners to express their mating preference, may help long-term population viability (Wedekind 2002; Asa et al. 2011; Pélabon et al. 2014). Female mate choice is a key component of sexual selection and is the area where most attention is currently being focused in captive management (Asa et al. 2011). Because of this, we will leave the potentially important male component of sexual selection (i.e., male–male competition and male mate choice) to future analysis; however we will discuss the importance of sexual conflict.

Sexual selection refers to the process of nonrandom mate choice that arises as a consequence of interindividual competition for sexual partners. This results in the evolution of sexually selected traits (e.g., mate choice preferences) that arise through direct benefits to females (e.g., increased fecundity or parental care, Andersson 1994) or indirect benefits to offspring (e.g., the inheritance of alleles that increase attractiveness, 'sexy sons', Fisher 1930) or viability (good genes, compatible genes, Candolin and Heuschele 2008). The link between mate choice and increased population viability can potentially be made for three mechanisms driving sexual selection: (i) direct benefits to females through increased female fecundity, (ii) increased genetic quality of offspring through additive genetic variation in fitness (good genes; e.g., Chargé et al. 2011), and (iii) increased genetic quality of offspring through nonadditive genetic variation (compatible genes) (Candolin and Heuschele 2008; see Box 2). Ideally, to ensure the long-term success of captive breeding and release programs, genetic diversity, and genetic quality have to both be maintained along generations in captivity. Any benefit of mate choice will depend on the specific program's goals. We explore the possible benefits and costs associated with incorporating mate choice below, as well as highlighting questions and assumptions we feel need to be addressed. Box 2 gives an overview of the main hypotheses that explain the costs and benefits of nonrandom mate choice in animals. Table 1 and Fig. 1 give an overview of the complex interactions between possible genetic benefits and risks associated with the integration of the main sexual selection theories (Fig. 1b) into the current genetic management of captive populations (Fig. 1a).

Direct benefits and differential maternal investment

Equally relevant for guiding the choice of enforcing monogamy in captive breeding is the observation that multiply mated females usually adjust the investment they make into offspring, affecting progeny quality and survival. Multiple lines of evidence show that females adjust their investment in offspring depending on the male they are mated to (Gil 1999); for example, when mated to preferred males (i) female mallards lay more eggs (Cunningham and Russell 2000), (ii) female house mice produce

---

Box 2: Fitness benefits associated with females mate choice

The utility of male attributes selected via female mate choice is species-specific and likely to include one or a combination of the following:

Direct benefits: Females can attain direct fitness benefits from choosing mates that improve their own fecundity; such as the male's ability to nest build, rear offspring, courtship feed, or provide other valuable resources within the territory (Norris 1990; Møller 1994; Brown 1997). Female choice may also work to limit negative fitness consequences of pairing by avoiding unhealthy males, sexually transmitted diseases, or male infertility (e.g., by selecting feather brightness in birds;.(Hamilton and Zuk 1982; Kokko et al. 2002; Matthews et al. 1997; Pitcher and Evans 2001).

Differential maternal investment: Females may adjust their investment in offspring depending on male attractiveness. In mallards (*Anas platyrhynchos*), females laid more eggs when mated with preferred males (Cunningham and Russell 2000). In the Houbara bustard (*Chlamydotis undulata*), artificially inseminated females that were visually stimulated by attractive males had better hatching success and increased chick growth compared to those stimulated with less attractive males (Loyau and Lacroix 2010).

Good genes: Females may choose male phenotypes indicative of 'good genes', which improve the fitness of their progeny (Andersson 1994; Møller 1994; Neff and Pitcher 2005). The parasite-mediated sexual selection theory predicts that these good genes play a crucial role in parasite resistance (Hamilton and Zuk 1982), with offspring being more resistant to local pathogens, and thus conferring higher fitness (Buchholz 1995; Kirkpatrick and Ryan 1991; Penn and Potts 1999). There is increasing support for the degree of male ornamentation (and female preference for it) being correlated with genetic quality (see 'the handicap principle'; Zahavi 1975).

Compatible genes:

Inbreeding avoidance: Females may choose males based on the degree of relatedness to limit inbreeding depression (Kempenaers 2007). In guppies (*Poecilia reticulate*), females prefer to mate with males newly introduced or with rare phenotypes (Hugues 1991); in chickens, *Gallus gallus*, females hold less sperm after insemination by one of their brothers (Pizzari et al. 2004).

Outbreeding avoidance: Females may avoid outbreeding in order to maintain local adaptations, to select males to optimize the degree of relatedness, or simply to increase the representation of genes identical by descent (Höglund et al. 2002; Puurtinen 2011). For instance, house sparrow (*Passer domesticus*) males failed to form breeding pairs with females too dissimilar at major histocompatibility complex (MHC) loci (Bonneaud et al. 2006). Peron's tree frog (*Litoria peronei*) males that were genetically similar to the female achieved higher siring success than less genetically similar males (Sherman et al. 2008). In the three-spines sticklebacks, female seems to be able to 'count' the number of MHC alleles in the sexual partner and choose males that share an optimum number of alleles with them (Aeschlimann et al. 2003).

Heterozygote advantage in offspring: Females may also seek to maximize heterozygosity in the offspring at key loci or at many loci (Brown 1997). For instance, in the domestic sheep, homozygous ewes inheriting mutant alleles from both parents have lower fecundity compare to heterozygous individuals for the same loci expressing increased ovulation rate (Gemmell and Slate 2006). Females may also try to maximize the offspring heterozygosity at key loci such as at MHC genes (reviewed in Penn 2002). In mice (and in humans), females prefer to mate with males carrying dissimilar MHC alleles than their own (Wedekind and Furi 1997; Penn and Potts 1998) which may enhance offspring immunocompetence. Although there is little evidence from tests of single parasites to support this hypothesis, MHC-heterozygous offspring may be resistant to multiple parasites (Penn and Potts 1999 and references within).

Sexy sons: Females may express a preference for heritable attractiveness in males, regardless of the utility of the trait. This may occur if the genes for the female preference become associated in linkage disequilibrium with genes for the trait underlying males attractiveness; females will select males that also carry the genes for the female preference of that male trait. This produces a positive feedback 'runaway' loop that is assumed to lead to the extravagance of male traits until the costs of such secondary sexual traits in terms of survival exceed the benefits in term of reproductive success (Fisher 1930; Weatherhead and Robertson 1979). Empirical evidence comes from studies on fruit flies and European starlings (Gwinner and Schwabl 2005; Taylor et al. 2007).

larger litter sizes (Drickamer et al. 2000), and (iii) female birds, insects, and crustacea deposit more testosterone in their eggs (Gil 1999; Kotiaho et al. 2003; Galeotti et al. 2006; Loyau et al. 2007). More recently, differential maternal investment has been investigated in supportive captive breeding of the endangered Houbara bustard. Artificially inseminated females visually stimulated by attractive males increased their hatching success as well as the allocation of androgens in their eggs and increased growth rate in chicks (Loyau and Lacroix 2010). Here, it was emphasized that using artificial insemination for species conservation without appropriate stimulation of the breeding females may lower their breeding performance with negative impact on the population viability. Thus, maximizing parental effort by allowing free mate choice in captive-bred populations might increase offspring quality and help in the long-term viability of captive and reinforced wild populations (Asa et al. 2011). However, while this expectation seems reasonable, it is unlikely to be this straightforward (Kokko and Brooks 2003). In a recent review of current progress in implementing mate choice in captive breeding programs (Asa et al. 2011), the zoo community's initial steps are primarily focusing on the direct benefits of

female mate choice to improve the probability of successful mating in valuable animals. While there is a general perception that giving animals choice should improve female fecundity (and consequently improve population persistence), there are a number of issues that need to be clarified from a captive breeding perspective. First is the general problem with female choice increasing the variance in reproductive success, thereby decreasing effective population size and increasing any imbalance in founder representation (Wedekind 2002). Thus, including mate choice in breeding management appears to directly conflict with current management goals that aim to minimize the loss of genetic variation and adaptation to captivity (Asa et al. 2011): adding an additional level of complexity in determining the best breeding strategy for captive populations (see above). Second is the idea that females in captivity are able to make accurate choices about male quality. Managers need to be clear on whether they are providing real choice for females to find the best mates or simply providing a 'simulation' of natural breeding to 'trick' females into increasing their reproductive investment accordingly. If we want females to make informed mate choice decisions, this makes a very strong assumption that male quality under captive conditions can be differentiated by females, even when limiting resources have been provided for. For example, if male coloration in the wild is a cue for health, territory quality, or foraging ability (e.g., Wolfenbarger 1999; Saks et al. 2003; Karino et al. 2005), how is it expressed under captive conditions where veterinary care is ongoing, food is provided *ad libitum*, and housing is standardized? Thus, the expected fecundity benefits in captivity may be much smaller (or even absent) compared to studies from wild populations where female choice is correlated with a limiting resource being provided by males. Third is the possibility that reproduction and survival (or current versus future reproduction) are traded-off against each other (Saino et al. 1999). Thus, it is possible that by promoting current reproductive output via direct benefits, future reproductive potential may be compromised; however, these effects are predicted in wild populations, and it is generally unknown how such relationships are affected by captive environments where key resources may not be limiting.

Benefits of sexual selection for population fitness and adaptation rate

Sexual selection can be a powerful force contributing to purge deleterious mutations from the genome, and theoretical work has shown that this can produce a net benefit that can improve population mean fitness and the rate of adaptation (Agrawal 2001; Siller 2001; Lorch et al. 2003; Whitlock and Agrawal 2009). Testing the benefit of sexual

Table 1. Potential benefits (B) and risks (R) from integrating theories of female mate choice into captive breeding programs for the viability of the captive population and that of any cohorts released into the wild. See Box 2 for a summary of each theory. When no benefit or risk was obvious, we indicate it by '?'; however, it suggests that more research is needed rather than implies that no risks can be safely assumed.

Theory	Impact on captive population	Impact on released cohort
Direct benefits/Maternal investment		
B	Increase female fecundity, lifespan, and offspring viability	Maintain males secondary sexual traits
	Select healthier males that afford expressing strong parental effort	
R	Decrease lifespan reproductive success if trade-off with parental effort	Select males adapted to captivity
Good genes		
B	Purge deleterious alleles	Select resistant individuals (e.g. if similar pathogens in the wild and in captivity)
R	Loss of genetic variance	Loss of genetic variance
	Decrease female fitness in case of sexual conflict	Select males adapted to captivity
	Decrease some fitness traits in males (e.g. if trade-off between immunity, reproduction, and lifespan)	Decrease female fitness in case of sexual conflict
Compatible genes		
Inbreeding avoidance		
B	Minimize inbreeding depression	Minimize inbreeding depression
R	Misled mate choice between kinship and familiarity	Loss of local adaptationMisled mate choice between kinship and familiarity
Outbreeding avoidance/'(k)inbreeding selection		
B	?	Maintain local adaptation
R	Increase risks of inbreeding depression	Increase risks of inbreeding depression
Maximizing heterozygosity in the offspring		
B	Minimize inbreeding depression	Minimize inbreeding depression
	Improve offspring viability	
	(heterozygous advantage)	Improve offspring viability (heterozygous advantage)
R	?	?
Sexy sons		
B	Maintain male ornamentation and female preferences	Maintain male ornamentation and female preferences
R	Decrease female fitness in case of sexual conflict	Select males adapted to captivity
		Decrease female fitness in case of sexual conflict

selection for population mean fitness, and the rate of adaptation has been achieved through experimental evolution approaches where females were either forced to mate under a monogamous regime or were allowed to mate with several males. For obvious reasons linked to generation time and laboratory facilities, this approach has mostly involved insects and other invertebrates, with a couple of notable exceptions involving guppies (*Poecilia reticulata*) and house mice (*Mus domesticus*; see examples below and Holman and Kokko 2013 for a recent overview of the topic).

In an elegant experiment, Almbro and Simmons (2014) exposed dung beetles (*Onthophagus taurus*) to a mutagenic treatment with ionizing radiation and then selected beetles under either enforced monogamy or sexual selection. After only two generations of sexual selection regime, the expression of male strength, a sexually selected trait, of irradiated beetles was almost twice as large as for the monogamous lines, and almost recovered the values of nonirradiated control individuals. In guppies, Pélabon et al. (2014) conducted an experimental evolution study where 19 popula-

tions of guppies were exposed to an enforced monogamous or a polygamous mating system for nine generations. Offspring size decreased across generations in both regimes, but the decrease was more pronounced in the enforced monogamy treatment. Therefore, despite being held in a benign (captive) environment for only nine generations, preventing mate choice and sexual selection resulted in the reduction in the expression of a trait that is potentially correlated with fitness (both sexual and nonsexual) in the wild. In the only mammalian species where the effect of mating system has been investigated, the house mouse, females that were free to mate with preferred mates produced (i) more litters, (ii) socially dominant sons, (iii) offspring with a better survival compared to females forced to mate with nonpreferred males (Drickamer et al. 2000). In addition to this, an experimental evolution approach where house mice were either polygamously or monogamously mated during 14 generations showed that offspring viability was improved when they were sired by males that had experienced the polygamous selection regime (Firman and Simmons 2012). Therefore, in the only study where divergent selection lines for mating system have been used in a mammal, sexual selection appears to confer a long-term fitness benefits to males and females, suggesting concordant effect on sexual and nonsexual traits.

Ultimately, if sexual selection produces a net benefit on population mean fitness, this should reduce the population extinction risk. Jarzebowska and Radwan (2009) used small populations (five males and five females) of the bulb mites (*Rhizoglyphus robini*) facing either enforced monogamy or sexual selection and looked at the extinction probability of each line. They found that 49% of the lines in the monogamy treatment went extinct versus 27% in the sexual selection group. In a similar experiment using the same species, Plesnar-Bielak et al. (2012) showed that the extinction probability of lines selected under enforced monogamy or sexual selection markedly differed when exposed to a harsh environment (a temperature stress): 100% of monogamous lines went extinct when reared at high temperature versus 0% for lines experiencing sexual selection.

Costly sexual traits and sexual conflict

Despite some studies providing supportive evidence that sexual selection promotes population mean fitness, this is not always the case as several examples show sexual selection does not purge deleterious mutation nor improve population fitness (in *Drosophila melanogaster*, Arbuthnott and Rundle 2012; Hollis and Houle 2011) or on the rate of adaptation to a novel environment (in the yeast, *Saccharomyces cerevisiae*, Reding et al. 2013). Moreover, sexual selection can favor the evolution of traits that have fitness costs and are instead associated with mating success (sexy sons, Fisher 1930; signal honesty, Zahavi 1975 or sexually antagonistic coevolution, Holland and Rice 1998). The Fisher–Zahavi traits evolve so that the benefits to the male in terms of mating success from female preferences are balanced by the costs of the traits. Because there are no population benefits involved, sexual selection primarily driven by these processes might be a burden when conditions change. This occurs because sexual selection is expected to exert its strongest negative effects on population viability under rapidly changing conditions when there is not enough time available for the costs of sexual traits to be adjusted to the new conditions (Candolin and Heuschele 2008). This is particularly relevant to understanding the possible role of sexual selection on the adaptation of captive populations to a novel environment, but the effect for most populations is currently unknown.

Sexual selection through antagonistic selection is a widespread phenomenon (Cox and Calsbeek 2009) that has been well documented and its associated theoretical framework intensively tackled (reviewed in Bonduriansky and

Chenoweth 2009; Cox and Calsbeek 2009; Van Doorn 2009). Two main forms of sexual conflict can be distinguished: the antagonistic interactions between the sexes (i.e., interlocus sexual conflict) and the genetic trade-offs for fitness between males and females (i.e., intralocus sexual conflict). Interlocus sexual conflict occurs when traits coded by alleles at different loci evolved so that it enhances the reproductive success in individuals from one sex at the cost of the fitness of their mating partners (Chapman et al. 2003; Bonduriansky and Chenoweth 2009). Behavioral examples include sexual coercion, mate guarding or mating plug, physical or physiological harassment of the partner, evasion of parental care, and resistance against mating (Chapman et al. 2003; Van Doorn 2009 and references within). By contrast, intralocus sexual conflict arises when the same set of fitness-related loci between sexes is subject to opposing selection pressures, preventing males and females from reaching their optima independently (Lande 1980; Chippindale et al. 2001). For instance, some secondary sexual traits in males improve male–male competition and mating success but are costly to produce for females, like horn phenotype in the Soay sheep, *Ovis aries* (Robinson et al. 2006) and red bill color in zebra finches, *Taeniopygia guttata* (Price and Burley 1994). In red deer, *Cervus elaphus*, selection favors males that carry low breeding values for female fitness resulting in the situation where males with relatively high fitness sired daughters with relatively low fitness (Foerster et al. 2007). Intralocus sexual conflict is controversial because such conflict is thought to be resolvable through the evolution of sex-specific gene expression, sex-linkage, and sexual dimorphism, enabling each sex to reach its adaptive optima (Bonduriansky and Chenoweth 2009; Cox and Calsbeek 2009; Stewart et al. 2010). But recent studies have shown that the conflict was not so easily resolved (Harano et al. 2010; Poissant et al. 2010; Tarka et al. 2014).

Such sexual conflicts may be relevant to population persistence, population genetics, and adaptation. When sexual conflict favors males, female fecundity is often reduced which may affect in turn population demography, mean population fitness, and increase extinction risks (Kokko and Brooks 2003; Rice et al. 2006; Morrow et al. 2008; Bonduriansky and Chenoweth 2009). It is thus important to account for sexual conflicts in the captive breeding programs to predict long-term outcomes of sexual selection on captive and reinforced population viability. For instance, in the lizard *Lacerta vivipara*, male sexual behavior is harmful and male-skewed sex ratios can threaten population persistence (Le Gaillard et al. 1998; see also Low 2005).

Because inter- and intralocus conflict have different genetic consequences, it is important to distinguish the evolutionary outcomes from both strategies. Although evolutionary outcomes of sexual conflicts are not yet fully understood, we briefly synthesize current knowledge. Interlocus sexual conflict generates coevolutionary arms races which have been thought to accelerate evolution of traits, particularly the antagonistic evolution of reproductive traits (Van Doorn 2009; Arbuthnott and Rundle 2012); this opposes the goal of captive breeding programs to maintain genetic diversity and prevent (response to) selection. Interlocus sexual conflict resulting in direct harm to females could be compensated by indirect genetic benefits (good genes or sexy sons, Cox and Calsbeek 2009). However, several empirical studies failed to show that costs related to sexual conflict were counterbalanced by good genes (Stewart et al. 2008), sexy sons (Rice et al. 2006), or compatible genes (Garner et al. 2010).

The evolutionary importance of intralocus sexual conflict is still debated (Chapman et al. 2003; Cox and Calsbeek 2009), with current evidence suggesting that when intralocus sexual conflict occurs across multiple loci, the so-called tug-of-war can neutralize benefits from sexual selection (Cox and Calsbeek 2009 and references within) and reduce population mean fitness (Bonduriansky and Chenoweth 2009). Paradoxically, theory also suggests the potential role of intralocus sexual conflict in maintaining genetic variation, although this idea has received little attention so far (Foerster et al. 2007). Antagonistic selection may maintain substantial levels of genetic variation in life history traits despite the directional selection to which they are subject (Kruuk et al. 2000); data from red deer natural populations show that sexually antagonistic selection could maintain heritable genetic variance in reproductive traits and fitness variation. Similarly in *Drosophila melanogaster*, gender-specific selection on loci expressed in both sexes may contribute to the maintenance of high levels of genetic variance for fitness within each sex (Chippindale et al. 2001). Sexual conflict could thus maintain genetic variation for fitness despite strong selection (Foerster et al. 2007). This genetic outcome may be of particular interest for the management of captive populations, but a detailed understanding of the strength of intralocus sexual conflict and its contribution to the maintenance of genetic variation will clearly require careful consideration (Foerster et al. 2007; Bonduriansky and Chenoweth 2009; Cox and Calsbeek 2009).

The net benefit of allowing sexual selection to operate likely depends on the relative importance of costs induced by sexual conflicts and benefits induced by the purging of mutational load. In some cases, environmental condition and population history can strongly modulate the net benefit of sexual selection. For instance, if populations are exposed to the arrival of newly maladapted alleles, the benefit of purging these alleles might outweigh the potential cost due to sexual conflicts. Long et al. (2012) have recently tested this idea using experimental populations of *Drosoph-ila melanogaster* that were either well adapted to their environment (cadmium-adapted populations), either pushed away from their adaptive peak by the income of migrant, maladapted, alleles. For each of these populations, they identified sexually successful and nonsuccessful males and used them to sire offspring. In agreement with the predictions, they found that sexually successful males sired unfit daughters in well-adapted populations, which corroborate the finding that sexual conflict produces a mismatch between sexual and nonsexual fitness in this species. However, sexually successful males sired fitter daughters in the populations where adaptation was prevented by the income of migrant alleles. This suggests that in unstable populations, the net benefit of purging deleterious alleles outweighs the cost of sexual conflicts. These results are mirrored by those reported by another recent study where the outcome of exposure to a regime of enforced monogamy versus polyandry depends on environmental quality (Grazer et al. 2014). Flour beetles (*Triboleum castaneum*) were maintained for 39 generation either under enforced monogamy or polyandry. Beetles from these selection lines were exposed to a poor or a good environment in terms of food quality. Reproductive success of pairs formed by males from the sexual selection lines and females from the enforced monogamy was low when reared in the good environment, again suggesting that sexual conflict incurs cost. However, when sexually selected males were mated with enforced monogamous females in the poor-quality environment, their reproductive success was improved suggesting that the benefits of sexual section outweighed the cost of sexual conflicts under stressful conditions. Despite the evidence of a net benefit of female choice to population viability from many of these studies, and hence suggesting that captive population management would benefit by incorporating female choice, these 'benefits' have generally not been considered within the complex framework of interactions and conflicting goals for long-term population persistence (e.g., Fig. 1). Thus, we encourage caution before female choice measures are adopted in captive breeding programs (see below).

Conclusions

To date, the main genetic focus of captive breeding programs has been on preserving genetic diversity, while genetic integrity is often neglected because of difficulty in measuring progress and conflicts with other genetic guidelines (on the basis of N_e). One means of preserving genetic integrity is incorporating female choice for male traits in captive breeding management. Based on current limited theoretical and empirical evidence, it appears that some mechanisms for mate choice may be safer to exploit than others. On the safer side are female preferences for compatible genes, general heterozygosity or allelic diversity at specific locus (e.g., major histocompatibility complex (MHC) genes), and differential maternal investment based on male's attractiveness. At the riskier end of the spectrum is selection for good genes in the presence of sexual conflict, as this could favor adaptation to captivity in males while decreasing female fitness by creating unbalanced selection pressures with sexual selection on males, while natural selection is lifted on females. One possibility we have not explored in our review is sexual selection acting on females through male choice and female–female competition; if and where this occurs, it could help balancing selection on both sexes and potentially obtain better results in terms of fitness for both. Another area still unexplored is the potential for integrating male–male competition; however, the risks of favoring males best adapted to captivity would likely be the same as in the good genes hypothesis.

Although there has been increasing attention focused on mate choice as a potential way of improving captive population management, its impact on genetic variability and adaptation to captivity is complex (Fig. 1). Incorporating mate choice into captive breeding recommendations presents a huge challenge, both in terms of the logistics of offering mate choice in captive settings and in implementing choice in a way that augments rather than hinders population management goals (Asa et al. 2011; see Fig. 1). Despite this, progress is possible, and a first step is identifying the key questions that need to be asked before considering implementing mate choice into a breeding program. First is assessing how long the captive population is expected to persist, how often and from where it may be reinforced, and if, when and how it may act as a source for reinforcing other populations (Lacy 2013). This should be a first step in any decision regarding the genetic management of captive populations because it determines how genetic adaptation to captivity needs to be considered, especially if mate choice accelerates adaptation to captivity. Second is identifying the mechanism (or sexual selection theory) driving mate choice in the system of interest. Is it likely that mate choice is linked to improved population persistence, and if so, are the expected benefits likely to be via improved fecundity of breeding females or the genetics of offspring? Also, is it reasonable to expect that phenotypic traits in males that females select on are still valid cues for genetic quality in captivity? Third is identifying whether sexual conflicts exist in the mating system. Fourth is considering the potential for conflicts with other genetic management goals. Because mate choice increases variation in mating success, this will generally reduce effective population size and erode genetic diversity in the captive population; thus, the benefits of incorporating mate choice will need to be balanced against any costs.

Currently, we need specific questions to be asked that link directly to the needs of captive management and then specific studies implemented (both empirical and via simulation studies) to look at specific management approaches, such has been successfully achieved in identifying ways to manage genetic adaptation to captivity through population fragmentation (Margan et al. 1998; Frankham 2008). It is only then that we will begin to seriously contribute to the genetic health of captive populations and the success of reintroductions. Thus, the goal of this review has not been to provide definitive answers and recommendations on the benefits (and costs) of mate choice and sexual selection in the management of captive populations, but rather to highlight the complexity of the relationships between mate choice, population fitness, and the current genetic goals of maintaining small populations in captivity. From this, we hope to encourage clear goal-oriented research and critical thinking into the role of mate choice and sexual selection in an area where its application and study are currently in its infancy.

Acknowledgements

We thank Christophe Eizaguirre and Miguel Soares for inviting us to contribute to this special issue, and two anonymous referees for helpful comments on earlier draft of this manuscript. We would like to thank Mikael Puurtinen for valuable discussion about the revised version of the manuscript and his great advice for the mind map. RC was funded by the Academy of Finland and the Centre of Excellence in Biological Interaction Research. CT was funded by the Agence Nationale de la Recherche (grant ANR-12-ADAP-0006-02-PEPS), and ML was supported by the Swedish Research Council (VR 2013-3634).

Literature cited

Aeschlimann, P. B., M. A. Häberli, T. B. H. Reusch, T. Boehm, and M. Milinski 2003. Female sticklebacks *Gasterosteus aculeatus* use self-reference to optimize MHC allele number during mate selection. Behavioral Ecology and Sociobiology **54**:119–126.

Agrawal, A. F. 2001. Sexual selection and the maintenance of sexual reproduction. Nature **411**:692–695.

Allendorf, F. W. 1993. Delay of adaptation to captive breeding by equalizing family size. Conservation Biology 7:416–419.

Allendorf, F. W., and G. Luikart 2007. Conservation and the Genetics of Populations. Blackwell Publishing, Oxford, UK.

Almbro, M., and L. W. Simmons 2014. Sexual selection can remove an experimentally induced mutation load. Evolution **68**:295–300.

Andersson, M. 1994. Sexual Selection. Princeton University Press, Princeton, NJ.

Arbuthnott, D., and H. D. Rundle 2012. Sexual selection is ineffectual or inhibits the purging of deleterious mutations in *Drosophila melanogaster*: sexual selection and deleterious mutations. Evolution **66**:2127–2137.

Asa, C. S., K. Traylor-Holzer, and R. C. Lacy 2011. Can conservation-breeding programmes be improved by incorporating mate choice? The International Zoo Yearbook 45:203–212.

Ballou, J. D., and R. C. Lacy 1995. Identifying Genetically Important Individuals for Management of Genetic Variation in Pedigreed Populations. Columbia University Press, New York, NY.

Barnosky, A. D., N. Matzke, S. Tomiya, G. O. U. Wogan, B. Swartz, T. B. Quental, C. Marshall, et al. 2011. Has the Earth's sixth mass extinction already arrived? Nature 471:51–57.

Bonduriansky, R., and S. F. Chenoweth 2009. Intralocus sexual conflict. Trends in Ecology & Evolution 24:280–288.

Bonneaud, C., O. Chastel, P. Federici, H. Westerdahl, and G. Sorci 2006. Complex Mhc-based mate choice in a wild passerine. Proceedings of the Royal Society. B, Biological sciences 273:1111–1116.

Brown, W. D. 1997. Courtship feeding in tree crickets increases insemination and female reproductive life span. Animal Behaviour 54:1369–1382.

Buchholz, R. 1995. Female choice, parasite load and male ornamentation in wild turkeys. Animal Behaviour 50:929–943.

Candolin, U., and J. Heuschele 2008. Is sexual selection beneficial during adaptation to environmental change? Trends in Ecology & Evolution 23:446–452.

Chapman, T., G. Arnqvist, J. Bangham, and L. Rowe 2003. Sexual conflict. Trends in Ecology & Evolution 18:41–47.

Chargé, R., G. Sorci, Y. Hingrat, F. Lacroix, and M. Saint Jalme 2011. Immune-mediated change in the expression of a sexual trait predicts offspring survival in the wild. PLoS ONE 6:e25305.

Chargé, R., G. Sorci, M. Saint Jalme, L. Lesobre, Y. Hingrat, F. Lacroix, and C. Teplitsky 2014. Does recognized genetic management in supportive breeding prevent genetic changes in life-history traits? Evolutionary Applications 7:521–532.

Charlesworth, D., and J. H. Willis 2009. The genetics of inbreeding depression. Nature Reviews Genetics 10:783–796.

Chippindale, A. K., J. R. Gibson, and W. R. Rice 2001. Negative genetic correlation for adult fitness between sexes reveals ontogenetic conflict in Drosophila. Proceedings of the National Academy of Sciences of the United States of America 98:1671–1675.

Cox, R. M., and R. Calsbeek 2009. Sexually antagonistic selection, sexual dimorphism, and the resolution of intralocus sexual conflict. The American Naturalist 173:176–187.

Cunningham, E. J. A., and A. F. Russell 2000. Egg investment is influenced by male attractiveness in the mallard. Nature 404:74–77.

De Mestral, L. G., and C. M. Herbinger 2013. Reduction in antipredator response detected between first and second generations of endangered juvenile Atlantic salmon Salmo salar in a captive breeding and rearing programme: antipredator response of Salmo salar fry. Journal of Fish Biology 83:1268–1286.

Drickamer, L. C., P. A. Gowaty, and C. M. Holmes 2000. Free female mate choice in house mice affects reproductive success and offspring viability and performance. Animal Behaviour 59:371–378.

Earnhardt, J. M. 1999. Reintroduction programmes: genetic trade-offs for populations. Animal Conservation 2:279–286.

Ewen, J. G. 2012. Reintroduction Biology Integrating Science and Management. Wiley-Blackwell, Hoboken, NJ.

Fa, J. E., S. M. Funk, and D. O'Connell 2011. Zoo Conservation Biology. Cambridge University Press, Cambridge.

Firman, R. C., and L. W. Simmons 2012. Male house mice evolving with post-copulatory sexual selection sire embryos with increased viability: post-copulatory sexual selection and embryo viability in mice. Ecology Letters 15:42–46.

Fischer, J., and D. Lindenmayer 2000. An assessment of the published results of animal relocations. Biological Conservation 96:1–11.

Fisher, R. A. 1930. The Genetical Theory of Natural Selection. Clarendon Press, Oxford.

Foerster, K., T. Coulson, B. C. Sheldon, J. M. Pemberton, T. H. Clutton-Brock, and L. E. B. Kruuk 2007. Sexually antagonistic genetic variation for fitness in red deer. Nature 447:1107–1110.

Ford, M. J. 2002. Selection in captivity during supportive breeding may reduce fitness in the wild. Conservation Biology 16:815–825.

Frankham, R. 1995. Conservation genetics. Annual Review of Genetics 29:305–327.

Frankham, R. 2008. Genetic adaptation to captivity in species conservation programs. Molecular Ecology 17:325–333.

Frankham, R., and D. A. Loebel 1992. Modeling problems in conservation genetics using captive Drosophila populations: rapid genetic adaptation to captivity. Zoo Biology 11:333–342.

Frankham, R., H. Manning, S. H. Margan, and D. A. Briscoe 2000. Does equalization of family sizes reduce genetic adaptation to captivity? Animal Conservation 4:357–363.

Fraser, D. J. 2008. How well can captive breeding programs conserve biodiversity? A review of salmonids. Evolutionary Applications 1:1–52.

Gaillard, J.-M., M. Festa-Bianchet, and N. G. Gilles Yoccoz 1998. Population dynamics of large herbivores: variable recruitment with constant adult survival. Trends in Ecology & Evolution 13:58–63.

Galeotti, P., D. Rubolini, G. Fea, D. Ghia, P. A. Nardi, F. Gherardi, and M. Fasola 2006. Female freshwater crayfish adjust egg and clutch size in relation to multiple male traits. Proceedings of the Royal Society. B, Biological sciences 273:1105–1110.

Garner, S. R., R. N. Bortoluzzi, D. D. Heath, and B. D. Neff 2010. Sexual conflict inhibits female mate choice for major histocompatibility complex dissimilarity in Chinook salmon. Proceedings of the Royal Society. B, Biological sciences 277:885–894.

Gemmell, N. J., and J. Slate 2006. Heterozygote advantage for fecundity. PLoS ONE 1:e125.

Gil, D. 1999. Male attractiveness and differential testosterone investment in zebra finch eggs. Science 286:126–128.

Grahn, M., A. Langefors, and T. von Schantz 1998. The Importance of Mate Choice in Improving Viability in Captive Populations, pp. 341–363. Oxford University Press, Oxford, UK.

Grazer, V. M., M. Demont, Ł. Michalczyk, M. J. Gage, and O. Y. Martin 2014. Environmental quality alters female costs and benefits of evolving under enforced monogamy. BMC Evolutionary Biology 14:21.

Griffith, B., J. M. Scott, J. W. Carpenter, and C. Reed 1989. Translocation as a species conservation tool: status and strategy. Science 245:477–480.

Gwinner, H., and H. Schwabl 2005. Evidence for sexy sons in European starlings (Sturnus vulgaris). Behavioral Ecology and Sociobiology 58:375–382.

Hamilton, W. D., and M. Zuk 1982. Heritable true fitness and bright birds: a role for parasites? Science 218:384–387.

Harano, T., K. Okada, S. Nakayama, T. Miyatake, and D. J. Hosken 2010. Intralocus sexual conflict unresolved by sex-limited trait expression. Current Biology 20:2036–2039.

Höglund, J., S. B. Piertney, R. V. Alatalo, J. Lindell, A. Lundberg, and P. T. Rintamäki. 2002. Inbreeding depression and male fitness in black grouse. Proceedings of the Royal Society. B, Biological Sciences 269:711–715.

Holland, B., and W. R. Rice 1998. Perspective: chase-away sexual selection: antagonistic seduction versus resistance. Evolution 52:1.

Hollis, B., and D. Houle 2011. Populations with elevated mutation load do not benefit from the operation of sexual selection: sexual selection and mutation load. Journal of Evolutionary Biology 24:1918–1926.

Holman, L., and H. Kokko 2013. The consequences of polyandry for population viability, extinction risk and conservation. Philosophical Transactions of the Royal Society of London. Series B, Biological Sciences 368:20120053.

Hugues, A. L. 1991. MHC polymorphism and the design of captive breeding programs. Conservation Biology 5:249–252.

Jarzebowska, M., and J. Radwan 2009. Sexual selection counteracts extinction of small populations of the bulb mites. Evolution 64:1283–1289.

Karino, K., T. Utagawa, and S. Shinjo 2005. Heritability of the algal-foraging ability: an indirect benefit of female mate preference for males' carotenoid-based coloration in the guppy, Poecilia reticulata. Behavioral Ecology and Sociobiology 59:1–5.

Kempenaers, B. 2007. Mate choice and genetic quality: a review of the heterozygosity theory. Advances in the Study of Behavior 37:189–278.

Kirkpatrick, M., and M. J. Ryan 1991. The evolution of mating preferences and the paradox of the lek. Nature 350:33–38.

Kokko, H., and R. Brooks 2003. Sexy to die for? Sexual selection and the risk of extinction. Annales Zoologici Fennici 40:207–219.

Kokko, H., E. Ranta, G. Ruxton, and P. Lundberg 2002. Sexually transmitted disease and the evolution of mating systems. Evolution 56:1091–1100.

Kotiaho, J. S., L. W. Simmons, J. Hunt, and J. L. Tomkins 2003. Males influence maternal effects that promote sexual selection: a quantitative genetic experiment with dung beetles Onthophagus taurus. The American Naturalist 161:852–859.

Kruuk, L. E. B., T. H. Clutton-Brock, J. Slate, J. M. Pemberton, S. Brotherstone, and F. E. Guinness 2000. Heritability of fitness in a wild mammal population. Proceedings of the National Academy of Sciences of the United States of America 97:698–703.

Lacy, R. C. 1987. Loss of genetic diversity from managed populations: interacting effects of drift, mutation, immigration, selection, and population subdivision. Conservation Biology 1:143–158.

Lacy, R. C. 1994. Managing genetic diversity in captive populations of animals. In M. L. Bowles, and C. J. Whelan, eds. Restoration and Recovery of Endangered Plants and Animals, pp. 63–89. Cambridge University Press, Cambridge.

Lacy, R. C. 2000. Should we select genetic alleles in our conservation breeding programs? Zoo Biology 19:279–282.

Lacy, R. C. 2013. Achieving true sustainability of zoo populations: achieving true sustainability of zoo populations. Zoo Biology 32:19–26.

Laikre, L. 1999. Hereditary defects and conservation genetic management of captive populations. Zoo Biology 18:81–99.

Laikre, L., M. K. Schwartz, R. S. Waples, and N. Ryman 2010. Compromising genetic diversity in the wild: unmonitored large-scale release of plants and animals. Trends in Ecology & Evolution 25:520–529.

Lande, R. 1980. Sexual dimorphism, sexual selection, and adaptation in polygenic characters. Evolution 34:292.

Long, T. A. F., A. F. Agrawal, and L. Rowe 2012. The effect of sexual selection on offspring fitness depends on the nature of genetic variation. Current Biology 22:204–208.

Lorch, P. D., S. Proulx, L. Rowe, and T. Day 2003. Condition-dependent sexual selection can accelerate adaptation. Evolutionary Ecology Research 5:867–881.

Low, M. 2005. Female resistance and male force: context and patterns of copulation in the New Zealand stitchbird Notiomystis cincta. Journal of Avian Biology 36:436–448.

Loyau, A., and F. Lacroix 2010. Watching sexy displays improves hatching success and offspring growth through maternal allocation. Proceedings of the Royal Society. B, Biological Sciences 277:3453–3460.

Loyau, A., M. Saint Jalme, R. Mauget, and G. Sorci 2007. Male sexual attractiveness affects the investment of maternal resources into the eggs in peafowl (Pavo cristatus). Behavioral Ecology and Sociobiology 61:1043–1052.

Lynch, M., and B. Walsh 1998. Genetics and Analysis of Quantitative Traits. Sinauer Associates, Sunderland, MA.

Margan, S. H., R. K. Nurthen, M. E. Montgomery, L. M. Woodworth, E. H. Lowe, D. A. Briscoe, and R. Frankham 1998. Single large or several small? Population fragmentation in the captive management of endangered species. Zoo Biology 17:467–480.

Matthews, I. M., J. P. Evans, and A. E. Magurran 1997. Male display rate reveals ejaculate characteristics in the Trinidadian guppy Poecilia reticulata. Proceedings of the Royal Society. B, Biological Sciences 264:695–700.

Milot, E., C. Perrier, L. Papillon, J. J. Dodson, and L. Bernatchez 2013. Reduced fitness of Atlantic salmon released in the wild after one generation of captive breeding. Evolutionary Applications 6:472–485.

Møller, A. P. 1994. Symmetrical male sexual ornaments, paternal care, and offspring quality. Behavioral Ecology 5:188–194.

Morrow, E. H., A. D. Stewart, and W. R. Rice 2008. Assessing the extent of genome-wide intralocus sexual conflict via experimentally enforced gender-limited selection. Journal of Evolutionary Biology 21:1046–1054.

Neff, B. D., and T. E. Pitcher 2005. Genetic quality and sexual selection: an integrated framework for good genes and compatible genes. Molecular Ecology 14:19–38.

Neff, B. D., S. R. Garner, and T. E. Pitcher 2011. Conservation and enhancement of wild fish populations: preserving genetic quality versus genetic diversity. Canadian Journal of Fisheries and Aquatic Sciences 68:1139–1154.

Norris, K. J. 1990. Female choice and the quality of parental care in the great tit Parus major. Behavioral Ecology and Sociobiology 27:275–281.

Pélabon, C., L.-K. Larsen, G. H. Bolstad, Å. Viken, I. A. Fleming, and G. Rosenqvist. 2014. The effects of sexual selection on life-history traits: an experimental study on guppies. Journal of Evolutionary Biology 27:404–416.

Penn, D. J. 2002. The scent of genetic compatibility: sexual selection and the major histocompatibility complex. Ethology 108:1–21.

Penn, D., and W. K. Potts 1998. Untrained mice discriminate MHC-determined odors. Physiology & Behavior 63:235–243.

Penn, D. J., and W. K. Potts 1999. The evolution of mating preferences and major histocompatibility complex genes. The American Naturalist 153:145–163.

Pimm, S. L., G. J. Russell, J. L. Gittleman, and T. M. Brooks 1995. The future of biodiversity. Science 269:347–350.

Pitcher, T. E., and J. P. Evans 2001. Male phenotype and sperm number in the guppy (Poecilia reticulata). Canadian Journal of Zoology 79:1891–1896.

Pizzari, T., P. Jensen, and C. K. Cornwallis 2004. A novel test of the phenotype-linked fertility hypothesis reveals independent components of fertility. Proceedings of the Royal Society. B, Biological Sciences 271:51–58.

Plesnar-Bielak, A., A. M. Skrzynecka, Z. M. Prokop, and J. Radwan 2012. Mating system affects population performance and extinction risk under environmental challenge. Proceedings of the Royal Society. B, Biological Sciences 279:4661–4667.

Poissant, J., A. J. Wilson, and D. W. Coltman 2010. Sex-specific genetic variance and the evolution of sexual dimorphism: a systematic review of cross-sex genetic correlations. Evolution 64:97–107.

Price, D. K., and N. T. Burley 1994. Constraints on the evolution of attractive traits: selection in male and female zebra finches. The American Naturalist 144:908.

Puurtinen, M. 2011. Mate choice for optimal (k)inbreeding. Evolution 65:1501–1505.

Quader, S. 2005. Mate choice and its implications for conservation and management. Current Science 89:1220–1229.

Ralls, K., J. D. Ballou, B. A. Rideout, and R. Frankham 2000. Genetic management of chondrodystrophy in California condors. Animal Conservation 3:145–153.

Reding, L. P., J. P. Swaddle, and H. A. Murphy 2013. Sexual selection hinders adaptation in experimental populations of yeast. Biology Letters 9:20121202.

Rice, W. R., A. D. Stewart, E. H. Morrow, J. E. Linder, N. Orteiza, and P. G. Byrne 2006. Assessing sexual conflict in the Drosophila melanogaster laboratory model system. Philosophical Transactions of the Royal Society of London. Series B, Biological Sciences 361:287–299.

Robinson, M. R., J. G. Pilkington, T. H. Clutton-Brock, J. M. Pemberton, and L. E. B. Kruuk 2006. Live fast, die young: trade-offs between fitness components and sexually antagonistic selection on weaponry in soay sheep. Evolution 60:2168–2181.

Saino, N., S. Calza, P. Ninni, and A. P. Møller 1999. Barn swallows trade survival against offspring condition and immunocompetence. Journal of Animal Ecology 68:999–1009.

Saks, L., I. Ots, and P. Hõrak 2003. Carotenoid-based plumage coloration of male greenfinches reflects health and immunocompetence. Oecologia 134:301–307.

Saura, M., A. Pérez-Figueroa, J. Fernández, M. A. Toro, and A. Caballero 2008. Preserving population allele frequencies in ex situ conservation programs. Conservation Biology 22:1277–1287.

Sherman, C. D. H., E. Wapstra, T. Uller, and M. Olsson 2008. Males with high genetic similarity to females sire more offspring in sperm competition in Peron's tree frog Litoria peronii. Proceedings of the Royal Society. B, Biological Sciences 275:971–978.

Siller, S. 2001. Sexual selection and the maintenance of sex. Nature 411:689–692.

Soulé, M. E. ed. 1980. Thresholds for Survival: Maintaining Fitness and Evolutionary Potential. Sinauer Associates, Sunderland, MA.

Stewart, A. D., A. M. Hannes, A. Mirzatuny, and W. R. Rice 2008. Sexual conflict is not counterbalanced by good genes in the laboratory Drosophila melanogaster model system. Journal of Evolutionary Biology 21:1808–1813.

Stewart, A. D., A. Pischedda, and W. R. Rice 2010. Resolving intralocus sexual conflict: genetic mechanisms and time frame. Journal of Heredity 101:S94–S99.

Tarka, M., M. Åkesson, D. Hasselquist, and B. Hansson 2014. Intralocus sexual conflict over wing length in a wild migratory bird. The American Naturalist 183:62–73.

Taylor, M. L., N. Wedell, and D. J. Hosken 2007. The heritability of attractiveness. Current Biology 17:R959–R960.

Van Doorn, G. S. 2009. Intralocus sexual conflict. Annals of the New York Academy of Sciences 1168:52–71.

Wang, J., and N. Ryman 2001. Genetic effects of multiple generations of supportive breeding. Conservation Biology 15:1619–1631.

Weatherhead, P. J., and R. J. Robertson 1979. Offspring quality and the polygyny threshold: "The sexy son hypothesis". The American Naturalist 113:201.

Wedekind, C. 2002. Sexual selection and life-history decisions: implications for supportive breeding and the management of captive populations. Conservation Biology 16:1204–1211.

Wedekind, C., and S. Furi 1997. Body odour preferences in men and women: do they aim for specific MHC combinations or simply heterozygosity? Proceedings of the Royal Society. B, Biological Sciences 264:1471–1479.

Whitlock, M. C., and A. F. Agrawal 2009. Purging the genome with sexual selection: reducing mutation load through selection on males. Evolution 63:569–582.

Wielebnowski, N. C., N. Fletchall, K. Carlstead, J. M. Busso, and J. L. Brown 2002. Noninvasive assessment of adrenal activity associated with husbandry and behavioral factors in the North American clouded leopard population. Zoo Biology 21:77–98.

Williams, S. E., and E. A. Hoffman 2009. Minimizing genetic adaptation in captive breeding programs: a review. Biological Conservation 142:2388–2400.

Wolf, C. M., B. Griffith, C. Reed, and S. A. Temple 1996. Avian and mammalian translocations: update and reanalysis of 1987 survey data. Conservation Biology 10:1142–1154.

Wolfenbarger, L. L. 1999. Red coloration of male northern cardinals correlates with mate quality and territory quality. Behavioral Ecology 10:80–90.

Woodworth, L. M., M. E. Montgomery, D. A. Briscoe, and R. Frankham 2002. Rapid genetic deterioration in captive populations: causes and conservation implications. Conservation Genetics 3:277–288.

Zahavi, A. 1975. Mate selection – a selection for a handicap. Journal of Theoretical Biology 53:205–214.

16

The implications of small stem cell niche sizes and the distribution of fitness effects of new mutations in aging and tumorigenesis

Vincent L. Cannataro,[1] Scott A. McKinley[2] and Colette M. St. Mary[1]

1 Department of Biology, University of Florida, Gainesville, FL, USA
2 Department of Mathematics, Tulane University, New Orleans, LA, USA

Keywords

aging, biomedicine, evolutionary theory, fitness, population genetics-theoretical, stem cells, tumorigenesis.

Correspondence

Vincent L. Cannataro, Department of Biology, University of Florida, 876 Newell Dr. room 220 Bartram Hall Gainesville, FL, 32611-8525, USA.

e-mail: vcannataro@ufl.edu

Abstract

Somatic tissue evolves over a vertebrate's lifetime due to the accumulation of mutations in stem cell populations. Mutations may alter cellular fitness and contribute to tumorigenesis or aging. The distribution of mutational effects within somatic cells is not known. Given the unique regulatory regime of somatic cell division, we hypothesize that mutational effects in somatic tissue fall into a different framework than whole organisms; one in which there are more mutations of large effect. Through simulation analysis, we investigate the fit of tumor incidence curves generated using exponential and power-law distributions of fitness effects (DFE) to known tumorigenesis incidence. Modeling considerations include the architecture of stem cell populations, that is, a large number of very small populations, and mutations that do and do not fix neutrally in the stem cell niche. We find that the typically quantified DFE in whole organisms is sufficient to explain tumorigenesis incidence. Further, deleterious mutations are predicted to accumulate via genetic drift, resulting in reduced tissue maintenance. Thus, despite there being a large number of stem cells throughout the intestine, its compartmental architecture leads to the accumulation of deleterious mutations and significant aging, making the intestinal stem cell niche a prime example of Muller's Ratchet.

Introduction

Evolution in somatic tissue

The epithelial tissues within many animals are continually replenished by populations of stem cells that divide throughout the organism's lifetime. For instance, the epithelial lining of the intestinal tract is replaced weekly by millions of independent populations of stem cells located in intestinal crypts [reviewed in Barker (2014)]. This continual division provides an opportunity for mutation, resulting in the accumulation of mutant lineages and somatic evolution (Lynch 2010). Stem cell lineages with decreased fitness, or a diminished ability to divide and survive, will represent a failure in this tissue renewal process and the aging of tissues and multicellular organisms as a whole (López-Otín et al. 2013; Moskalev et al. 2013). Lineages with increased fitness, or faster division rates and an increased propensity to survive, will

result in the accumulation of cells and neoplasia (Merlo et al. 2006). Although considered premalignant at the onset, the accumulation of cells into a polyp, in which cells continually divide and accumulate subsequent mutations, can develop a cancerous phenotype over time (Winawer 1999).

Distribution of fitness effects

The effect that a new mutation will have on an individual's fitness can be characterized by a distribution of fitness effects (DFE). The DFE of several organisms have been experimentally estimated using mutation accumulation experiments or directed mutagenesis experiments in the laboratory (Eyre-Walker and Keightley 2007; Halligan and Keightley 2009). The majority of random mutations to a genome that affect fitness have a deleterious effect on fitness, while a small subset increase fitness

(Eyre-Walker and Keightley 2007). Additionally, many mutations that affect fitness have a small effect, while few have a large effect. In general, both beneficial (Imhof and Schlotterer 2001; Orr 2003; Kassen and Bataillon 2006) and deleterious (Elena *et al.* 1998) mutational fitness effects can be described well using an exponential distribution. We note that certain beneficial DFE may not be exponentially distributed, and are better classified as having either a bounded or heavier-than-exponential tail (Rokyta *et al.* 2008; Bank *et al.* 2014; Levy *et al.* 2015), and compound distributions or distributions with more parameters may better fit empirical measures of DFE (Sanjuán *et al.* 2004).

By understanding the mutational DFE in somatic tissue, we can predict the evolutionary trajectories of tissues within multicellular organisms as they age. Absolute fitness is typically measured as the reproductive success of a genetically identical lineage, which can be measured empirically as the growth rate of a population and interpreted ecologically as the death rate of individuals in a population subtracted from the birth rate. Within stem cell populations, this is analogous to the differentiation rate of the stem cell lineage subtracted from the division rate. However, the total growth rate of the healthy stem cell population is necessarily zero to insure tissue homeostasis. As we describe in the next section, the stem cells exist in two populations: a static niche population with cells that are undergoing division and migrating into the second population, containing cells that are undergoing division and differentiation. Therefore, although only mutations to stem cell division rate would confer a change in selection pressure within the stem cell niche, fixed mutations to both division and differentiation rate will alter the rate of growth of the total stem cell lineage, the expected size of the stem cell population as a whole, and contribute to the probability of a tumorigenesis event. Hence, mutations to division and differentiation rate affect the reproductive success of stem cell lineages, that is, fitness, and we consider distributions of mutational effects on these two rates in this work. Although there has been no direct measurement of the distribution of fitness effects in somatic tissue [but see Vermeulen *et al.* (2013), Snippert *et al.* (2014) for estimations of the selective advantage for some known cancer drivers], the evolution of cancer progression has been previously modeled using discrete (Beerenwinkel *et al.* 2007; Bozic *et al.* 2010; McFarland *et al.* 2013) and continuous (Foo *et al.* 2011) fitness effects. Here, we differ from these previous models by investigating different mutational effect frameworks using parameters derived from whole organisms to explore mutation accumulation in crypts initialized at their measured healthy size in mice and humans and quantify both aging and tumorigenesis.

When quantifying tumor incidence, we are concerned with the moment that the regulatory regime in the intestinal crypt breaks down: when the stem cell division rate exceeds its differentiation rate. We call this point the tumorigenesis threshold. The resulting population will accumulate stem cells without bound, which is thought to be the cause of crypt fission and the main mechanism of polyp or adenoma growth (Loeffler and Grossmann 1991; Wong *et al.* 2002). We investigate the full spectrum of deleterious and beneficial mutational effects on the progression of a healthy crypt to tumor initiation using empirically measured rates of division.

The evolution of multicellularity has necessitated the evolution of regulatory systems that hold somatic stem cells at a relatively low fitness (when compared to their maximum potential) in order to ensure the cooperation of the different cellular systems constituting a whole organism. As such, in addition to the beneficial mutations that would be accounted for by a typical DFE for whole organisms [which are commonly assumed to already be highly fit (Orr 2010)], we expect mutations of large effect in somatic tissue as regulatory processes become dysfunctional, such as the deactivation of tumor suppressor genes or the activation of oncogenes. It is reasonable to hypothesize that a heavy-tailed distribution could better classify mutational effects that have a beneficial effect in somatic stem cells by capturing both the mutations of small effect and also having a nontrivial probability of capturing the mutations of large effect often associated with cancer.

We evaluate whether or not the DFE estimated in whole organisms can explain known tumor incidence in the intestine. Further, we explore whether or not tumor incidence is better explained by a heavy-tailed distribution for mutations beneficial to fitness. Thus, we create a model of an evolving intestinal stem cell pool and implement alternate DFE and compare the resultant incidence curves to known tumor incidence curves.

Materials and methods

Description of the model

Crypt population structure

The base of each intestinal crypt harbors a population of symmetrically dividing cells expressing markers associated with the stem cell phenotype (Lopez-Garcia *et al.* 2010; Snippert *et al.* 2010). Within this population, there exists a subpopulation niche that is responsible for maintaining tissue homeostasis (Kozar *et al.* 2013; Vermeulen *et al.* 2013). We model the stem cells of the intestinal crypt as two populations of cells, the first being this stem cell niche, which consists of a fixed population of stem cells, N, and the second consisting of the stem cells displaced from this niche but not yet committed to differentiation. The sum of these populations represent the total number of stem cells within the crypt, N_T. In our model, cells within the niche divide at rate λ and displace their neighbors through overcrowding, as proposed by Lopez-Garcia et al. (2010) and revealed by in vivo live imaging by Ritsma et al. (2014). This population of cells experiences genetic drift and selection; cells that have a higher division rate are more likely to push their neighbors out of the niche [as demonstrated by Snippert et al. (2014)] and cells with lower division rates are more likely to be displaced. Mutations may occur at division with mutation rate μ and result in either a lineage with a new division rate or a lineage with a new rate of committing to differentiation. Displaced stem cells divide at the rate of their progenitor cells in the niche and commit to differentiation at rate v, hereafter referred to as the differentiation rate. We assume that once a lineage commits to differentiation, it is destined to be expelled from the crypt. We define tumorigenesis in the crypt as the moment a lineage of stem cells with a division rate greater than its differentiation rate has become fixed in the niche, resulting in exponential population growth. We note that, although a stem cell's propensity to commit to differentiation in healthy tissue is partially dependent on external signaling queues, such as Wnt signals from Paneth cells in the small intestinal crypt stem cell niche (Clevers 2013), the ability of a stem cell to interpret and respond to, or even gain independence from, external signals is an intrinsic and heritable property of the stem cell (Reya and Clevers 2005). The parameters λ, N, and N_T have been previously estimated (Kozar et al. 2013; Vermeulen et al. 2013), and we calculate v according to eqn 4 in the Appendix, where $v = \left(1 + \frac{N}{\bar{y}(t)}\right)\lambda$ and $\bar{y}$ is the average number of stem cells outside of the niche.

Distribution of fitness effects

We first describe our model of mutations that affect the division rate of stem cells and address mutations that affect differentiation rate later in section "Mutations that alter the differentiation rate of stem cells result in rapid aging and tumorigenesis" When mutations occur, the new division rate is greater than the previous rate with probability P_B, and the mean positive change of rate is s_+. We consider positive and negative changes that are exponentially distributed for deleterious effects and exponentially or Pareto distributed for beneficial effects, see the Appendix. The mean negative change is s_-. We define the exponential DFE in eqn 1 and the power-law DFE in eqn 2.

$$m(\lambda;\lambda_0)_{\exp} = \begin{cases} (1-P_B)\frac{\beta}{\lambda_0}e^{-\beta\left(1-\frac{\lambda}{\lambda_0}\right)} & \lambda<\lambda_0 \\ P_B\frac{\alpha}{\lambda_0}e^{-\alpha\left(\frac{\lambda}{\lambda_0}-1\right)} & \lambda>\lambda_0 \end{cases} \quad (1)$$

$$m(\lambda;\lambda_0)_{Pareto} = \begin{cases} (1-P_B)\frac{\beta}{\lambda_0}e^{-\beta\left(1-\frac{\lambda}{\lambda_0}\right)} & \lambda<\lambda_0 \\ P_B\frac{\alpha-1}{\lambda_0}\left(\frac{\lambda}{\lambda_0}\right)^{-\alpha} & \lambda>\lambda_0 \end{cases} \quad (2)$$

The power-law distribution is well defined if $\alpha > 1$ and is considered to be heavy-tailed (having infinite variance) if $1 < \alpha < 3$.

Selection assumptions

We are concerned with the mutations that arise and reach fixation within the stem cell niche. Due to drift, all stem cells with the same division rate as the background population have an equal probability of reaching fixation, commonly referred to as neutral drift dynamics (Lopez-Garcia et al. 2010; Snippert et al. 2010). Following Wodarz and Komarova (2005), we use a Moran model to estimate the probability that a mutant lineage fixes in the stem cell niche:

$$p_{\text{fix}}(\lambda;\lambda_{\text{old}}) = \frac{1-\lambda_{\text{old}}/\lambda}{1-(\lambda_{\text{old}}/\lambda)^N} \quad (3)$$

where N is the number of cells in the niche. The mutation rate is low relative to the division rate, so we assume that there are at most two competing division rates at any given time.

Using the above formula (3), we can use Bayes' theorem to compute the probability density $\Phi(\lambda|\lambda_{\text{old}})$ of a new fixed division rate λ given that the previous division rate is λ_{old}:

$$\Phi(\lambda|\lambda_{\text{old}}) = \frac{p_{\text{fix}}(\lambda;\lambda_{\text{old}})m(\lambda;\lambda_{\text{old}})}{\int_0^\infty p_{\text{fix}}(\ell;\lambda_{\text{old}})m(\ell;\lambda_{\text{old}})d\ell}. \quad (4)$$

As described above, tumorigenesis occurs when the division rate λ is greater than differentiation rate v and we define the point at which this happens to be the tumorigenesis threshold. In our modeling framework, each new fixed mutation presents a new possibility that the division rate exceeds the threshold for tumorigenesis. From (4), we can iteratively derive the sequence of functions $\{f_n\}$ that represent the density of the distribution of the stem cell division rates conditioned that n mutations have fixed in the stem cell niche and tumorigenesis has not occurred as of mutation $n-1$. If we let λ_0 denote the initial stem cell division rate, then $f_1(\lambda) = \Phi(\lambda|\lambda_{\text{old}} = \lambda_0)$. For each n, let p_n denote the probability that tumorigenesis occurs due to the nth mutation (given that n mutations have occurred). Then, $p_1 = \int_v^\infty f_1(\lambda)d\lambda$. From there, we can write the recursive formulae

$$f_{n+1}(\lambda) = \frac{1}{1-p_n} \int_0^v \Phi(\lambda|\lambda_{\text{old}} = \ell)f_n(\ell)d\ell \quad \text{and}$$

$$p_{n+1} = \int_v^\infty f_{n+1}(\lambda)d\lambda. \tag{5}$$

From this, we have a recursive formula for the probabilities, $\{q_n\}$, that tumorigenesis has not occurred given n fixed mutations: $q_1 = 1 - p_1$ and

$$q_{n+1} = (1-p_{n+1})q_n. \tag{6}$$

To translate this result to an individual's lifetime, we model the time-dependent arrival of new mutations as a Poisson process with fixed rate parameter μ mutations per cell division. We keep track of the time-dependent number $M(t)$ of mutations that fix in the stem cell niches by time t. Then, using T to denote the time that tumorigenesis occurs in a given crypt, we can write the probability that tumorigenesis has not occurred as of time t by the equation

$$\mathbb{P}(T > t) = \sum_{n=1}^\infty q_n \mathbb{P}(M(t) = n) \tag{7}$$

Depending on the species, an individual has hundreds of thousands or even millions of crypts. The probability that an individual has at least one crypt that has undergone tumorigenesis can be calculated by considering the distribution of fixed mutations that have accumulated among the individual's crypts and the probability that these mutations result in tumorigenesis. This can then be extrapolated to the incidence rate of tumors among a population of individuals (Fig. 1). Let $\mathcal{T}$ represent the time that tumorigenesis first occurs in any of an individual's crypts. We use the following estimate to calculate tumorigenesis incidence data reported in the Results section. In the Supporting Information, we describe the full calculation and a few simplifying assumptions we make to develop a computationally tractable model.

$\mathbb{P}(\text{No tumorigenesis at time } t)$

$$\approx \prod_{n=1}^{\hat{n}} \mathbb{P}(\text{No crypt tumorigenesis at time}$$

$$t|n \text{ mutations})^{\mathbb{E}(\# \text{ of crypts with n mutations})}$$

i.e.,

$$\mathbb{P}(\mathcal{T} > t) \approx \prod_{n=1}^{n_{\max}} (q_n)^{C(\hat{\mu}t)^n e^{-\hat{\mu}t}/n!}. \tag{8}$$

In the above, C is the number of crypts in the length of intestine being investigated, $n_{\max}$ is the maximum number of mutations simulated, and $\hat{\mu} = N\mu\lambda_0 \int_0^\infty p_{\text{fix}}(\lambda; \lambda_0)m(\lambda; \lambda_0)d\lambda$, where, as above, N

is the number of cells in the stem cell niche, μ is the mutation rate per cell division, and p_{fix} is defined by eqn 3.

Parameter choices

Some estimates of crypt dynamics parameters have shifted over time, for example, the stem cell division rate in the mouse was formerly thought to be once every 1–1.5 days (Lopez-Garcia et al. 2010), but more recent estimates indicate they divide once every 3–10 days (Kozar et al. 2013). Kozar et al. (2013) demonstrated that the division rate of stem cells in the stem cell niche of mice varied from approximately 0.1 to 0.2 to 0.3 divisions per day along the proximal small intestine, distal small intestine, and colon, respectively. Likewise the estimated number of stem cells within the mouse stem cell niche varies from approximately five to six to seven, respectively. The total number of cells in crypts expressing stem cell markers has been reported to be 14–16 in mice (Lopez-Garcia et al. 2010; Snippert et al. 2010; Clevers 2013). For the analysis of our mouse model, we chose the middle value of these parameter ranges, a crypt with 15 total cells expressing stem cell markers, with 6 of the cells constituting the stem cell niche dividing 0.20 times per stem cell per day. To estimate the differentiation rate of stem cells outside the stem cell niche, we used a continuous time Markov chain, described in the Appendix. According to this model, in order for the total stem cells in the crypt of a mouse to stay at a constant population size, the differentiation rate of stem cells outside of the stem cell niche must be 0.333 per stem cell per day.

The parameters associated with crypt dynamics in mice have been well described; however, we were unable to obtain any data on population incidence of intestinal polyps or tumors in wild type mice. On the other hand, while crypt dynamics in humans have not been as well studied, there exist incidence data for large intestine polyps (Chapman 1963). To parameterize the human colon crypt system, we considered a few sources. Nicolas et al. (2007) analyzed the methylation patterns within the human colon crypt and their Bayesian analysis suggests a posterior density mode between 15 and 20 stem cells maintaining homeostasis and constituting the stem cell niche within the crypt. Their posterior density provides more support for numbers of stem cells larger than this mode than for numbers smaller, so we chose 20 as an initial value for the number of stem cells within the stem cell niche. Bravo and Axelrod (2013) report an average of 35.7 quiescent stem cells within the human colon crypt through a staining experiment, so we assume there are 36 total stem cells within the human colon crypt. Human colon stem cells divide about once every 7 days (Potten et al. 2003), which would mean they would have to differentiate at a rate of

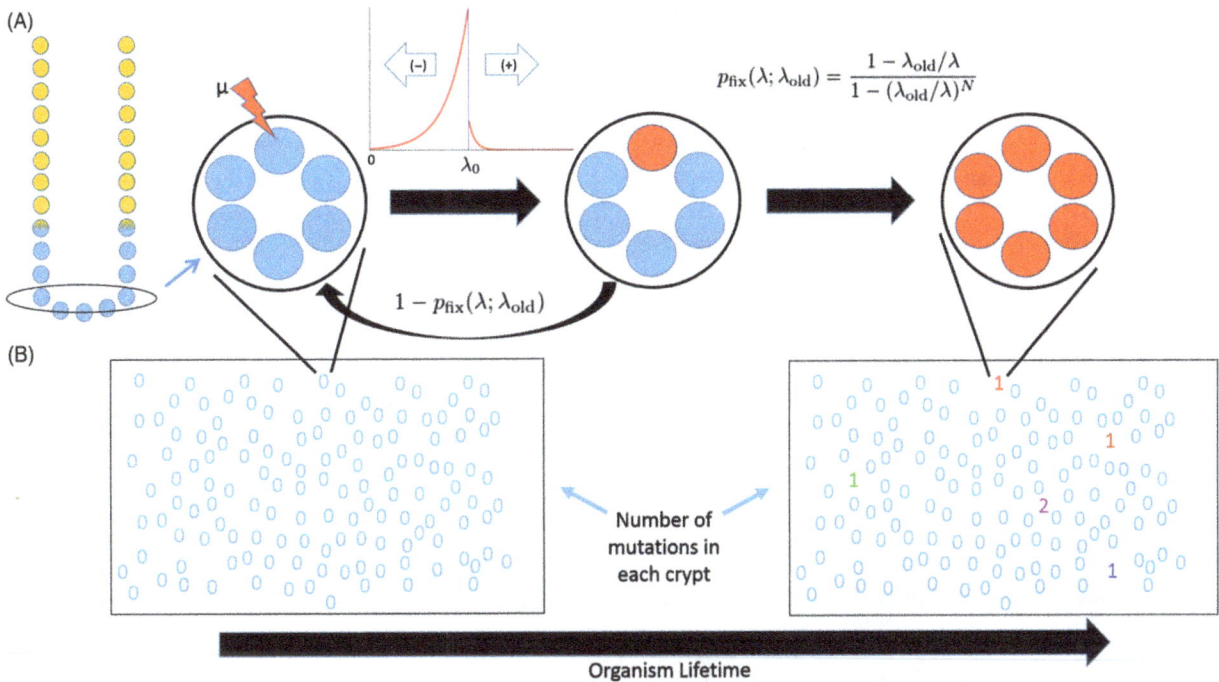

Figure 1 A representation of our model. (A) A cross section of an intestinal crypt, blue circles at the base of the crypt represent stem cells, while yellow circles represent cells that have committed to differentiation. The oval cross section at the base encompasses the stem cell niche, while stem cells above this niche are destined to commit to differentiation. Taking a top-down look at the oval, large circles represent a cross section of the intestinal crypt base, which houses the intestinal stem cells, represented by smaller blue and red circles. Mutations may occur to a single cell in the stem cell niche. These mutations alter the fitness of the cell according to a specified distribution of fitness effects. Given the new fitness, the mutated lineage has a certain probability, $p_{fix}(\lambda; \lambda_{old})$, of reaching fixation within the stem cell niche. (B) Here, the rectangles represent a cross section of the intestinal epithelium with the numbers representing the locations of individual crypts and describing the number of fixed mutations for each crypt. An organism accumulates fixed mutations over its lifetime.

about 0.321 per day to maintain homeostasis at the assumed initial parameters.

We parameterized the initial DFE based on those measured in whole organisms to evaluate whether they can account for known tumorigenesis incidence. The distribution of fitness effects has been estimated in mutation accumulation experiments and directed mutagenesis experiments. We consider the DFE proposed by Joseph and Hall (2004) in a mutation accumulation study because they report the expected effect size of deleterious and beneficial mutations, as well as the mutation rate and the proportion of mutations that were beneficial in a diploid eukaryotic system (*Saccharomyces cerevisiae*). They report an average beneficial heterozygous fitness effect of 0.061, which is slightly lower but within an order of magnitude of the effect of average beneficial mutation measured for vesicular stomatitis virus of 0.07 (Sanjuán *et al.* 2004) and *E.coli* of 0.087 (Kassen and Bataillon 2006). They found that 5.75% of accumulated mutations were beneficial and that the overall mutation rate to alleles that alter fitness was 6.3×10^{-5} mutations per haploid genome per generation. This would result in a diploid beneficial mutation rate of

$2 \times 6.3 \times 10^{-5} \times 0.0575 = 7.245 \times 10^{-6}$. This is within an order of magnitude of the beneficial mutation rate reported for *E. coli* (Wiser *et al.* 2013).

Mutation accumulation experiments may not capture the true distribution of fitness effects because they rely on observing the mutations of lineages that survive and persist in a population. Because of this, they are biased against mutations of large deleterious effect. Additionally, the random passaging of individuals to repopulate new generations may result in drastically different estimates of average mutational effect size for the same species. For instance, average deleterious effect of mutations in *Saccharomyces cerevisiae* has been estimated to be 0.061 (Joseph and Hall 2004), 0.086 (Wloch *et al.* 2001), and 0.217 (Zeyl and DeVisser 2001). Directed mutagenesis of random genome targets in an RNA virus revealed an average nonlethal deleterious fitness effect of 0.244 (Sanjuán *et al.* 2004). It is likely that the inherent average effect size of a mutation of deleterious effect would be better reflected by the larger estimates because mutations of large deleterious effect may be lost in mutation accumulation experiments.

After building our model with DFE parameters estimated from whole organisms, we describe overall patterns of mutation accumulation and risk of tumorigenesis and then we utilized least squares analysis to explore the best fit among a series of plausible choices of μ and the expected value of s_+ for the human incidence curves and compared to data from Chapman (1963) (best fit figures available as Figs S1–S3). For the division rate scenario, both for the exponential and for power-law DFE, we vary the expected value of s_+ from 0.041 to 0.07 and μ from 2.5×10^{-5} to 7×10^{-4}. For the differentiation rate scenario, we vary s_+ from 0.041 to 0.07 and μ from 2.5×10^{-7} to 7×10^{-4}.

The model described above was executed using R version 3.1.1. R scripts developed for this study are available at https://github.com/vcannataro/Somatic-Evo-DFE.

Results

Mutations result in both aging and tumorigenesis within the intestine

Because stem cell niche populations are small, it is possible for mutant lineages with a fitness disadvantage to fix in the niche. This, coupled with the fact that the vast majority of mutations that occur will have a deleterious effect on stem cell fitness, results in the expected value of the probability density describing the new division rates to move away from the tumorigenesis threshold with subsequent fixed mutations (Fig. 2A,C,E). In general, the accumulation of fixed mutations within crypts results in impaired stem cell maintenance and lower stem cell production, contributing to the aging of the tissue and organism.

The probability that a particular fixed mutation will result in tumorigenesis in the crypt, p_n (eqn 5), is equal to the area under these densities that crosses the tumorigenesis threshold (Fig. 2B,D). For the initial parameterization in mice and humans, this increases at first, but then decreases with subsequent fixed mutations as the probability densities describing division rate move away from the tumorigenesis threshold.

Predicted incidence curves in mice and humans using DFE derived from a whole organism

Using the model described in "Selection-assumptions", we determined the cumulative probability distribution of tumorigenesis within a population of crypts in an individual organism. For mice, using the initial parameters in Table 1 and exponentially distributed beneficial fitness effects, we find that the incidence of tumorigenesis is predicted to increase linearly with age, with close to nine percent of mice experiencing tumorigenesis at 3 years of age (Fig. 3A). Human tumorigenesis incidence in the large intestine is pre-

dicted to be approximately 36% at 80 years of age (Fig. 3B), using an exponentially distributed beneficial fitness effects and the initial null parameters from Table 1.

The only incidence data for early tumors or polyps were found for the large intestine in humans. The predicted incidence curve derived from an exponentially distributed DFE follows the same qualitative dynamics as the tumor incidence data. Incidence curves that are derived from a power-law distribution using the initial parameters in Table 1 predict nearly 100% tumorigenesis by 80 years of age and do not follow the incidence data dynamics. Hence, we performed a least squares analysis, varying parameters that have not been characterized for human somatic tissue, to find the parameter set in our exploratory space with the best fit to the observed incidence curve to the data.

Altering the expected beneficial fitness effects and the mutation rate provides better fits for both exponential and power-law derived incidence curves

The expected mean fitness effects (s_+, s_-) of the DFE and the mutation rate (μ) per division of a mutation that alters the stem cell fitness were inferred from whole organisms as an initial parameter choice (Table 1). A parameter space around the initial choices was explored, and a least squares analysis was performed to find a better fit to the data (additional information in the Appendix, Figs S1–S3). Just the mutation rate (μ) and expected beneficial fitness effect (s_+) are presented because changes to the expected deleterious fitness effect (s_-) had little effect on the resultant tumorigenesis incidence curves. We found that both the exponential and power-law scenarios can provide similarly good fits to the data, however, with distinctly different parameters. The exponential DFE derived curve provided the best fit with the same mutation rate as our initial choice (Table 1), with a slightly larger expected beneficial fitness effect $(\mathbb{E}[s_+] = 0.064$, Fig. 4A, red dashed line). Interestingly, assuming the same expected beneficial fitness effect as in Table 1 and varying the mutation rate provides a reasonable fit with a slightly larger mutation rate $(\mu = 1.75 \times 10^{-4}$, 4A, blue dashed line). The power-law DFE provided a similarly good fit to the incidence curve, but for a parameter space that assumes a much smaller expected beneficial fitness effect and a large mutation rate $(\mathbb{E}[s_+] = 0.044$, $\mu = 5 \times 10^{-4}$, Fig. 4B red dashed line).

Mutations that alter the differentiation rate of stem cells result in rapid aging and tumorigenesis

Mutations affecting differentiation rate influence the lifetime of a stem cell lineage. Mutations that increase differentiation rate will decrease the fitness of the lineage, while mutations that decrease differentiation rate increase fitness.

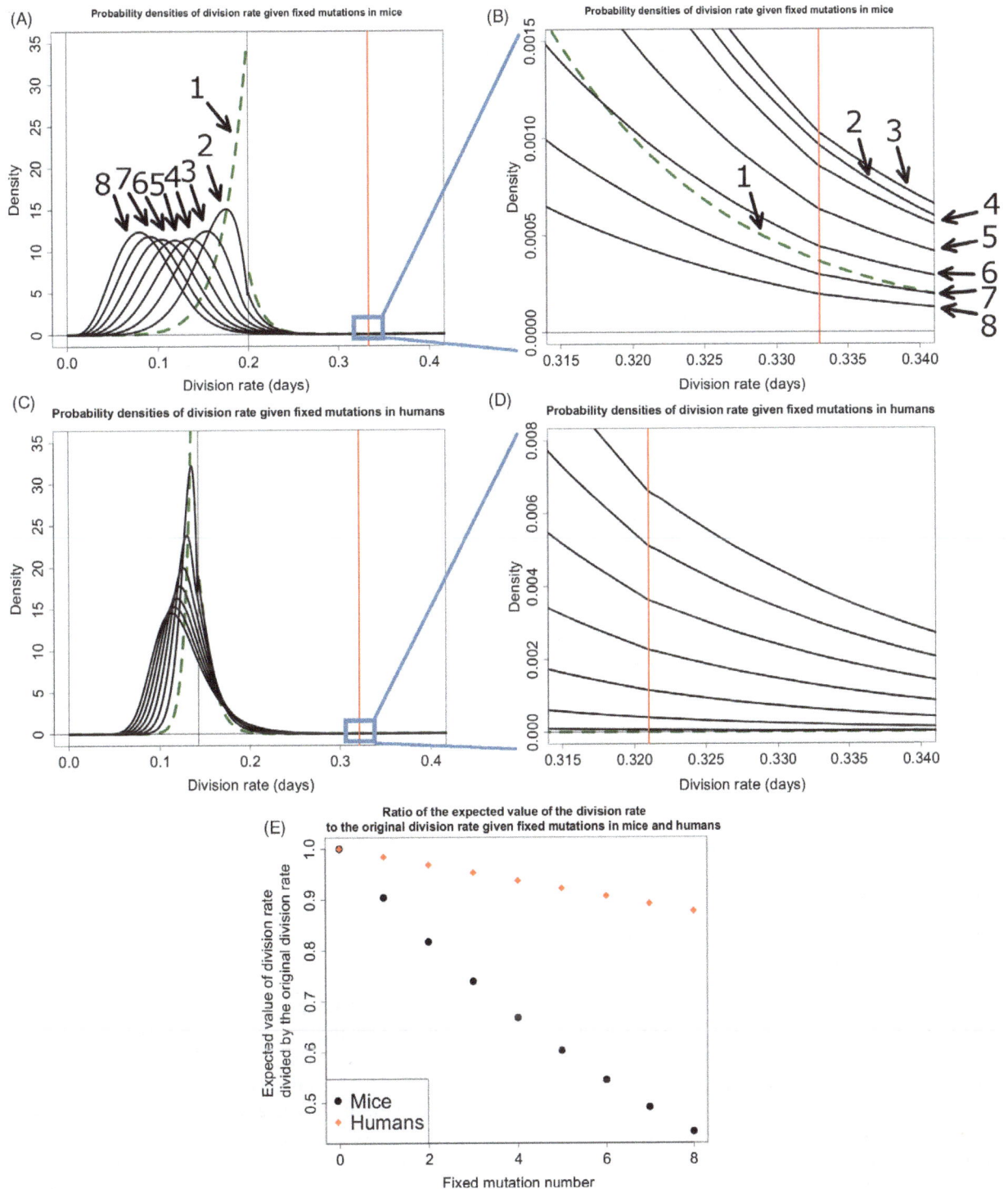

Figure 2 The accumulation of probability densities describing stem cell division rate. (A) Exponentially distributed fitness effects on division rate using the parameters in Table 1 for the mouse. The first density is a green dashed line. Each probability density represents the division rate of a fixed lineage after n fixed mutations, with n indicated by an arrow. (B) Zooming in on the tumorigenesis threshold, we see that the area of the division rate density that is over the tumorigenesis threshold increases at first and then decreases with subsequent mutation. There is a change in slope of the densities at the tumorigenesis threshold because subsequent densities are calculated from the previous density which has had the area to the right of the tumorigenesis threshold removed and the area to the left renormalized to 1. (C,D) are the same as (A) and (B), respectively, but are for the human scenario. The larger population size decreases the strength of drift. Order of mutations in (C) proceeds as in (A) and proceeds from 1 through 8 from bottom to top in (D). (E) The expected values of the probability densities in (A) and (B) divided by their original values over subsequent fixed mutations.

Table 1. Initial model parameters, combining whole organism DFE with organismal crypt parameters. See text above for reasoning behind initial parameter choices.

Parameter	Description	Value in mouse (Ref)	Value in human (Ref)
P_B	Percent of mutations with a beneficial effect	0.0575 (Joseph and Hall 2004)	0.0575 (Joseph and Hall 2004)
s_+	Effect size of a mutation of beneficial effect	0.061 (Joseph and Hall 2004)	0.061 (Joseph and Hall 2004)
s_-	Effect size of a mutation of deleterious effect	0.217 (Zeyl and DeVisser 2001)	0.217 (Zeyl and DeVisser 2001)
μ	Mutation rate per genes influencing fitness per division	$2 \times 6.3 \times 10^{-5}$ (Joseph and Hall 2004)	$2 \times 6.3 \times 10^{-5}$ (Joseph and Hall 2004)
λ_0	Normal stem cell division rate per day	0.2 (Kozar et al. 2013)	0.143 (Potten et al. 2003)
v_0	Normal stem cell differentiation rate per day	0.333 (this study)	0.321 (this study)
N	Number of stem cells in the stem cell niche at the base of the crypt	6 (Kozar et al. 2013)	20 (Nicolas et al. 2007)
N_T	Total number of stem cells expressing stem cell markers in the crypt	15 (Clevers 2013)	36 (Bravo and Axelrod 2013)
Crypts	Number of crypts in the small and large intestine, respectively	7.5×10^5, 4.5×10^5 (Potten et al. 2003)	5×10^7, 2×10^7 (Potten et al. 2003)

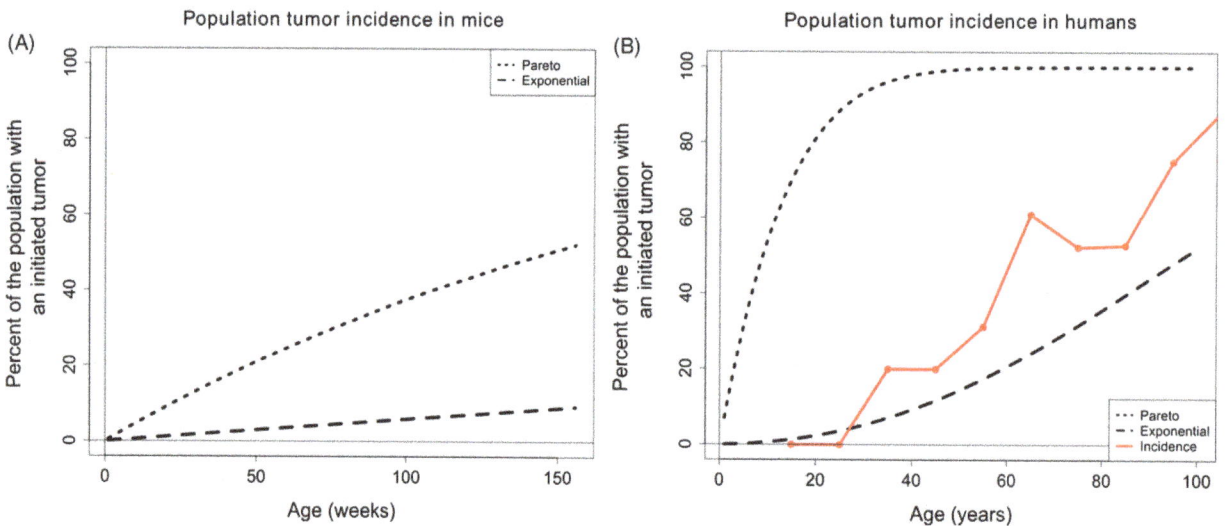

Figure 3 Tumorigenesis incidence in mice and humans using whole organism DFE parameters. (A) The population incidence of tumorigenesis throughout the entire intestinal tract of the mouse. (B) The population incidence of tumorigenesis throughout the large intestine in humans. The black dashed lines are generated from the species specific parameters listed in Table 1. The solid red line connects large intestine polyp incidence data found during autopsy (Chapman 1963).

Mutations affecting differentiation rate neutrally drift to fixation in the stem cell niche because the differentiation phenotype is not expressed in the niche, hence all cells divide at the same rate. Thus, the probability of fixation of mutations to differentiation rate is $(1/N)$, regardless of mutational effect. We only considered an exponential mutational effect distribution because the distinction between exponential and power-law distributions is only significant in prevalence of large deviations from the mean, and beneficial mutational effects in this scenario exist between v_0 and zero. Because all mutations that solely affect differentiation rate drift neutrally, and the majority of mutations decrease fitness (by increasing differentiation rate), the majority of fixed mutations move stem cell pools away from the tumorigenesis threshold (Fig. 5).

Mutational effects are typically described as a proportion of the phenotype they are affecting, and as such, the same DFE applied to a larger rate will have a larger absolute expected effect. The differentiation rate of stem cells displaced from the niche is necessarily larger than the intrinsic division rate because only a subpopulation of the entire stem cell population is exposed to committing to differentiation; however, all cells are dividing (Ritsma et al. 2014), and the stem cell population is maintained at a steady-state equilibrium. Thus, mutations affecting differentiation rate in our model have a larger absolute effect for the same proportional change in rate when compared to the previous analysis on mutations to division rate. Hence, given a fixed mutation, we see a high incidence of tumorigenesis when the mutation affects differentiation rate (Fig. 6A,B). Fitting

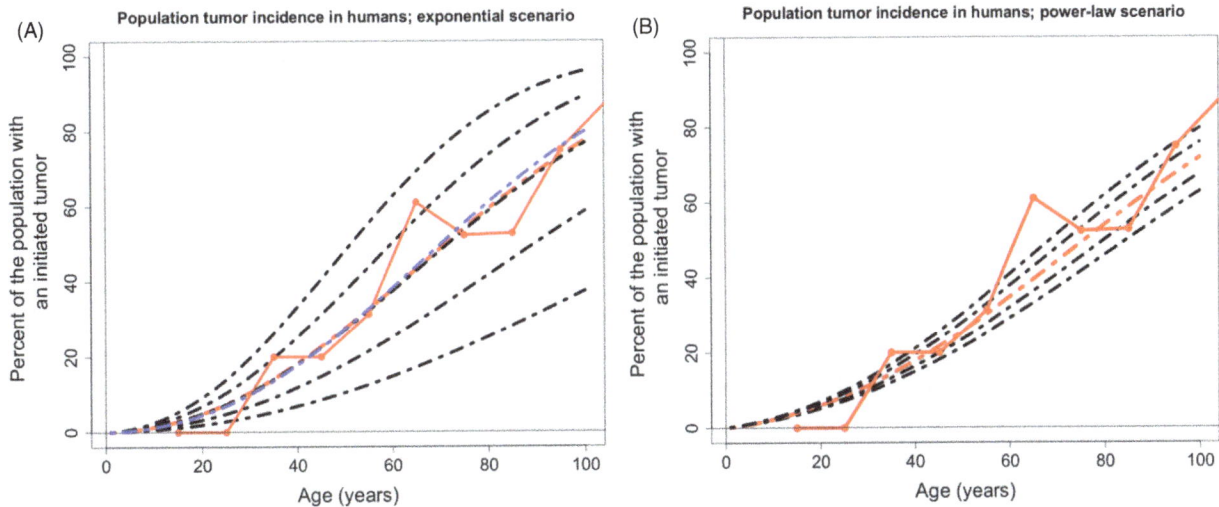

Figure 4 Tumorigenesis incidence curves resulting from least squares parameter fitting. (A) Incidence curves derived from the assumption of an exponential beneficial DFE. The best fit to the data out of the explored parameter space has the same μ as the yeast reported in Table 1 and $\mathbb{E}[s_+] = 0.064$ (red dashed line). Black dashed lines derived from $\mathbb{E}[s_+] = 0.064$ and, from bottom to top, $\mu = 7.5 \times 10^{-5}$ to 1.75×10^{-4} by 2.5×10^{-5}. Blue dashed line is the predicted incidence curve with the best fit with $\mathbb{E}[s_+] = 0.061$ (initial DFE derived from yeast reported in Table 1), which had $\mu = 1.75 \times 10^{-4}$. (B) Incidence curves derived from the assumption of a power-law beneficial DFE. All parameters are the same as in Table 1, except $\mathbb{E}[s_+] = 0.044$ for each curve and, ranging from top to bottom, μ ranges from 4.5×10^{-4} to 5.5×10^{-4} by 2.5×10^{-5}, with 5×10^{-4} providing the best fit.

analyses along a range of plausible parameter space revealed a poorer fit when compared to mutations that alter division rate because mutations that alter differentiation rate will always result in large tumor incidence at early age.

Discussion

Whole organism DFE are sufficient to explain tumorigenesis

We hypothesized that mutations in somatic tissues would differ in their distribution, compared to unicellular whole organisms, because of the regulatory processes that control cell division and differentiation rates in multicellular organisms. However, we found that whole organism DFE were sufficient to account for patterns of tumorigenesis in the intestines. This suggests that somatic evolution is not unique, but instead is based on the same patterns of mutation that we see in whole organisms. Hence, the differences in evolutionary patterns between somatic tissues and whole organisms, such as the tendency of tissues to age via the accumulation of deleterious mutations while populations of whole organisms instead evolve to greater mean fitness in benign environments, arise as a consequence of the small populations of stem cells within multicellular organisms and the asexual nature of cell division. Somatic aging via mutation is thus akin to the action of Muller's ratchet, the accumulation of deleterious mutations in organisms that cannot eliminate them via recombination. Indeed, the ratchet acts more strongly than in populations of

organisms as a result of the relative importance of drift versus selection in very small stem cell populations (i.e., niches). This raises the interesting question of why somatic tissues are organized in this way and whether small stem cell pools predominate to minimize tumorigenesis at the expense of aging, as has been suggested by Michor *et al.* (2003).

Given the role of well-known large effect mutations in cancer, it is tempting, from a mathematical modeling point of view, to adopt a heavy-tailed (infinite variance) distribution for the DFE. In contrast to the DFE employed for modeling populations of whole organisms (e.g., an exponential distribution), which tend to exhibit small incremental changes, a heavy-tailed regime enables a significant contribution from "one-shot" large mutations. To probe this possibility, we included in our simulations a power-law (Pareto) distribution which, through its shape parameter α, can be either heavy-tailed ($1 < \alpha \leq 3$) or not ($\alpha > 3$). It is noteworthy, then, that the best fit parameters were very far from the heavy-tailed regime ($\alpha \approx 16$). The prevalence of large outlier mutations for such a distribution is comparable to what would be seen from an exponential distribution, meaning that the heavy-tailed regime is not an appropriate modeling framework to explain the data.

Small populations and genetic drift lead to aging

One of our primary findings is that mutation effects drive crypt aging as much, if not more so, than tumorigenesis.

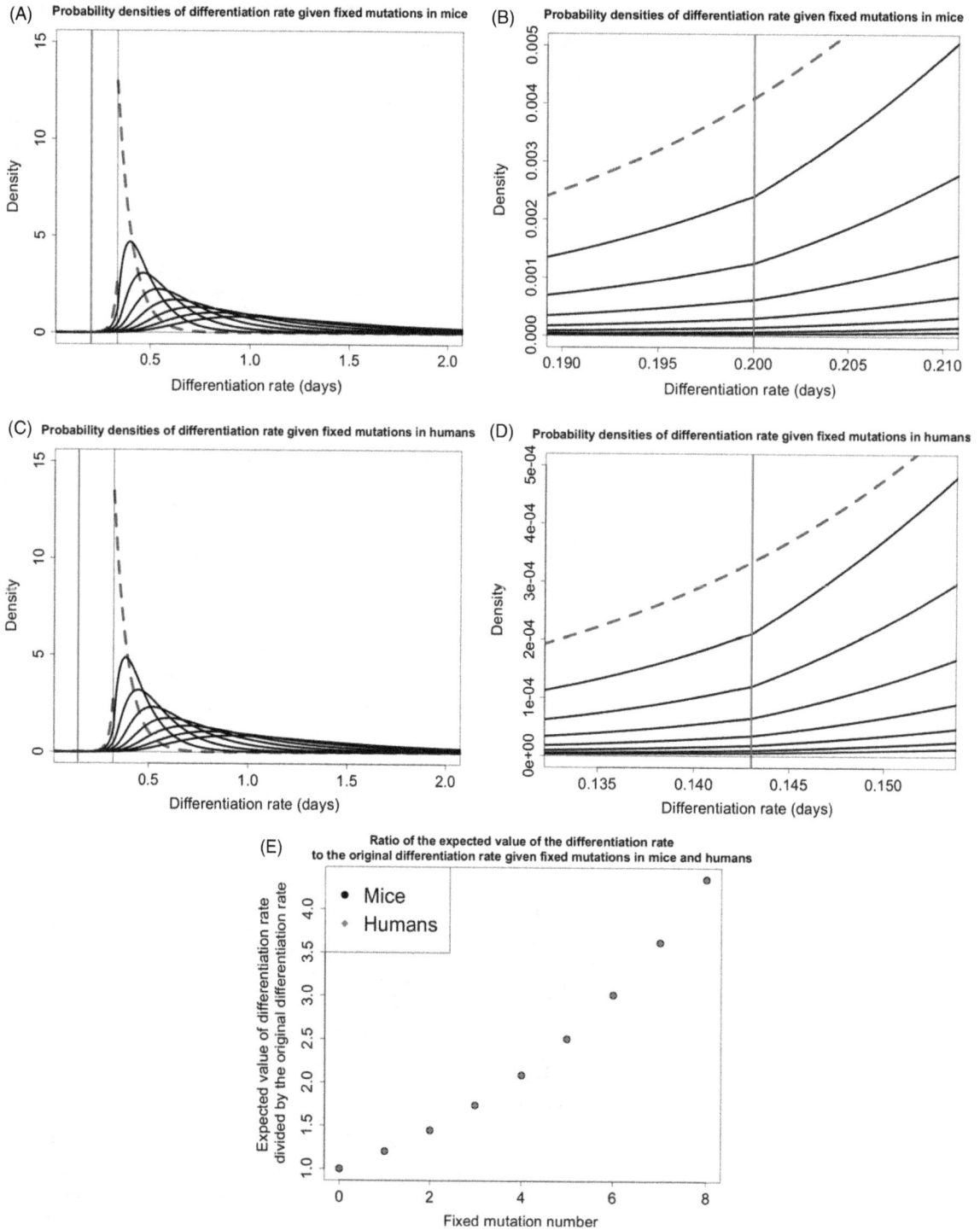

Figure 5 The accumulation of probability densities describing stem cell differentiation rate. (A) Exponentially distributed fitness effects on differentiation rate using the parameters in Table 1 for the mouse. The first density is a green dashed line. Each probability density represents the differentiation rate of a fixed lineage after *n* fixed mutations, with subsequent mutations traveling away from the original differentiation rate. (B) Zooming in on the tumorigenesis threshold, we see that the area of the differentiation rate density that is over the tumorigenesis threshold decreases with subsequent mutation. There is a change in slope of the densities at the tumorigenesis threshold because subsequent densities are calculated from the previous density which has had the area to the left of the tumorigenesis threshold removed and the area to the right renormalized to 1. (C,D) are the same as (A) and (B), respectively, but are for the human scenario. Order of mutations in (C) proceeds as in (A). (E) The expected values of the probability densities in (A) and (B) divided by their original values over subsequent fixed mutations.

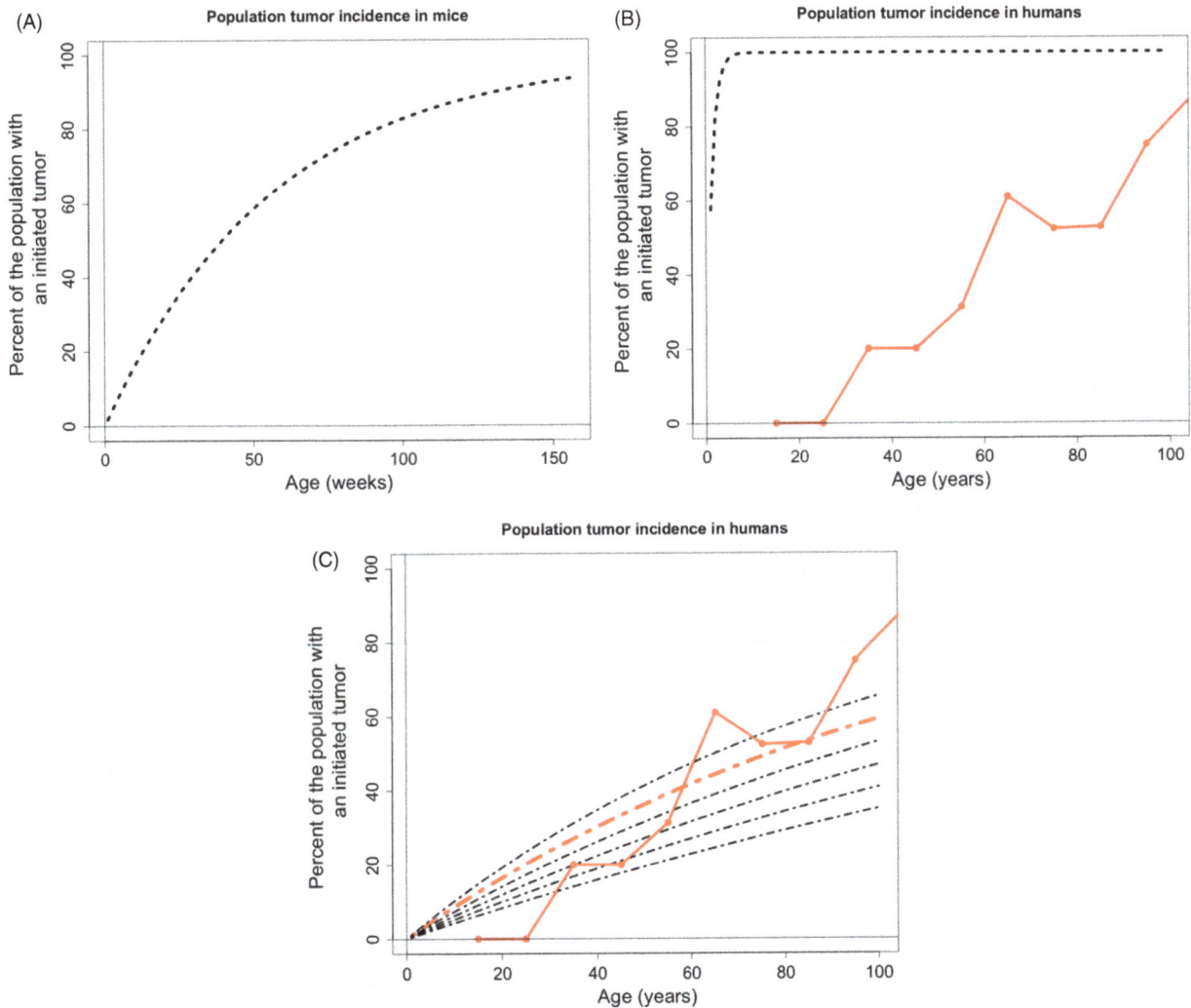

Figure 6 The tumorigenesis incidence resulting from stem cell mutational effects on differentiation rate. Calculations presented for tumor incidence in (A) mice, (B) humans, and (C) Best fit incidence curve in red; expected beneficial fitness effect of 0.057 and mutation rate of 2.5×10^{-6}. The other curves have the same mutation rate but vary around the expected beneficial fitness effect by increments of 0.001.

Tumor formation and aging are two manifestations of the accumulation of cellular genetic damage. This damage is especially relevant to the aging process when it affects the functional competence of stem cells and compromises their ability to replenish the various cell populations of their constituent tissue (López-Otín et al., 2013). Mutations to stem cells that result in aging have been associated with diminishing the stem cell's potential to proliferate (Rossi et al. 2007), competitively exclude healthy stem cells (Nijnik et al. 2007), and self-renew or differentiate (Jones and Rando 2011; Moskalev et al. 2013). These effects on stem cell dynamics would decrease the number of functional stem and nonstem cells in tissues, thus resulting in tissue aging, as defined in aging reviews and experimental work (above) and previous mathematical models (Wodarz 2007). As mutations become fixed in the intestinal stem cell

niche, the expected value of the probability density describing new stem cell lineage division rates decreases when we consider mutations that affect division rate, and the expected value for differentiation rate increases when we consider mutations that affect differentiation rate, and thus, crypts are predominately aging. The intestinal stem cell niche is maintained at a population size smaller than the effective population sizes of whole organisms and our findings derive directly from this population structure.

Our study, which emphasizes small healthy crypt populations, contrasts with previous studies that have investigated the accumulation of deleterious mutations in somatic tissue. These studies have looked at larger initial population sizes and in effect model hyperplasia or growing tumors. For instance, McFarland et al. (2013) modeled populations with an initial population size of approximately 1000 cells based on estimates from hyperplasia in mice 2 weeks after APC deletion. Similarly, McFarland *et al.* (2014) investigated mutation accumulation in models of hyperplasia and growing cancers. Datta *et al.* (2013) modeled deleterious mutations in housekeeping genes in an exponentially growing tumor initialized at 1×10^6 cells. Beckman and Loeb (2005) assumed their population of cells was sufficiently large to ensure a deleterious mutation of any strength could not become fixed. These approaches are useful to describe tumor growth in initiated tumors but fail to capture the relative importance of drift in evolving stem cell niches and the process of tumorigenesis from healthy stem cell niches, which exist as very small populations.

We find that crypts with fixed mutations are distributed along a range of both aging and tumor formation. The expected value of the division rate density moving away from the tumorigenesis threshold causes the probability of tumorigenesis per fixed mutation to eventually decrease. For example, in mice, there is a smaller probability that the fourth fixed mutation in a stem cell niche will result in tumorigenesis when compared to the third mutation in the exponential beneficial DFE scenario. The human intestinal crypt stem cell niche consists of a larger number of stem cells so drift plays a smaller role in the evolutionary trajectory of these crypts. Nonetheless, the mode of the distribution of division rate still moves away from the tumorigenesis threshold, albeit at a slower rate.

Although our model assumes that the size of the stem cell niche (N) remains constant and mutations only change the division rate or differentiation rate of lineages, it is possible that mutations could alter the niche size. If mutations altered the size of a crypt's stem cell niche, they would change the probability of fixation of subsequent mutations.

Mutations that only affect differentiation rate do not match incidence data curves

Analyses of colon cancer genomes from different individuals reveals that a small number of genes, associated with large fitness advantage, are commonly mutated among cancers (Wood *et al.* 2007). For instance, many colon cancers contain cells that have mutations in genes involved in the Wnt-signaling cascade responsible for maintaining "stemness" (Clevers and Nusse 2012). A study by Smith *et al.* (2002) found that 56% of 106 sequenced tumors had mutations in the APC gene, which, when nonfunctional, results in the activation of the Wnt cascade (Reya and Clevers 2005). Additionally, cancers that have a mutation in the APC gene tend to have the mutation distributed throughout the tumor, suggesting the mutations occurred early in tumor growth (Sottoriva *et al.* 2015). Because the Wnt-signaling cascade is involved with maintaining a stem cell phenotype, mutations in this cascade would influence the propensity for stem cells to differentiate. Additionally, Smith *et al.* (2002) also found that 61.3% of colorectal cancers had mutations in p53, involved in regulating apoptosis, and 27.4% of colorectal cancers had mutations in K-ras, thought to drive cancer growth by accelerating stem cell division and leading to enhanced crypt fission (Snippert *et al.* 2014).

Our modeling scenario of mutations only having an effect on the differentiation rate of stem cells and having effect sizes equal to those measured in whole organisms results in rapid tumorigenesis, with nearly 100% of human individuals having a polyp in their large intestine at young age. Indeed, individuals with familial adenomatous polyposis (FAP), who already have a germline mutation in one copy of their APC gene and only need one mutational hit on the other to form an adenoma, regularly develop adenomas as teenagers (Bozic *et al.* 2010). In our model, the large tumorigenesis incidence associated with mutations solely affecting differentiation rate is due to both the mutations having a larger absolute effect toward the tumorigenesis threshold and there being a higher overall probability of fixation of new mutations among all the crypts due to the mutations fixing through neutral drift. Even when we decrease the expected beneficial mutational effect size and decrease the mutation rate in an attempt to better fit the tumorigenesis incidence data, we find that mutations only affecting differentiation rate still result in more tumorigenesis than predicted at young age. However, the data were derived from autopsies on individuals >10 years of age, so data for tumorigenesis are lacking in this age group. Additionally, we modeled scenarios where mutations only affect division or differentiation, nature is certainly more complex, and mutation in both differentiation and division rates are likely to co-occur within a crypt population. Indeed, the APC protein discussed above contributes directly or indirectly to cellular division, differentiation, migration, cell orientation, and apoptosis (Dikovskaya *et al.* 2007; McCartney and Näthke 2008).

We model stem cell dynamics and mutational effects on those dynamics as a property that is controlled by an individual stem cell's genome, that is, a stem cell's heritable ability to produce or respond to internal signals, or respond to external signals, to divide or differentiate. The external signals regulating stem cell phenotype, such as Wnt signals produced by Paneth cells in the small intestine (Clevers 2013), are produced by cells differentiated from stem cells. Mutations to the stem cell genome may eventually influence the production of these signals in daughter cells. These mutations would drift neutrally in the niche, as they are not expressed until after stem cell differentiation, and, unless the lineage harboring the mutation reaches fixation in the niche, would eventually be lost from the crypt because Paneth cells die in approximately 20 days (Bry *et al.* 1994). Thus, mutations that result in differential signaling output by daughter cells can be modeled as neutrally fixed mutations acting intrinsically in the stem cells.

The influence of organism specific factors on somatic evolution

We find less tumor incidence in mice than humans throughout their respective lifetimes using the same DFE parameters. Mice only live a few years and have an order of magnitude fewer crypts in their entire intestine than humans have in just their large intestine (Potten *et al.* 2003). They also have smaller numbers of stem cells within their crypts, although those stem cells are dividing at a faster rate than human stem cells. Overall, this results in a lower chance of mutant lineages reaching fixation within crypts during the shorter mouse lifetime, and therefore, a reduction in the overall number of crypts with fixed mutations is lower. For instance, using the distribution of fixed mutations derived in the Appendix, at 2 years old, a mouse is expected to have about 75 crypts with two mutations, and only about 28% of mice will have a single crypt with three mutations. At 85 years old, a human is expected to have about 44 crypts with five mutations, one crypt with six mutations, and about four percent of humans at 85 years old will have a crypt with seven fixed mutations. As humans age they experience more fixed mutations, each of which confers a higher probability of tumorigenesis than the previous, whereas mice are expected to experience the accumulation of fewer mutations, possibly explaining the near linearity of the mouse incidence curve and the upwards curvature of the human incidence curve. Of note, given that a tumorigenesis event has occurred, it is likely the product of one mutation in the mouse model, whereas multiple mutations may contribute to the initiation of a tumor in the human model (the Appendix, Fig. S4).

The incidence of polyps at autopsy reported by Chapman (1963) was based on visual observations of discernible elevations of the mucosa in the entire large intestine during autopsy. It would take time for an initiated tumor to grow to a visible mass, so the true tumorigenesis incidence curve may lie in front of the data recorded in this study, with a lag time of growth before the tumor is visible. This lag time would be a function of the individual mutational spectrum of the initiated tumor and the tumor's environment.

Overall, we have shown that small homeostatic populations of stem cells, typical of somatic tissues in multicellular organisms, accumulate mutations that affect cellular fitness, contributing both to aging and tumorigenesis over an organism's lifetime. We show that the evolution of intestinal stem cell populations under the assumption of an organismal DFE, as opposed to the assumption of a heavy-tailed beneficial DFE, best predicted early tumor formation. However, aging, rather than tumorigenesis, predominated among crypts in the intestine. Our modeling approach emphasizes tumorigenesis in the context of aging, and vice versa, and demonstrates the importance of mutational processes within very small populations in both these phenomena.

Acknowledgement

We thank Charlie Baer, Andres J. Garcia, and Jake M. Ferguson for useful discussion. This research was partially supported by the National Science Foundation under Grant No. 0801544 in the Quantitative Spatial Ecology, Evolution and Environment Program at the University of Florida.

Literature Cited

Bank, C., R. T. Hietpas, A. Wong, D. N. Bolon, and J. D. Jensen 2014. A bayesian MCMC approach to assess the complete distribution of fitness effects of new mutations: uncovering the potential for adaptive walks in challenging environments. Genetics **196**:841–852.

Barker, N. 2014. Adult intestinal stem cells: critical drivers of epithelial homeostasis and regeneration. Nature Reviews Molecular Cell Biology **15**:19–33.

Beckman, R. A. and L. A. Loeb 2005. Negative clonal selection in tumor evolution. Genetics **171**:2123–2131.

Beerenwinkel, N., T. Antal, D. Dingli, A. Traulsen, K. W. Kinzler, V. E. Velculescu, B. Vogelstein, and M. A. Nowak 2007. Genetic progression and the waiting time to cancer. PLoS Computational Biology **3**:e225.

Bozic, I., T. Antal, H. Ohtsuki, H. Carter, D. Kim, S. Chen, R. Karchin, K. W. Kinzler, B. Vogelstein, and M. A. Nowak 2010. Accumulation of driver and passenger mutations during tumor progression. Proceedings of the National Academy of Sciences of the United States of America **107**:18545–18550.

Bravo, R. and D. E. Axelrod 2013. A calibrated agent-based computer model of stochastic cell dynamics in normal human colon crypts useful for in silico experiments. Theoretical Biology and Medical Modelling **10**:66.

Bry, L., P. Falk, K. Huttner, A. Ouellette, T. Midtvedt, and J. I. Gordon 1994. Paneth cell differentiation in the developing intestine of normal and transgenic mice. Proceedings of the National Academy of Sciences of the United States of America **91**:10335–10339.

Chapman, I. 1963. Adenomatous polypi of large intestine: incidence and distribution. Annals of Surgery **157**:223–226.

Clevers, H. 2013. The intestinal crypt, a prototype stem cell compartment. Cell **154**:274–284.

Clevers, H. and R. Nusse 2012. Wnt-catenin signaling and disease. Cell **149**:1192–1205.

Datta, R. S., A. Gutteridge, C. Swanton, C. C. Maley, and T. Graham 2013. Modelling the evolution of genetic instability during tumour progression. Evolutionary Applications **6**:20–33.

Dikovskaya, D., D. Schiffmann, I. P. Newton, A. Oakley, K. Kroboth, O. Sansom, T. J. Jamieson, V. Meniel, A. Clarke, and I. S. Näthke 2007. Loss of APC induces polyploidy as a result of a combination of defects in mitosis and apoptosis. Journal of Cell Biology **176**:183–195.

Elena, S. F., L. Ekunwe, N. Hajela, S. A. Oden, and R. E. Lenski 1998.

Distribution of fitness effects caused by random insertion mutations in *Escherichia coli*. Genetica 102-103:349–358.

Eyre-Walker, A. and P. D. Keightley 2007. The distribution of fitness effects of new mutations. Nature Reviews Genetics 8:610–618.

Foo, J., K. Leder, and F. Michor 2011. Stochastic dynamics of cancer initiation. Physical Biology 8:015002.

Gillespie, D. T. 1977. Exact Stochastic Simulation of couple chemical reactions. The Journal of Physical Chemistry 81:2340–2361.

Halligan, D. L. and P. D. Keightley 2009. Spontaneous mutation accumulation studies in evolutionary genetics. Annual Review of Ecology, Evolution, and Systematics 40:151–172.

Imhof, M. and C. Schlotterer 2001. Fitness effects of advantageous mutations in evolving *Escherichia coli* populations. Proceedings of the National Academy of Sciences of the United States of America 98:1113–1117.

Jones, D. L. and T. A. Rando 2011. Emerging models and paradigms for stem cell ageing. Nature Cell Biology 13:506–512.

Joseph, S. B. and D. W. Hall 2004. Spontaneous mutations in diploid *Saccharomyces cerevisiae*: more beneficial than expected. Genetics 168:1817–1825.

Kassen, R. and T. Bataillon 2006. Distribution of fitness effects among beneficial mutations before selection in experimental populations of bacteria. Nature Genetics 38:484–488.

Kozar, S., E. Morrissey, A. M. Nicholson, M. van der Heijden, H. I. Zecchini, R. Kemp, S. Tavaré, L. Vermeulen, and D. J. Winton 2013. Continuous clonal labeling reveals small numbers of functional stem cells in intestinal crypts and adenomas. Cell Stem Cell 13:626–633.

Levy, S. F., J. R. Blundell, S. Venkataram, D. A. Petrov, D. S. Fisher, and G. Sherlock 2015. Quantitative evolutionary dynamics using high-resolution lineage tracking. Nature 519:181–186.

Loeffler, M. and B. Grossmann 1991. A stochastic branching model with formation of subunits applied to the growth of intestinal crypts. Journal of Theoretical Biology 150:175–191.

Lopez-Garcia, C., A. M. Klein, B. Simons, and D. J. Winton 2010. Intestinal stem cell replacement follows a pattern of neutral drift. Science 330:822–825.

López-Otín, C., M. A. Blasco, L. Partridge, M. Serrano, and G. Kroemer 2013. The hallmarks of aging. Cell 153:1194–1217.

Lynch, M. 2010. Evolution of the mutation rate. Trends in Genetics 26:345–352.

McCartney, B. M. and I. S. Näthke 2008. Cell regulation by the Apc protein. Apc as master regulator of epithelia. Current Opinion in Cell Biology 20:186–193.

McFarland, C. D., K. S. Korolev, G. V. Kryukov, S. R. Sunyaev, and L. A. Mirny 2013. Impact of deleterious passenger mutations on cancer progression. Proceedings of the National Academy of Sciences of the United States of America 110:2910–2915.

McFarland, C. D., L. Mirny, and K. S. Korolev 2014. A tug-of-war between driver and passenger mutations in cancer and other adaptive processes. Proceedings of the National Academy of Sciences of the United States of America 111:15138–15143.

Merlo, L. M. F., J. W. Pepper, B. J. Reid, and C. C. Maley 2006. Cancer as an evolutionary and ecological process. Nature Reviews Cancer 6:924–935.

Michor, F., S. Frank, R. M. May, Y. Iwasa, and M. A. Nowak 2003. Somatic selection for and against cancer. Journal of Theoretical Biology 225:377–382.

Moskalev, A. A., M. V. Shaposhnikov, E. N. Plyusnina, A. Zhavoronkov,

A. Budovsky, H. Yanai, and V. E. Fraifeld 2013. The role of DNA damage and repair in aging through the prism of Koch-like criteria. Ageing Research Reviews 12:661–684.

Nicolas, P., K.-M. Kim, D. Shibata, and S. Tavaré 2007. The stem cell population of the human colon crypt: analysis via methylation patterns. PLoS Computational Biology 3:e28.

Nijnik, A., L. Woodbine, C. Marchetti, S. Dawson, T. Lambe, C. Liu, N. P. Rodrigues, T. L. Crockford, E. Cabuy, A. Vindigni, T. Enver, J. I. Bell, P. Slijepcevic, C. C. Goodnow, P. A. Jeggo, and R. J. Cornall 2007. DNA repair is limiting for haematopoietic stem cells during ageing. Nature 447:686–690.

Orr, H. A. 2003. The distribution of fitness effects among beneficial mutations. Genetics 163:1519–1526.

Orr, H. A. 2010. The population genetics of beneficial mutations. Philosophical Transactions of the Royal Society of London. Series B, Biological Sciences 365:1195–1201.

Potten, C. S., C. Booth, and D. Hargreaves 2003. The small intestine as a model for evaluating adult tissue stem cell drug targets. Cell Proliferation 36:115–129.

Reya, T. and H. Clevers 2005. Wnt signalling in stem cells and cancer. Nature 434:843–850.

Ritsma, L., S. I. J. Ellenbroek, A. Zomer, H. J. Snippert, F. J. de Sauvage, B. D. Simons, H. Clevers, and J. van Rheenen 2014. Intestinal crypt homeostasis revealed at single-stem-cell level by in vivo live imaging. Nature 507:362–365.

Rokyta, D. R., C. J. Beisel, P. Joyce, M. T. Ferris, C. L. Burch, and H. A. Wichman 2008. Beneficial fitness effects are not exponential for two viruses. Journal of Molecular Evolution 67:368–376.

Rossi, D., D. Bryder, J. Seita, A. Nussenzweig, J. Hoeijmakers, and I. L. Weissman 2007. Deficiencies in DNA damage repair limit the function of haematopoietic stem cells with age. Nature 447:725–729.

Sanjuán, R., A. Moya, and S. F. Elena 2004. The distribution of fitness effects caused by single-nucleotide substitutions in an RNA virus. Proceedings of the National Academy of Sciences of the United States of America 101:8396–8401.

Smith, G., F. A. Carey, J. Beattie, M. J. V. Wilkie, T. J. Lightfoot, J. Coxhead, R. C. Garner, R. J. C. Steele, and C. R. Wolf 2002. Mutations in APC, Kirsten-ras, and p53-alternative genetic pathways to colorectal cancer. Proceedings of the National Academy of Sciences of the United States of America 99:9433–9438.

Snippert, H. J., L. G. van der Flier, T. Sato, J. H. van Es, M. van den Born, C. Kroon-Veenboer, N. Barker, A. M. Klein, J. van Rheenen, B. Simons, and H. Clevers 2010. Intestinal crypt homeostasis results from neutral competition between symmetrically dividing Lgr5 stem cells. Cell 143:134–144.

Snippert, H. J., A. G. Schepers, J. H. van Es, B. Simons, and H. Clevers 2014. Biased competition between Lgr5 intestinal stem cells driven by oncogenic mutation induces clonal expansion. EMBO Reports 15:62–69.

Sottoriva, A., H. Kang, Z. Ma, T. Graham, M. P. Salomon, J. Zhao, P. Marjoram, K. Siegmund, M. F. Press, D. Shibata, and C. Curtis 2015. A Big Bang model of human colorectal tumor growth. Nature Genetics 47:209–216.

Vermeulen, L., E. Morrissey, M. van der Heijden, A. M. Nicholson, A. Sottoriva, S. Buczacki, R. Kemp, S. Tavare, and D. J. Winton 2013. Defining stem cell dynamics in models of intestinal tumor initiation. Science 342:995–998.

Winawer, S. J. 1999. Natural history of colorectal cancer. The American Journal of Medicine 106: 3S–6S; discussion 50S–51S.

Wiser, M. J., N. Ribeck, and R. E. Lenski 2013. Long-term dynamics of adaptation in asexual populations. Science **342**:1364–1367.

Wloch, D. M., K. Szafraniec, R. H. Borts, and R. Korona 2001. Direct estimate of the mutation rate and the distribution of fitness effects in the yeast *Saccharomyces cerevisiae*. Genetics **159**:441–452.

Wodarz, D. 2007. Effect of stem cell turnover rates on protection against cancer and aging. Journal of Theoretical Biology **245**:449–458.

Wodarz, D. and N. L. Komarova 2005. Computational Biology of Cancer. World Scientific, Singapore.

Wong, W.-M., N. Mandir, R. A. Goodlad, B. C. Y. Wong, S. B. Garcia, S.-K. Lam, and N. A. Wright 2002. Histogenesis of human colorectal adenomas and hyperplastic polyps: the role of cell proliferation and crypt fission. Gut **50**:212–217.

Wood, L., D. Parsons, and S. Jones 2007. The genomic landscapes of human breast and colorectal cancers. Science **318**:1108–1113.

Zeyl, C. and J. DeVisser 2001. Estimates of the rate and distribution of fitness effects of spontaneous mutation in *Saccharomyces cerevisiae*. Genetics **157**:53–61.

Appendix

Description of the mathematical methodology

In this section, we describe the mathematical model underlying our research approach. It is a multiscale model, including the dynamics of the stem cell niche, the consequences for the larger stem cell population and a crypt, the dynamics in a population of crypts that comprise an individual's colon, and the dynamics of tumorigenesis among many individuals in a population. Due to the very large number of crypts in the colon and the desire to analyze population level incidence curves, we used several principles of rare-event analysis in our numerical computations and also introduced a few approximations to make computations tractable.

Population dynamics within the stem cell niche

First, we develop a model for the population dynamics of a crypt immediately after a mutation has occurred. Suppose that there are N cells in the stem cell niche and let $X(t)$ represent the number of cells that are descended from the original mutated cell at time t. We model $X(t)$ as a continuous time Markov chain (CTMC) that takes its values in the set $\{0, \ldots, N\}$ with $X(0) = 1$.

In accordance with the Markov process assumption, the time between divisions of a given stem cell is independent of all other cells and exponentially distributed with rate parameters λ_{old} or λ for the old and the new lineages, respectively. When a cell divides, we assume that there is crowding in the stem cell niche and an old cell is forced out. [In fact, the actual order of events remains unclear. It has also been hypothesized that a cell may leave the niche first, then triggering a cell division to replace it (Lopez-Garcia *et al.* 2010, but see Ritsma *et al.* 2014).] Whether or not the value of the process $X(t)$ changes depends on whether the cell that has been forced out is from the same lineage as the one that divided. The assumption that leads to the simplest mathematical model is nearest neighbor displacement. There are two cases (i) the dividing cell is of the same lineage as both of its neighbors, and (ii) the dividing cell is adjacent to a cell of the opposing lineage. In the first case, the value of $X(t)$ does not change as a result of the cell division. In the latter case, there is a one-half probability that a cell of the opposing lineage will be displaced. As such, for $X(t) \in \{1, 2, \ldots, N-1\}$, the Markov transition rates are given by

$$\text{Nearest neighbor displacement:} \begin{cases} X(t) \rightarrow X(t) + 1 \\ \quad \text{at rate } \lambda \\ X(t) \rightarrow X(t) - 1 \\ \quad \text{at rate } \lambda_0. \end{cases} \quad (9)$$

(The rate of one of the two border cells dividing is 2λ for the new lineage and $2\lambda_{old}$ for the old lineage and then each is multiplied by the one-half probability of displacing an opposing lineage cell.) An alternate hypothesis is that after division, any other cell in the crypt might be displaced. The corresponding transition rates would be

$$\text{Nonlocal displacement:} \begin{cases} X(t) \rightarrow X(t) + 1 \text{ at rate } \frac{1}{N}X(t)(N - X(t))\lambda \\ X(t) \rightarrow X(t) - 1 \text{ at rate } \frac{1}{N}X(t)(N - X(t))\lambda_{old} \end{cases} \quad (10)$$

The probability of fixation is actually the same for both models (though the expected time until fixation will differ). Let $\{t_1, t_2, \ldots\}$ be the sequence of times when $X(t)$ changes values. Disregarding the role of time in the process, we track the values with the process $\{X_n\}_{n \geq 0}$ defined by $X_n := X(t_n)$. The probability of a transition $X \rightarrow X+1$ is the rate at which the size of the mutant lineage increases divided by the total rate of change in lineage count. For both models, this probability of an increase in the mutant lineage size is $p = \lambda/(\lambda + \lambda_{old})$. Using the classical theory of hitting probabilities for biased random walks (Wodarz and Komarova 2005), one can readily derive the probability $p_{fix}(\lambda; \lambda_{old})$ recorded in eqn 3 in the main text.

The intervals between mutations that fix in the stem cell niche

The DFEs used in this work are both considered in terms of percentage increase or decrease, rather than in terms of absolute quantities of change. In mathematical terms, this means that the densities can be expressed in terms of the ratio λ/λ_{old}. A remarkable consequence of this assumption is that the probability of a new lineage fixing in the niche is independent of the prevailing division rate λ_{old}. To see this, consider the probability of that a new lineage fixes after a mutation drawn from the exponential DFE. Recalling eqn 3, we note that the probability of fixation formula can be written in terms of the ratio of the new to the old division rate, $r = \lambda/\lambda_{old}$,

$$p_{fix}(\lambda; \lambda_{old}) = p_{fix}(r) = \frac{1 - r^{-1}}{1 - r^{-N}}.$$

We can then write

have not yet differentiated. Assuming, for the moment, that all members of the stem cell niche have a division rate λ, the CTMC $Y(t)$ is defined by the transition rates

$$Y(t) \rightarrow Y(t) + 1 \text{ at rate } (N + Y(t))\lambda$$
$$Y(t) \rightarrow Y(t) - 1 \text{ at rate } Y(t)v.$$

The form of the rate of increase follows from the observation that $Y(t)$ increases anytime a stem cell divides, whether that stem cell is in the crypt or not. On the other hand, because stem cells in the niche are assumed to not differentiate, the total rate of decrease is proportional to the number of stem cells outside the niche. Because the population size is so small, there is high variability and we note that $Y(t)$ can regularly hit the value zero. Because the niche is protected by unrelated biological processes, this does not constitute extinction of the full stem cell population. As soon as another

$$\hat{p} = \mathbb{P}\{\text{Fixation } |\lambda_{old}\} = \int_0^\infty p_{fix}(\lambda; \lambda_{old}) m(\lambda; \lambda_{old}) d\lambda$$

$$= \int_0^{\lambda_{old}} p_{fix}\left(\frac{\lambda}{\lambda_{old}}\right)(1 - P_B)\frac{\beta}{\lambda_{old}}e^{-\beta\left(1 - \frac{\lambda}{\lambda_{old}}\right)} d\lambda + \int_{\lambda_{old}}^\infty p_{fix}\left(\frac{\lambda}{\lambda_{old}}\right)P_B\frac{\alpha}{\lambda_{old}}e^{-\alpha\left(1 - \frac{\lambda}{\lambda_{old}}\right)} d\lambda$$

$$= \int_0^1 p_{fix}(r)(1 - P_B)\beta e^{-\beta(1-r)} dr + \int_1^\infty p_{fix}(r)P_B\alpha e^{-\alpha(r-1)} dr,$$

which is independent of the choice of value λ_{old}. A similar result holds for the power-law DFE. Generally, this property holds for any DFE that can be expressed in terms of the ratio λ/λ_{old}. It follows that the number of mutations that must occur in order for a new division rate to fix is distributed Geometrically with success probability $\hat{p}$. By standard properties of CTMC, we can then say that the time between the arrivals of "successful" mutations is exponentially distributed with rate parameter $\mu\hat{p}\lambda_{old}N$.

Population dynamics outside the stem cell niche

Once outside the niche, a stem cell can either divide (at rate λ), or it can differentiate into transient amplifying cells (at rate v). For the purposes of this model, we consider differentiated cells to be dead. There is a chance that the lineage of a stem cell outside the niche can undergo sufficiently many mutations to cause tumorigenesis, but we found by way of numerical investigations that this does not significantly contribute to overall incidence of these cancers. As such, let $Y(t)$ denote the number of stem cells outside the niche that

stem cell in the niche divides, the population outside the niche is renewed again. A typical trace for $Y(t)$ can be seen in Fig. S5.

The law of this CTMC, $y_n(t) = \mathbb{P}\{Y(t) = n\}$, satisfies the system of master equations

$$\frac{d}{dt}y_n(t) = (N + (n - 1))\lambda y_{n-1}(t)1_{n \geq 1}(n)$$
$$+ (n + 1)vy_{n+1}(t) - ((N + n)\lambda + nv)y_n(t). \tag{11}$$

One can then show that the mean $\bar{y}(t) = \sum_{n=1}^\infty ny_n(t)$ satisfies the ODE

$$\frac{d}{dt}\bar{y}(t) = N\lambda + (\lambda - v)\bar{y}(t). \tag{12}$$

If $\lambda < v$, this ODE converges to a steady-state value $N\lambda/(v-\lambda)$. Otherwise the mean diverges to infinity with exponential growth. For this reason, we consider this threshold to be the initiation of tumorigenesis.

An alternate way to view the dynamics is to note that each time a stem cell in the niche divides it creates a new independent lineage outside the crypt. Let $Y^j(t)$ be the number of living stem cells outside the crypt that are descended from (and include) the product of the jth stem cell division in the niche. As such $Y(t) = \sum_{j=1}^{\infty} Y^j(t)$. Each process $Y^j(t)$ can be understood as a branching process, with an offspring distribution that is geometrically distributed with "success probability" $q = v/(v + \lambda)$. (The number of offspring is determined by the number of times the cell divides before differentiating. This is a sequence of independent trials where the probability of having another offspring, rather than differentiating, is $\lambda/(v + \lambda)$.) As long as the mean of this offspring distribution is less than or equal to one, these lineages will eventually go extinct. Therefore, the critical stem cell division rate corresponds to when the mean of the offspring distribution (which can be shown to be $(1 - q)/q = \lambda/v$) is less than one. In other words, the critical division rate λ_* is simply $\lambda_* = v$.

Population dynamics in the crypt

Of course, tumorigenesis in a given crypt is exceedingly unlikely, even over the lifetime of an individual. We model the colon as a collection of $C \approx 10^7$ individual crypts that are mathematically identical and independent. The number of fixed mutations in the ith crypt at time t is denoted $M^i(t)$, and let $\{\lambda_0^i, \lambda_1^i \ldots\}$ denote the sequence of division rates that become fixed in the ith crypt at times $\{0, \tau_1^i, \tau_2^i \ldots\}$, respectively. It follows that the inter-arrival times are independent and distributed as

$$\tau_{k+1}^i - \tau_k^i \sim \text{Exp}(\hat{p}\mu\lambda_k^i N). \tag{13}$$

Whether tumorigenesis has occurred in the ith crypt will be tracked by the function $\chi^i(t)$, defined by

$$\chi^i(t) = \begin{cases} 1, & \text{if } v \leq \max\left(\lambda_k^i : k \leq M^i(t)\right) \\ 0, & \text{otherwise.} \end{cases}$$

That is to say, $\chi^i(t) = 1$ if tumorigenesis has occurred before time t. It follows that the time of first tumorigenesis in the colon is given by the time

$$\mathcal{T} := \inf\left\{t > 0 : \sum_{i=1}^{C} \chi^i(t) \geq 1\right\}. \tag{14}$$

The per capita population incidence curves are then just the cumulative distribution function of the random variable $\mathcal{T}$, which can be expressed in terms of the individual crypt dynamics as follows:

$$\mathbb{P}\{\mathcal{T} > t\} = \mathbb{P}\{\chi^i(t) = 0 \text{ for all } i \in \{1, \ldots, C\}\}.$$

To prepare for our numerical approximation of this quantity, we introduce one last bit of notation,

$\{N_m(t)\}_{m=0}^{\infty}$, which represents the number of crypts that have seen the arrival of m new fixed lineages as of time t. Then

$$\mathbb{P}\{\mathcal{T} > t | N_m(t) = n_m \text{ for all } m\}$$
$$= \prod_{m=0}^{\infty} \mathbb{P}\{\chi(t) = 0 | M(t) = m\}^{n_m}. \tag{15}$$

These dynamics can be simulated by Gillespie's method (Gillespie 1977), but such an approach is computationally intensive. For this reason, we introduced a few simplifying assumptions. For example, we model the arrival rates of new fixed lineages in the crypts as being constant over time [having fixed rate $\hat{\mu} = \hat{p}\mu\lambda_0 N$, rather than a sequence of rates given in eqn (13)]. This allows us to assume that the number of mutations in each crypt at time t is Poisson distributed with mean $\hat{\mu}t$. With such a tremendously large number of crypts in the colon, it is in turn reasonable to assume that the number of crypts takes the form $n_m \approx C\mathbb{P}\{M(t) = m\} \approx Ce^{-\hat{\mu}t}(\hat{\mu}t)^m/m!$. To complete the derivation of eqn (8) in the main text, we truncate the infinite product in eqn (15) and note that in the notation of the main text, $\mathbb{P}\{\chi(t) = 0 | M(t) = m\} = q_m$.

DFE equations

To define the parameters of the system, we specified the probability P_B of a beneficial (versus deleterious) mutation and the respective means s_+ and s_- of the DFE conditioned on the event that the mutation is beneficial or deleterious. The form of the mean of the DFE depends on whether it is exponential or heavy tailed. The exponential form DFE has mean

$$\mathbb{E}(\lambda_{\text{new}} | \lambda_{\text{old}})_{\exp} = \lambda_{\text{old}}\left(1 + \frac{P_B}{\alpha} - \frac{(1 - P_B)}{\beta}\right)$$

while the heavy-tailed DFE has mean

$$\mathbb{E}(\lambda_{\text{new}} | \lambda_{\text{old}})_{\text{Pareto}} = P_B\left(\frac{\alpha - 1}{\alpha - 2}\right)\lambda_{\text{old}} + (1 - P_B)\left(\lambda_{\text{old}} - \frac{\lambda_{\text{old}}}{\beta}\right).$$

The conditional means have the form

$$s_+ := \mathbb{E}(\lambda_{\text{new}} | \lambda_{\text{old}}, \textbf{beneficial})_{\exp} = \lambda_{\text{old}}\left(1 + \frac{1}{\alpha}\right)$$

$$s_- := \mathbb{E}(\lambda_{\text{new}} | \lambda_{\text{old}}, \textbf{deleterious})_{\exp} = \lambda_{\text{old}}\left(1 - \frac{1}{\beta}\right)$$

for the exponential DFE, and

$$s_+ := \mathbb{E}(\lambda_{\text{new}} | \lambda_{\text{old}}, \textbf{beneficial})_{\text{Pareto}} = \lambda_{\text{old}}\left(\frac{\alpha - 1}{\alpha - 2}\right)$$

$$s_- := \mathbb{E}(\lambda_{\text{new}} | \lambda_{\text{old}}, \textbf{deleterious})_{\text{Pareto}} = \lambda_{\text{old}}\left(1 - \frac{1}{\beta}\right)$$

for the heavy-tailed DFE.

Least squares analysis

We generated tumor incidence curves and used a least squares analysis to determine which set of these parameters best fit the tumor incidence data described in Chapman (1963). The best fit has the smallest sum of squared residuals of the parameter space explored in Figs S1–S3.

Tumor mutational profile

Figure S4 contains the probabilities that each individual mutational profile was the culprit in tumorigenesis given that tumorigenesis occurred in a single crypt. They were calculated using Bayes' theorem to compute the probability that a certain mutational load fixed in the crypt given that a tumorigenesis event happened,

$$
\mathbb{P}(M(T) = n \mid T = t)
$$

$$
= \frac{f_T(t \mid M(T) = n)\mathbb{P}(M(T) = n)}{\sum_{j=1}^{\hat{n}} f_T(t \mid M(T) = j)\mathbb{P}(M(T) = j)}
$$

where we recall that $M(t)$ is the number of fixed mutations as of time t, T is the precise time that tumorigenesis occurs in the crypt, and f_T refers to density of the random variable T. The quantity $\mathbb{P}(M(T) = n)$ is the probability that tumorigenesis occurs exactly on the nth mutation, a quantity we defined earlier as p_n and gave a recursive formula

for in eqn 5 in the main text. To compute the quantity $f_T(t \mid M(T) = n)$, note that because the arrival time of the nth mutation is independent of the event that it causes tumorigenesis, we have that $f_T(t \mid M(T) = n) = f_{\tau_n}(t)$, where τ_n is the arrival time of the nth mutation. By hypothesis, τ_n is Poisson distributed with mean $\hat{\mu}t$ with $\hat{\mu}$ being defined after eqn 8 in the main text. The distributions of mutational profiles given the tumorigenesis event occurred at a certain point in time throughout an organism's lifetime are given in Fig. S4.

Permissions

The contributors of this book come from diverse backgrounds, making this book a truly international effort. This book will bring forth new frontiers with its revolutionizing research information and detailed analysis of the nascent developments around the world.

We would like to thank all the contributing authors for lending their expertise to make the book truly unique. They have played a crucial role in the development of this book. Without their invaluable contributions this book wouldn't have been possible. They have made vital efforts to compile up to date information on the varied aspects of this subject to make this book a valuable addition to the collection of many professionals and students.

This book was conceptualized with the vision of imparting up-to-date information and advanced data in this field. To ensure the same, a matchless editorial board was set up. Every individual on the board went through rigorous rounds of assessment to prove their worth. After which they invested a large part of their time researching and compiling the most relevant data for our readers.

The editorial board has been involved in producing this book since its inception. They have spent rigorous hours researching and exploring the diverse topics which have resulted in the successful publishing of this book. They have passed on their knowledge of decades through this book. To expedite this challenging task, the publisher supported the team at every step. A small team of assistant editors was also appointed to further simplify the editing procedure and attain best results for the readers.

Apart from the editorial board, the designing team has also invested a significant amount of their time in understanding the subject and creating the most relevant covers. They scrutinized every image to scout for the most suitable representation of the subject and create an appropriate cover for the book.

The publishing team has been an ardent support to the editorial, designing and production team. Their endless efforts to recruit the best for this project, has resulted in the accomplishment of this book. They are a veteran in the field of academics and their pool of knowledge is as vast as their experience in printing. Their expertise and guidance has proved useful at every step. Their uncompromising quality standards have made this book an exceptional effort. Their encouragement from time to time has been an inspiration for everyone.

The publisher and the editorial board hope that this book will prove to be a valuable piece of knowledge for researchers, students, practitioners and scholars across the globe.

List of Contributors

Tracy Arcella and Glen R. Hood
Department of Biological Sciences, University of Notre Dame, Notre Dame, IN, USA

Thomas H. Q. Powell
Department of Biological Sciences, University of Notre Dame, Notre Dame, IN, USA
Department of Entomology and Nematology, University of Florida, Gainesville, FL 32611, USA

Sheina B. Sim
Department of Biological Sciences, University of Notre Dame, Notre Dame, IN, USA
USDA-ARS US PBARC, 64 Nowelo Street, Hilo, HI 96720, USA

Wee L. Yee
USDA-ARS, Yakima Agricultural Research Laboratory, Wapato, WA, USA

Dietmar Schwarz
Department of Biology, Western Washington University, Bellingham, WA, USA

Scott P. Egan
Department of Biological Sciences, University of Notre Dame, Notre Dame, IN, USA
Advanced Diagnostics and Therapeutics, University of Notre Dame, Notre Dame, IN, USA
Department of Biosciences, Rice University, Houston, TX 77005, USA

Robert B. Goughnour
Washington State University Extension, Vancouver, WA, USA

James J. Smith
Departments of Entomology and Lyman Briggs College, Michigan State University, E. Lansing, MI, USA

Jeffrey L. Feder
Department of Biological Sciences, University of Notre Dame, Notre Dame, IN, USA
Advanced Diagnostics and Therapeutics, University of Notre Dame, Notre Dame, IN, USA
Environmental Change Initiative, University of Notre Dame, Notre Dame, IN, USA

Jessica Hua and Jason T. Hoverman
Department of Forestry and Natural Resources, Purdue University, West Lafayette, IN, USA

Devin K. Jones, Brian M. Mattes and Rick A. Relyea
Department of Biological Sciences, Rensselaer Polytechnic Institute, Troy, NY, USA

Rickey D. Cothran
Department of Biological Sciences, Southwestern Oklahoma State University, Weatherford, OK, USA

Mieke Jansen, Joost Vanoverbeke, Melissa Schepens and Luc De Meester
Laboratory of Aquatic Ecology, Evolution and Conservation, KU Leuven, Leuven, Belgium

Anja Coors
ECT Oekotoxikologie GmbH, Flörsheim A.M., Germany
Biodiversity and Climate Research Centre (BiK-F), Frankfurt A.M., Germany

Pim De Voogt
Institute for Biodiversity and Ecosystem Dynamics (IBED), Universiteit Amsterdam, Amsterdam, The Netherlands

Karel A. C. De Schamphelaere
Laboratory for Environmental Toxicology and Aquatic Ecology, Environmental Toxicology Unit (GhEnToxLab), Ghent University, Ghent, Belgium

Gregory Brazzola and Claus Wedekind
Department of Ecology and Evolution, Biophore, University of Lausanne, Lausanne, Switzerland

Nathalie Chèvre
Institute of Earth Surface Dynamics, University of Lausanne, Lausanne, Switzerland

Anna Kuparinen
Department of Environmental Sciences, University of Helsinki, Helsinki, Finland

Jeffrey A. Hutchings
Department of Biology, Dalhousie University, Halifax, NS, Canada

Department of Biosciences, Centre For Ecological and Evolutionary Synthesis, University of Oslo, Oslo, Norway
Department of Natural Sciences, University of Agder, Kristiansand, Norway

Robin S. Waples
National Marine Fisheries Service, National Oceanic and Atmospheric Administration, Northwest Fisheries Science Center, Seattle, WA, USA

Alex J. Malvezzi, Christopher J. Gobler and Demian D. Chapman
School of Marine and Atmospheric Sciences, Stony Brook University, Stony Brook, NY, USA

Christopher S. Murray and Hannes Baumann
Department of Marine Sciences, University of Connecticut, Groton, CT, USA

Kevin A. Feldheim
Pritzker Laboratory for Molecular Systematics and Evolution, Field Museum of Natural History, Chicago, IL, USA

Joseph D. DiBattista
Red Sea Research Center, King Abdullah University of Science and Technology, Thuwal, Saudi Arabia

Dany Garant
Département de Biologie, Université de Sherbrooke, Sherbrooke, QC, Canada

Lise Marty
IFREMER, Laboratoire Ressources Halieutiques, Unité Halieutique Manche-Mer du Nord, Boulogne-sur-mer, France

Ulf Dieckmann
IIASA, Evolution and Ecology Program, Laxenburg, Austria
International Institute of Applied Systems Analysis, Laxenburg, Austria

Bruno Ernande
IFREMER, Laboratoire Ressources Halieutiques, Unité Halieutique Manche-Mer du Nord, Boulogne-sur-mer, France
IIASA, Evolution and Ecology Program, Laxenburg, Austria

Gabriel Pigeon and Fanie Pelletier
Département de Biologie and Centre d'Études Nordiques, Université de Sherbrooke, Sherbrooke, QC, Canada

Département de Biologie, Canada Research Chair in Evolutionary Demography and Conservation, Université de Sherbrooke, Sherbrooke, QC, Canada

Marco Festa-Bianchet
Département de Biologie and Centre d'Études Nordiques, Université de Sherbrooke, Sherbrooke, QC, Canada

David W. Coltman
Department of Biological Sciences, University of Alberta, Edmonton, AB, Canada

Lindsey W. Sargent
Department of Biological Sciences, University of Notre Dame, Notre Dame, IN, USA

David M. Lodge
Department of Biological Sciences and Environmental Change Initiative, University of Notre Dame, Notre Dame, IN, USA

Tam T. Tran
Institute of Aquaculture, Nha Trang University, Nha Trang, Vietnam
Laboratory of Aquatic Ecology, Evolution and Conservation, University of Leuven, Leuven, Belgium

Lizanne Janssens, Lin Op de Beeck and Robby Stoks
Laboratory of Aquatic Ecology, Evolution and Conservation, University of Leuven, Leuven, Belgium

Khuong V. Dinh
Institute of Aquaculture, Nha Trang University, Nha Trang, Vietnam
National Institute of Aquatic Resources, Technical University of Denmark, Copenhagen, Denmark

Silva Uusi-Heikkilä
Department of Biology and Ecology of Fishes, Leibniz-Institute of Freshwater Ecology and Inland Fisheries, Berlin, Germany
Division of Genetics and Physiology, Department of Biology, University of Turku, Turku, Finland

Andrew R. Whiteley
Department of Environmental Conservation, University of Massachusetts, Amherst, MA, USA

Shuichi Matsumura
Faculty of Applied Biological Sciences, Gifu University, Gifu, Japan

Paul A.Venturelli
Department of Fisheries, Wildlife, and Conservation Biology, University of Minnesota, St Paul, MN, USA

Christian Wolter, David Bierbach and Giovanni Polverino
Department of Biology and Ecology of Fishes, Leibniz-Institute of Freshwater Ecology and Inland Fisheries, Berlin, Germany

Jon Slate
Department of Animal and Plant Sciences, University of Sheffield, Western Bank, Sheffield, UK

Craig R. Primmer
Division of Genetics and Physiology, Department of Biology, University of Turku, Turku, Finland

Thomas Meinelt
Department of Ecophysiology and Aquaculture, Leibniz-Institute of Freshwater Ecology and Inland Fisheries, Berlin, Germany

Shaun S. Killen
Institute of Biodiversity, Animal Health and Comparative Medicine, College of Medical, Veterinary and Life Sciences, University of Glasgow, Glasgow, UK

Arne Ludwig
Department of Evolutionary Genetics, Leibniz-Institute for Zoo and Wildlife Research, Berlin, Germany

Robert Arlinghaus
Department of Biology and Ecology of Fishes, Leibniz-Institute of Freshwater Ecology and Inland Fisheries, Berlin, Germany
Chair of Integrative Fisheries Management, Faculty of Life Sciences, Albrecht-Daniel-Thaer Institute of Agricultural and Horticultural Sciences, Humboldt-Universität zu Berlin, Berlin, Germany

Maren Wellenreuther
Department of Biology, University of Lund, Lund, Sweden
Institute for Plant and Food Research, Lund, New Zealand

Sarah Otto
Department of Zoology and Biodiversity Research Centre, University of British Columbia, Vancouver, BC, Canada

Neala W. Kendall
School of Aquatic and Fishery Sciences, University of Washington, Seattle, WA, USA
Washington Department of Fish and Wildlife, Olympia, WA, USA

Mikko Heino
International Institute of Applied Systems Analysis, Laxenburg, Austria
Department of Biology, University of Bergen, Bergen, Norway
Institute of Marine Research, Bergen, Norway

André E. Punt and Thomas P. Quinn
School of Aquatic and Fishery Sciences, University of Washington, Seattle, WA, USA

Khuong Van Dinh
Institute of Aquaculture, Nha Trang University, Nha Trang, Vietnam
Laboratory of Aquatic Ecology, Evolution and Conservation, University of Leuven, Leuven, Belgium

Lizanne Janssens, Lieven Therry and Hajnalka A. Gyulavári
Laboratory of Aquatic Ecology, Evolution and Conservation, University of Leuven, Leuven, Belgium

Lieven Bervoets
Systemic, Physiological and Ecotoxicological Research Group, University of Antwerp, Antwerp, Belgium

Robby Stoks
Institute of Aquaculture, Nha Trang University, Nha Trang, Vietnam
Laboratory of Aquatic Ecology, Evolution and Conservation, University of Leuven, Leuven, Belgium

Rémi Chargé
Department of Biological and Environmental Science, Centre of Excellence in Biological Interactions, University of Jyväskylä, Jyväskylä, Finland

Céline Teplitsky
Centre d'Ecologie et de Sciences de la Conservation UMR 7204 CNRS/MNHN/UPMC, Muséum National d'Histoire Naturelle, Paris, France

Gabriele Sorci
Biogéosciences, UMR CNRS 6282, Université de Bourgogne, Dijon, France

Matthew Low
Department of Ecology, Swedish University of Agricultural Sciences, Uppsala, Sweden

Vincent L. Cannataro and Colette M. St. Mary
Department of Biology, University of Florida, Gainesville, FL, USA

Scott A. McKinley
Department of Mathematics, Tulane University, New Orleans, LA, USA

Index

www.ingramcontent.com/pod-product-compliance
Lightning Source LLC
Chambersburg PA
CBHW082042190326
41458CB00010B/3435